Embedded Machine Learning for Cyber-Physical, IoT, and Edge Computing

Sudeep Pasricha • Muhammad Shafique
Editors

Embedded Machine Learning for Cyber-Physical, IoT, and Edge Computing

Hardware Architectures

 Springer

Editors
Sudeep Pasricha
Colorado State University
Fort Collins, CO, USA

Muhammad Shafique
New York University Abu Dhabi
Abu Dhabi, United Arab Emirates

ISBN 978-3-031-19570-9   ISBN 978-3-031-19568-6   (eBook)
https://doi.org/10.1007/978-3-031-19568-6

This Springer imprint is published by the registered company Springer Nature Switzerland AG
The registered company address is: Gewerbestrasse 11, 6330 Cham, Switzerland

Paper in this product is recyclable.

Preface

Machine learning (ML) has emerged as a prominent approach for achieving state-of-the-art accuracy for many data analytic applications, ranging from computer vision (e.g., classification, segmentation, and object detection in images and video), speech recognition, language translation, healthcare diagnostics, robotics, and autonomous vehicles to business and financial analysis. The driving force of the ML success is the advent of neural network (NN) algorithms, such as deep neural networks (DNNs)/deep learning (DL) and spiking neural networks (SNNs) with support from today's evolving computing landscape to better exploit data and thread-level parallelism with ML accelerators.

Current trends show an immense interest in attaining the powerful abilities of NN algorithms for solving ML tasks using embedded systems with limited compute and memory resources, i.e., so-called *Embedded ML*. One of the main reasons is that embedded ML systems may enable a wide range of applications, especially the ones with tight memory and power/energy constraints, such as mobile systems, Internet of Things (IoT), edge computing, and cyber-physical applications. Furthermore, embedded ML systems can also improve the quality of service (e.g., personalized systems) and privacy as compared to centralized ML systems (e.g., based on cloud computing). However, *state-of-the-art NN-based ML algorithms are costly in terms of memory sizes and power/energy consumption, thereby making it difficult to enable embedded ML systems.*

This book explores and identifies the most challenging issues that hinder the implementation of embedded ML systems. These issues arise from the fact that, to achieve better accuracy, the development of NN algorithms has led to state-of-the-art models with higher complexity with respect to model sizes and operations, the implications of which are discussed below:

- *Massive Model Sizes*: Larger NN models usually obtain higher accuracy than the smaller ones because they have a larger number of NN parameters that can learn the features from the training dataset better. However, a huge number of parameters may not be fully stored on chip, hence requiring large-sized off-chip memory to store them and intensive off-chip memory accesses during run time.

Furthermore, these intensive off-chip accesses are significantly more expensive in terms of latency and energy than on-chip operations, hence exacerbating the overall system energy.

- *Complex and Intensive Operations*: The complexity of operations in NN algorithms depends on the computational model and the network architecture. For instance, DNNs and SNNs have different complexity of operations since DNNs typically employ multiply-and-accumulate (MAC) while SNNs employ more bio-plausible operations like leaky-integrate-and-fire (LIF). Besides, more complex neural architectures (e.g., residual networks) may require additional operations to accommodate the architectural variations. These complex architectures with a huge number of parameters also lead to intensive neural operations (e.g., a large number of MAC operations in DNNs), thereby requiring high computing power/energy during model execution.

In summary, *achieving acceptable accuracy for the given ML applications while meeting the latency, memory, and power/energy constraints of the embedded ML systems is not a trivial task.*

This volume of the book focuses on addressing these challenges from a hardware perspective, with multiple solutions towards the design of efficient accelerators, memory, and emerging technology substrates for embedded ML systems. A brief outline of the book along with the section structure is as follows.

1. *Efficient Hardware Acceleration*: To improve the performance efficiency of NN algorithms, ML-focused hardware acceleration has been considered an effective approach. Therefore, the first part of the book focuses on hardware acceleration techniques for embedded ML systems.

   - Chapter "Massively Parallel Neural Processing Array (MPNA): A CNN Accelerator for Embedded Systems" develops a convolutional neural network (CNN) accelerator that employs efficient computing architecture coupled with dataflows that exploit parameter/data reuse on-chip.
   - Chapter "Photonic NoCs for Energy-Efficient Data-Centric Computing" discusses how an approximate computing paradigm can be used in photonic-based network-on-chip (NoC) systems to achieve energy-efficient data movement during the execution of NN and other data-centric applications.
   - Chapter "Low- and Mixed-Precision Inference Accelerators" describes the design choices and the implications of implementing low- and mixed-precision DNNs on the flexibility and energy efficiency of the inference accelerators.
   - Chapter "Designing Resource-Efficient Hardware Arithmetic for FPGA-Based Accelerators Leveraging Approximations and Mixed Quantizations" explains the designs of resource-efficient hardware arithmetic for field programmable gate array (FPGA)-based DNN accelerators by leveraging approximation and quantization.

- Chapter "Efficient Hardware Acceleration of Emerging Neural Networks for Embedded Machine Learning: An Industry Perspective" provides a comprehensive review of an industry perspective on the efficient hardware acceleration for emerging neural networks targeting embedded applications.

2. *Memory Design and Optimizations*: Oftentimes, memories are one of the biggest bottlenecks when processing NN algorithms due to frequent accesses to them (to load and store parameters and activations during execution) and their physical limitations of high latency- and energy-per-access. Hence, the second part of the book explores techniques for memory design and optimizations for embedded ML systems.

- Chapter "An Off-Chip Memory Access Optimization for Embedded Deep Learning Systems" discusses optimization techniques that exploit data reuse for reducing the number of DRAM memory accesses and DRAM energy-per-access for DNN hardware accelerators.
- Chapter "In-Memory Computing for AI Accelerators: Challenges and Solutions" explains the challenges of designing energy-efficient in-memory computing (IMC) for DNN hardware accelerators and then describes the recent advances to address these challenges.
- Chapter "Efficient Deep Learning Using Non-Volatile Memory Technology in GPU Architectures" describes how non-volatile memory (NVM) technologies can be used in graphic processing unit (GPU) architectures for deep learning acceleration.
- Chapter "SoC-GANs: Energy-Efficient Memory Management for System-On-Chip Generative Adversarial Networks" discusses the on-chip memory management for achieving energy-efficient generative adversarial network (GAN) acceleration on system-on-chip architecture.
- Chapter "Using Approximate DRAM for Enabling Energy-Efficient, High-Performance Deep Neural Network Inference" presents how to leverage approximate DRAM with reduced voltage and reduced latency for achieving energy-efficient and high-performance DNN inference.

3. *Emerging Substrates*: To improve the efficiency of NN acceleration, recent efforts have also explored new device technologies for the corresponding hardware accelerators, such as silicon photonics, and NVM technologies like resistive random access memory (ReRAM), phase change memory (PCM), and spin-transfer torque magnetic RAM (STT-MRAM). The fourth part of the book focuses on emerging substrates for embedded ML systems.

- Chapter "On-Chip DNN Training for Direct Feedback Alignment in FeFET" studies the benefits of using a ferroelectric field-effect transistor (FeFET) for DNN training on-chip leveraging the direct feedback alignment (DFA) algorithm.
- Chapter "Platform-Based Design of Embedded Neuromorphic Systems" describes how platform-based design methodologies can be employed to

develop neuromorphic systems considering different manufacturing processes/NVM technologies.
- Chapter "Light Speed Machine Learning Inference on the Edge" presents a fast silicon photonic-based BNN accelerator by employing microring resonator (MR)-based optical devices for light-speed computing.
- Chapter "Low-Latency, Energy-Efficient In-DRAM CNN Acceleration with Bit-Parallel Unary Computing" explains how to enable low-latency and energy-efficient CNN acceleration in DRAM by leveraging bit-parallel unary computing.

We hope this book provides a comprehensive review and useful information on the recent advances in embedded machine learning for cyber-physical, IoT, and edge-computing applications.

Fort Collins, CO, USA   Sudeep Pasricha
Abu Dhabi, UAE   Muhammad Shafique
October 25, 2022

Acknowledgments

This book would not be possible without the contributions of many researchers and experts in the field of embedded systems, machine learning, IoT, edge platforms, and cyber-physical systems. We would like to gratefully acknowledge the contributions of Rachmad Putra (Technische Universität Wien), Muhammad Abdullah Hanif (New York University, Abu Dhabi), Febin Sunny (Colorado State University), Asif Mirza (Colorado State University), Mahdi Nikdast (Colorado State University), Ishan Thakkar (University of Kentucky), Maarten Molendijk (Eindhoven University of Technology), Floran de Putter (Eindhoven University of Technology), Henk Corporaal (Eindhoven University of Technology), Salim Ullah (Technische Universität Dresden), Siva Satyendra Sahoo (Technische Universität Dresden), Akash Kumar (Technische Universität Dresden), Arnab Raha (Intel), Raymond Sung (Intel), Soumendu Ghosh (Purdue University), Praveen Kumar Gupta (Intel), Deepak Mathaikutty (Intel), Umer I. Cheema (Intel), Kevin Hyland (Intel), Cormac Brick (Intel), Vijay Raghunathan (Purdue University), Gokul Krishnan (Arizona State University), Sumit K. Mandal (Arizona State University), Chaitali Chakrabarti (Arizona State University), Jae-sun Seo (Arizona State University), Yu Cao (Arizona State University), Umit Y. Ogras (University of Wisconsin, Madison), Ahmet Inci (University of Texas, Austin), Mehmet Meric Isgenc (University of Texas, Austin), and Diana Marculescu (University of Texas, Austin), Rehan Ahmed (National University of Sciences and Technology, Islamabad), Muhammad Zuhaib Akbar (National University of Sciences and Technology, Islamabad), Lois Orosa (ETH Zürich, Skanda Koppula (ETH Zürich), Konstantinos Kanellopoulos (ETH Zürich), A. Giray Yağlikçi (ETH Zürich), Onur Mutlu (ETH Zürich), Saideep Tiku (Colorado State University), Liping Wang (Colorado State University), Xiaofan Zhang (University of Illinois Urbana-Champaign), Yao Chen (University of Illinois Urbana-Champaign), Cong Hao (University of Illinois Urbana-Champaign), Sitao Huang (University of Illinois Urbana-Champaign), Yuhong Li (University of Illinois Urbana-Champaign), Deming Chen (University of Illinois Urbana-Champaign), Alexander Wendt (Technische Universität Wien), Horst Possegger (Technische Universität Graz), Matthias Bittner (Technische Universität Wien), Daniel Schnoell (Technische Universität Wien), Matthias Wess (Technische Universität Wien),

Dušan Malić (Technische Universität Graz), Horst Bischof (Technische Universität Graz), Axel Jantsch (Technische Universität Wien), Floran de Putter (Eindhoven University of Technology), Alberto Marchisio (Technische Universitat Wien), Fan Chen (Indiana University Bloomington), Lakshmi Varshika Mirtinti (Drexel University), Anup Das (Drexel University), Supreeth Mysore Shivanandamurthy (University of Kentucky), Sayed Ahmad Salehi (University of Kentucky), Biresh Kumar Joardar (University of Houston), Janardhan Rao Doppa (Washington State University), Partha Pratim Pande (Washington State University), Georgios Zervakis (Karlsruhe Institute of Technology), Mehdi B. Tahoori (Karlsruhe Institute of Technology), Jörg Henkel (Karlsruhe Institute of Technology), Zheyu Yan (University of Notre Dame), Qing Lu (University of Notre Dame), Weiwen Jiang (George Mason University), Lei Yang (University of New Mexico), X. Sharon Hu (University of Notre Dame), Jingtong Hu (University of Pittsburgh), Yiyu Shi (University of Notre Dame), Beatrice Bussolino (Politecnico di Torino), Alessio Colucci (Technische Universität Wien), Vojtech Mrazek (Brno University of Technology), Maurizio Martina (Politecnico di Torino), Guido Masera (Politecnico di Torino), Ji Lin (Massachusetts Institute of Technology), Wei-Ming Chen (Massachusetts Institute of Technology), Song Han (Massachusetts Institute of Technology), Yawen Wu (University of Pittsburgh), Yue Tang (University of Pittsburgh), Dewen Zeng (University of Notre Dame), Xinyi Zhang (University of Pittsburgh), Peipei Zhou (University of Pittsburgh), Ehsan Aghapour (University of Amsterdam), Yujie Zhang (National University of Singapore), Anuj Pathania (University of Amsterdam), Tulika Mitra (National University of Singapore), Hiroki Matsutani (Keio University), Keisuke Sugiura (Keio University), Soonhoi Ha (Seoul National University), Donghyun Kang (Seoul National University), Ayush Mittal (Colorado State University), Bharath Srinivas Prabakaran (Technische Universität Wien), Ganapati Bhat (Washington State University), Dina Hussein (Washington State University), Nuzhat Yamin (Washington State University), Rafael Makrigiorgis (University of Cyprus), Shahid Siddiqui (University of Cyprus), Christos Kyrkou (University of Cyprus), Panayiotis Kolios (University of Cyprus), Theocharis Theocharides (University of Cyprus), Anil Kanduri (University of Turku), Sina Shahhosseini (University of California, Irvine), Emad Kasaeyan Naeini (University of California, Irvine), Hamidreza Alikhani (University of California, Irvine), Pasi Liljeberg (University of Turku), Nikil Dutt (University of California, Irvine), Amir M. Rahmani (University of California, Irvine), Sizhe An (University of Wisconsin-Madison), Yigit Tuncel (University of Wisconsin-Madison), Toygun Basaklar (University of Wisconsin-Madison), Aditya Khune (Colorado State University), Rozhin Yasaei (University of California, Irvine), Mohammad Abdullah Al Faruque (University of California, Irvine), Kruttidipta Samal (University of Nebraska, Lincoln), Marilyn Wolf (University of Nebraska, Lincoln), Joydeep Dey (Colorado State University), Vipin Kumar Kukkala (Colorado State University), Sooryaa Vignesh Thiruloga (Colorado State University), Marios Pafitis (University of Cyprus), Antonis Savva (University of Cyprus), Yue Wang (New York University), Esha Sarkar (New York University), Saif Eddin Jabari (New York University Abu Dhabi), Michail Maniatakos (New York University Abu Dhabi), Mahum Naseer (Technische Universität Wien), Iram

Tariq Bhatti (National University of Sciences and Technology, Islamabad), Osman Hasan (National University of Sciences and Technology, Islamabad), Hao Fu (New York University), Alireza Sarmadi (New York University), Prashanth Krishnamurthy (New York University), Siddharth Garg (New York University), Farshad Khorrami (New York University), Priyadarshini Panda (Yale University), Abhiroop Bhattacharjee (Yale University), Abhishek Moitra (Yale University), Ihsen Alouani (Queen's University Belfast), Stefanos Koffas (Delft University of Technology), Behrad Tajalli (Radboud University), Jing Xu (Delft University of Technology), Mauro Conti (University of Padua), and Stjepan Picek (Radboud University).

This work was partially supported by the National Science Foundation (NSF) grants CCF-1302693, CCF-1813370, and CNS-2132385; by the NYUAD Center for Interacting Urban Networks (CITIES), funded by Tamkeen under the NYUAD Research Institute Award CG001, Center for Cyber Security (CCS), funded by Tamkeen under the NYUAD Research Institute Award G1104, and Center for Artificial Intelligence and Robotics (CAIR), funded by Tamkeen under the NYUAD Research Institute Award CG010; and by the project "eDLAuto: An Automated Framework for Energy-Efficient Embedded Deep Learning in Autonomous Systems," funded by the NYUAD Research Enhancement Fund (REF). The opinions, findings, conclusions, or recommendations presented in this book are those of the authors and do not necessarily reflect the views of the National Science Foundation and other funding agencies.

Contents

Part I
Efficient Hardware Acceleration for Embedded Machine Learning

Massively Parallel Neural Processing Array (MPNA): A CNN Accelerator for Embedded Systems

Rachmad Vidya Wicaksana Putra, Muhammad Abdullah Hanif, and Muhammad Shafique

1 Introduction

Machine Learning (ML) algorithms have rapidly proliferated into different field of applications, ranging from object recognition, automotive, healthcare to business [22, 33]. The field of ML encompasses several algorithms, and the most influential ones in recent years are the Deep Neural Networks (DNNs) or Deep Learning [13, 14]. The reason is that DNNs have achieved state-of-the-art accuracy and even surpassed humans' accuracy, especially through Convolutional Neural Networks (CNNs) [35]. In recent years, larger and deeper CNNs have been proposed in the literature since they can achieve higher accuracy than the smaller ones, thereby becoming the key enabler for many applications (e.g., advanced vision processing). Such large CNN models typically require a huge memory footprint, intensive computations, and energy consumption [5]. Furthermore, recent trends show that many ML applications are moving toward mobile and embedded platforms, such as Cyber-Physical System (CPS) and IoT-Edge devices, due to performance, privacy, and security reasons. These embedded platforms typically employ the pretrained CNN models for performing inferences. However, performing such an inference is challenging because the embedded platforms are resource- and power/energy-constrained. For instance, the ResNet-152 model needs more than 200MB of memory footprint and 11.3 billion operations to perform an inference for a single input image [16]. Such a high amount of processing is infeasible

R. V. W. Putra (✉)
Embedded Computing Systems, Institute of Computer Engineering, Technische Universität Wien, Vienna, Austria
e-mail: rachmad.putra@tuwien.ac.at

M. A. Hanif · M. Shafique
Division of Engineering, New York University Abu Dhabi, Abu Dhabi, UAE
e-mail: mh6117@nyu.edu; muhammad.shafique@nyu.edu

© The Author(s), under exclusive license to Springer Nature Switzerland AG 2024
S. Pasricha, M. Shafique (eds.), *Embedded Machine Learning for Cyber-Physical, IoT, and Edge Computing*, https://doi.org/10.1007/978-3-031-19568-6_1

to be performed by embedded platforms in an efficient manner. Therefore, *it is necessary to design a specialized hardware accelerator that efficiently performs CNN inferences for embedded systems.*

1.1 State of the Art and Their Limitations

A significant amount of works has been carried out for proposing specialized CNN accelerators. Some of the accelerators aim at accelerating the *un-structurally sparse networks* by exploiting sparse weights and/or activations to decrease the computational requirements, which is expected to improve the performance and energy efficiency [1, 11, 12, 17, 19, 24, 29, 38]. However, recent studies show that employing sparsity does not directly lead to energy savings, and it requires more complex and sophisticated accelerator designs to achieve high performance which incur considerably high power/energy and area consumption [9, 37]. Moreover, since these accelerators typically employ the Rectified Linear Unit (ReLU) to convert all the negative activations to zeros, they cannot efficiently handle the advanced activation functions that do not result in high sparsity (e.g., Leaky ReLU [15, 27]), thereby decreasing their efficiency benefits. Meanwhile, the other accelerators aim at accelerating *dense networks* for achieving high performance and energy efficiency [6, 7, 10, 18, 21, 25, 26, 34, 36]. They can also be used for accelerating the *structurally sparse networks* by tailoring the dataflows to the respective accelerator architectures [2, 37]. However, they also employ ReLU operations which make them inefficient for computing the advanced activation functions and hence decreasing their efficiency gains. Furthermore, most of these accelerators consume relatively large area and high power/energy which are not suitable for embedded applications [10, 18]. Moreover, despite showing a good performance for the convolutional (CONV) layers, many of these accelerators offer limited acceleration for the fully connected (FC) layers, as we will show with the help of a motivational case study in Sect. 1.2.

1.2 Motivational Case Study and Research Challenges

Motivational Case Study To obtain high performance and energy efficiency, state-of-the-art CNN accelerators exploit the reuse of weights and activations (including partial sums), hence reducing the number of off-chip memory (DRAM) accesses [5, 30, 31]. In this respect, the conventional systolic array-based designs (like Google's TPU [18]) are very effective, as each Processing Element (PE) in the Systolic Array (SA) engine performs three key tasks, as follows:

- It receives data from their upstream neighbor.
- It performs basic DNN operations, i.e., multiply and accumulate (MAC).
- It passes the data along with the partial sum to their downstream neighbor.

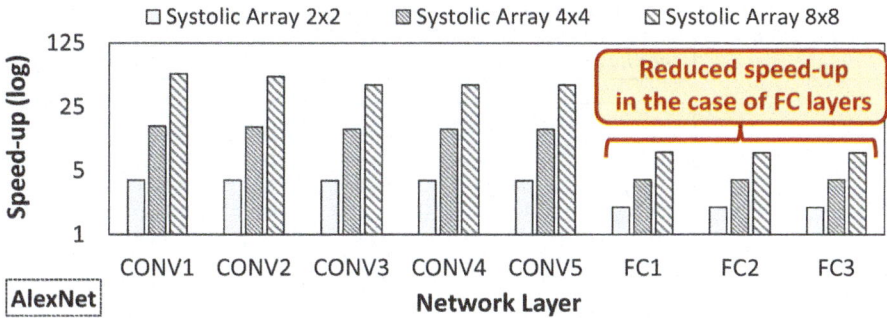

Fig. 1 Speed-up values for CONV and FC layers of AlexNet for different sizes of systolic array engines, which are normalized to the 1 × 1 systolic array engine

Therefore, the overall speed-up of the SA engine is significant for computations that involve weight and activation reuse (i.e., CONV layers), as shown in Fig. 1 for AlexNet [20]. However, if only activation reuse happens (i.e., a single input is used for multiple computations, while the weights are used only once), the speed-up of SA engines is very limited, as shown in Fig. 1. Such operations are found in the FC layers, and their conventional dataflow on an SA engine is shown in Fig. 2. These observations indicate that the conventional SA-based engines can provide a high speed-up for the CONV layers, but it does not provide comparable speed-up for the FC layers. This significantly limits the overall speed-up of CNN acceleration, especially when the networks are dominated by FC layers. Therefore, *there is a significant need for a CNN accelerator that can expedite both the CONV and FC layers to obtain a high speed-up for the complete CNN model while consuming a low operational power/energy*. However, designing such an accelerator bears a wide range of challenges, as discussed in the following.

Associated Research Challenges From results in Figs. 1 and 2, we identify the following research challenges:

- First, specialized SA-based designs need to be developed to facilitate accelerating both the CONV and FC layers without consuming significant area and power/energy overheads as compared to the conventional SA designs.
- Second, the SA designs should consider diverse dataflows of both the CONV and FC layers, while fully utilizing the available memory bandwidth. For instance, the acceleration of CONV layers requires simple, fast, yet massively parallel PEs to maximally the reuse of weights and activations (including partial sums). Meanwhile, the acceleration of FC layers should maximize the activation reuse in a single-sample batch processing. Note, the acceleration of FC layers can only exploit weight reuse in a multi-sample batch processing, which is not suitable for latency-sensitive/real-time embedded applications as targeted in this work.

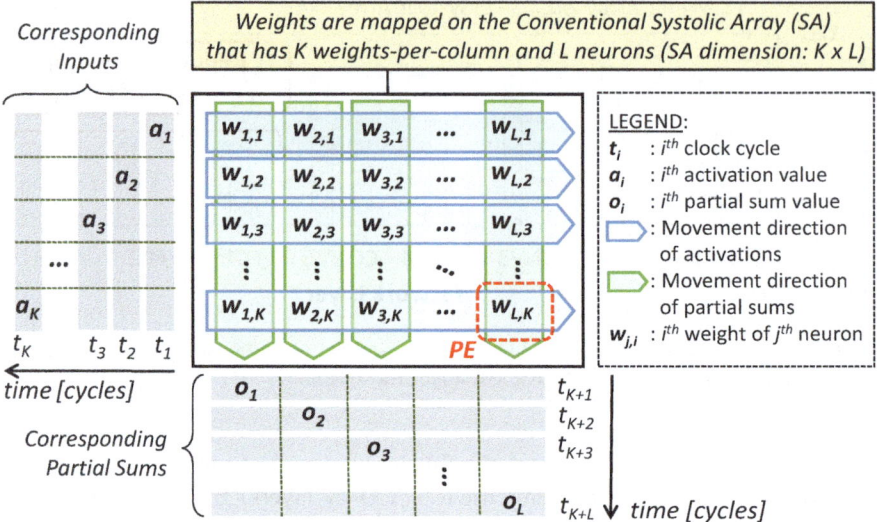

Fig. 2 The dataflow for the FC layer execution using a conventional SA engine. The input activations are fed to the SA from the left side and then shifted one step to the right toward the adjacent processing element (PE) at each cycle. The partial sums start appearing in the SA output at $K+1$ cycles

- Third, the dataflows should consider minimizing the off-chip memory (DRAM) accesses to optimize the energy consumption, since the DRAM access energy typically dominates the total CNN acceleration system energy [4, 30, 31, 35, 38].

1.3 Our Novel Contributions

To address the above research challenges, we propose a novel **Massively Parallel Neural Processing Array (MPNA)** accelerator through the following key steps:

- **A Design Methodology:** The MPNA architecture is designed using a methodology that systematically integrates heterogeneous SA designs, specialized on-chip memory, and other necessary components, while exploring different dataflow patterns to maximize data reuse and jointly accelerate the CONV and FC layers
- **Optimized Dataflows:** We propose different optimizations of dataflow patterns for enabling efficient processing on heterogeneous SA engine and maximally exploiting data reuse, thereby improving the overall processing efficiency.
- **Hardware Implementation and Evaluation:** We perform functional validation of the MPNA accelerator architecture and synthesize it using the ASIC design tools for a 28-nm CMOS technology library. Our experimental results show that the MPNA architecture offers up to $2\times$ performance speed-up and 51% energy saving as compared to the baseline accelerator. Our MPNA achieves 149.7GOP-S/W performance efficiency at 280 MHz and incurs 239 mW of operational power.

2 Preliminaries

2.1 Convolutional Neural Networks (CNNs)

In this section, we explain the fundamentals of CNN processing, which are necessary to understand the contributions of this work. A neural network is composed of a number of layers that are connected in a cascade fashion. Each layer receives inputs from the preceding layer, performs certain operations, and passes the results to the succeeding layer. A CNN, a particular type of neural networks, typically consists of four types of processing layers: convolutional (CONV) layer for extracting features, fully connected (FC) layer for classification, activation layer for introducing non-linearity, and pooling layer for sub-sampling. Each layer of CONV processing is illustrated in Fig. 3a and can be represented using for loops line in Fig. 3b. Furthermore, the FC layer processing can also be represented using the same loops

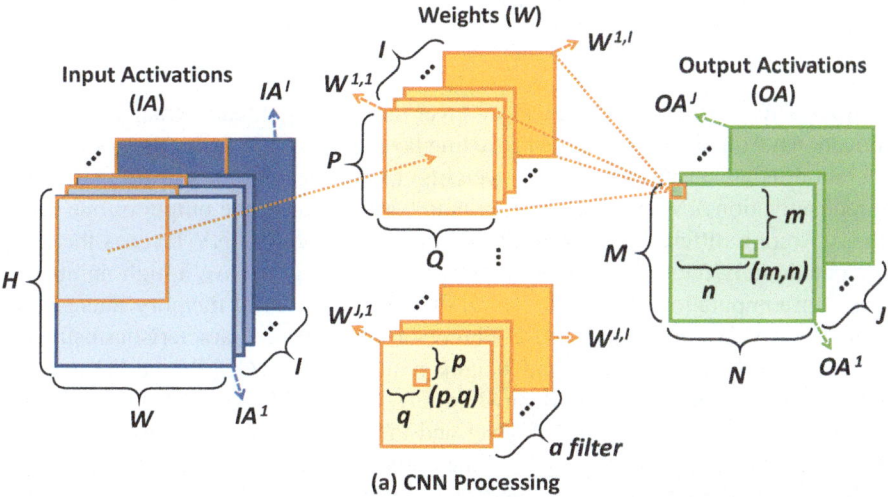

(a) CNN Processing

```
for j = 1 : J { % Loop for the output activations
  for i = 1 : I { % Loop for the input activations
    for m = 1 : M { % Loop for the rows of output activations
      for n = 1 : N { % Loop for the columns of output activations
        for p = 1 : P { % Loop for the rows of the filter weights
          for q = 1 : Q { % Loop for the columns of the filter weights
            OA^j (m, n) = OA^j (m, n) + W^{j,i}(p, q) x IA^i (m + p - 1, n + p - 1);
}}}}}} % Loops end
```

(b) Pseudocode of the CONV processing

Fig. 3 (a) Illustration of a CONV layer processing, i.e., a set of input activations are convolved with the weights for generating a set of output activations. (b) Pseudocode of the CONV layer processing. An FC layer processing can be represented using the same loops with $H = W = P = Q = 1$

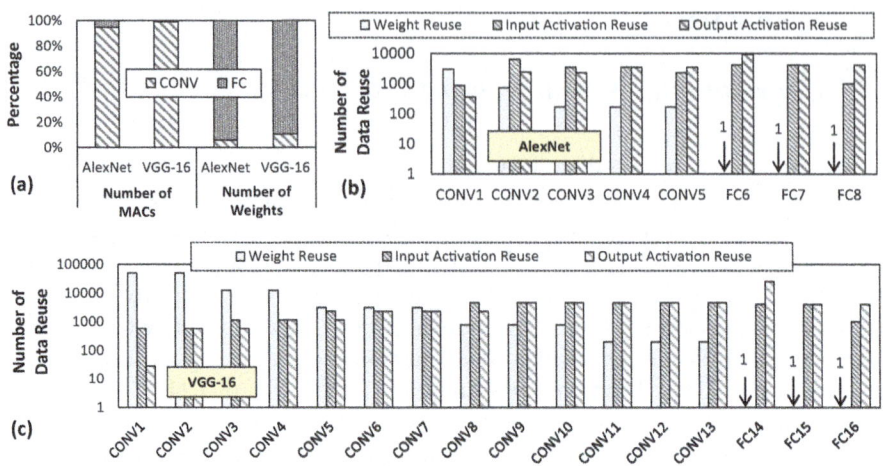

Fig. 4 (**a**) Percentage of the number of MAC operations and filter weights for the AlexNet and the VGG-16. The number of reuse for different data types, i.e., weights, input activations, and output activations for (**b**) the AlexNet and (**c**) the VGG-16

with $H = W = P = Q = 1$. A CONV layer receives input feature maps (i.e., input activations) from the inputs or the preceding layer and then performs the convolution operation using several filters (i.e., weights) to produce output feature maps (i.e., output activations), where each feature map corresponds to the output of one set of filters. Among different layers in a CNN processing, the CONV layer is the most computationally intensive as input activations and weights have a high number of reuse for computations. Meanwhile, the FC layer is the most memory intensive as weights have a low number of reuse. Figure 4a shows these characteristics using the percentage of MAC operations and weights required for the CONV and FC layers in the AlexNet and the VGG-16. Meanwhile, Fig. 4b and c shows the number of reuse for different data types in the AlexNet and the VGG-16, respectively. IA^i is the input activations at channel-i, OA^j is the output activations at channel-j, and $W^{j,i}$ is the weight filters between IA^i and OA^j. Furthermore, $OA^j(m, n)$ denotes the activation at location (m, n) in the j-th output activations. Meanwhile, $W^{j,i}(p, q)$ denotes the weight at location (p, q) in the filter kernel between IA^i and OA^j. We consider CONV stride $= 1$, unless stated otherwise. The FC layers can be considered as a special case of CONV layers where the input and the output have a 1-dimensional array, and hence they can be represented using the above terminologies.

2.2 Systolic Array (SA)

In SA-based computations, first, weights are accessed from the weight memory and then are loaded to the PEs in the array. The weights are held stationary in the PEs in

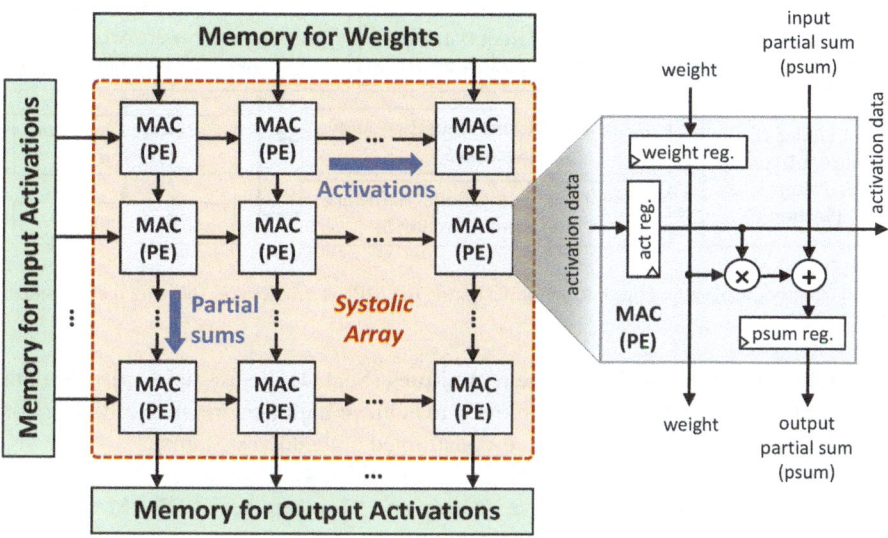

Fig. 5 The conventional SA architecture that shows the processing elements (PEs), the connections among PEs as well as between the array and memories. Note that we refer the on-chip memories to as the on-chip buffers to distinguish them from the off-chip memory (DRAM)

the manner that the same column of the array holds the weights from the same filter or neuron. During the processing, the input activations are accessed from the input activation memory, and then they are streamed into the array. These activations are passed on neighboring PEs from left to right of the array at each clock cycle, and the respective partial sums are moved downstream on neighboring PEs from top to bottom of the array. The input activations are aligned so that each input activation reaches a specific PE at the same time when its respective partial sum also reaches the same PE, hence producing a correct output partial sum. If the number of weights of a filter is larger than the number of rows in the systolic array, the output partial sums are divided into multiple sets (portions). Therefore, accumulators are required to hold the generated partial sums when the rest of the partial sums are computed in the array. A detailed description of the SA-based computations can be found in [18] (Fig. 5).

3   Our Design Methodology

We develop a novel methodology for designing an optimized CNN accelerator for embedded systems, as shown in Fig. 6. It consists of the following key steps, which are explained in detail in the subsequent sections.

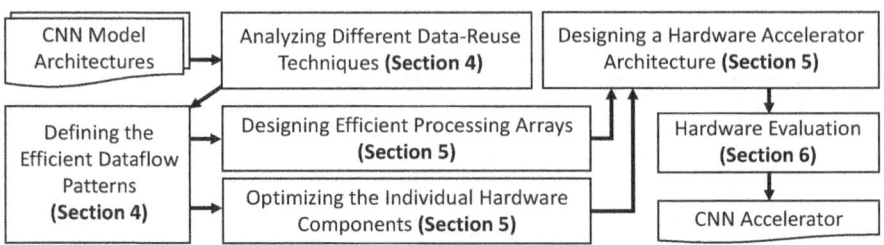

Fig. 6 Overview of our methodology for designing the MPNA accelerator, showing the key steps

1. **Analyzing different data reuse techniques (Sect. 4).** It aims at identifying data reuse techniques that can be exploited to achieve high performance efficiency of the given CNN targeting resource-constrained embedded systems.
2. **Defining the efficient dataflow patterns (Sect. 4).** It identifies the dataflow patterns that offer high data reuse on-chip and a low number of DRAM accesses, while considering the SA-based computations.
3. **Designing efficient processing arrays (Sect. 5).** The processing arrays are designed to support the selected dataflow patterns for executing the complete layers of a CNN, thereby efficiently processing the respective operations.
4. **Optimizing the individual hardware components (Sect. 5).** It aims at minimizing the latency, area, and power/energy consumption of the elementary functions that lead to the optimized system-level design.
5. **Designing a hardware accelerator architecture (Sect. 5).** We determine the key architectural parameters like the size of processing arrays, interconnect of components, and memory organization to judiciously integrate different hardware components into an MPNA architecture.
6. **Hardware evaluation (Sect. 6).** We evaluate the functionality of the MPNA architecture through functional simulations and synthesize it using the ASIC design flow with 28-nm technology library for obtaining the characteristics of area, performance, and power consumption.

4   Dataflow Optimization

4.1   Data Reuse Analysis

The CNN complexity can be estimated using the number of MAC operations and weights required by the CONV and FC layers. Table 1 provides the number of MAC operations and weights of the AlexNet and VGG-16 networks for inferring one input sample. These indicate that the CONV layers are computationally intensive due to their high number of MAC operations, while the FC layers are memory intensive due to their high number of weights that need to be accessed from memories, as

Table 1 Number of MACs and weights in the AlexNet and the VGG-16 for inferring one sample

Layer	Number of MACs		Number of weights	
	AlexNet	VGG-16	AlexNet	VGG-16
CONV layers	1.07 B	15.34 B	3.74 M	14.71 M
FC layers	58.62 M	123.63 M	58.63 M	123.64 M

indicated by Fig. 4a. Moreover, the CONV layers and the FC layers have different reuse factors for different data types (i.e., weights, input activations, and output activations), as shown in Fig. 4b and c. The *reuse factor* defines the number of MAC operations that are performed for a specific data type [31]. In CONV layers, all data types typically have comparable reuse factor, while in FC layers, weights have a significantly lower reuse factor than the activations. Furthermore, we observe that different layers in a network have different reuse factor characteristics, which is in line with previous studies [23, 30, 31]. For instance, the order of reuse factor for the AlexNet-CONV1 is weights, input activations, and output activations, while the order for the AlexNet-FC6 is output activations, input activations, and weights, respectively. This order of reuse factor is proportional to the significance of each data type to be stored longer in the on-chip memory and used for multiple computations, while avoiding costly DRAM accesses. *These observations are then exploited for determining the efficient dataflow patterns to maximally benefit from the data reuse*, thereby minimizing the number of DRAM accesses.

4.2   Proposed Dataflow Patterns

To effectively use the (off-chip and on-chip) memories and the compute capabilities of our SA-based architecture, we propose a set of dataflow patterns (as shown in Fig. 7) that can be employed by both the CONV and FC layers.

Before devising efficient dataflow patterns, we present different possible types of data reuse schemes and their dependencies on different data types as follows:

- **Weight Reuse:** It is defined by the number of times a specific weight used in the MAC operations of a given layer, which equals the size of output activations in a specific channel-j [31]. Hence, to maximally exploit the weight reuse, all input activations in a specific channel-i and the respective output activations should be available on-chip.
- **Input Activation Reuse:** It is defined by the number of times a specific input activation used by the same filter multiplied by the number of filters for the MAC operations of a given layer [31]. Hence, to maximally exploit the input activation reuse, all the weights from a specific channel-i across all filters and the respective output activations should be available on-chip.
- **Output Activation Reuse:** It is defined by the number of times partial sums accumulated to generate a specific output activation. It equals the size of a filter of a given layer [31]. Hence, to maximally exploit the output activation reuse, all

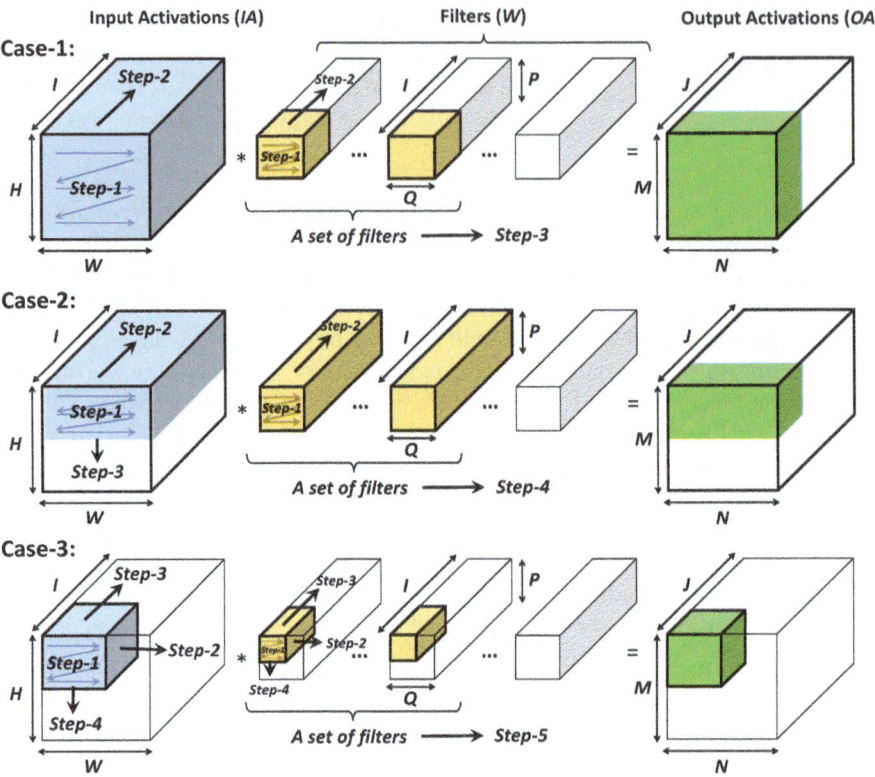

Fig. 7 Different possible dataflow patterns that are considered in this work. Portions of data stored in the on-chip buffers are highlighted with the bold boxes

input activations and weights that correspond to a specific output activation should be available on-chip.

In embedded applications, the operational power is typically limited, hence leading to limited hardware resources that can be designed and used for performing the CNN inference at a time. Therefore, *we consider a higher priority to generate a set of final output activations on-chip before starting other computations that generate different sets of partial sums (output activations)*. In this manner, the size of the accumulator units and the on-chip activation buffers are optimized. Furthermore, we leverage the data reuse observations (like the ones in Fig. 4) and the data tiling approach[1] to devise different possible dataflow patterns for embedded applications, as explained in the following:

---

[1] The data tiling defines the portion of the weights and the input activations that need to be accessed from the off-chip DRAM, stored in the on-chip memories (buffers), and then computed together to generate a portion of output activations at one time.

- **Case-1:** It has the following conditions: (1) all input and output activations can be stored in the on-chip buffer, (2) the weights can only be partially stored in the on-chip buffer, and (3) the complete output activations in a specific channel-j can be stored in a single accumulation unit. In this case, *we aim at maximally reusing the weights while considering the output activation reuse*. To do this, we define a set of weight filters where the number of filters per set equals the SA column (L), and the number of weights per filter equals a multiple of the SA row (K). This dataflow pattern is shown in Fig. 7a.
- **Case-2:** It has the following conditions: (1) all input and output activations can be stored in the on-chip buffer, (2) the weight filters can only be partially stored in the on-chip buffer, and (3) the complete output activations in a specific channel-j cannot be stored in a single accumulation unit. In this case, *we aim at maximally reusing the output activation reuse while considering the weight reuse*. To do this, we define a portion of input activations so that it includes activations from all channels. We also define a set of weight filters where the number of filters per set equals L and include all weights from each filter. This dataflow pattern is shown in Fig. 7b.
- **Case-3:** For other cases, the best configuration for data partitioning and scheduling is selected using the ROMANet methodology [31] with the following constraints. First, the number of filters per set (as a tile of weights) should be a multiple of L. Second, the number of weights per filter should be a multiple of K. This dataflow pattern is shown in Fig. 7c.

5   The MPNA Architecture

5.1   Overview of Our MPNA Architecture

Figure 8a presents the top-level view of our MPNA accelerator architecture, showing its detailed components, which are explained in the subsequent subsections. Our MPNA architecture consists of heterogeneous systolic arrays (SAs), accumulation block, pooling-and-activation block, on-chip buffers, and a control unit. The arrays receive weights and input activations from the respective on-chip buffers, perform MAC operations, and forward the partial sums to the accumulation block. Each SA is designed to support specific types of dataflow patterns and data parallelism for accelerating CONV and FC layers while incurring minimum overheads. This accumulation block holds the generated partial sums while their remaining partial sums are being computed on the arrays and then accumulates them together to generate the updated partial sums or final output activations. Afterward, the accumulator block forwards these partial sums (or final output activations) to the subsequent block for performing pooling-and-activation operations or sending them to the on-chip buffer. These data are then either used for further computations or moved to the DRAM until the rest of operations are completed.

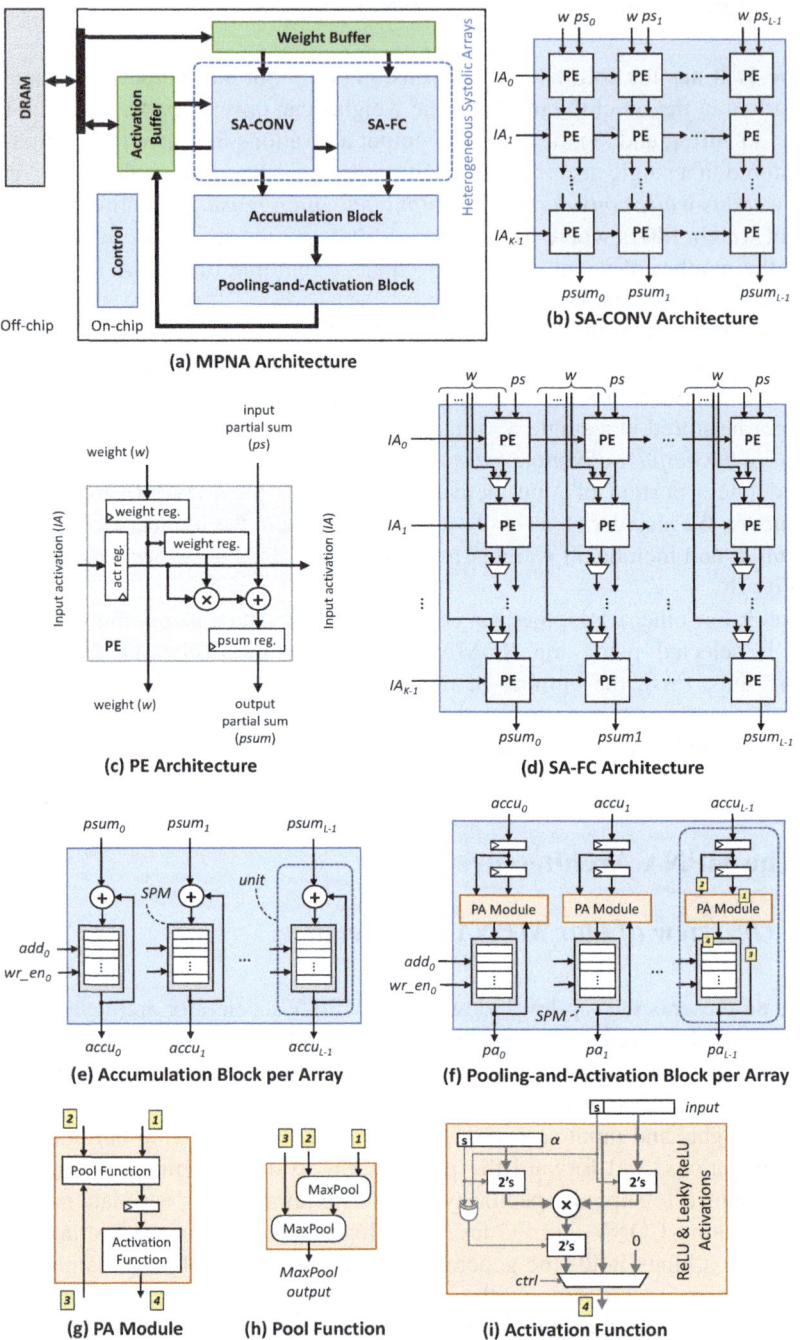

Fig. 8 MPNA architecture. (**a**) Top-level view of our MPNA architecture showing different components and their interconnections. The heterogeneous systolic arrays, (**b**) the SA-CONV, (**c**) the PE, and (**d**) the SA-FC. (**e**) Accumulation block per array for accumulating and storing the partial sums generated by the arrays. The micro-architecture of (**f**) a Pooling-and-Activation block per array and (**g**) a Pooling-and-Activation module for performing. (**h**) MaxPooling and (**i**) ReLU and Leaky-ReLU activation functions

5.2 Heterogeneous Systolic Arrays (SAs)

Based on our observations in Fig. 4, we propose for utilizing two different SAs (i.e., heterogeneous SAs) that can process different types of layers in a given network, that is, an array that targets only accelerating CONV layers (i.e., SA-CONV) and an array that targets accelerating FC layers (i.e., SA-FC).

Systolic Array for CONV Layers (SA-CONV) For CONV layers, *we employ the SA design to exploit the weight, input activation, and partial sum (output activation) reuse*, following the design of [18] (i.e., so-called SA-CONV). Figure 8b shows that our SA-CONV integrates a massively parallel array of Processing Elements (PEs) for MAC operations. Each PE receives an input activation from its left-adjacent PE and a weight and a partial sum from the top-adjacent PE and passes a generated partial sum to its bottom-adjacent PE. The left-most PEs in the array receive input from the input activation buffer, while the top-most PEs receive weights from the weight buffer. The generated partial sums are then passed to the accumulation block by the bottom-most PEs. To support such a processing dataflow on the array, weights from the same filter (or neuron) are mapped on the same column of the array. Meanwhile, weights that need to be multiplied with the same input activation are mapped on parallel columns. In this manner, activation reuse and weight reuse are maximized. Furthermore, we also include an additional register that holds a weight that is being used for MAC operation, while moving new weights (which will be used in the next iteration) to their respective locations, as shown in Fig. 8c. In this manner, the initialization time for weight loading on the array can be significantly reduced.

Systolic Array for FC Layers (SA-FC) The SA-CONV can provide a significant throughput for batch processing with large batch size due to the weight reuse in CONV layer processing. However, it can significantly affect the latency of the overall CNN inference which is an important parameter for many real-world applications. The reason is that FC layer processing in a CNN inference has low weight reuse, thereby making the SA-CONV inefficient for accelerating FC layers and decreasing the benefit of batch processing with large batch size, as shown in Fig. 2. Toward this, *we propose a novel systolic array architecture that can expedite both the CONV and FC layers* (i.e., so-called SA-FC). However, the overall bandwidth required for accelerating FC layers is huge, especially for larger CNNs. Therefore, our proposed SA-FC is designed so that it can be multiplexed for processing both the bandwidth-intensive FC and computation-intensive CONV layers. In this manner, the SA-FC can also be used for batch processing while incurring minimum area and power overheads as compared to the SA-CONV. Figure 8d shows that, unlike the SA-CONV, the SA-FC has dedicated connections from the weight buffer to each PE. It enables the system to update the weights in PEs at every clock cycle, hence providing a matching data throughput to support high-performance execution of the FC layers. The supporting dataflow for the SA-FC is shown in Fig. 9.

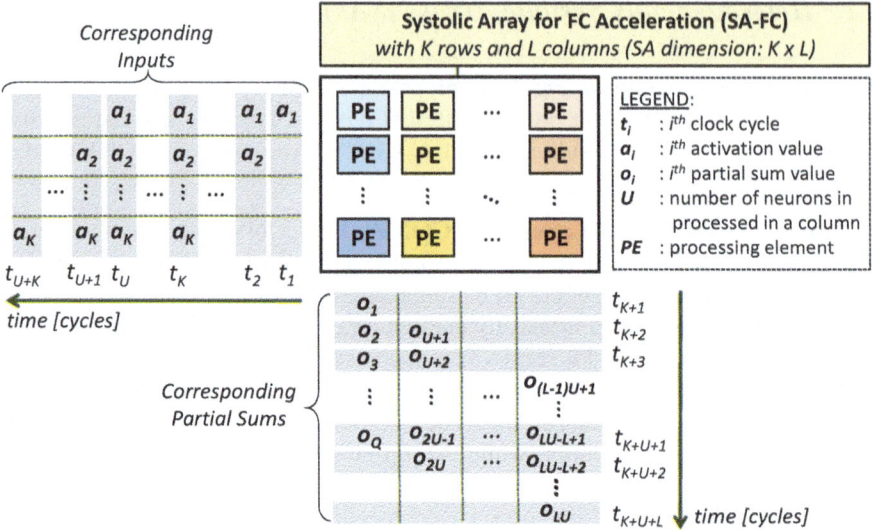

Fig. 9 The dataflow for the FC layer execution using our SA-FC engine. The input activations are fed to the array from the left side. Meanwhile, each PE will receive and store a new weight each clock cycle to generate a new partial sum accordingly cycle for FC layer processing

Integration of the SA-CONV and the SA-FC To determine the processing array design, several aspects need to be considered. First, the SA-CONV design is not efficient for processing FC layers especially in the real-time or latency-sensitive applications. Second, the SA-FC has area and power overheads over the SA-CONV, thereby limiting its array size for efficient processing. Third, the available data for SA computations are limited by the memory bandwidth. Toward this, *we propose to integrate the SA-CONV and the SA-FC as heterogeneous systolic arrays for providing a better design* with respect to the area, performance, and power/energy efficiency as compared to employing the individual design of SA-CONV or SA-FC.

5.3   Accumulation Block

The accumulation block consists of several accumulation units, whose number equals the total number of columns in the SA-CONV and the SA-FC. Each accumulation unit consists of (1) a Scratch-Pad Memory (SPM) for storing the partial sums of the output activations that are generated by the arrays and (2) an adder for accumulating the incoming partial sums with the other partial sums in the SPM. Once the final output activations are computed, the values are passed to the pooling-and-activation block for further processing. The accumulation block is shown in Fig. 8e

5.4 Pooling-and-Activation Block

After the CONV and FC layer processing, an activation function is typically employed and followed by a pooling layer to reduce the size of feature maps for subsequent layers. For these activation and pooling operations, our MPNA also provides specialized hardware. Our MPNA considers the state-of-the-art MaxPooling function, which is used by almost all modern CNNs. Furthermore, since the state-of-the activation functions are typically monotonically increasing, they can be moved after the pooling operation to reduce the number activation functions and the hardware complexity. The pooling-and-accumulation block consists of several pooling-and-accumulation units, whose number equals the total number of columns in the SA-CONV and the SA-FC. Each pooling-and-activation unit consists of (1) an SPM to hold the intermediate pooling results and (2) a pooling and activation computation module. Our MPNA architecture currently supports two state-of-the-art activation functions that are commonly used in CNNs, i.e., ReLU and Leaky-ReLU [32]. The pooling-and-activation block is shown in Fig. 8f

5.5 Hardware Configuration

We analyze the characteristics of the workloads such as the AlexNet [20] to determine the hardware configuration for our MPNA architecture. A summary of the hardware configuration for our MPNA architecture is provided in Table 2. For the AlexNet case, we make the following observations:

- The output activations of the last three CONV layers (i.e., CONV3 until CONV5) should fit in the SPM of the accumulation and pooling-and-activation blocks. Since the size of output activations in these layers is 13×13, we select the SPM size that can store up to 256 elements. In this manner, pooling and activations can be efficiently performed with local data in the blocks, thereby avoiding accessing data from the buffers or even the DRAM.

Table 2 The hardware configuration of our MPNA architecture

Module	Description
Systolic arrays	Size of SA-CONV = 8×8 of PEs
	Size of SA-FC = 8×8 of PEs
SPM	Size of SPM in each accumulation unit
	and each pooling-and-activation unit = 256 B
Weight buffer	Size of weight buffer = 32 KB
Activation buffer	Size of weight buffer = 256 KB
DRAM	Size of DRAM = 2 Gb
	Bandwidth of DRAM = 12.8 GB/s [28]

- For holding the input and output activations of the CONV3-CONV5 layers on-chip, we select a 256-KB activation buffer for two arrays. The reason is that this buffer is greater than four times of the $13 \times 13 \times 384$, i.e., the biggest size of input activations across CONV3-CONV5 layers (i.e., CONV4). In this manner, we can maximize the activation reuse.
- We select the size of 8×8 of PEs for each processing array, as it provides high parallelism while requiring relatively low off-chip memory bandwidth, as compared to the SA-FC design with the same number of PEs.

6   Evaluation Methodology

For evaluating the MPNA architecture, we build the MPNA design using RTL codes and then perform the logic simulation through the ModelSim for functional and timing validations. Afterward, we synthesize the design for a 28-nm technology using the Synopsys Design Compiler to extract critical path delay, area, and power. We also employ the CACTI 7.0 [3] for modeling the off-chip and on-chip memories and then estimating the respective latency, area, and power/energy. We compare our SA-FC with the SA-CONV to evaluate the overheads. Afterward, we compare our MPNA design with the conventional SA-based accelerators (as the baselines) across different array sizes (i.e., 2×2, 4×4, and 8×8). Furthermore, we compare our MPNA accelerator with several well-known CNN accelerators such as Eyeriss [8], SCNN [29], and FlexFlow [25]. In this evaluation, we consider the AlexNet [20] as the workload.

7   Experimental Results and Discussion

7.1   Systolic Array Evaluation

We first evaluate our SA-CONV and SA-FC designs to obtain their profiles on area and power, and the results are shown in Fig. 10a–b. This figure shows that our SA-FC architecture incurs insignificant overheads as compared to the SA-CONV architecture, i.e., 2.1% area and 4.4% power overheads on average across different sizes of arrays. The reason is that our SA-FC design employs a relatively simple additional modifications for each PE (i.e., multiplexer and wires), thereby consuming significantly smaller area and power as compared to the combined components in a PE (i.e., registers, multiplier, addition, and wires).

In terms of performance, the experimental results are shown in Fig. 10c. This figure indicates that the SA-FC 8×8 achieves $8.1\times$ speed-up as compared to the SA-CONV 8×8 when accelerating the FC layers. This performance improvement is due to the micro-architectural enhancements in the SA-FC (i.e., multiplexer

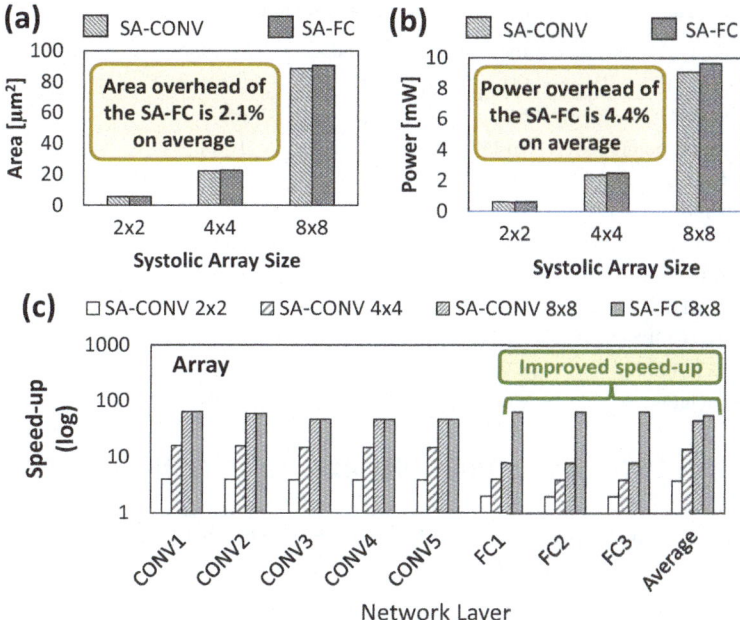

Fig. 10 Comparisons of the SA-CONV with the SA-FC in terms of (**a**) area and (**b**) power. (**c**) Performance speed-up of different array designs that are normalized to the SA-CONV 1 × 1 for the AlexNet workload

and wires) that can provide data (i.e., weights and activations) timely to PEs for producing results each clock cycle.

7.2 Comparison with Other CNN Accelerators

The comparisons of our MPNA accelerator with the state-of-the-art accelerators are summarized in Table 3. This table shows that, in general, our MPNA accelerator achieves competitive characteristics as compared to other accelerators for a full CNN acceleration (i.e., for both the CONV and FC layers).

7.2.1 Performance Evaluation

Figure 11 shows the performance comparison between our MPNA accelerator and the conventional SA-based accelerators (i.e., SA-CONV-based designs). Our MPNA design achieves up to 2× speed-up for expediting all layers of the AlexNet as compared to the SA-CONV 8 × 8-based accelerators. The speed-up on CONV layers is due to the higher parallelism of computations offered by the heterogeneous

Table 3 Comparisons to the state-of-the-art CNN accelerators

Evaluated aspects	Eyeriss [8]	SCNN [29]	FlexFlow [25]	MPNA (this work)
Technology (nm)	65	16	65	28
Precision (fixed-point)	16-bit	16-bit	16-bit	8-bit
Number of PEs (MACs)	168	64	256	128
On-chip memory (KB)	181.5	1024	64	288
Area (mm^2)	12.25	7.9	3.89	2.34
Power (mW)	278	NA	~1000	239
Frequency (MHz)	100–250	1000	1000	280
Performance (GOPS)	23.1	NA	420	35.8
Efficiency (GOPS/W)	83.1	NA	300–500	149.7
Acceleration target	CONV	CONV	CONV	CONV+FC

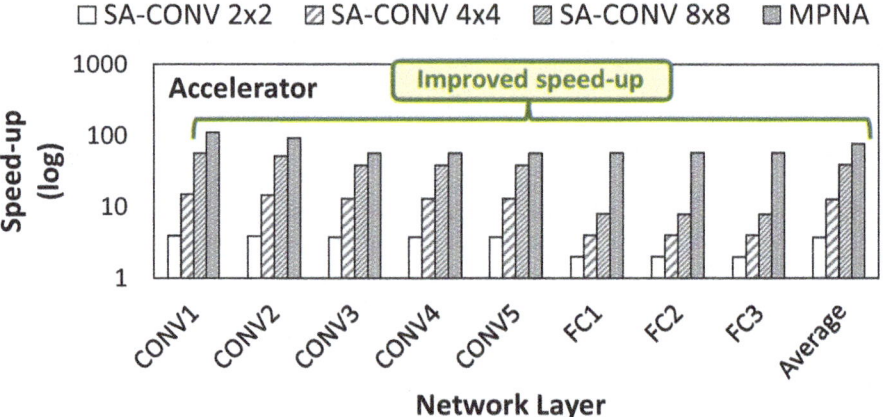

Fig. 11 Performance speed-up of different SA-based accelerators that are normalized to the accelerator with SA-CONV 1 × 1 for the AlexNet workload

arrays in the MPNA as compared to the SA-CONV-based designs with smaller array sizes (i.e., 2 × 2, 4 × 4, and 8 × 8). Meanwhile, the speed-up on FC layers is mainly due to the enhancements in SA-FC that enable the array generating output partial sums at each clock cycle, thereby providing a higher throughput as compared to the SA-CONV-based designs. Our MPNA design also achieves better performance than Eyeriss [8] especially for FC layers, since Eyeriss only prioritizes for accelerating CONV layers through the row stationary dataflow, and Eyeriss also does not disclose their speed-up for the FC layers. Furthermore, the MPNA can operate at 280 MHz with 35.8 GOPs, which is higher than Eyeriss, as shown in Table 3. Although the MPNA has lower performance (GOPs) and operating frequency than other designs (e.g., FlexFlow and SCNN), it offers other important benefits for embedded systems (e.g., power/energy and area), which will be discussed in the following subsections.

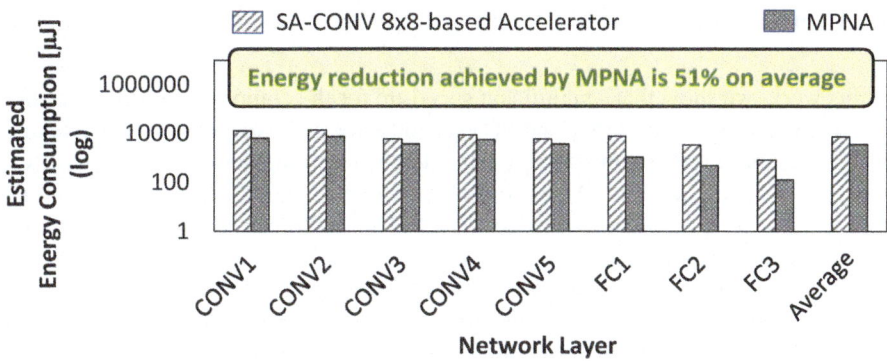

Fig. 12 Estimated energy consumption of the conventional SA-based accelerator with 8×8 array size and our MPNA for the AlexNet workload

7.2.2 Power and Energy Consumption

For operational power, our MPNA consumes 239 mW on average which is dominated by the Pooling-and-Activation block, mainly due to its local memory (SPM)
operations (i.e., memory accesses). This power number is comparable to the
operational power of Eyeriss, and however our MPNA offers better acceleration
for a complete CNN architecture (i.e., including both the CONV and FC layers).
Furthermore, power consumption of the MPNA is also significantly lower than
other designs (e.g., 1W power for FlexFlow). Our MPNA achieves a performance
efficiency of ~149GOPs/W, which is considered high for embedded systems with
power budgets, such as the battery-powered IoT devices. In terms of energy
consumption, our MPNA achieves about 51% of energy saving as compared to
the conventional SA-based accelerator, as shown in Fig. 12. The reason is that our
MPNA effectively exploits (1) data reuse through the optimized dataflows and (2)
high parallelism from the heterogeneous arrays that lead to reduced processing
latency, thereby decreasing the energy consumption.

7.2.3 Area Footprint

Our MPNA design occupies 2.34 mm^2 area which encompasses the computation
part (i.e., about 1.38 mm^2) and the on-chip memories (i.e., about 0.96 mm^2),
including both the activation and weight buffers. Furthermore, Table 3 shows that
our MPNA accelerator occupies a competitively small area as compared to other
state-of-the-art CNN accelerators. This characteristic is especially beneficial for
embedded applications which typically require small-sized hardware implementation to enable their use cases, such as mobile and wearable devices.

8 Conclusion

In this work, we show that a significant speed-up for the complete SNN architecture (i.e., including both the CONV and FC layers) can be achieved through a synergistic design methodology encompassing (1) the dataflow optimization that exploit different types of data reuse and (2) the MPNA architecture with heterogeneous systolic arrays and specialized on-chip buffers. The MPNA architecture is synthesized for a 28-nm technology through the ASIC design flow and evaluated for performance, power/energy, and area. The results show performance gain of our design as compared to the conventional systolic array-based accelerators. They also show that our MPNA achieves better power/energy and area than several state-of-the-art CNN accelerators. All these results suggest that our MPNA accelerator is suitable for various resource- and power-/energy-constrained embedded systems.

Acknowledgments This work was partly supported by the Indonesia Endowment Fund for Education (IEFE/LPDP) Graduate Scholarship Program, Ministry of Finance, Republic of Indonesia, under Grant PRJ-1477/LPDP.3/2017.

References

1. Albericio, J., Judd, P., Hetherington, T., Aamodt, T., Jerger, N.E., Moshovos, A.: Cnvlutin: Ineffectual-neuron-free deep neural network computing. In: Proceedings of the 43rd International Symposium on Computer Architecture, pp. 1–13 (2016)
2. Anwar, S., Hwang, K., Sung, W.: Structured pruning of deep convolutional neural networks. J. Emerg. Technol. Comput. Syst. **13**(3) (2017). https://doi.org/10.1145/3005348
3. Balasubramonian, R., Kahng, A.B., Muralimanohar, N., Shafiee, A., Srinivas, V.: Cacti 7: New tools for interconnect exploration in innovative off-chip memories. ACM Trans. Archit. Code Optim. **14**, 1–25 (2017)
4. Capra, M., Peloso, R., Masera, G., Ruo Roch, M., Martina, M.: Edge computing: a survey on the hardware requirements in the internet of things world. Future Int. **11**(4), 100 (2019). https://doi.org/10.3390/fi11040100. https://www.mdpi.com/1999-5903/11/4/100
5. Capra, M., Bussolino, B., Marchisio, A., Shafique, M., Masera, G., Martina, M.: An updated survey of efficient hardware architectures for accelerating deep convolutional neural networks. Future Int. **12**(7), 113 (2020)
6. Chen, T., Du, Z., Sun, N., Wang, J., Wu, C., Chen, Y., Temam, O.: Diannao: A small-footprint high-throughput accelerator for ubiquitous machine-learning. In: 19th International Conference on Architectural Support for Programming Languages and Operating Systems, pp. 269–284 (2014). https://doi.org/10.1145/2541940.2541967
7. Chen, Y.H., Emer, J., Sze, V.: Eyeriss: A spatial architecture for energy-efficient dataflow for convolutional neural networks. In: 2016 ACM/IEEE 43rd Annual International Symposium on Computer Architecture, pp. 367–379 (2016). https://doi.org/10.1109/ISCA.2016.40
8. Chen, Y.H., Krishna, T., Emer, J.S., Sze, V.: Eyeriss: An energy-efficient reconfigurable accelerator for deep convolutional neural networks. IEEE J. Solid-State Circuits **52**(1), 127–138 (2016)
9. Chen, Y., Yang, T.J., Emer, J., Sze, V.: Understanding the limitations of existing energy-efficient design approaches for deep neural networks. In: 2018 Proceedings of SysML Conference (2018)

10. Fowers, J., Ovtcharov, K., Papamichael, M., Massengill, T., Liu, M., Lo, D., Alkalay, S., Haselman, M., Adams, L., Ghandi, M., Heil, S., Patel, P., Sapek, A., Weisz, G., Woods, L., Lanka, S., Reinhardt, S.K., Caulfield, A.M., Chung, E.S., Burger, D.: A configurable cloud-scale DNN processor for real-time AI. In: 2018 ACM/IEEE 45th Annual International Symposium on Computer Architecture (ISCA), pp. 1–14 (2018). https://doi.org/10.1109/ISCA.2018.00012

11. Gondimalla, A., Chesnut, N., Thottethodi, M., Vijaykumar, T.: Sparten: A sparse tensor accelerator for convolutional neural networks. In: Proceedings of the 52nd Annual IEEE/ACM International Symposium on Microarchitecture, pp. 151–165 (2019)

12. Han, S., Liu, X., Mao, H., Pu, J., Pedram, A., Horowitz, M.A., Dally, W.J.: EIE: Efficient inference engine on compressed deep neural network. In: 2016 ACM/IEEE 43rd Annual International Symposium on Computer Architecture, pp. 243–254 (2016). https://doi.org/10.1109/ISCA.2016.30

13. Hanif, M.A., Khalid, F., Putra, R.V.W., Rehman, S., Shafique, M.: Robust machine learning systems: Reliability and security for deep neural networks. In: 2018 IEEE 24th International Symposium on On-Line Testing And Robust System Design, pp. 257–260 (2018). https://doi.org/10.1109/IOLTS.2018.8474192

14. Hanif, M.A., Khalid, F., Putra, R.V.W., Teimoori, M.T., Kriebel, F., Zhang, J.J., Liu, K., Rehman, S., Theocharides, T., Artusi, A., et al.: Robust computing for machine learning-based systems. In: Dependable Embedded Systems, pp. 479–503. Springer, Cham (2021)

15. He, K., Zhang, X., Ren, S., Sun, J.: Delving deep into rectifiers: Surpassing human-level performance on ImageNet classification. In: Proceedings of the IEEE International Conference on Computer Vision, pp. 1026–1034 (2015)

16. He, K., Zhang, X., Ren, S., Sun, J.: Deep residual learning for image recognition. In: Proceedings of the IEEE Conference on Computer Vision and Pattern Recognition, pp. 770–778 (2016)

17. Hegde, K., Yu, J., Agrawal, R., Yan, M., Pellauer, M., Fletcher, C.: UCNN: Exploiting computational reuse in deep neural networks via weight repetition. In: 2018 ACM/IEEE 45th Annual International Symposium on Computer Architecture (ISCA), pp. 674–687. IEEE (2018)

18. Jouppi, N.P., Young, C., Patil, N., Patterson, D., Agrawal, G., Bajwa, R., Bates, S., Bhatia, S., Boden, N., Borchers, A., Boyle, R., Cantin, P., Chao, C., Clark, C., Coriell, J., Daley, M., Dau, M., Dean, J., Gelb, B., Ghaemmaghami, T.V., Gottipati, R., Gulland, W., Hagmann, R., Ho, C.R., Hogberg, D., Hu, J., Hundt, R., Hurt, D., Ibarz, J., Jaffey, A., Jaworski, A., Kaplan, A., Khaitan, H., Killebrew, D., Koch, A., Kumar, N., Lacy, S., Laudon, J., Law, J., Le, D., Leary, C., Liu, Z., Lucke, K., Lundin, A., MacKean, G., Maggiore, A., Mahony, M., Miller, K., Nagarajan, R., Narayanaswami, R., Ni, R., Nix, K., Norrie, T., Omernick, M., Penukonda, N., Phelps, A., Ross, J., Ross, M., Salek, A., Samadiani, E., Severn, C., Sizikov, G., Snelham, M., Souter, J., Steinberg, D., Swing, A., Tan, M., Thorson, G., Tian, B., Toma, H., Tuttle, E., Vasudevan, V., Walter, R., Wang, W., Wilcox, E., Yoon, D.H.: In-datacenter performance analysis of a tensor processing unit. In: 2017 ACM/IEEE 44th Annual International Symposium on Computer Architecture, pp. 1–12 (2017). https://doi.org/10.1145/3079856.3080246

19. Kim, D., Ahn, J., Yoo, S.: Zena: Zero-aware neural network accelerator. IEEE Design Test **35**(1), 39–46 (2017)

20. Krizhevsky, A., Sutskever, I., Hinton, G.E.: ImageNet classification with deep convolutional neural networks. In: Advances in Neural Information Processing Systems, pp. 1097–1105 (2012)

21. Kwon, H., Samajdar, A., Krishna, T.: Maeri: Enabling flexible dataflow mapping over DNN accelerators via reconfigurable interconnects. In: 23rd International Conference on Architectural Support for Programming Languages and Operating Systems, pp. 461–475 (2018). https://doi.org/10.1145/3173162.3173176

22. LeCun, Y., Bengio, Y., Hinton, G.: Deep learning. Nature **521**(7553), 436–444 (2015)

23. Li, J., Yan, G., Lu, W., Jiang, S., Gong, S., Wu, J., Li, X.: SmartShuttle: Optimizing off-chip memory accesses for deep learning accelerators. In: 2018 Design, Automation Test in Europe Conference Exhibition, pp. 343–348 (2018). https://doi.org/10.23919/DATE.2018.8342033
24. Li, J., Jiang, S., Gong, S., Wu, J., Yan, J., Yan, G., Li, X.: SqueezeFlow: a sparse CNN accelerator exploiting concise convolution rules. IEEE Trans. Comput. **68**(11), 1663–1677 (2019)
25. Lu, W., Yan, G., Li, J., Gong, S., Han, Y., Li, X.: FlexFlow: A flexible dataflow accelerator architecture for convolutional neural networks. In: 2017 IEEE International Symposium on High Performance Computer Architecture, pp. 553–564 (2017). https://doi.org/10.1109/HPCA.2017.29
26. Luo, T., Liu, S., Li, L., Wang, Y., Zhang, S., Chen, T., Xu, Z., Temam, O., Chen, Y.: DaDianNao: A neural network supercomputer. IEEE Trans. Comput. **66**(1), 73–88 (2016)
27. Maas, A.L., Hannun, A.Y., Ng, A.Y., et al.: Rectifier nonlinearities improve neural network acoustic models. In: Proceedings of the 30 th International Conference on Machine Learning, vol. 30, p. 3. Citeseer (2013)
28. Malladi, K.T., Nothaft, F.A., Periyathambi, K., Lee, B.C., Kozyrakis, C., Horowitz, M.: Towards energy-proportional datacenter memory with mobile dram. In: 2012 39th Annual International Symposium on Computer Architecture (ISCA), pp. 37–48. IEEE (2012)
29. Parashar, A., Rhu, M., Mukkara, A., Puglielli, A., Venkatesan, R., Khailany, B., Emer, J., Keckler, S.W., Dally, W.J.: SCNN: An accelerator for compressed-sparse convolutional neural networks. In: 2017 ACM/IEEE 44th Annual International Symposium on Computer Architecture, pp. 27–40 (2017). https://doi.org/10.1145/3079856.3080254
30. Putra, R.V.W., Hanif, M.A., Shafique, M.: DRMap: A generic dram data mapping policy for energy-efficient processing of convolutional neural networks. In: 2020 57th ACM/IEEE Design Automation Conference, pp. 1–6 (2020). https://doi.org/10.1109/DAC18072.2020.9218672
31. Putra, R.V.W., Hanif, M.A., Shafique, M.: ROMANet: Fine-grained reuse-driven off-chip memory access management and data organization for deep neural network accelerators. IEEE Trans. Very Large Scale Integr. Syst. **29**(4), 702–715 (2021). https://doi.org/10.1109/TVLSI.2021.3060509
32. Redmon, J., Divvala, S., Girshick, R., Farhadi, A.: You only look once: Unified, real-time object detection. In: Proceedings of the IEEE Conference on Computer Vision and Pattern Recognition, pp. 779–788 (2016)
33. Shafique, M., Marchisio, A., Putra, R.V.W., Hanif, M.A.: Towards energy-efficient and secure edge ai: A cross-layer framework ICCAD special session paper. In: 2021 IEEE/ACM International Conference On Computer Aided Design, pp. 1–9 (2021). https://doi.org/10.1109/ICCAD51958.2021.9643539
34. Shin, D., Lee, J., Lee, J., Lee, J., Yoo, H.J.: DNPU: An energy-efficient deep-learning processor with heterogeneous multi-core architecture. IEEE Micro **38**(5), 85–93 (2018). https://doi.org/10.1109/MM.2018.053631145
35. Sze, V., Chen, Y., Yang, T., Emer, J.S.: Efficient processing of deep neural networks: a tutorial and survey. Proc. IEEE **105**(12), 2295–2329 (2017). https://doi.org/10.1109/JPROC.2017.2761740
36. Tu, F., Yin, S., Ouyang, P., Tang, S., Liu, L., Wei, S.: Deep convolutional neural network architecture with reconfigurable computation patterns. IEEE Trans. Very Large Scale Integr. Syst. **25**(8), 2220–2233 (2017)
37. Yu, J., Lukefahr, A., Palframan, D., Dasika, G., Das, R., Mahlke, S.: Scalpel: customizing DNN pruning to the underlying hardware parallelism. ACM SIGARCH Comput. Architect. News **45**(2), 548–560 (2017)
38. Zhang, S., Du, Z., Zhang, L., Lan, H., Liu, S., Li, L., Guo, Q., Chen, T., Chen, Y.: Cambricon-X: An accelerator for sparse neural networks. In: 2016 49th Annual IEEE/ACM International Symposium on Microarchitecture, pp. 1–12 (2016). https://doi.org/10.1109/MICRO.2016.7783723

Photonic NoCs for Energy-Efficient Data-Centric Computing

Febin P. Sunny, Asif Mirza, Ishan G. Thakkar, Mahdi Nikdast, and Sudeep Pasricha

1 Introduction

To match the increasing demand in processing capabilities of modern applications, the core count in emerging manycore systems has been steadily increasing. For example, Intel Xeon processors today have up to 56 cores [1], while NVIDIA's GPU's have reported over 10,000 shader cores [2]. Emerging application-specific processors are pushing these numbers to new highs, e.g., the Cerebras AI accelerator has over 400,000 lightweight cores [3]. The increasing number of cores creates greater core-to-core and core-to-memory communication.

Electrical networks-on-chip (ENoCs), which employ conventional metallic interconnects, already dissipate very high power to support the high bandwidths and low-latency requirements of data-driven parallel applications today and are unlikely to scale to meet the demands of future applications [4]. Fortunately, chip-scale silicon photonics technology has emerged in recent years as a promising development to enhance multicore systems with light speed photonic links that can overcome the bottlenecks of slow and noise-prone electrical links. Silicon photonics can enable photonic NoCs (PNoCs) with a promise of much higher bandwidths and lower latencies than ENoCs [5].

Typical PNoC architectures employ several photonic devices such as photonic waveguides, couplers, splitters, and multiwavelength laser sources, along with mod-

F. P. Sunny (✉) · A. Mirza · M. Nikdast · S. Pasricha
Department of Electrical and Computer Engineering, Colorado State University, Fort Collins, CO, USA
e-mail: Febin.Sunny@colostate.edu; asifmirz@rams.colostate.edu; Mahdi.Nikdast@colostate.edu; sudeep@colostate.edu

I. G. Thakkar
Department of Electrical and Computer Engineering, Lexington, KY, USA
e-mail: igthakkar@uky.edu

© The Author(s), under exclusive license to Springer Nature Switzerland AG 2024
S. Pasricha, M. Shafique (eds.), *Embedded Machine Learning for Cyber-Physical, IoT, and Edge Computing*, https://doi.org/10.1007/978-3-031-19568-6_2

ulators, detectors, and switches, devised using devices such as microring resonators (MRs) and Mach-Zehnder interferometers (MZIs) [5]. PNoCs employ a laser source (either off-chip or on-chip) to generate light with one or more wavelengths, which is coupled by an optical coupler to an on-chip photonic waveguide. This waveguide guides the input optical power of potentially multiple carrier wavelengths (referred to as wavelength-division-multiplexed (WDM) transmission), via a series of optical power splitters, to the individual nodes (e.g., processing cores) on the chip. Each wavelength serves as a carrier for a data signal. Typically, multiple data signals are generated at a source node in the electrical digital domain as sequences of logical 0 and 1 voltage levels. These input electrical data signals can be modulated onto the wavelengths using a group (bank) of modulator MRs (e.g., 64-bit data modulated on 64 wavelengths), typically using on-off keying (OOK) modulation. Subsequently, the carrier wavelengths are routed over the PNoC till they reach their destination node, where the wavelengths are filtered and dropped into the waveguide by a bank of filter MRs that redirect the wavelengths to photodetectors to recover the data in the electrical domain. Each node in a PNoC can communicate to multiple other nodes through such WDM-enabled photonic waveguides in PNoCs.

Unfortunately, optical signals accumulate losses and crosstalk noise as they traverse PNoCs. This accumulation of losses necessitates high optical input power from the laser for signal-to-noise ratio compensation and to guarantee that the signal can be received at the destination node with sufficient power to enable error-free recovery of the transmitted data. Moreover, the sensitivity of an MR to the wavelength it is intended to couple with is related to its physical properties (e.g., radius, width, thickness, refractive index of the device material) that can vary with fabrication process and thermal variations. To rectify these problems, MRs require active "tuning" components to correct the impact of these variations. MRs can be tuned either by free-carrier injection (electro-optic tuning) or thermally tuning the device (thermo-optic tuning), both these techniques aim at affecting the effective refractive index of MR material, thereby the changing optical properties to counteract the impact of variations. Such tuning entails energy and power overheads, which can become significant as the number of MRs in PNoCs increases. Novel solutions are therefore urgently needed to reduce these power overheads, so that PNoCs can serve as a viable replacement to ENoCs in emerging and future manycore architectures.

One promising direction towards this goal is to utilize approximate computing in conjunction with silicon photonic communication. As computational complexity and data volumes increase for emerging applications, ensuring fault-free computing for them becomes increasingly difficult, for various reasons including the following: (i) traditional redundancy-based fault tolerance require additional resources which is hard to allocate among the increasing resource demands for big-data processing, and (ii) the ongoing scaling of semiconductor devices makes them increasingly sensitive to variations, e.g., due to imperfect fabrication processes. Approximate computing trades off "acceptable errors" during execution for reduced energy consumption and runtime and is a potential solution to both these challenges [6]. With diminishing performance-per-watt gains from Dennard scaling, leveraging such aggressive techniques to achieve higher energy-efficiency is becoming increasingly important.

In this chapter, we explore how to leverage data approximation to benefit the energy and power consumption footprints of PNoC architectures. To achieve this goal, we analyze how data approximation impacts the output quality of various applications and how that will impact energy and power requirements for laser operation, transmission, and MR tuning. The framework discussed in this chapter, called ARXON [7], extends a previous work (LORAX [8]) to implement an aggressive loss-aware approximated-packet-transmission solution that reduces power overheads due to the laser, crosstalk mitigation, and MR tuning. The main contributions of this work are as follows:

- Developed an approach that relies on approximating a subset of data transfers for applications, to reduce energy consumption in PNoCs while still maintaining acceptable output quality for applications
- Proposed a strategy that adaptively switches between two modes of approximate data transmission, based on the photonic signal loss profile along the traversed path
- Evaluated the impact of utilizing multilevel signaling (pulse-amplitude modulation) instead of conventional on-off keying (OOK) signaling along with approximate transfers for achieving significantly better energy-efficiency
- Explored how adapting existing approaches towards MR tuning and crosstalk mitigation can help further reduce power overheads in PNoCs
- Evaluated ARXON framework on multiple applications and show its effectiveness over the best-known prior work on approximating data transfers over PNoC architectures

2 Related Work

By carefully relaxing the requirement for computational correctness, so as not to impact the quality of service (QoS) of the application, it has been shown that many applications can execute with a much lower energy consumption and without significantly impacting application output quality. Some examples for approximation-tolerant applications that can save energy through this approach include audio transcoding, image processing, encoding/decoding during video streaming [9, 10], and big-data applications [11, 12]. The fast-growing repository of machine learning (ML) applications represents a particularly promising target for approximation because of the exhibited resilience to errors in parameter values by ML applications. As an example, it is possible to approximate the weights (e.g., from 32-bit floating-point to 8-bit fixed point) in convolutional and deep neural networks and with negligible degradation in the output classification accuracy [10]. Many other approaches have been proposed for ML algorithm-level approximations [13–15]. With ML applications becoming increasingly prevalent in resource-constrained environments such as mobile and IoT platforms, there is growing interest in utilizing approximated versions of ML applications for faster and lower-energy inference [16].

In general, the approximate computing solutions can be broadly categorized into four types based on their scope [17]: hardware, storage, software, and systems. The approximation of hardware components allows for a reduction in their complexity, and consequently their energy overheads [18]. For instance, an approximate full adder can utilize simpler approximated components such as XOR/XNOR-based adders and pass transistor-based multiplexers, for reduced energy consumption and operational latency [19, 20]. Additional reduction in circuit complexity and power dissipation can be enabled by avoiding XOR operations [21]. Techniques for storage approximation can include reducing refresh rates in DRAM [22, 23], which results in a deterioration of stored data, but at the advantage of increased energy-efficiency. Approaches for software approximation include algorithmic approximation that leverages domain specific knowledge [23–25]. They may also refer to approximating annotated data, variables, and high-level programming constructs (e.g., loop iterations), via annotations in the software code [26, 27]. At the system level, approximation involves modification of architectures to support imprecise operations. Attempts to design an approximate NoC architecture fall under the system level approximation category.

Several strategies have been proposed to approximate data transfers over ENoC architectures by using strategies that reduce the number of bits or packets being transmitted to reduce NoC utilization and thus reduce communication energy. An approximate ENoC for GPUs was presented in [28], where similar data packets were coalesced at the memory controller, to reduce the packets that traverse over the network. A hardware-data-approximation framework with an online data error control mechanism, which facilitates approximate matching of data patterns within a controllable value range, for ENoCs, was presented in [29]. In [30], traffic data was approximated by dropping values from a packet before it is sent on to the ENoC, at a set interval. The data is then recreated at the destination nodes using a linear interpolator-based predictor. A dual voltage ENoC is proposed in [31], where lower-priority bits in a packet are transferred at a lower voltage level, which can save energy at the cost of possible bit flips. In contrast, the higher priority bits of the packet, including header bits, are transmitted with higher voltage, ensuring a lower bit-error rate (BER) for them.

The approaches discussed so far focused on approximations for ENoCs. The complex and unique design space of approximation techniques for PNoCs remains relatively unexplored. There is a wide body of work which discusses strategies to make PNoCs more efficient and overall viable [32–54]. However, the use of approximate data communication in PNoCs for the first time was explored in [55]. The authors explored different levels of laser power for transmission of bits across a photonic waveguide, with a lower level of laser power used for bits that could be approximated, but at the cost of higher BER for these bits. The work focused specifically on approximation of floating-point data, where the least significant bits (LSBs) were transmitted at a lower laser-power level. However, the specific number of these bits to be approximated as well as the laser-power levels were decided in an application-independent manner, which ignores application-specific sensitivity to approximation. Moreover, the laser-power level is set statically and without considering the dynamic optical loss that photonic signals encounter as they traverse the network. LORAX framework [8] improved upon the work in [55] by using a loss- and QoS aware approach to adapt laser power at runtime for approximate communication in PNoCs. LORAX took into consideration the impact of adaptive approximation, varying laser-power levels, and the use of 4-pulse amplitude modulation (PAM4) on application output quality, to maximize application-specific energy savings in an acceptable manner. There may be apparent similarities between these approaches and works in say [55] or [56], but the design considerations, modeling, and implementation in hardware required for PNoC are very different from an ENoC. These differences stem from the differences in physical operation of the basic components in ENoCs (transistors operating in digital domain) and PNoCs (photonic devices such as MZIs or MRs operating in the optical analog domain).

The ARXON (AppRoXimation framework for On-chip photonic Networks) framework discussed in this chapter improves upon LORAX in multiple ways through:

(i) Considering integer data for approximation in addition to floating-point data (LORAX only considered floating-point data)
(ii) Integrating the impact of fabrication-process variations (FPV) and thermal variations (TV) on MR tuning and leveraging it for energy savings
(iii) Approximating error correction techniques, which are commonly used in PNoCs, to save more energy
(iv) Analyzing the potential for approximation for a much broader set of applications and across multiple PNoC architectures.

Section 5 of this chapter discusses the ARXON framework in detail, with evaluation results presented in Sect. 6.

3   Data Formats and Approximations

3.1   Floating-Point Data

In many applications, floating-point data can be safely considered for approximation and without impacting the QoS of the approximation, as explored and demonstrated in [8, 55]. This compatibility of floating-point data to approximation is in large part due to the way in which it is represented. The IEEE-754 standard defines a standardized floating-point data representation, which consists of three parts: sign (S), exponent (E), and mantissa (M), as shown in Fig. 1. The value of the data stored is

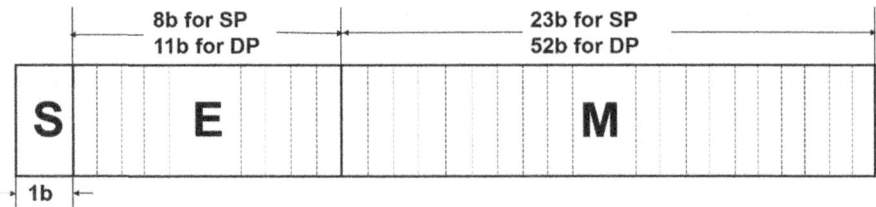

Fig. 1 IEEE 754 floating-point representation

$$X = (-1)^S \times 2^{E-\text{bias}} \times (1 + M) \tag{1}$$

where X is the resulting floating-point value. The bias values are fixed values, 127 and 1203, respectively, for single and double precision representation, and are used to ensure that the exponent is always positive, thereby eliminating the need to store the exponent sign bit. The single precision (SP) and double precision (DP) representations vary in the number of bits allocated to the E and M (see Fig. 1). E is 8 bits for SP and 11 bits for DP, while M is 23 bits for SP and 52 bits for DP. S remains 1 bit for both cases. From Eq. (1), we can observe that the S and E values notably affect the value of X. But X is typically less sensitive to alterations in M in many cases. M also takes up a significant portion of the floating-point data representation. We consider S and E as MSBs that may not be altered, whereas M makes up the LSBs that are more suitable for approximation to save energy during photonic transmission.

3.2 Integer Data

Integer data approximation is significantly more challenging, as it does not have exploitable separations in its representation, like those present in the IEEE754 standard for floating-point data. An integer data value is usually represented as an N-bit chunk of bits that can represent signed or unsigned integer data. If unsigned, the N-bits of data can be used to represent an integer value in the range from 0 to 2^{N-1}. If signed, the most significant bit represents the sign bit, and the remaining $N-1$ bits represent an integer value in the range from $-(2^{N-1}-1)$ to $+(2^{N-1}-1)$. Moreover, the number of bits, N, in an integer data word can change depending on the usage or application. N is usually in the range from 8 to 64 bits in today's platforms. Therefore, devising a generalized approach to approximate the integer data values is challenging. As a result, we have opted for an application-specific approach, where we identify possible integer variables that have larger than required size, depending on the values they handle. We deem the size of an unsigned integer variable as larger than required, if the MSBs of the variable are not holding any useful information, i.e., consisting of 0-bits. We approximate such unnecessarily large unsigned-integer variables by truncating their MSBs. We also consider LSB approximation for integer packets, when viable. However, we observed that integer data is generally not as tolerant to LSB approximation as floating-point data, so this approximation approach cannot utilized be as aggressive as LSB approximation in floating-point data and is thus used sparsely in the proposed framework.

3.3 Applications Considered for Approximation

To establish the effectiveness of this approach that focuses on approximating floating-point LSB data and integer MSB data, we evaluate the breakdown of integer and floating-point data usage across multiple applications. We selected the ACCEPT benchmark suite [25], which consists of several applications, including some from the well-known benchmark suite PARSEC [56], that exhibit a relatively strong potential for data approximations. While the applications in this suite may be executed on a single core, to adapt these to a PNoC-based multicore platform with 64 cores, we used a multi-application simulation approach where the applications were replicated across the 64 cores to emulate multi-application workloads on real systems. Along with the applications from [25], we also considered several convolutional neural-network applications from the tinyDNN [57] benchmark suite to see how the ARXON framework would perform for ML applications. The multi-application simulation approach was adopted, as this would simulate multiple applications will be running and competing for on-chip resources simultaneously, as in a real many-core system.

To count the total number of integer and floating-point packets in transit across the memory hierarchy during the simulations, we used the gem5 [58] full system-level simulator and the Intel PIN tool [59] in tandem. Figure 2 shows the breakdown of the floating-point and integer packets across the applications for large input workloads. We considered all floating-point data packets as candidates for approximation. As for integer packets, we identified specific variables and a subset of their bits ("approximable integer packets") that can be approximated safely. The goal while selecting floating-point and integer packets for approximation was to keep application-specific error to below 10% of the original output. It can be observed that while a majority of the applications have integer packets that cannot be approximated without hurting output quality significantly, most of these applications have a nontrivial percentage of their overall packet count that can be approximated. This is a promising observation that establishes the validity of the framework. However, before we describe the framework in detail (Sect. 5), we briefly cover challenges in PNoCs related to crosstalk and signal loss (Sect. 4), which this approximation approach leverages for energy savings.

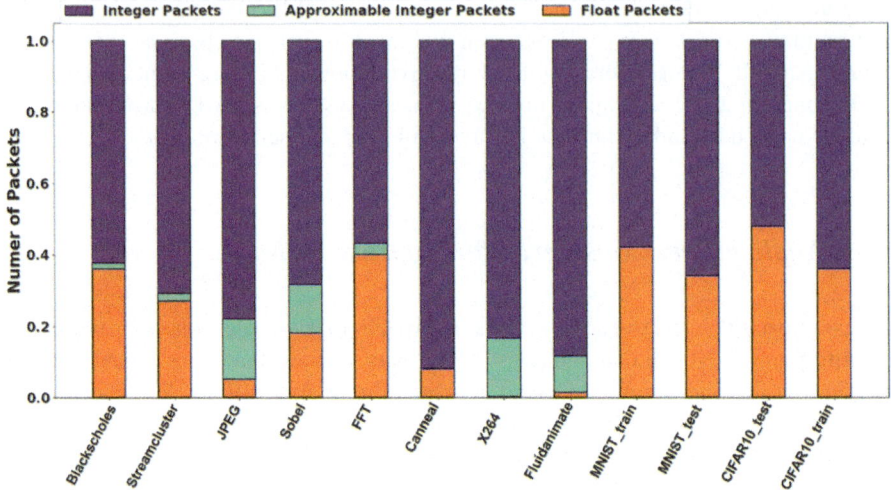

Fig. 2 Characterization of applications considered for evaluation, as presented in [7]

4   Crosstalk and Optical Loss in PNoCs

The overall data movement on the chip increases as the number of on-chip processing elements increases and applications utilize more data. To meet the demands of the increased communication, a larger number of photonic waveguides, wavelengths, and MR devices are necessary. However, using a larger number of photonic components makes it challenging to maintain acceptable BER and achieve sufficient signal-to-noise ratio (SNR) in any PNoC architecture due to optical signal loss and crosstalk noise accumulation in photonic building blocks [60].

Light wave propagation in photonic interconnects relies significantly on the precise geometry adjustment of photonic components. Any distortion in waveguide geometries and shape can notably affect the optical power and energy-efficiency in waveguides. For instance, sidewall roughness due to inevitable lithography and etching-process imperfections can result in scattering and hence optical losses in waveguides [61]. Such losses experienced by the optical signal as it passes through a waveguide is called propagation losses. In addition to propagation loss, there is optical loss whenever a waveguide bends (i.e., bending loss) or when a wavelength passes (i.e., passing loss) or drops (i.e., drop loss) into an MR device. To compensate for the losses and ensure appropriate optical-power levels at destination nodes where the signals are detected, increased laser power is required.

Crosstalk is another inherent phenomenon in photonic interconnects that degrades energy-efficiency and reliability. Crosstalk in PNoCs occurs due to variations in MR geometry or refractive index and imperfect spectral properties of MRs, which can cause an MR to couple optical power from another optical channel/wavelength in addition to its own optical channel (i.e., resonant wavelength). This leakage of power from one optical channel to other essentially becomes crosstalk noise in photonic interconnects. Such crosstalk noise is of concern when multiple optical channels exist within close (<1 nm) spectral proximity (referred to as channel spacing). This is especially evident in dense-wavelength-division multiplexed (DWDM) waveguides, which is necessary to support a higher bandwidth for emerging manycore platforms. In such DWDM systems, not only will optical signals in each channel suffer from optical loss, but inter- and intra-channel crosstalk [62] accumulating on optical signals can severely reduce SNR and increase BER. Reducing crosstalk is challenging, and techniques to minimize crosstalk (e.g., [43, 63, 64]) introduce further power and latency overheads.

It should be noted that the optical-power loss and crosstalk noise from a single silicon photonic device (e.g., MR) can be very small, and hence negligible [65]. However, in PNoCs integrating a large number of such devices (e.g., hundreds of thousands of MRs), the small power loss and crosstalk noise at the device-level accumulate to a point that they can severely reduce the performance and energy-efficiency in such architectures. In ARXON framework, as we are considering approximated data packets, we can intelligently relax crosstalk-mitigation mechanisms and optical loss compensation for the approximated bits, to aggressively reduce power and energy consumption overheads.

5 ARXON Framework: Overview

This section of this chapter discusses the components of the ARXON framework. Section 5.1 provides an overview of our loss-aware laser-power optimization strategy. Sections 5.2 and 5.3 discuss how crosstalk mitigation and tuning can be relaxed to save power during approximate bit transfers. Finally, Sect 5.4 describes the integration of multilevel signaling to reduce power dissipation further during approximate communication in PNoCs.

5.1 Loss-Aware Power Management for Approximation

Optical signals transmitted over a waveguide (photonic link) experiences various optical losses along the path from a source to a destination, as discussed in Sect 4. To express how these optical losses tie in with the required input laser power provisioned to the optical signals in the waveguide, we can use the following model [66]:

$$P_{laser} - S_{detector} \geq P_{photoloss} + 10 \times \log_{10} N_\lambda \qquad (2)$$

Here, P_{laser} is the laser power in dBm, $S_{detector}$ is the receiver sensitivity, and N_λ is the number of unique wavelengths, i.e., optical channels in the link. Also, $P_{photoloss}$ is the total optical loss accumulated on the optical signal during its transmission, which includes propagation, crossing, and bending losses in the waveguides, through- and drop-port losses of MR modulators and filters and modulating loss in modulator MRs due to imperfect modulation [63]. P_{laser} thus depends on the link bandwidth in terms of N_λ and the total loss $P_{photoloss}$ encountered by each optical signal traversing the network. A signal can only be accurately recovered at the destination node if the received signal power is higher than $S_{detector}$, after encountering $P_{photoloss}$. Ensuring this requires a high-enough P_{laser} to compensate for all optical losses.

To approximate data transmission for floating-point data transfers [55], use lower P_{laser} for transmitting LSBs while keeping P_{laser} unchanged for MSBs. However, this technique fails to take into account the higher losses encountered by a packet if the destination node is relatively farther along a waveguide from a source node. The higher losses encountered can drive the signal power at the detector MRs lower than $S_{detector}$, which would result in detecting logic "0" for all the approximated signals at the destination node (e.g., with OOK modulation). On the other hand, in the scenario where the destination is closer to the source, it may be possible to detect the approximated signals accurately, even with the reduced P_{laser} for the approximated bits, as the losses encountered are low enough that the signal power at the detector MRs would be higher than $S_{detector}$. For each data transfer on a waveguide, if we are aware of the distance of the destination from the source, it is possible to calculate the losses encountered for the signals, which can allow us to determine whether the signals can be recovered accurately, or if they will be detected as "0s." In this scenario, it is more efficient to simply truncate all the approximated bits (i.e., reduce P_{laser} to 0 for optical channels carrying the approximated signals) when the destination is farther along the waveguide and there is no likelihood of the signal being recovered accurately. Moreover, in the cases where the destination is closer to the source, we can transmit the approximated signals with a lower P_{laser} and still retrieve the correct data. This intelligent distance-aware transmission model for data approximation allows for some of the data to be detected accurately at the destination, while approximating other data depending on its content and distance to the destination.

Figure 3 shows the operational details of the distance-aware transmission model in our framework, on a single-writer-multiple-reader (SWMR) waveguide that is part of a PNoC architecture. Note that while the framework is to illustrate an SWMR waveguide, it is also applicable (with minimal changes) to multiple-writer-multiple reader (MWMR) and multiple-writer-single reader (MWSR) waveguides that are also used in many PNoCs. In Fig. 3, only one sender node is active per data transmission phase, and there is one receiver node (out of three nodes in the figure) that acts as the destination for the transmission. In a pre-transmission phase (called receiver-selection phase), the sender notifies the rest of the nodes about the destination for the upcoming data transmission, and only the destination node will activate its MR banks, whereas the other nodes will power down their MR banks

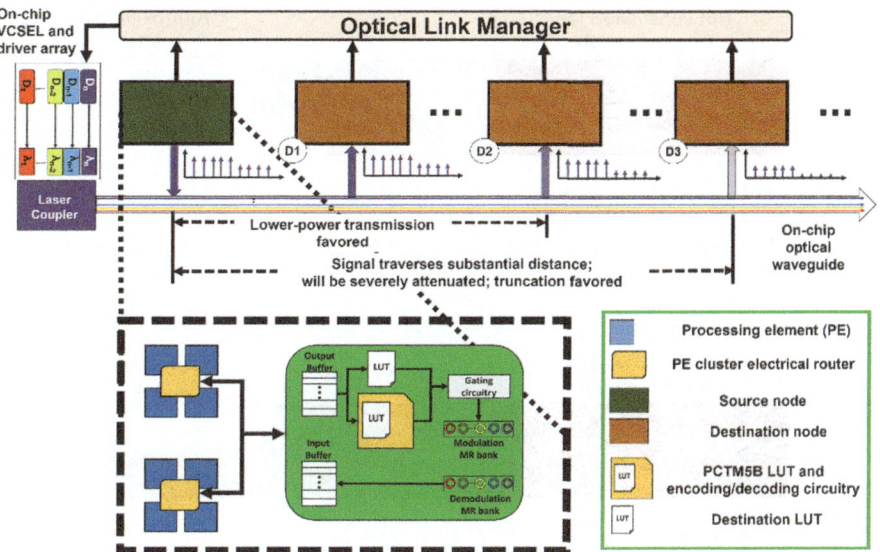

Fig. 3 Overview of the ARXON framework

to save power. As shown in Fig. 3, if the destination node is close to the sender node (e.g., D1), we can transmit the approximated bit signals with a lower P_{laser}. Otherwise, if the destination node is farther away from the sender node (e.g., D3), we determine that it would not be possible to detect the approximated signals at that destination due to the greater losses the signals will encounter. In this second scenario, we can dynamically turn off P_{laser}, essentially truncating the bits.

We consider both integer and floating-point data for approximation. For floating-point data, we perform distance aware transmission of the LSBs (M bits of the floating-point data representation) of the data in a controlled manner, so as not to impact the overall output quality of the application. Figure 4 shows how transmission of data will conform to the distance aware transmission policy of our framework. In the case where substantial losses are expected to be encountered between a source and destination, we adopt the strategy shown in Fig. 4a, where the data is truncated, as the approximated bits would have been lost during transmission anyway. When the signal can have enough power to be successfully received at the destination node, we adopt the strategy shown in Fig. 4b, where the LSB of the data is transmitted at a lowered laser power than its nonapproximated counterparts. The power at which the bits can be transmitted, and the number of the approximated bits will depend on the application, as discussed in Sect. 6.

For approximating integer variables, we take a different approach. From our analysis, we observe that indiscriminate approximation of integer data in an application can significantly reduce output quality. Therefore, we instead profile applications and log the range of values stored in each integer variable. If the range of values is smaller than the bit size allotted to the variable (e.g., the case where

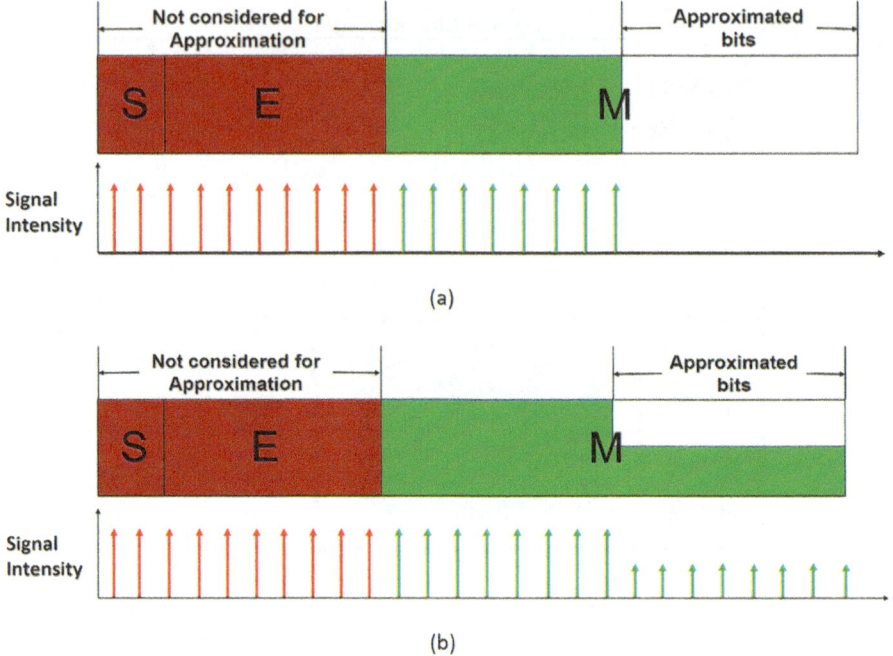

Fig. 4 Floating-point data transmission on a photonic waveguide via (**a**) truncation and (**b**) lowering laser power

a 32-bit integer variable only stores values up to 24 bits throughout the run of the application), we consider it a candidate for approximation. We can remove or truncate the MSBs that are unused in such variables that will otherwise take up input laser power, modulation/demodulation energy, and tuning energy necessary for transmission. We can also try and approximate the LSBs of the integer packets, and this approach can work in integer variables that store very large values where slight errors in the LSBs have minimal impact. But integer variables amenable to LSB approximation without significantly reducing output quality are rare. Nonetheless, for any such amenable integer data, the distance aware transmission model is applied to approximated LSB bits as well. Figure 5 summarizes our approximation strategy for integer packets.

To implement these strategies, we require the following: (i) a dynamic control mechanism for the laser power being injected into the on-chip waveguides and (ii) a mechanism to annotate approximable variables in the application source code, for runtime adaptation of transfers involving these variables.

We use an on-chip laser array with vertical-cavity surface-emitting lasers (VCSELs) [67], which can be directly controlled using on-chip laser drivers. With the laser drivers, we control the power fed into each individual VCSEL, thus controlling the power of the laser output for a particular wavelength corresponding to that VCSEL. The gateway interface (GWI) that interfaces the electrical layer

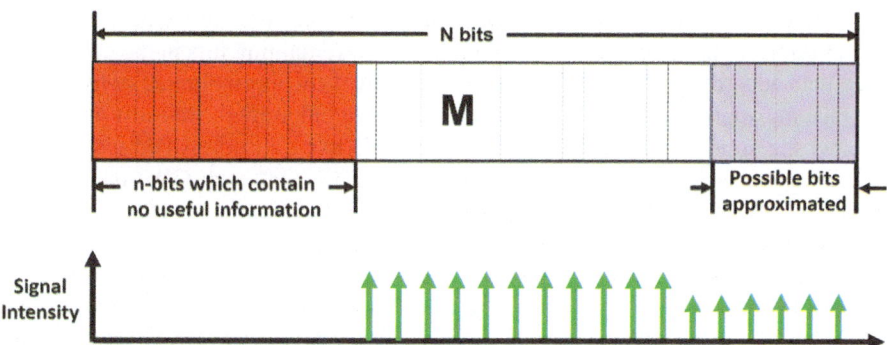

Fig. 5 The adopted approximated integer data transmission method

of the chip to the PNoC (see Fig. 3) communicates the desired P_{laser} power level (including 0 for truncation) to the drivers, via an optical link manager, similar in structure to the one proposed in [68].

To generate necessary flags for data that is approximable, identification of candidate packets to be approximated is done at the processing-element level, via source-code annotations [25]. To allow for proper decoding of approximated or truncated packets at the destination flags indicating the type of data and the amount of data being truncated/approximated is necessary. For this, the two additional flags can be included in the packet header, at the processing-element level. The first (1 bit) flag indicates whether the approximable packet contains integer or float data and the second (1 bit) flag indicates whether the approximation is to be done for LSBs or MSBs. The number of bits that can be safely approximated or truncated are determined offline for each application and stored in lookup tables (LUTs) at the network interface (NI) which connects processing elements to routers that are in turn connected to GWIs. The number of bits approximated/truncated in a packet is also passed as part of the header flit of the packet to the GWI. This information can be used to gate (i.e., prevent) those bits from being passed into encoding/decoding circuitry. We also add six bits to the header flit to convey the number of bits truncated/approximated in the data packet, which is necessary information for decoding the data at the destination GWI. These six bits represent the number of approximated/truncated bits in the range from 0 to 32 bits, which is the range of approximation/truncation in ARXON.

Usually the header flit of the packet contains the routing information, which can just be the destination address. We consider a flit size of 64 bits, i.e., 64 bits from a packet are transmitted per transmission cycle. The number of used bits in the header flit in a NoC do not exceed 16 bits (for the destination and source addresses), thus making it possible to incorporate the 8 necessary bits containing the two bits for the necessary flags and 6 bits for the approximation/truncation size information without causing any additional latency overheads. Once the header flit is received at the destination GWI, the flags and the approximated/truncated bits information are used to select the appropriate LSB/MSB to be not considered for decoding. The packet ID from the flit can then be used to track the remaining flits in the packet and treat them accordingly, depending on whether they were approximated or truncated.

Once the approximable bits in a packet have been identified, we must determine whether the approximation during their transfer is to be accomplished via reduced power transmission or truncation. This requires a LUT at each GWI (see Fig. 3) populated with the destination IDs to which the loss values are sufficiently large enough to warrant truncation. The values can be easily calculated post-fabrication at design time, as the distance between nodes as well as the cumulative loss to their GWI from the source does not change at runtime. Once the decision to truncate or transmit at a lower laser power is made, depending on the destination node, the required power levels for the wavelengths are communicated to the VCSEL drivers via the optical-link manager. We discuss the overheads of the tables and the application specific tuning of P_{laser} for the approximated signals in Sect. 6.

5.2 Relaxed Crosstalk Mitigation Strategy

Due to the challenges with signal crosstalk outlined in Sect. 4, PNoCs must utilize one or more crosstalk mitigation strategies to reduce BER and achieve high SNR. We consider a state-of-the-art crosstalk mitigation strategy from [43] that can be applied at the link level in PNoCs. Analyses from [43] showed that a "1" carried by the wavelengths in the DWDM wavelength group adjacent to the resonant wavelength ($\lambda \_ MR$) of an MR causes higher crosstalk in that MR. Hence, an encoding strategy to avoid two consecutive optical channels to carry "1," by replacing adjacent "1s" with "0s" was proposed, to reduce interchannel crosstalk. This technique essentially helps by reducing the optical signal-strength of immediate nonresonant wavelengths and improving SNR. Two encoding techniques were proposed that encoded nibbles (4 bits) of data. The PCTM5B technique encoded the nibble to 5-bit data, while the PCTM6B technique encoded the nibble to 6-bit data. Table 1 shows the code words used in these encoding techniques. Note that to implement PCTM5B on a photonic link with 64-bit word parallel transfers, 16 additional bits are required, which increases the number of MRs by 25%. Similarly, for PCTM6B, 32 additional bits are required for a 64-bit data word, and this increases the number of MRs by 50%. We consider the lower-overhead PCTM5B technique to be integrated into the PNoCs considered for analyses (Sect .6) by default, to meet BER goals.

We assume the baseline configuration of the PNoC to have implemented PCTM5B, for crosstalk mitigation. This means the encoder/decoder circuitry and the LUT, containing the data word-code word pairs, are incorporated into the GWI. Using these additions, the incoming packets from the processing elements can be encoded to PCTM5B code words before they are transmitted to their destination, and at the destination, the packets are decoded using the LUTs. In the approximation scheme employed in ARXON, applying crosstalk mitigation

Table 1 Data word to code word conversion [43]

Data word	Code word	Data word	Code word
Code words for PCTM5B technique			
0000	00000	1000	01000
0001	00001	1001	01001
0010	00010	1010	01010
0011	10101	1011	10100
0100	00100	1100	01100
0101	00101	1101	10010
0110	00110	1110	10001
0111	10110	1111	10000
Code words for PCTM6B technique			
0000	000000	1000	001000
0001	000001	1001	001001
0010	000010	1010	001010
0011	100000	1011	010100
0100	000100	1100	100010
0101	000101	1101	010010
0110	010101	1110	010001
0111	100001	1111	010000

via PCTM5B technique to the approximated bits is an unnecessary overhead as it does not provide any benefits towards BER. By relaxing crosstalk mitigation for the truncated or approximated bits, it is possible to reduce the energy costs of the mitigation strategy. We do this by leveraging the approximation information gathered using our offline analysis of applications, where we consider that some LSB/MSB of the data can be approximated/truncated. We do not consider these approximable bits for encoding, by gating their access to the encoder. Similarly, at the destination, when an approximated/truncated packet is received, the information from our LUTs are used to gate the approximated/truncated bits from being passed into the decoder circuitry.

5.3 Relaxed MR Tuning Strategy

Thermo-optic or electro-optic tuning of MRs in a PNoC is crucial for ensuring reliable communication, by counter-acting the effects of FPV and TV. We assume the use of thermo-optic tuning in PNoCs, due to its better range of resonant wavelength shift ($\Delta\lambda_{MR}$) correction. Electro-optic tuning can provide a tuning range of at most 1.5 nm [69]. In contrast, thermo-optic tuning can provide a tuning range of about 6.6 nm corresponding to the temperature range of up to 60 K [70] at 0.11 nm/K sensitivity [71]. This comes at the price of higher energy consumption (~mW/nm) and slower operation (in units of μs). In our framework, we aim to reduce the overhead of tuning the MRs associated with truncated bits. We relax the tuning requirement for MRs associated with the truncated bits, by turning off the tuning mechanism for those MRs. We do not consider approximated bits for relaxed MR tuning, as the added noise this approach generates, due to thermal drift of λ_{MR}, may render the approximated bits unreadable at the destination GWI.

5.4 Integrating Multilevel Signaling

The discussion in the previous subsections assumes the use of conventional on-off keying (OOK) signal modulation, where each photonic signal can have one of two amplitude levels: high or on (when transmitting a "1"), and low or off (when transmitting a "0"). In contrast, multilevel signaling is a signal-modulation scheme where more than two levels of voltage can be used to modulate multiple bits of data simultaneously in each optical signal. The obvious benefit with such multilevel signaling is an increase in the bandwidth. But for PNoCs multilevel signaling also provides the added benefit of reduced power consumption (compared to an OOK-based PNoC), by reducing the number of optical channels needed to obtain the same bandwidth. Leveraging this technique in the photonic domain has, however, traditionally been a cumbersome process with high overheads, e.g., when using the signal superposition techniques from [72]. But with advances such as the introduction of optical digital to analog converter (ODAC) circuits [73] that are much more compact and faster than MZIs used in techniques involving superimposition [72], multilevel signaling has been shown to be more energy efficient than OOK [66]. The overall reduction in power and energy makes multilevel signaling a promising candidate for more aggressive energy savings in silicon photonic networks.

Four-level pulse amplitude modulation (PAM4) is a multilevel signal modulation scheme where two extra levels of voltage (or optical signal power in case of optical modulation) are added in between the "0" and "1" levels of OOK. This allows PAM4 to transmit two bits per modulator as opposed to one bit per modulator in OOK. This in turn increases the communication bandwidth of PAM4 when compared to OOK. We are interested in evaluating the impact of using PAM4 in PNoCs and how its use will impact the effectiveness of the discussed approximation strategies in ARXON. While PAM4 promises better energy-efficiency than OOK, it is prone to higher BER due to having multiple levels of the signal close to each other in optical power. Thus, we cannot reduce the laser-power level of the LSB bits to the level used in OOK, as it would significantly reduce the likelihood of accurate data recovery even when destination nodes are relatively close to the source. Thus, when PAM4 is used, we need to increase the laser power compared to OOK. We used an empirically determined value of $1.5\times$ the laser power that was used for OOK, to prevent the degradation of approximated signals transmitted with PAM4. This may seem like a backward step in conserving energy, but the reduced-operational cost per modulation, reduced modulator and demodulator losses, and the reduced wavelength count for achieving the same bandwidth as OOK may reduce the overall laser power and energy consumption. Also, while it is possible to add more signaling levels (e.g., to use a PAM8 modulation scheme [74]), as the number of amplitude levels increases, the optical signal becomes increasingly susceptible to noise and causes increase in BER [75]. To ensure reliable communication when using PAM8, the bandwidth and speed of operation must be sacrificed [74]. Considering these constraints, we limit the extent of multilevel signaling integration in our framework to PAM4. The experimental results in the next section quantify the impact and trade-off when using PAM4 signaling in our framework.

6 ARXON Evaluation and Simulation Results

6.1 *Simulation Setup*

To evaluate our ARXON framework, we implement it in Clos [76] and SwiftNoC PNoC architectures [77] for a 64-core processor, with baseline OOK signaling, PCTM5B crosstalk mitigation, and thermo-optic tuning in MRs.

The Clos PNoC, shown in Fig. 6a, has an 8-ary three-stage topology for a 64-core system with eight clusters and eight cores per cluster. The PNoC is used for communication between clusters. It utilizes an optical crossbar topology with point-to-point photonic links utilizing SWMR waveguides for inter-cluster communication. Each cluster has two concentrators, and a group of four cores is connected to each concentrator, where concentrators communicate with each other using an electrical router.

For the SwiftNoC PNoC, as shown in Fig. 6b, we have again considered a 64-core system. Each node here has four cores and communication within the node happens through a 5×5 router, with the fifth port of the router connected to a GWI, which facilitates transfers between the CMOS-electrical layer and the photonic layer. Each GWI connects four nodes (16 processing cores). The architecture utilizes eight waveguide groups with four MWMR waveguides per group in a crossbar topology. In order to support the MWMR communication, SwiftNoC utilizes a concurrent token stream arbitration that provides multiple simultaneous tokens and increases channel utilization.

We performed a simulation-based analysis to visualize the impact of losses on laser power, modeled using Eq. (2). These losses are critical in ARXON's loss-aware approximation/truncation strategy. The Clos PNoC has a waveguide length of 4.5 cm, and the SwiftNoC PNoC has a waveguide length of 8.3 cm over the considered 400 mm^2 chip. In both PNoCs, the first MR is encountered at ~1 cm, and the last MR is encountered at ~3.8 cm for Clos PNoC and ~7.8 cm for SwiftNoC. This relationship is visualized in Fig. 7, where the sudden jumps in power indicate a new GWI with the optical devices being encountered along the waveguide.

The considered PNoC architectures were modeled and simulated using an in-house SystemC-based cycle-accurate simulator. A combination of gem5 full-system simulator [58] and Intel PIN toolkit [59] was used to generate traces for the

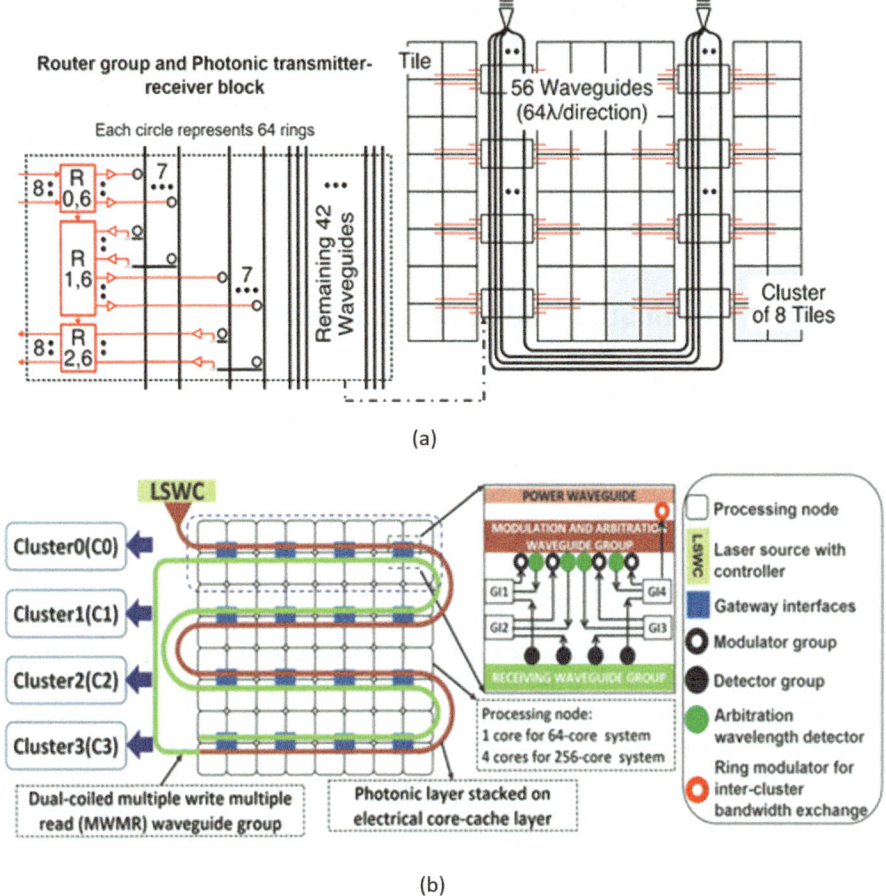

(a)

(b)

Fig. 6 PNoC architectures considered for analyses. (**a**) 8-ary 3-stage Clos architecture with 64 cores [76] and (**b**) schematic overview of SwiftNoC architecture [77]

considered application; these were replayed on the PNoC simulators to determine the effectiveness of ARXON framework on the PNoC. The PIN tool was used to obtain the addresses of the variables we deemed suitable for approximation from our profiling analysis of applications and then to track accesses to them. Using this information in gem5 simulation, we track the relevant data flow at various levels of the simulated system (processor level, memory controller level, DRAM level, and cache level). The information generated while the simulation is running was consolidated, and custom python scripts were created to extract the necessary information about the data packets (e.g., timestamp at origin, their source, destination, data values, and control values from the packet header) and to generate the traces necessary for our cycle accurate simulator to simulate the applications on these PNoC architectures. Then, details of the approximate data communication

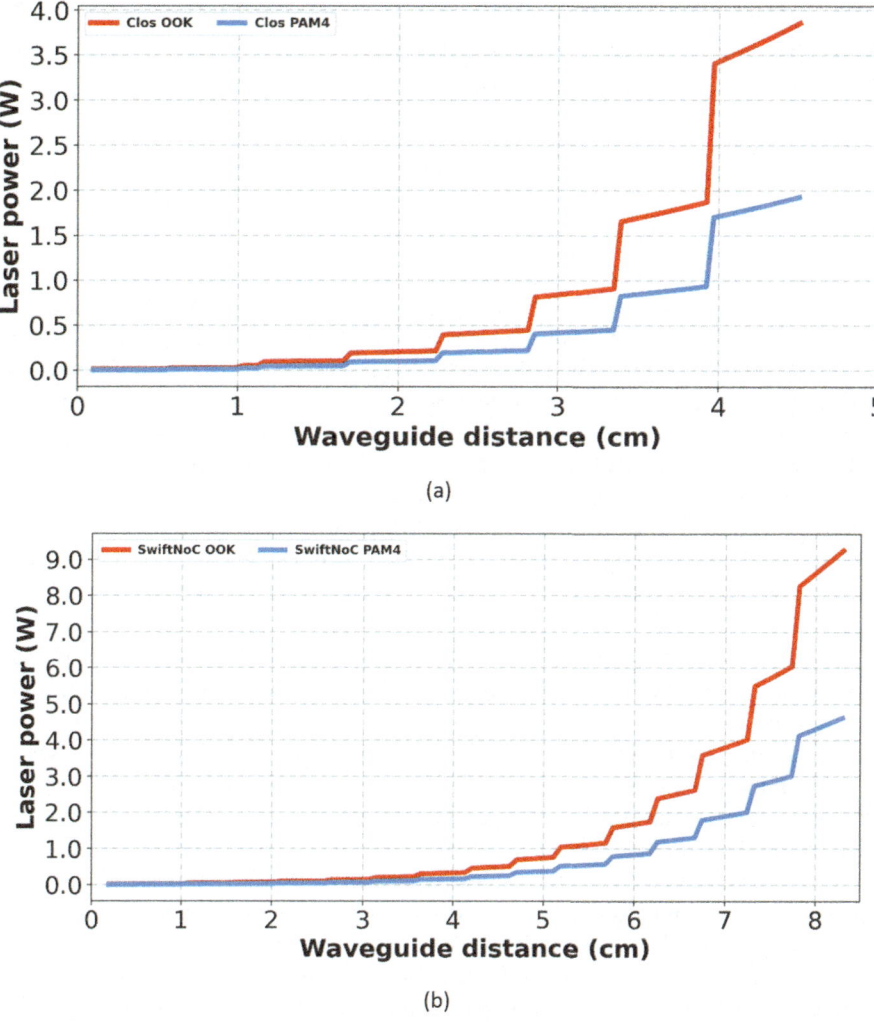

Fig. 7 Laser-power consumption behavior over the length of the waveguide in (**a**) Clos PNoC and (**b**) SwiftNoC. (Taken from Sunny et al. [7])

(i.e., whether a packet was truncated or transmitted at lower power) were used to modify data packets in a subsequent gem5 simulation, to estimate the impact of the approximation strategy on output quality for the application being considered. Table 2 shows gem5 architectural simulation parameters considered in our experiments. We have based our simulations on x86 cores, but these simulations and our approach is applicable to systems having other types of cores as well, e.g., ARM cores. Twelve applications, ten from the ACCEPT benchmark suite and two from tinyDNN benchmark suite, were used in our evaluations. The performance was evaluated at

Table 2 64-core architecture configuration

Simulated component	Specification
No. of cores, processor type	64, ×86
DRAM	8 GB, DDR3
Memory controllers	8
L1 I/D cache, line size	128 KB each, direct mapped, 64 B
L2 cache, line size, coherence	2 MB, two-way set associative, 64 B, MESI

the 22 nm CMOS node for 400 mm^2 chips, with cores and routers operating at 5 GHz clock frequency. DSENT [78] was used to calculate the energy consumption of routers and the GWI at each node. Each GWI holds two LUTs for our framework; these are one which holds the information regarding which destination addresses are preferred for truncation and another for PCTMB5 encoding scheme. The size of both the LUTs at GWI level is fixed and is application independent, as the information one store is hardware dependent and the other stores a fixed set of encoding-decoding information. The PCTM5B LUT takes up only 144 bits for storing encoding decoding information at each GWI. The destination ID LUT can take up a maximum of 32 bits at each GWI for Clos PNoC and 64 bits for SwiftNoC variants.

The table containing information regarding number of bits to be approximated/truncated for integer/float approximable packets is stored at the network interface (NI) of each processor. The maximum number of bits required in these LUTs for the worst case (application with the highest number of approximable variables) is a few hundred bits for the applications we considered. CACTI v6.5 [79] and scaling equations from [80] were used to evaluate the power, area, and delay for the lookup tables in NIs and GWIs. These values were found to be 0.236 mm^2 for the area consumption for all the tables, with a total power overhead, for reading from and writing into the tables, of 0.135 mW for Clos and 0.472 mm^2 and 0.27 mW respectively, for SwiftNoC. The combined power and area consumption of associated circuitry necessary for accessing information in the LUTs, calculated using gate-level analysis, is 0.0274 mm^2 and 4.224 mW for Clos and 0.0548 mm^2 and 8.448 mW for SwiftNoC. LUTs in both Clos and SwiftNoC have the same number of entries as both architectures have the same number of processing elements. The encoding/decoding scheme is the same and the approximations done depend on the output error quality of the application and not the architecture, while SwiftNoC has double the number of GWIs, and hence double the number of LUTs. The access time for scratchpad RAMs designed with 22-nm technology node was under 1 cycle from synthesis estimates.

For implementing the dynamic P_{laser} control, ARXON needs a VCSEL control unit. The VCSEL control in ARXON was modeled after the optical link manager in [68], where the channel management for their PNoC design was described. However, since we are considering PNoCs from prior works with their own channel management systems in place for our analysis, we only adopt the approach for

Table 3 Loss and power parameters considered for PNoC simulations

Parameters considered	Standard values	Aggressive values
Receiver sensitivity	−20 dBm [81]	−23.4 dBm [82]
MR through loss	0.02 dB [83]	0.02 dB [83]
MR drop loss	0.7 dB [84]	0.5 dB [85]
Propagation loss	1 dB/cm	0.25 dB/cm [86]
Bending loss	0.01 dB/90° [87]	0.005 dB/90° [88]
Thermo-optic tuning	6.67 mW/nm [89]	240 μW/nm [71]

VCSEL control from [68]. The VCSEL control described in [68] uses a combination of MRs and PDs, but we only require the MR-based switching mechanism for the VCSEL output. From the data available in [68], we calculated the area overhead necessary for implementing the VCSEL control, which was 0.093 mm^2 for OOK variants and 0.047 mm^2 for PAM4 variants of both the architectures.

Clos and SwiftNoC PNoC architectures with PCTM5B are used as baselines for our analyses in this work. We have also considered a two-cycle overhead for PCTM5B encoding and decoding of the signals, as calculated in [43]. We considered $N_\lambda = 64$ for OOK, which would enable 64-bit transmission across a waveguide per cycle. For PAM4, we only need to consider $N_\lambda = 32$ to achieve the same bandwidth as with OOK modulation. Table 3 shows the energy values for losses and power dissipation in different photonic devices, which we have used in our modeling efforts. We use a "standard" set of values for these parameters from existing prototyping efforts, and a more "aggressive" set of values as per future projections from various research efforts. Our approach sacrifices reliability of approximated bits in floating point data LSB and selected integer variable data, for EPB and laser-power savings, as discussed in Sects. 5.2, 5.3, and 5.4.

We use the standard values for most of our simulations and use the aggressive values in Sect. 6.4. These values are used to calculate laser power from Eq. (2) and total power after considering tuning and lookup-table overheads. We consider a laser efficiency of 10% for our on-chip VCSELs, which is midway, the initial and worst-case efficiencies mentioned in [67]. We additionally consider a PAM4-induced-signaling loss of 5.8 dB in P_{pho} for laser-power calculations for PAM4 [66]. To compensate for the increased sensitivity of PAM4 to bit errors, we also consider laser-power levels that are 1.5× than those used for OOK signaling. For ensuring reliable communication, we have considered a BER of 10^{-9} in our designs. Finally, we calculated application output error for the non-machine learning applications due to our approximation approach as:

$$\text{Percentage (Output) Error} = \frac{|approximated\ value - exact\ value|}{exact\ value} \times 100 \tag{3}$$

The "exact value" refers to the original output values, which can be can be pixel values of output images/frames, like in the case of JPEG, Sobel or X264, or a set of values presented in the output files, like in the case of Blackscholes. The "approximated value" refers to the value of these outputs once the approximation approach is applied to the applications. For our analysis, we assume an error threshold of 10% output error, which was seen experimentally to be the limit at which the errors became apparent in the outputs of the majority of the applications [8]. For example, artifacts become noticeable in JPEG output as we cross the 10% error threshold. Thus, we want to ensure that none of the approximation strategies degrade output quality by more than 10%. For our machine-learning applications (convolution neural networks for MNIST and CIFAR10 classification), we have considered the drop in classification accuracy to measure the impact of our framework, and we have set the threshold as 10% drop in the accuracy.

6.2   Impact of ARXON on Considered Applications

Our first set of experiments involve analyzing the sensitivity of an application to varying degrees of approximation of their floating-point data. We are interested in studying the impact of our approximation strategies on output error due to (i) approximating a number of bits in the packets carrying data deemed approximable and (ii) varying levels of lowered laser power for those approximated bits.

Figure 8 shows the results of our comprehensive study for the applications we considered (as depicted earlier in Fig. 2). The z-axis shows the percentage error (PE) in application output or drop in accuracy for ML applications, as a function of the reduction in P_{laser} level for the photonic signals that carry the approximated bits (x-axis; varying from 0% to 100%, where 100% refers to truncation), and the number of bits that were considered for approximation (y-axis; with the number of approximated float and integer bits given in [float, integer] format). The subset of combination of these values were selected for enabling viable trade-offs between output quality and power consumption. It should be noted that not all applications consider both floating-point and integer data for approximation. For example, Fluidanimate only considers integers for approximation while the ML applications (convolution neural network-based classifiers for CIFAR10 and MNIST datasets) only considers floating-point data. This selection of datatypes to be approximated was made after profiling the application and determining the datatypes that do not have adequate impact on the traffic (e.g., floating-point data in Fluidanimate and X264) or the functionality of the application (integers in the case of the ML applications considered). This is a more comprehensive version of the experiments in our earlier work, presented in [8].

In those experiments in [8], we had determined how much floating-point approximation can be tolerated by the applications from ACCEPT benchmark. Here we not only consider a larger number and variety of applications, but also use more comprehensive analyses to determine thresholds than in [8] to explore how approximating the integer bits along with the float bits affects the output quality. It is clear from our analyses that not all applications can tolerate the same level

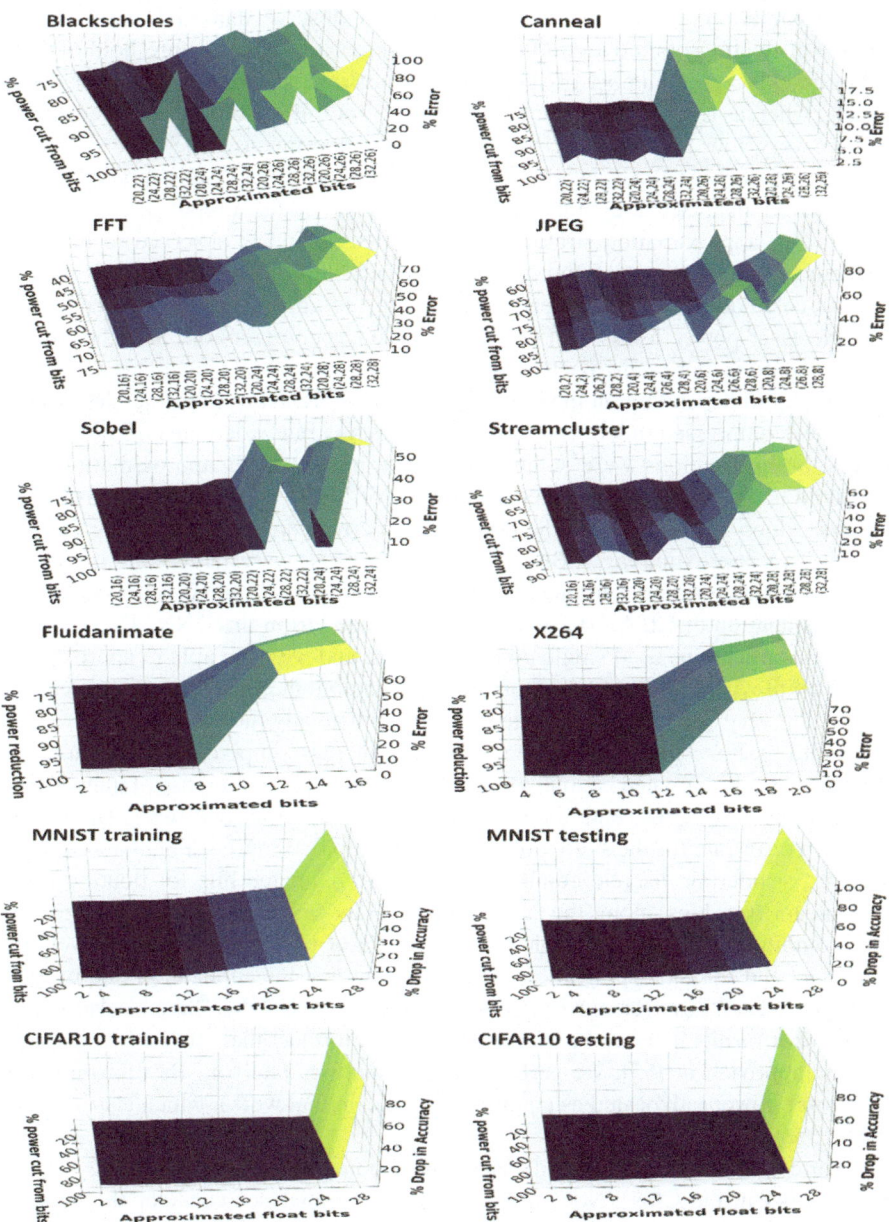

Fig. 8 Percentage error (PE)/drop in accuracy in application output as a function of the number of approximated bit signals (*y* axis) and reduction in laser power (*x* axis) for the approximated signals, for Blackscholes, canneal, fft, jpeg, sobel, streamcluster, fluidanimate, and X264 benchmarks with large input workloads and MNIST (training and testing) and CIFAR10 (training and testing) models

of approximation. From the PE values, we can observe that FFT with a large volume of floating-point data traffic (see Fig. 2) reaches the error threshold of 10% rather quickly as the number of approximated bits increases and laser power levels reduce, whereas Canneal with a lower floating-point traffic-volume observed seems to have very low PE values across the various experiments. The edge detection algorithm Sobel performs well in approximated conditions, possibly owing to the lowered data accuracy requirements to construct the output. Streamcluster involves an approximation strategy for data streams and is observed to be quite resilient to greater levels of approximation. Blackscholes, which performs market options calculations, is particularly sensitive to the approximated number of bits and the laser-power levels. JPEG performs image compression, and the output image quality is more sensitive to approximation. Fluidanimate generates a video of flowing liquid depending on the input data provided. X264 is a video codec, which generates compressed video from the input, which is raw video data. Fluidanimate and X264 applications were subjected to only integer MSB approximation, and threshold is quickly breached after the amount of MSBs approximated start taking up bits, which contain values; the quick rise in error can be explained by the fact that we are approximating MSBs which would cause very large shift in values. Moreover, we considered implementations of deep convolutional neural networks for classification of CIFAR-10 and MNIST datasets, from tinyDNN. The machine learning applications used single precision floats, and we were able to approximate till the point where we encroached on the exponent, but the decay of output accuracy ramped up very quickly once we tried to approximate any further.

From Fig. 8 we can see that there is a sharp increase in percentage output error (PE), as we approximated beyond a certain number of bits, in the case of many of the applications considered, e.g., the applications in the bottom two rows. The erratic jumps in error rate for the six applications in the top two rows of Fig. 8 are because we are considering discrete combinations of approximated bits for floating point and integer variables, along the "approximated bits" axis. Table 4 summarizes the best combination of approximable bits and the laser-power-transmission levels for these bits and for each application while ensuring that the application output error does not exceed 10% for our proposed framework (ARXON). Table 4 also shows the number of bits that can be truncated, selected to meet the <10% PE constraint. For the approach in [55], we perform approximations on 16 LSBs transmitted at 20% laser power (advocated as an optimal choice in that work), which also satisfies the <10% PE constraint.

Figure 9 shows the EPB and laser power comparison results for the various frameworks in the Clos PNoC architecture. These analyses consider the benefits from distance-aware transmission and the relaxed encoding technique for approximated packets for ARXON. Figure 9a shows that using ARXON-OOK results in lower EPB than the previous approaches, including our previous framework LORAX-OOK. The better EPB for LORAX and ARXON can be attributed the fact that they avoid wasteful transmission at lower laser power when it is unlikely

Table 4 Number of bits considered for approximation and laser-transmission-power level for the corresponding signals across benchmarks and frameworks considered

Application name	Truncation	[55]	LORAX [8]		ARXON		
	Truncated bits (float)	16 bits approximated, with 20% power reduction	Approximated bits (float)	% Power reduction	Approximated bits in floating-point packets	Approximated bits in integer packets	% power reduction
Blackscholes	12		32	90	32	24	90
Canneal	32		32	100	32	24	100
FFT	8		32	50	32	20	50
JPEG	20		24	80	22	4	80
Sobel	32		32	100	32	20	100
Streamcluster	12		28	80	28	20	80
Fluidanimate	–		–	–	–	8	100
X264	–		–	–	–	12	100
MNIST_train	24		24	100	24	–	100
MNIST_test	24		24	100	24	–	100
CIFAR10_train	24		24	100	24	–	100
CIFAR10_test	24		24	100	24	–	100

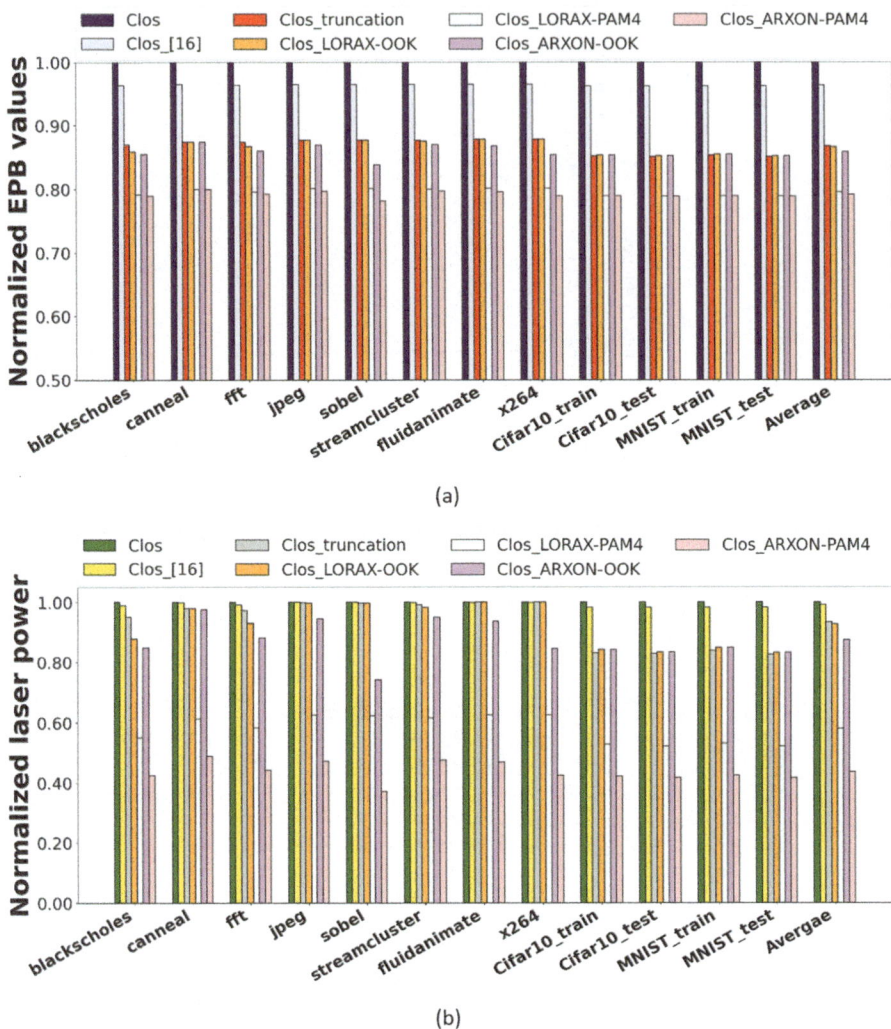

Fig. 9 (**a**) Energy per bit (EPB) and (**b**) laser power comparison across different frameworks for Clos PNoC architecture

that the destination can recover the transmitted data due to high optical losses. Also, [55] has noticeably higher EPB values for which we are not considering the benefits of relaxed encoding and distance-aware transmission for the framework to be consistent with the framework presented in that chapter. The ARXON-OOK framework improves upon LORAX-OOK [55] and truncation, by adaptively switching between truncation and an application-specific laser-power-intensity level for approximated bits of both floating-point and integer packets. The ARXON-PAM4 variant of our framework achieves the largest reduction in EPB, even though it uses $1.5\times$ higher laser-power levels for the approximated bits. The use of fewer wavelengths in PAM4 allows for more energy savings, despite greater losses and the use of more laser power per wavelength than OOK variant.

On average, ARXON-PAM4 shows 21%, 17.2%, 9.7%, 9.2%, and 1.2% lower EPB compared to the baseline Clos, [55], truncation, LORAX-OOK, and LORAX-PAM4 approaches, respectively. ARXON-OOK exhibits lower EPB on average while having a 6% higher EPB than the LORAX-PAM4 approach. In the best-case scenarios for the Blackscholes and Sobel applications, ARXON-PAM4 has 21.2% and 23.5% lower EPB than the Clos baseline; and 17.4% and 15.6% lower EPB than [55]; 9.8% and 11.5% lower EPB when compared to truncation; 8.6% and 10.25% lower EPB than LORAX-OOK; and 1.24% and 2.5% lower EPB than LORAX-PAM4 for these two applications.

Figure 9b shows the laser power reduction. On average, ARXON-PAM4 uses 50.45%, 49.5%, 43.2%, 42.5%, and 7.7% lower laser power compared to the baseline Clos, [55], truncation, LORAX-OOK, and LORAX-PAM4, respectively. ARXON-OOK exhibits lower average laser-power consumption on average while exhibiting 28% higher laser power consumption than LORAX-PAM4. For the best case Blackscholes and Sobel applications, laser power for ARXON-PAM4 is 51.7% and 59.2% lower than the Clos baseline and 50.8% and 57.9% lower than [55], while against truncation it is 51% and 58.5% lower; against LORAX-OOK, we see 38% and 57% lowered laser-power utilization; and against LORAX-PAM4, we have 6.5% and 20% lower laser-power utilization.

Figure 10 shows the same analyses but done for the frameworks implemented on the SwiftNoC architecture. The larger data rate and the larger number of GWIs in the architecture have impacted the packets and their distance aware transmission profile, creating more avenues to truncate the packets, yielding better EPB results in this architecture. The general trend in EPB and laser-power savings is similar to that for the Clos architecture, with Blackscholes and Sobel applications again exhibiting the best EPB and laser-power saving values. From Fig. 10a, ARXON-PAM4 exhibits 36%, 23.8%, 13.5%, 12.9%, and 1.8% lower EPB on average than baseline SwiftNoC, [55], truncation, LORAX-OOK, and LORAX-PAM4, respectively.

The results for SwiftNoC show the same trend as the Clos architecture for normalized laser power (Fig. 10b), albeit with lower laser power across applications with average laser power consumption for ARXON-PAM4 at 57.2%, 56.4%, 50.8%, 49.3%, and 15.7% better than baseline SwiftNoC, [55], truncation, LORAX-OOK, and LORAX-PAM4, respectively.

These results highlight the promise of our ARXON framework, as it improves upon the ability LORAX exhibited to trade-off output correctness with energy-efficiency and laser-power savings in PNoC architectures executing selected applications.

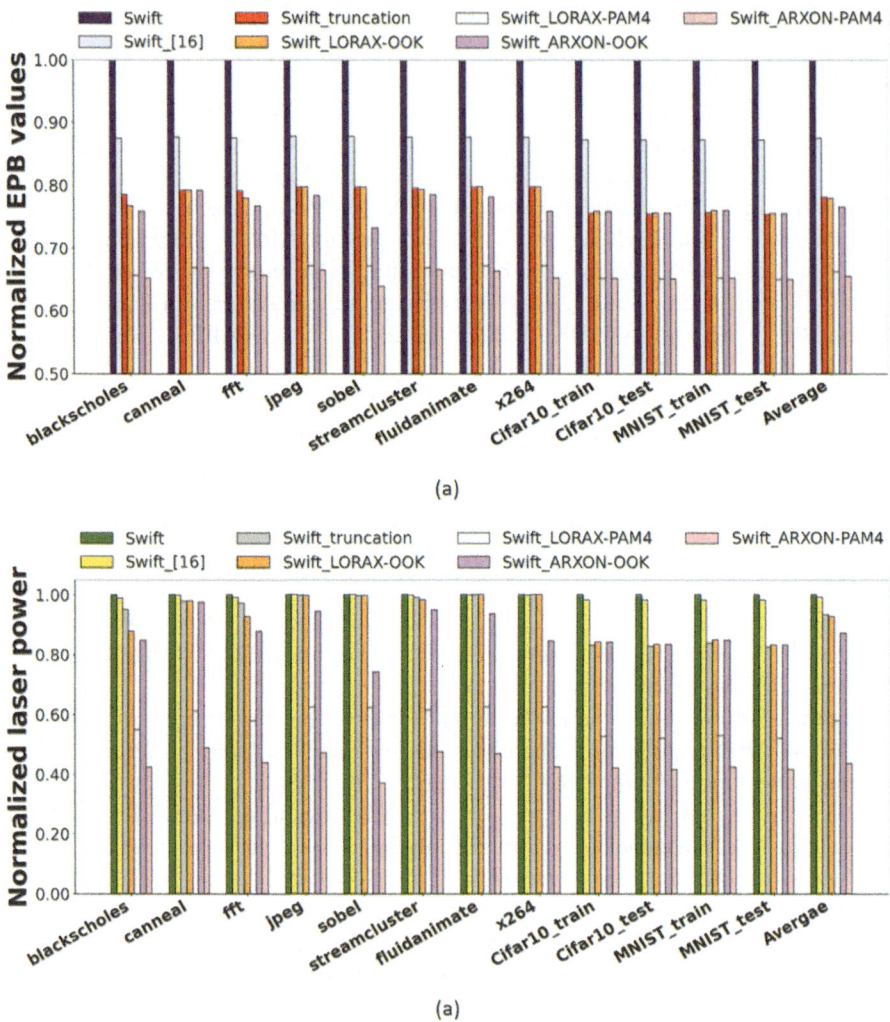

Fig. 10 (**a**) Energy per bit (EPB) and (**b**) laser power comparison across different frameworks for SwiftNoC architecture

6.3 MR Tuning Relaxation-Based Analyses

We also consider the potential for relaxed thermo-optic tuning for truncated bits, in addition to distance-aware transmission for float and integer packets and relaxing crosstalk-mitigation encoding techniques. We have considered thermal MR tuning in our work for its larger range of operation over other tuning methods such as electro-optic tuning. However, thermal tuning strategies are much slower in operation when compared to electro-optic tuning (microseconds for operation as opposed to nano to picoseconds for electro-optic tuning). However, this overhead cannot be avoided, as using just the electro-optic tuning method will not offer sufficient coverage for FPV and TV encountered by MRs, the effect of which must be mitigated for robust operation of the PNoC.

However, with the increasing maturity of silicon photonics, we envision faster thermo-optic tuning strategies or a combination of different tuning strategies to reduce this tuning latency. Therefore, in this section we explore the potential of energy savings due to relaxed MR tuning, i.e., by turning off the tuning mechanism for MRs associated with truncated bits. For this experiment, we utilize thermal and process-variation information. For TV, we have referred to the study conducted in [90] and have adopted the worst-case TV induced shift to be 6.5 nm. For analysis of FPV, we utilized the FPV analysis method as described in [90], where FPV is considered as a Gaussian random distribution. As the granularity of the method is at 30 nm, we have opted for analyzing FPV at the GWI level rather than for individual MR devices. We have generated FPV maps for the architectures using the method from [91] and have selected locations corresponding to the GWIs in the layouts. We took the average of device variations (i.e., width and thickness) in that location. This was repeated over 100 different FPV maps.

Utilizing the FPV and TV information obtained, we implement the tuning-relaxation approach, where we turn off thermo-optic tuning for all truncated bits. We use a gating mechanism similar to the one utilized for the encoding strategy, as mentioned in Sect. 5.2, to implement the control necessary for relaxing the tuning. With this mechanism, we can power gate the tuning circuits to the MR, as per the information from LUTs, turning them off for the transmission cycle, again similar to the description in Sect. 5.2. From our analysis, this had a substantial impact on the EPB values of our ARXON framework, as shown in Fig. 11. Our observations in Fig. 11a for Clos PNoC and Fig. 11b for SwiftNoC, show that the ARXON variants have substantial savings over the other frameworks considered, a trend maintained even while using the aggressive values as it was with standard values. To reiterate, the standard values are from existing prototyping efforts, and the aggressive values are as per future projections from various research efforts. The savings exhibited by ARXON is because the tuning-based approach is again dependent on the traffic profile of the applications, with higher truncated packets meaning better savings. So, we see Blackscholes and Sobel as the best performing applications again. We do not consider laser-power savings in this scenario, as the tuning relaxation approach does not impact the laser power. On average, ARXON-PAM4 has 38.1%, 36.1%, 26.8%, 26.4%, and 19.2% better EPB values than baseline Clos, [55], truncation, LORAX-OOK, and LORAX-PAM4. When implemented in SwiftNoC, ARXON-PAM4 exhibits 48.6%, 39.3%, 29%, 28.5%, and 16.9% better EPB than baseline, [55], truncation, LORAX-OOK, and LORAX-PAM4, respectively. This only adds to the significant reduction in the overall laser power consumption achieved by ARXON, showing how our framework achieves better laser power and EPB values for all the applications considered in our analyses.

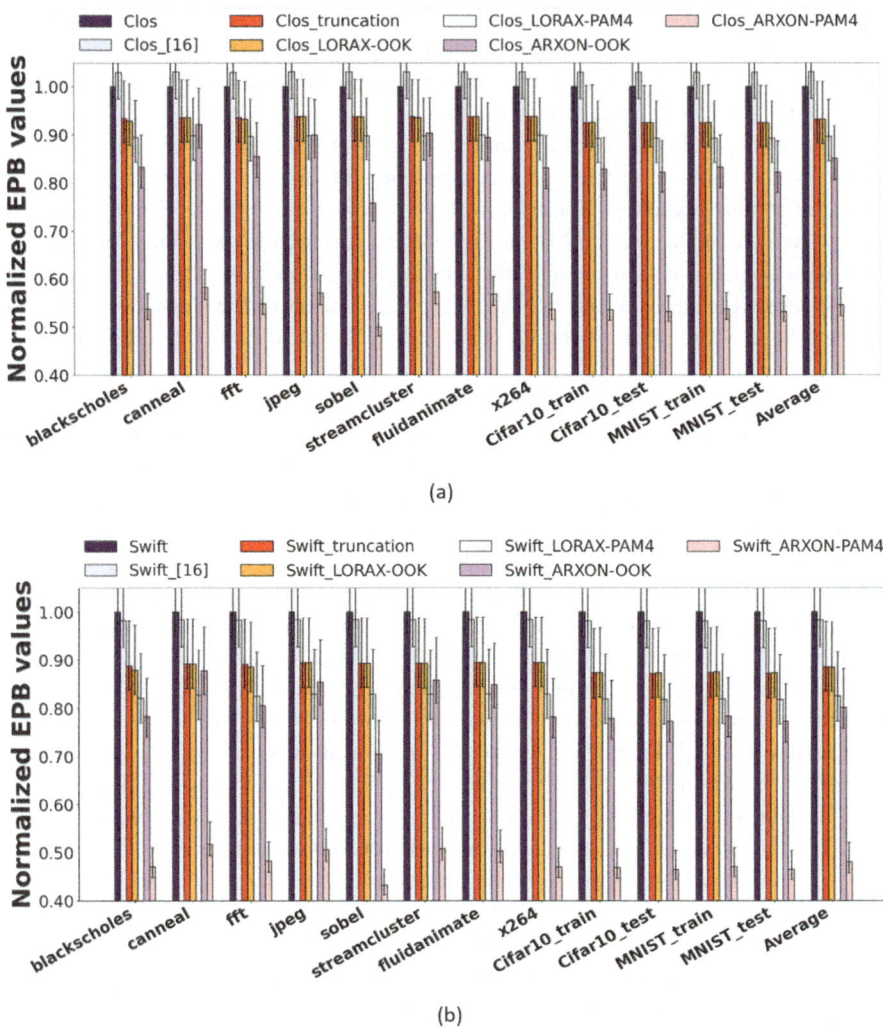

Fig. 11 EPB values for ARXON implemented on (**a**) Clos and (**b**) SwiftNoC while considering thermal-tuning relaxation

6.4 Power Dissipation Breakdown

We performed an experiment to determine how much more power can be saved as silicon photonics technology matures and devices with improved characteristics become available (aggressive values from Table 3). For this, we contrast the power dissipation with our framework on the Clos and SwiftNoC architectures, for the standard and aggressive values of parameters in Table 3. As the EPB and laser power, once normalized, follow the same trends, we decided to use a detailed

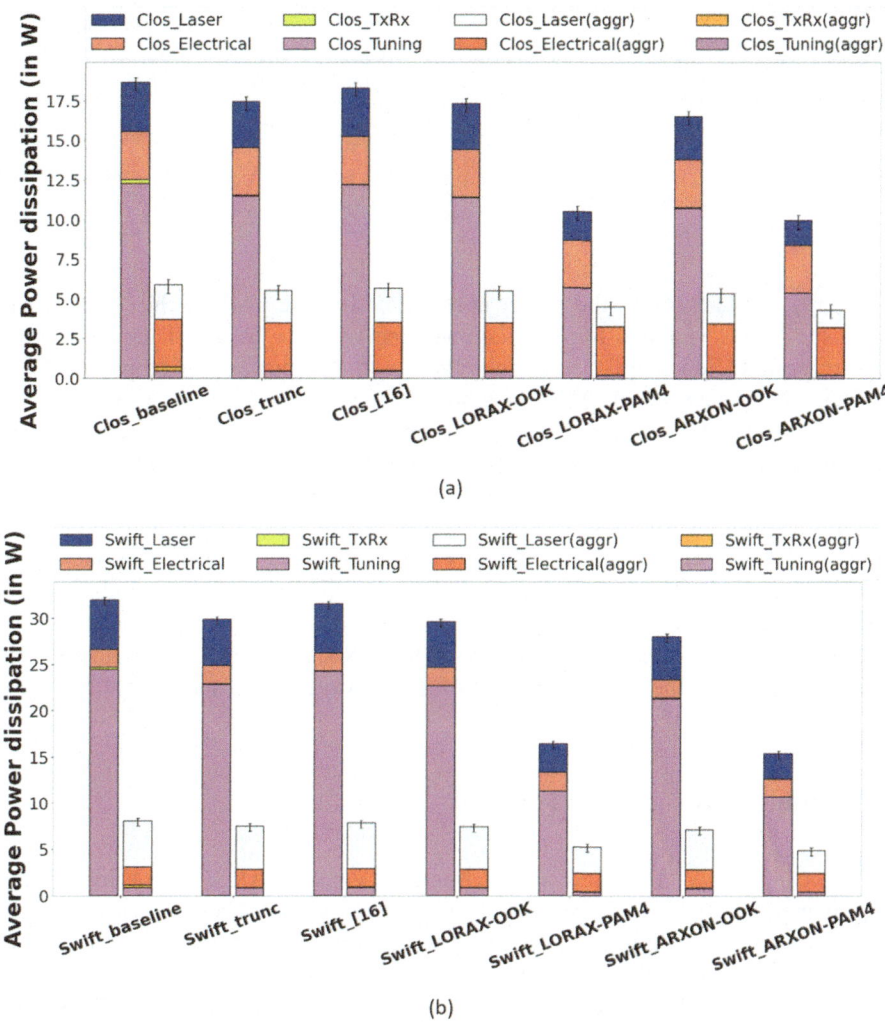

Fig. 12 Power dissipation breakdown for standard and aggressive values ("aggr" in the plots) for (**a**) Clos and (**b**) SwiftNoC PNoCs

power-dissipation breakdown to show how much ARXON improves the power consumption in PNoC and in which areas.

Figure 12 shows the detailed power breakdown for the framework applied on Clos and SwifNoC, averaged across the applications. From the figures we can clearly observe how ARXON impacts both laser power and tuning-power dissipation, having the lowest power dissipation in both these categories and in total, be it while considering standard loss and power utilization values or while considering aggressive values.

Table 5 Comparison of power savings between systems implementing the discussed PNoC variants

PNoC variant	Total power (W)		Power savings (%)	
	Clos	SwiftNoC	Clos	SwiftNoC
Truncation	94.25	106.25	1.57	3.19
[55]	94.95	108.25	0.84	1.37
LORAX-OOK	94.0	106.0	1.83	3.42
LORAX-PAM4	86.25	93.25	9.92	15.03
ARXON-OOK	90.75	100.25	5.22	8.66

Finally, Table 5 shows the power consumption at the 64-core chip level when using the PNoC variants. For this comparison, we have assumed the individual core to be a 14 nm x86-64 core from Intel, with the power consumption of the 64-core chip being 77.75 W. This assumption includes power consumption of 128 KB private L1 caches, 2 MB L2 cache (shared between four cores), and memory controllers (shared between four cores) [92]. This assumption sets the total power consumption for the baseline Clos PNoC-based system at 95.75 W and for the baseline SwiftNoC PNoC-based system at 109.75 W. Table 5 considers power and loss values for PNoC variants calculated using the standard parameter values from Table 3. It can be seen that even at the entire chip granularity, the ARXON framework provides notable reduction in overall power consumption, with ARXON-PAM4-based Clos and SwiftNoC PNoCs saving 10.97% and 16.86% power compared to the Clos and SwiftNoC PNoC baselines, respectively.

7 Conclusion

In this chapter, we discussed a new framework called ARXON for loss-aware approximation of data communicated over PNoC architectures. We also studied how multilevel signaling can assist with the proposed approximation framework. We considered crosstalk mitigation strategies and dynamic MR tuning as avenues to save energy while our distance aware transmission technique is in effect. Our results indicate that using multilevel signaling as part of our framework can reduce laser-power consumption by up to 57.2% over a baseline PNoC architecture. Our framework also shows up to 56.4% lower laser power and up to 23.8% better energy-efficiency compared to the best-known prior work on approximating communication in PNoCs. These results highlight the potential of approximation in PNoC architectures to reduce energy and power consumption in emerging manycore platforms.

References

1. Intel Xeon Platinum Processor Family. [Online]: https://www.intel.com/content/www/us/en/products/processors/xeon/scalable/platinum-processors/platinum-8180.html
2. NVIDIA ampere GA102 GPU Architecture Whitepaper. [Online]: https://images.nvidia.com/aem-dam/en-zz/Solutions/geforce/ampere/pdf/NVIDIA-ampere-GA102-GPU-Architecture-Whitepaper-V1.pdf
3. Cerebras Wafer-Scale Engine. [Online]: https://www.cerebras.net/
4. Pasricha, S., Dutt, N.: On-Chip Communication Architectures. Morgan Kauffman (2008). ISBN 978-0-12-373892-9
5. Alexoudi, T., Terzenidis, N., Pitris, S., Moralis-Pegios, M., Maniotis, P., Vagionas, C., Mitsolidou, C., Mourgias-Alexandris, G., Kanellos, G.T., Miliou, A., Vyrsokinos, K.: Optics in computing: from photonic network-on-chip to chip-to-chip interconnects and disintegrated architectures. J. Lightwave Technol. **37**, 363–379 (2019)
6. Xu, Q., Mytkowicz, T., Kim, N.S.: Approximate computing: a survey. IEEE Des. Test. **33**, 8–22 (2016)
7. Sunny, F., Mirza, A., Thakkar, I., Nikdast, M., Pasricha, S.: ARXON: a framework for approximate communication over photonic networks-on-chip. IEEE Trans. Very Large Scale Integr. VLSI Syst. **29**(6), 1206–1219 (2021)
8. Sunny, F., Mirza, A., Thakkar, I., Nikdast, M., Pasricha, S.: LORAX: loss-aware approximations for energy-efficient silicon photonic networks-on-chip. In: ACM GLSVLSI. ACM (2020)
9. Qiao, F., Zhou, N., Chen, Y., Yang, H.: Approximate computing in chrominance cache for image/video processing. In: IEEE ICMBD, pp. 180–183 (2015)
10. Nyugen, D.T., Kim, H., Lee, H.J., Chang, I.J.: An approximate memory architecture for reduction of a reduction of refresh power consumption in deep learning applications. In: IEEE ISCAS. IEEE (2018)
11. Liu, H., Ong, Y.S., Shen, X., Cai, J.: When Gaussian process meets big data: A review of scalable GPs. IEEE Trans. Neural Netw. Learn. Syst. **31**, 4405–4423 (2020)
12. Ahmadvand, H., Goudarzi, M., Foroutan, F.: Gapprox: using gallup approach for approximation in big data processing. J Big Data. **6**, 20 (2019)
13. Younes, H., Ibrahim, A., Rizk, M., Valle, M.: Algorithmic level approximate computing for machine learning classifiers. In: IEEE ICECS. IEEE (2019)
14. Sen, S., Raghunathan, A.: Approximate computing for long short term memory (LSTM) neural networks. In: IEEE TCAD. IEEE (2018)
15. Van Leussen, M., Huisken, J., Wang, L., Jiao, H., De Gyvez, J.P.: Reconfigurable support vector machine classifier with approximate computing. In: IEEE ISVLSI. IEEE (2017)
16. Ibrahim, A., Osta, M., Alameh, M., Saleh, M., Chible, H., Valle, M.: Approximate computing methods for embedded machine learning. In: IEEE ICECS. IEEE (2018)
17. Yellu, P., Boskov, N., Kinsy, M.A., Yu, Q.: Security threats in approximate computing systems. In: ACM GLSVLSI, pp. 387–392. IEEE (2019)
18. Han, J., Orshansky, M.: Approximate computing: an emerging paradigm for energy-efficient design. In: IEEE ETS, pp. 1–6. IEEE (2013)
19. Chippa, V.K., Venkataramani, S., Chakradhar, S.T., Roy, K., Raghunathan, A.: Approximate computing: an integrated hardware approach. In: Asilomar Conference on Signals, Systems and Computers (ACSSC). IEEE (2013)
20. Yang, Z., Jain, A., Liang, J., Han, J., Lombardi, F.: Approximate XOR/XNOR-based adders for inexact computing. In: IEEE-Nano. IEEE (2013)
21. Ramasamy, M., Narmadha, G., Deivasigamani, S.: Carry based approximate full adder for low power approximate computing. In: ICSCC. IEEE (2019)
22. Raha, A., Sutar, S., Jayakumar, H., Raghunathan, V.: Quality configurable approximate DRAM. In: TC. IEEE (2017)
23. Venketaramani, S., Chippa, V.K., Chakradhar, S.T., Roy, K., Raghunathan, A.: Quality programmable vector processors for approximate computing. In: IEEE MICRO. IEEE (2013)

24. Esmaeilzadeh, H., Sampson, A., Ceze, L., Burger, D.: Neural acceleration for general purpose approximate programs. In: IEEE MICRO. IEEE (2013)
25. Sampson, A., Baixo, A., Ransford, B., Moreau, T., Yip, J., Ceze, L., Oskin, M.: ACCEPT: A Programmer-Guided Compiler Framework for Practical Approximate Computing, White Chapter. University of Washington (2014)
26. Sampson, A., Dietl, W., Fortuna, E., Gnanapragasam, D., Ceze, L., Grossman, D.: EnerJ: approximate data types for safe and general low-power computation. In: PLD. ACM (2011)
27. Park, J., Esmaeilzadeh, H., Zhang, X., Naik, M., Harris, W.: FlexJava: language support for safe and modular approximate programming. In: FSE. ACM (2015)
28. Raparti, Y., Pasricha, S.: DAPPER: data aware approximate NoC for GPGPU architectures. In: IEEE/ACM NOCS. IEEE (2018)
29. Boyapati, R., Huang, J., Majumder, P., Yum, K.H., Kim, E.J.: APPROX-NoC: a data approximation framework for network-on-chip architectures. In: ISCA. IEEE (2017)
30. Wang, L., Wang, X., Wang, Y.: ABDTR: approximation-based dynamic traffic regulation for networks-on-chip systems. In: IEEE ICCD. IEEE (2017)
31. Ahmed, A.B., Fujiki, D., Matsutani, H., Koibuchi, M., Amano, H.: AxNoC: low-power approximate network-on-chips using critical-path isolation. In: IEEE/ACM NOCS (2018)
32. Bahirat, S., Pasricha, S.: METEOR: hybrid photonic ring-mesh network-on-chip for multicore architectures. ACM Trans. Embed. Comput. Syst. **13**(3), 116:1–116:33 (Mar 2014)
33. Bahirat, S., Pasricha, S.: HELIX: design and synthesis of hybrid nanophotonic application-specific network-on-chip architectures. In: IEEE International Symposium on Quality Electronic Design (ISQED). IEEE (2014)
34. Bahirat, S., Pasricha, S.: 3D HELIX: design and synthesis of hybrid nanophotonic application-specific 3D network-on-chip architectures. In: Workshop on Exploiting Silicon Photonics for Energy Efficient Heterogeneous Parallel Architectures (SiPhotonics). IEEE (2014)
35. Bahirat, S., Pasricha, S.: A particle swarm optimization approach for synthesizing application-specific hybrid photonic networks-on-chip. In: IEEE International Symposium on Quality Electronic Design (ISQED). IEEE (2012)
36. Bahirat, S., Pasricha, S.: UC-PHOTON: a novel hybrid photonic network-on-chip for multiple use-case applications. In: IEEE International Symposium on Quality Electronic Design (ISQED). IEEE, Santa Clara (2010)
37. Bahirat, S., Pasricha, S.: Exploring hybrid photonic networks-on-chip for emerging chip multiprocessors. In: IEEE/ACM International Conference on Hardware/Software Codesign and System Synthesis (CODES+ISSS). IEEE, Grenoble (2009)
38. Chittamuru, S.V.R., Thakkar, I., Pasricha, S., Vatsavai, S.S., Bhat, V.: Exploiting process variations to secure photonic NoC architectures from snooping attacks. IEEE Trans. Comput. Aided Des. Integr. Circuits Syst. **40**, 850–863 (2021)
39. Chittamuru, S.V.R., Thakkar, I., Pasricha, S.: LIBRA: thermal and process variation aware reliability management in photonic networks-on-chip. IEEE Trans. Multi-Scale Comput. Syst. **4**(4), 758–772 (2018)
40. Chittamuru, S.V.R., Dharnidhar, D., Pasricha, S., Mahapatra, R.: BiGNoC: accelerating big data computing with application-specific photonic network-on-chip architectures. IEEE Trans. Parallel Distrib. Syst. **29**(11), 2402–2415 (2018)
41. Chittamuru, S.V.R., Thakkar, I., Pasricha, S.: HYDRA: heterodyne crosstalk mitigation with double microring resonators and data encoding for photonic NoC. IEEE Trans. Very Large Scale Integr. VLSI Syst. **26**(1), 168–181 (2018)
42. Chittamuru, S.V.R., Desai, S., Pasricha, S.: Swiftnoc: a reconfigurable silicon-photonic network with multicast enabled channel sharing for multicore architectures. ACM J. Emerg. Technol. Comput. Syst. **13**(4), 58:1–58:27 (2017)
43. Chittamuru, S.V.R., Pasricha, S.: Crosstalk mitigation for high-radix and low-diameter photonic NoC architectures. IEEE Des. Test. **32**(3), 29–39 (2015)
44. Thakkar, I., Chittamuru, S.V.R., Pasricha, S.: Mitigating the energy impacts of VBTI aging in photonic networks-on-chip architectures with multilevel signaling. In: IEEE Workshop on Energy-Efficient Networks of Computers (E2NC): From the Chip to the Cloud. IEEE (2018)

45. Pasricha, S., Chittamuru, S.V.R., Thakkar, I., Bhat, V.: Securing photonic NoC architectures from hardware trojans. In: IEEE/ACM International Symposium on Networks-on-Chip (NOCS). IEEE, Torino (2018)
46. Chittamuru, S.V.R., Thakkar, I., Pasricha, S.: SOTERIA: exploiting process variations to enhance hardware security with photonic NoC architectures. In: IEEE/ACM Design Automation Conference (DAC). IEEE, San Francisco (2018)
47. Thakkar, I., Chittamuru, S.V.R., Pasricha, S.: Improving the reliability and energy-efficiency of high-bandwidth photonic NoC architectures with multilevel signaling. In: IEEE/ACM International Symposium on Networks-on-Chip (NOCS). IEEE (2017)
48. Chittamuru, S.V.R., Thakkar, I., Pasricha, S.: Analyzing voltage bias and temperature induced aging effects in photonic interconnects for manycore computing. In: ACM System Level Interconnect Prediction Workshop (SLIP). IEEE (2017)
49. Dang, D., Chittamuru, S.V.R., Mahapatra, R.N., Pasricha, S.: Islands of heaters: a novel thermal management framework for photonic NoCs. In: IEEE/ACM Asia & South Pacific Design Automation Conference (ASPDAC). IEEE (2017)
50. Thakkar, I., Chittamuru, S.V.R., Pasricha, S.: A comparative analysis of front-end and back-end compatible silicon photonic on-chip interconnects. In: ACM/IEEE System Level Interconnect Prediction Workshop (SLIP). IEEE (2016)
51. Thakkar, I., Chittamuru, S.V.R., Pasricha, S.: Run-time laser power management in photonic NoCs with on-chip semiconductor optical amplifiers. In: IEEE/ACM International Symposium on Networks-on-Chip (NOCS). IEEE (2016)
52. Chittamuru, S.V.R., Thakkar, I., Pasricha, S.: PICO: mitigating heterodyne crosstalk due to process variations and intermodulation effects in photonic NoCs. In: IEEE/ACM Design Automation Conference (DAC). IEEE (2016)
53. Chittamuru, S.V.R., Thakkar, I., Pasricha, S.: Process variation aware crosstalk mitigation for DWDM based photonic NoC architectures. In: IEEE International Symposium on Quality Electronic Design (ISQED). IEEE (2016)
54. Chittamuru, S.V.R., Pasricha, S.: SPECTRA: a framework for thermal reliability management in silicon-photonic networks-on-chip. In: IEEE International Conference on VLSI Design (VLSI). IEEE (2016)
55. Lee, J., Killian, C., Le Beux, S., Chillet, D.: Approximate nanophotonic interconnects. In: IEEE/ACM NOCS. IEEE (2019)
56. Bieneia, C.: Benchmarking Modern Multiprocessors. Ph. D Thesis, Princeton University, January (2011)
57. https://github.com/tiny-dnn/tiny-dnn
58. Binkert, N., Beckmann, B., Black, G., Reinhardt, S.K., Saidi, A., Basu, A., Hestness, J., Hower, D.R., Krishna, T., Sardashti, S., Sen, R.: The gem5 simulator. Comp. Arch. News. **39**, 1–7 (2011)
59. Luk, C.K., Cohn, R., Muth, R., Patil, H., Klauser, A., Lowney, G., Wallace, S., Reddi, V.J., Hazelwood, K.: Pin: building customized program analysis tools with dynamic instrumentation. ACM Sigplan. **40**, 190–200 (2005)
60. Pasricha, S., Nikdast, M.: A survey of silicon photonics for energy-efficient manycore computing. IEEE Des. Test. **37**, 60–81 (2020)
61. Soref, R.I., Bennett, B.R.: Electrooptical effects in silicon. IEEE J. Quantum Electron. **23**, 123–129 (1987)
62. Thakkar, I., et al.: Mitigation of homodyne crosstalk noise in silicon photonic NoC architectures with tunable decoupling. In: CODES+ISSS. IEEE (2016)
63. Chittamuru, S.V.R., Thakkar, I., Pasricha, S.: HYDRA: heterodyne crosstalk mitigation with double microring resonators and data encoding for photonic NoCs. IEEE Trans. Very Large Scale Integr. VLSI Syst. **26**, 168–181 (2018)
64. Chittamuru, S.V.R., Thakkar, I., Pasricha, S.: PICO: mitigating heterodyne crosstalk due to process variations and intermodulation effects in photonic NoCs. In: IEEE/ACM DAC (2016)
65. Nikdast, M., et al.: Crosstalk noise in WDM-based optical networks-on-chip: a formal study and comparison. In: IEEE Trans. Very Large Scale Integr. VLSI Syst., vol. 23, pp. 2552–2565 (2015)

66. Thakkar, I., et al.: Improving the reliability and energy-efficiency of high bandwidth photonic noc architectures with multilevel signaling. In: IEEE/ACM NOCS. IEEE (2017)
67. Li, H., Fourmigue, A., Le Beux, S., Letartre, X., O'Connor, I., Nicolescu, G.: Thermal aware design method for VCSEL-based On-Chip Optical Interconnect. In: IEEE/ACM DATE. IEEE (2015)
68. Wu, X., Xu, J., Ye, Y., Wang, Z., Nikdast, M., Wang, X.: Suor: sectioned unidirectional optical ring for chip multiprocessor. ACM J. Emerg. Technol. Comput. Syst. **10**, 1–25 (2014)
69. Mohamed, M., Li, Z., Chen, X., Shang, L., Mickelson, A.R.: Reliability-aware design flow for silicon photonics on-chip interconnect. IEEE Trans. Very Large Scale Integr. VLSI Syst. **22**, 1763–1776 (2014)
70. Padmaraju, K., Bergman, K.: Resolving the thermal challenges for silicon microring resonator devices. Nanophotonics. **3**, 269–281 (2013)
71. Sun, C., Wade, M.T., Lee, Y., Orcutt, J.S., Alloatti, L., Georgas, M.S., Waterman, A.S., Shainline, J.M., Avizienis, R.R., Lin, S., Moss, B.R.: Single-chip microprocessor that communicates directly using light. Nature. **528**, 24–31 (2015)
72. Kao, T.J., Louri, A.: Optical multilevel signaling for high bandwidth and power-efficient on-chip interconnects. IEEE Photon. Technol. Lett. **27**(19), 2051–2054 (2015)
73. Roshan-Zamir, A., Wang, B., Telaprolu, S., Yu, K., Li, C., Seyedi, M.A., Fiorentino, M., Beausoleil, R., Palermo, S.: A 40 Gb/s PAM4 silicon microring resonator modulator transmitter in 65nm CMOS. In: OIC. IEEE (2016)
74. D-Demers, R., LaRochelle, S., Shi, W.: Ultrafast pulse-amplitude modulation with a femtojoule silicon photonic modulator. Optica. **3**(6), 622–627 (2016)
75. Wu, X., Dama, B., Gothoskar, P., Metz, P., Shastri, K., Sunder, S., der Spiegel, J.V., Wang, Y., Webster, M., Wilson, W.: A 20Gb/s NRZ/PAM-4 1V transmitter in 40nm CMOS driving a Si-photonic modulator in 0.13μm CMOS. In: ISSCC. IEEE (2013)
76. Joshi, A., Batten, C., Kwon, Y.-J., Beamer, S., Shamim, I., Asanovic, K., Stojanovic, V.: Silicon-photonic clos networks for global on-chip communication. In: IEEE/ACM NOCS. IEEE (2009)
77. Chittamuru, S.V.R., Deasi, S., Pasricha, S.: Swiftnoc: a reconfigurable silicon photonic network with multicast enabled channel sharing for multicore architectures. ACM J. Emerg. Technol. Comput. Syst. **13**, 1–27 (2017)
78. Sun, C., Chen, C.-H.O., Kurian, G., Wei, L., Miller, J., Agarwal, A., Peh, L.-S., Stojanovic, V.: DSENT a tool connecting emerging photonics with electronics for opto-electronic networks-on-chip modeling. In: IEEE/ACM NOCS. IEEE (2012)
79. Chen, K., Li, S., Muralimanohar, N., Ahn, J.H., Brockman, J.B., Jouppi, N.P.: CACTI-3DD: architecture-level modeling for 3D diestacked DRAM main memory. In: IEEE/ACM DATE. IEEE (2012)
80. Stillmaker, A., Baas, B.: Scaling equations for the accurate prediction of CMOS device performance from 180nm to 7nm. Integration. **58**, 74–81 (2017). https://doi.org/10.1016/j.vlsi.2017.02.002
81. Biberman, A., Preston, K., Hendry, G., Sherwood-Droz, N., Chan, J., Levy, J.S., Lipson, M., Bergman, K.: Photonic network-on-chip architectures using multilayer deposited silicon materials for high-performance Chip multiprocessors. ACM J. Emerg. Technol. Comput. Syst. **7**, 1–25 (2011)
82. Chen, H.T., Verbist, J., Verheyen, P., De Heyn, P., Lepage, G., De Coster, J., Absil, P., Yin, X., Bauwelinck, J., Van Campenhout, J., Roelkens, G.: High Sensitivity 10 Gb/s Si photonic receiver based on a low-voltage waveguide-coupled Ge avalanche photodetector. Opt. Express. **23**, 815–822 (2015)
83. Bahirat, S., Pasricha, S.: OPAL: A multi-layer hybrid photonic NoC for 3D ICs. In: ASPDAC. IEEE (2011)
84. Jayatileka, H., Caverley, M., Jaeger, N.A.F., Shekhar, S., Chrostowski, L.: Crosstalk limitations of microring-resonator based WDM demultiplexers on SOI. In: OIC. IEEE (2015)
85. Yahya, M.R., Wu, N., Fang, Z., Ge, F., Shah, M.H.: A low insertion loss 5×5 optical router for mesh photonic network-on-chip topology. In: IEEE CSUDET. IEEE (2019)

86. http://www.aimphotonics.com/pdk
87. Behadori, M., Nikdast, M., Cheng, Q., Bergman, K.: Universal design of waveguide bends in silicon-on-insulator photonics platform. J. Lightwave Technol. **37**, 3044–3054 (2019)
88. Grani, P., Bartolini, S.: Design options for optical ring interconnect in future client devices. ACM J. Emerg. Technol. Comput. Syst. **10**, 1–25 (2014)
89. Yu, K., Li, C., Li, H., Titriku, A., Shafik, A., Wang, B., Wang, Z.: A 25 gb/s hybrid-integrated silicon photonic source synchronous receiver with microring wavelength stabilization. IEEE J. Solid-State Circuits. **51**, 2129–2141 (2016)
90. Thakkar, I.G., Pasricha, S.: LIBRA: thermal and process variation aware reliability management in photonic networks-on-chip. In: TMSCS (2018)
91. Mirza, A., Sunny, F., Walsh, P., Hassan, K., Pasricha, S., Nikdast, M.: Silicon photonic microring resonators: a comprehensive design-space exploration and optimization under fabrication-process variations. In: IEEE Transactions on Computer-Aided Design of Integrated Circuits and Systems (2021)
92. Totoni, E., Behzad, B., Ghike, S., Torrellas, J.: Comparing the power and performance of Intel' SCC to state-of-the-art CPUs and GPUs. In: IEEE International Symposium on Performance Analysis of Systems & Software. IEEE (2012)

Low- and Mixed-Precision Inference Accelerators

Maarten J. Molendijk, Floran A. M. de Putter, and Henk Corporaal

1 Introduction

Neural Networks can solve increasingly more complex tasks in fields such as Computer Vision (CV) and Natural Language Processing (NLP). While these Neural Networks can perform complex tasks with increasingly higher accuracy, the sheer size of these networks often prevents deployment on edge devices that have limited memory capacity and are subject to severe energy constraints. To overcome the issues preventing the deployment of neural networks onto edge devices, efforts toward reducing the model size and reducing the computational costs have been made. These efforts are most often focused on either the algorithmic side, tailoring the neural network and its properties, or on the hardware side, creating efficient system designs and arithmetic circuitry.

In an effort to reduce the computational cost and model size of neural networks, several approaches are taken. One of these approaches is to automate the synthesis of the neural network architecture while taking into account the hardware resources, this is called hardware-aware neural architecture search (NAS) [23, 28]. Another way to increase the energy efficiency is by compressing the model size, applying either quantization [10] or pruning [4].

In parallel to research on model compression, research has been performed on creating highly specialized hardware that exploits the opportunities arising from model compression. ASICs that support neural network inference for operand precisions as low as 1 bit exploit the advantages extreme quantization brings: low memory size and bandwidth and simplified compute logic. In the pursuit of the most energy-efficient hardware design, several design choices regarding memory

M. J. Molendijk (✉) · F. A. M. de Putter · H. Corporaal
Eindhoven Artificial Intelligence Systems Institute and PARsE lab, Eindhoven University of Technology, Eindhoven, The Netherlands
e-mail: m.j.molendijk@tue.nl; f.a.m.d.putter@tue.nl; h.corporaal@tue.nl

hierarchy, hardware parallelization of operations, and data-flow are made that impact both the ASIC's efficiency and its flexibility.

For instance, many architectures have a fixed datapath; the movement of the data is fixed at design time which can impose limitations on the layer types, channel dimensions, and kernel dimensions. Furthermore, these architectures typically have limited programmability and configurability, which restricts the execution schedules that can (efficiently) be run.

In this chapter, a look will be taken at several different approaches of neural network accelerators specifically designed for inference with very low-precision operands. The efficiency (and origin thereof) of the architectures will be analyzed and compared to the flexibility that these architectures offer.

In short, the contributions of this work are:

- Overview of state-of-the-art low- and mixed-precision neural network accelerators, in Sect. 3.
- Analysis on the trade-off between the flexibility and the energy efficiency of accelerators, in Sect. 4.

The remainder of this chapter is structured as follows: in Sect. 2, background information on neural network architecture and quantization is presented. Thereafter, in Sect. 3, the low- and mixed-precision accelerators are presented and a comparison is presented in Sect. 4. Section 5 concludes this chapter.

2 Background: Extreme Quantization and Network Variety

Modern neural network architectures consist of many different layers with millions of parameters and operations. The storage required to store all parameters and features is not in line with the storage capacity typically found on embedded devices, leading to costly off-chip memory accesses. Next to the memory and bandwidth limitations, computational costs for full-precision (float32) operations require power-hungry compute blocks that quickly overtax the energy requirements of the embedded devices. To reduce both the computational cost and the cost of data access and transport, quantization can be applied.

Quantization leads to lower precision parameters and therefore induces information loss. Naturally, when weights and activations can represent fewer distinct values, the representational capabilities of the network decrease. This decrease may create an accuracy loss. In [14], Gholami et al. show, however, that quantization down to integer8 can be done without significant accuracy loss. But even when quantizing down to integer8, the memory requirements can still overtax the memory capacity typically found in embedded systems. Therefore, research has been done on extreme quantization, i.e., quantization below 8-bit precision.

In the next subsection, several frequently utilized building blocks for convolutional neural networks (CNNs) are listed. Thereafter, in Sects. 2.2 and 2.3, two forms of extreme quantization, namely `binary` and `ternary` quantization, are discussed. Finally, in Sect. 2.4, the need for mixed-precision is considered.

2.1 Neural Network Architecture Design Space

Neural network architectures have a great variety in the type of layers, the size of these layers, and the connectivity between these layers. Furthermore, with mixed-precision architectures, the precision can also be chosen on a per-layer basis. An example network is shown in Fig. 1. Some common building blocks are listed below:

- Convolutional Layer
- Fully connected Layer
- Depth-wise Convolutional Layer
- Residual Addition
- Requantization
- Pooling

The working horse of CNNs is the convolutional layer. Between different convolutional layers, there can be variety in the kernel size, number of input feature maps, output feature maps, etc. In Fig. 2, the different parameters of a convolutional layer are presented. These parameters will later on prove to be an important basis for designing efficient hardware. The goal of Sect. 3 is to show how these network parameters relate to hardware design, hardware efficiency, and hardware flexibility.

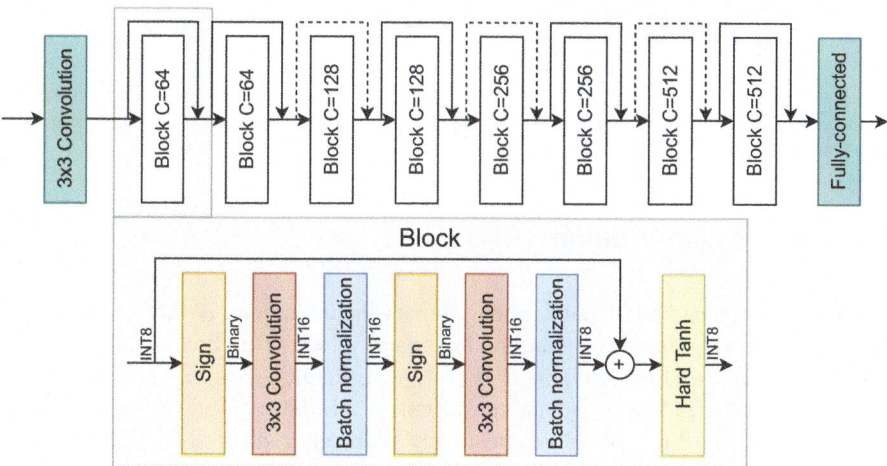

Fig. 1 Binary ResNet-18, an exemplary network containing several different building blocks and using different operand precisions. Note that the first layer and the "skip connections" have a higher than binary precision. Furthermore, the number of channels C can differ between the building blocks

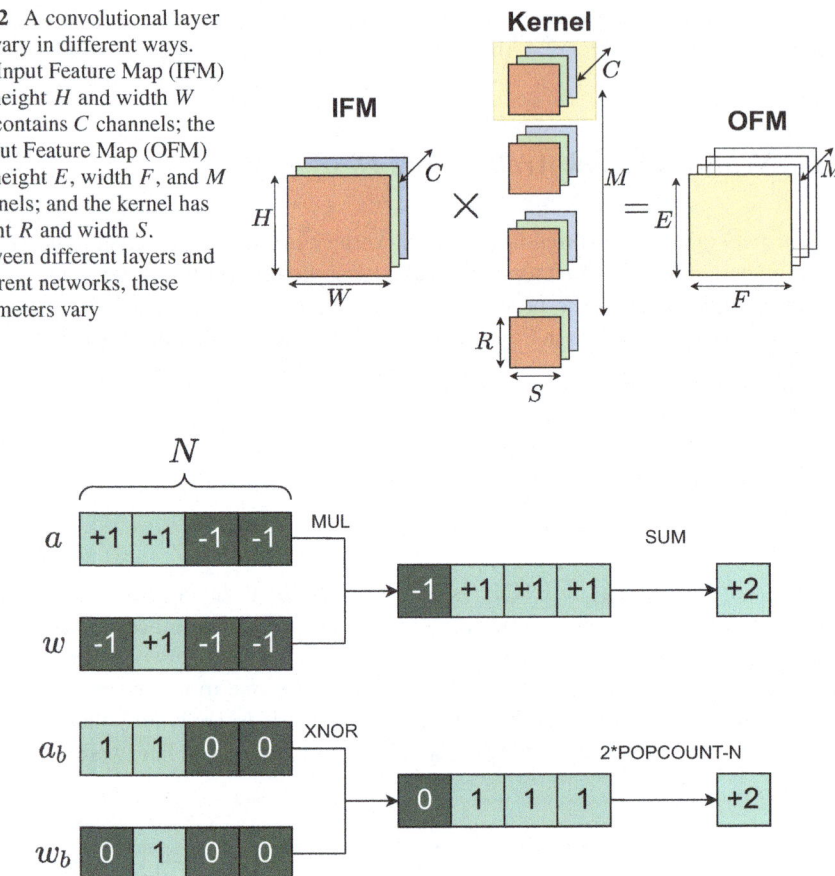

Fig. 2 A convolutional layer can vary in different ways. The Input Feature Map (IFM) has height H and width W and contains C channels; the Output Feature Map (OFM) has height E, width F, and M channels; and the kernel has height R and width S. Between different layers and different networks, these parameters vary

Fig. 3 Simplified arithmetic circuitry as a consequence of binary quantization. The top displays the default multiplication, while the bottom displays how binary quantization can replace it with XNOR and popcount. N is the number of bits of the input vector

2.2 Binary Quantization

On the extreme end of quantization is binary quantization. Binary quantization restricts both weights and activations to binary values. This means that the activations $a \in \{-1, +1\}$ and weights $w \in \{-1, +1\}$. Reduction of the precision of the operands introduces several advantages. First of all, the required storage capacity and bandwidth on the device are drastically reduced, compared to float32 by a factor of 32. Furthermore, the Multiply–Accumulate (MAC) operation, involving expensive multiplication hardware, can be replaced by the much more simple and cheaper XNOR and popcount operations [21]. An example of this simplified arithmetic is shown in Fig. 3.

The output value of the `popcount` produces a value that needs to be stored with a larger bit-width compared to the binary input value, e.g., `integer16`. Therefore, to feed the outputs into a new layer, the nonlinear activation function needs to requantize the values back to the binary bit-width. For this purpose, the *sign* function is used. The quantized operand can be derived from its unquantized form as follows:

$$X_{quant} = Sign(X) = \begin{cases} +1 & \text{if } X \geq 0 \\ -1 & \text{if } X < 0 \end{cases} \tag{1}$$

This function is non-differentiable; for training, a Straight-Through Estimator (STE) [3] can be used that passes gradients as is. By employing an STE, gradient descent is possible, and binary neural networks can be trained.

2.3 Ternary Quantization

Compared to binary quantization, ternary quantization allows for only one—*albeit very important*—extra value to be represented in the operands, namely zero. Ternary networks therefore have operands $w, a \in \{-1, 0, +1\}$ called *trits*. Next to the increased representational capabilities, the ability to represent zero also solves some issues found in binary networks. First of all, zero padding is not possible in binary networks since it lacks the ability to represent zero, and this is most often solved by employing on–off padding. Furthermore, the ability to represent zero introduces the capability to exploit sparsity, i.e., skipping computations when either the weights or activation is zero. As will be seen later on, this can have a significant impact on the efficiency of the computational hardware if the network itself is sparse.

The arithmetic circuitry required to perform multiply–accumulate (MAC) operations on ternary operands is very similar to that of binary networks. The MAC operation can be replaced by a `Gated-XNOR` [8] (XNOR and AND gates) combined with two `popcount` modules, one for the +1s and one for the -1s. The arithmetic is shown in Fig. 4.

Again, as with the binary `popcount`, the final result has a higher bit-width and needs to be requantized before being fed into the next layer. The quantization function typically uses a symmetric threshold value Δ:

$$X_{quant} = Ternarize(X) = \begin{cases} +1 & \text{if } X > \Delta \\ 0 & \text{if } |X| \leq \Delta \\ -1 & \text{if } X < -\Delta \end{cases} \tag{2}$$

During computation, each trit occupies 2 bits. However, this is a wasteful way to store them since theoretically $log_2(3) = 1.58$ bits are needed for each trit. Muller et

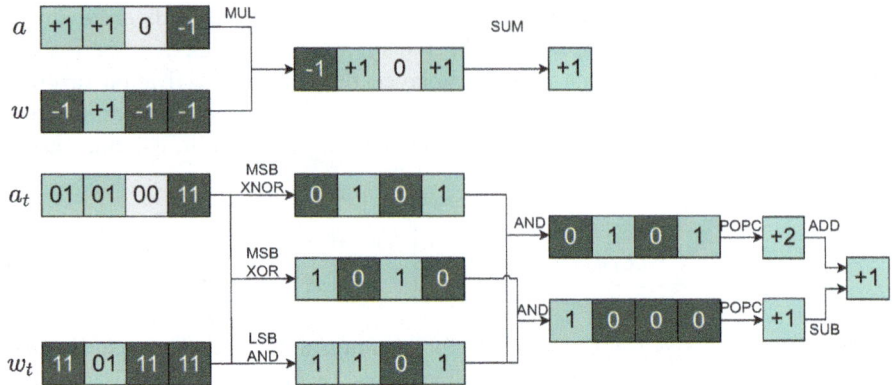

Fig. 4 Simplified arithmetic circuitry as a consequence of ternary quantization. The top displays the default multiplication, while the bottom displays the ternary simplified variant. Note that the ternary variant needs two popcount modules (one to count +1s and one to count −1s)

al. [18] derived an efficient mapping, compressing 5 trits into 8 bits, yielding a total storage of 1.6 bit per trit, close to the theoretical lower bound.

2.4 Mixed-Precision

Despite all the advantages of extreme quantization, binary and ternary quantization often induce severe accuracy loss, especially on more complex tasks. For example, there is a large gap in accuracy when comparing `integer8` quantization to `binary` and `ternary` [5, 10]. Moreover, the accuracy loss that is induced differs per layer in the network [11]; i.e., some layers are more resilient to extreme quantization than others. Therefore, a combination of different precisions in a per-layer fashion can give a good balance between accuracy and efficiency.

An overview of different data precisions typically found in neural network architectures is given in Fig. 5. The figure shows the width of different data formats and how the bits are allocated. Next to the data format, the range is displayed, i.e., the minimum and maximum value that can be attained using that data format. Note that the range for the floating-point number only displays the positive numbers, while it is able to represent negative numbers using the sign bit.

In the past, `float32` was used as the *de facto* standard for neural networks. Gradually, movements toward smaller data types like `float16` were made to save on storage and computational cost. Moreover, it was found that the dynamic range of the data types has a larger impact on the accuracy than the relative precision, leading to the creation of `bfloat16` [27] (Brain Floating Point) and `tf32` [15] (TensorFloat32), both trading off relative precision in favor of increased range. Using `integer8` precision completely gets rid of the expensive floating-

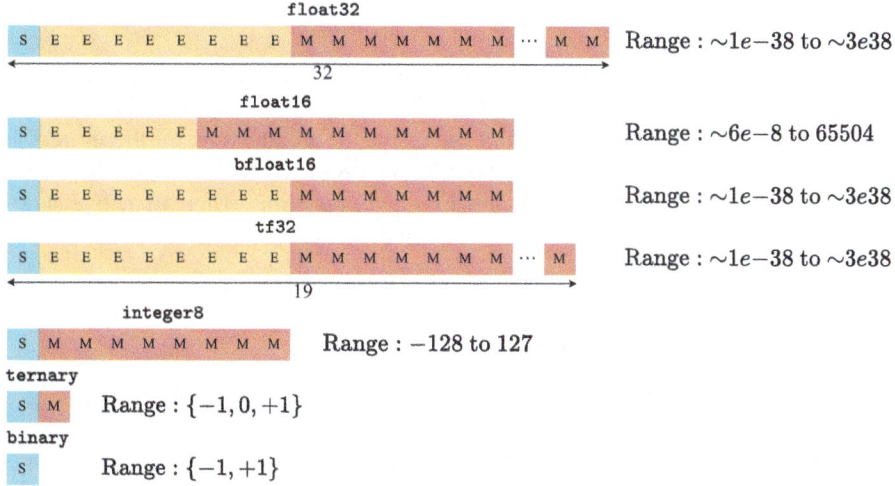

Fig. 5 Breakdown of the bit usage inside data formats commonly used in neural networks. S is sign, E is exponent, and M is mantissa. Floating point data formats specifically for neural networks prefer higher range over more precision

point arithmetic, vastly increasing the throughput and energy efficiency, with at the extreme end `binary` and `ternary` quantization.

By the nature of floating-point arithmetic units, exponents are added up together, while the mantissa bits are multiplied. Therefore, `bf16`, which has 3 less mantissa bits compared to `float16`, will have a *two times* smaller footprint, while compared to `float32` it will even have an *eight times* smaller area. This is because the area of the multiplier unit is roughly proportional to the square of the mantissa bits. In Sect. 3, accelerators that support `integer8` (which can also be used for fixed-point arithmetic), `binary`, and `ternary` precisions are discussed.

3 Accelerators for Low- and Mixed-Precision Inference

With the aim to get the energy per MAC operation as low as possible, several accelerators specifically designed for low-precision inference have been created. Some of these architectures also support different precisions on the same platform. The accelerators can be split into two groups: fully digital accelerators and mixed-signal/analog compute-in-memory (CIM) approaches. Although state-of-the-art CIM architectures [2, 26] and mixed-signal implementations [25] have the potential to achieve high energy efficiency, they also introduce new unique challenges. These challenges include longer design time and chip-to-chip variation due to CMOS process variation, which makes it more difficult to benchmark the actual performance of such a design, and no programmability, making it more difficult to use the accelerator. The further focus in this chapter will therefore be solely on fully digital implementations.

First, characterization criteria that are important to embedded neural network accelerators will be established, and these include key performance indicators to measure the efficiency (in both area and energy) of the architecture. Furthermore, the basis for the flexibility analysis is laid out, based on the robustness of architectures against different layer types, dimensionality, and precisions. Thereafter, five state-of-the-art digital inference accelerators will be discussed.

3.1 Characterization Criteria

The accelerators will be characterized according to both their flexibility and their energy efficiency. Defining flexibility as a quantitative metric can often be cumbersome, although some recent effort toward bringing structure has been made [12]. Next to the flexibility aspects, the most important quantitative performance evaluation criteria for neural network inference accelerators will be listed and motivated.

3.1.1 Flexibility

Before the characterization criteria are established, a closer look is taken at the nature of a convolution kernel. A convolution kernel can be described by 6 nested for-loops (7 when adding the batch dimension), and an exemplary schedule is shown in Listing 1. It is assumed that the target application is image processing, i.e., inputs are referred to as *pixels*.

In Listing 1, the for-loops are arranged in a so-called *output stationary* way, i.e., one output pixel is calculated as soon as possible. In other words, all calculations that are needed for a set of output pixels are performed before moving to the next set of output pixels. This avoids having to store and reload partially calculated output pixels.

```
for h in [0, H - R + 1]:                          Output feature map height
  for w in [0, W - S + 1]:                         Output feature map width
    for m in [0, M]:                                      Output channels
      acc = bias[m]
      for c in [0, C]:                                      Input channels
        for r in [0, R]:                                      Kernel height
          for s in [0, S]:                                    Kernel width
            acc += ifm[h + r][w + s][c] * weights[n][r][s][m]
      ofm[h][w][m] = acc
```

Listing 1 A naive convolutional layer with output stationary schedule and a stride of 1; *acc* is the temporary accumulated value, for simplicity, the IFM is assumed to be padded. The loop iterators are visualized in Fig. 2

Loop nest optimization (LNO) can be performed to increase data locality. Two important techniques, part of LNO, are loop tiling (also known as loop blocking), where a loop is split up into an inner and an outer loop, and loop interchange, where two loops are swapped in hierarchy level. The problem of finding the best combination of the two is called the **temporal mapping** problem (i.e., finding the best execution schedule). The temporal mapping greatly influences the number of memory accesses needed and therefore indirectly greatly influences the energy efficiency of an accelerator.

Next to the temporal mapping, the operations performed in the convolutional kernel can also be parallelized in hardware. The problem of finding the optimal parallelization dimensions is called the **spatial mapping** problem. Using optimal spatial mapping can increase data reuse in hardware and reduce memory traffic. A good example of this is the mapping on a systolic array. It is important to note that the spatial mapping should be carefully chosen, as it imposes constraints on the dimensions being parallelized.

Hardware parallelization over a dimension is called vectorization. Vectorization over any of the dimensions given in Fig. 2 will be denoted as the vectorization factor v_{param}, where param can be any of the dimensions in Fig. 2. For instance, when parallelizing over the C dimension using a vectorization factor of 32, it is denoted as $v_C = 32$. This vectorization factor also implies constraints: any convolutional network layer that does not have an input channel multiple of 32 will not run at 100% utilization. There will be a trade-off between the vectorization factor and the flexibility with respect to convolutional layers with certain layer dimensions being able to run at full utilization.

Research has been done on structurally exploring the temporal and spatial mapping design space [20, 29]. Most recently, the ZigZag framework [17] has been published aiming to fully co-design temporal mapping with hardware architecture finding the best spatial and temporal mappings available.

One other facet of flexibility is **programmability**. Programmability allows running different, possibly even non-DNN workloads on the accelerators. Especially, high-level programmability increases the usability of the device since it allows the workload to be configured while programming it via a high-level language, requiring less knowledge about the hardware implementation from a user perspective.

3.1.2 Performance Characteristics

To compare the performance of the several accelerators reviewed, some quantitative metrics that reflect the performance of the accelerator are established. First of all, the most widely promoted metric to compare accelerators is to compare the energy efficiency, defined as the energy per operation (either [pJ/op] or inversely in [TOPS/W]).

Secondly, the memory capacity plays an important role in the efficiency of the accelerator. Since off-chip memory access energy is much larger than the energy needed to compute, off-chip memory access should be avoided at all costs.

More on-chip memory means fewer external memory accesses, benefiting energy consumption. Two different ways to implement on-chip memory are SRAM and Standard-Cell Memory (SCM). While SRAM has a much higher memory density, it is less efficient in terms of energy usage for smaller sizes compared to SCM. Especially, when applying voltage–frequency scaling, the SCM can be scaled to a much lower voltage than SRAM. Therefore, SCM tends to be a popular choice to keep down the energy cost of the total system while sacrificing area and storage capacity. Other important metrics are throughput [GOPS] and area efficiency [GOPS/mm^2].

3.2 Five Low- and Mixed-Precision Accelerators Reviewed

Five state-of-the-art accelerators will be discussed and compared against one another. These accelerators were chosen because of their support for very low precisions (i.e., binary or ternary). These accelerators are:

- *XNOR Neural Engine* [6] is a binary neural network accelerator built into a programmable microcontroller unit. A full system on a chip (SoC), implemented in 22-nm technology, is presented including the accelerator, RISC host processor, and peripherals.
- *ChewBaccaNN* [1] is an architecture for binary neural network inference that exploits efficient data reuse by co-designing the memory hierarchy with the neural network ran on the architecture. The hard-wired kernel size allows efficient data reuse.
- *CUTIE* [22] is an accelerator for ternary neural networks. This is a massively parallel architecture, hard-coding all the network parameters into the hardware design. Furthermore, it exploits sparsity opportunities from ternary networks that are not present in binary networks.
- *Knag et al.* produced a binary neural network accelerator in 10-nm FinFet technology [16]. The design focuses on utilizing the compute near memory paradigm, minimizing the cost of data movement by interleaving memory and computational elements.
- *BrainTTA* is a flexible, fully programmable solution based on a Transport-Triggered Architecture. The architecture has support for mixed-precision and focuses, next to the energy efficiency objective also on flexibility, trying to minimize the concessions made while still pursuing energy efficiency.

A summary of these architectures is given in Table 1, and the strengths and weaknesses of the architectures are discussed in Sect. 4.

3.2.1 XNOR Neural Engine (XNE)

XNOR Neural Engine [6] is a binary accelerator exploiting the arithmetic simplifications introduced by binarizing the weights and activations (see Fig. 3). Conti et

Table 1 Comparison of performance, efficiency, and flexibility of the architectures discussed

	ChewBaccaNN [1]	CUTIE [22]		XNE [6]		10nm FinFet [16]	BrainTTA
Implementation characteristics							
Tech node [nm]	22	22		22		10	28
Supply voltage [V]	0.4	0.65		0.6	0.4	0.39	0.9
Inference precision[a]	b	b[b], t		b		b	b, t, i8
Memory technology	SCM	SRAM	SCM	SRAM	SCM	SRAM	SRAM
Key Performance Indicators							
Peak throughput [GOPS]	240	16,000		67	5	3400	880
Energy/op [fJ] binary	4.48/15.38[c]	–		115	21.6	1.62	101
Energy/op [fJ] ternary	–	2.19	1.70	–		–	188
Energy/op [fJ] 8-bit	–	–		–		–	1105
Core area [mm²]	0.7	7.5		2.32		0.39	3.6
Area efficiency [GOPS/mm²]	343	2133		28.88		8717	244.44
Memory capacity [kB]	153	1190	281	520	16	161	1024
Flexibility							
Full utilization for							
C multiple of	16	128		128		1024	32/16/4[d]
M multiple of	Any	128		128		128	32
R is	7	3		Any		2	Any
S is	7	3		Any		2	Any
Partial result support (for scheduling freedom)	Yes	No[e]		No		No	Yes
Residual layer support	Yes	No		No		No	Yes
Programmability	None	None		None		None	C-language

[a] b = binary, t = ternary, i8 = integer8
[b] Only estimates were provided, under the assumption that all ternary specific hardware is removed
[c] For 7 × 7 and 3 × 3 convolution, respectively
[d] For binary, ternary and integer8, respectively
[e] Partial result support is not needed since the output pixel computation is fully unrolled in hardware

al. present an SoC consisting of an accelerator core (XNE) inside a microcontroller unit (MCU) and peripheries. The accelerator can independently run simple network configurations but requires the programmable MCU to execute more complex

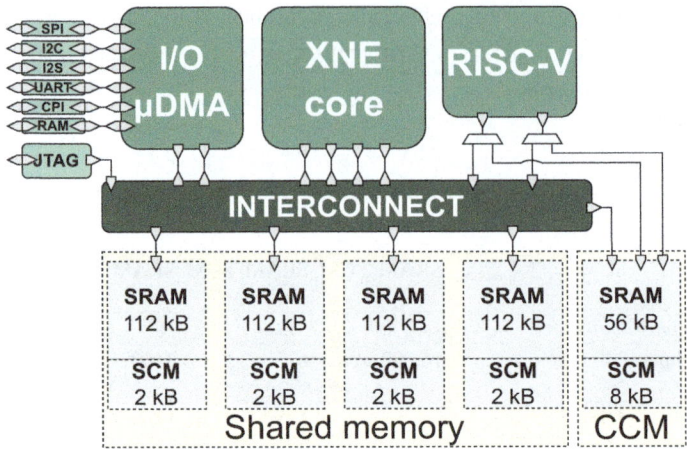

Fig. 6 Top-level view of the SoC with XNE inside the MCU. The memory is a hybrid of latch-based SCM and SRAM

layers. The MCU is programmed using some assembly dialect. The full system is shown in Fig. 6. It consists of:

- *XNE core*, where the binary MAC operations are performed; this core consists of a *streamer*, to stream feature maps and weights in and out of the architecture, a *controller* consisting of a finite-state machine, the programmable microcode processor, and a latch-based register file.
- *RISC-V host processor*, used to realize more complex layer behaviors than supported with the XNE core alone.
- *Shared Memory*, shared between the μDMA, RISC-V core, and XNE core. This memory is a hybrid of SRAM and SCM, allowing aggressive voltage scaling when the SRAMs are turned off.
- *Core-Coupled Memory (CCM)*, primarily for the RISC-V core, again composed of both SRAM and SCM.
- *μDMA*, which is an autonomous unit able to send and receive data via several communication protocols from and to the shared memory.

The accelerator core, XNE, is shown in Fig. 7. The throughput of the design can be chosen at design time by means of a throughput parameter TP. This throughput parameter can be described as follows: it takes the accelerator TP cycles to calculate TP output pixels. While doing this, the accelerator keeps the same input activations for TP cycles while loading TP weights each cycle (for a total of TP sets of TP weights). Therefore, this TP parameter essentially hard-wires the C and M dimension of the convolution dimensions shown in Fig. 2 into the design.

For instance, each accumulator in Fig. 7 contains the partial result of one output pixel (i.e., the number of accumulators is equal to the output feature map channel vectorization v_M). Therefore, all the inputs that are processed while a single accumulator is selected via the mux should contribute to the same output pixel.

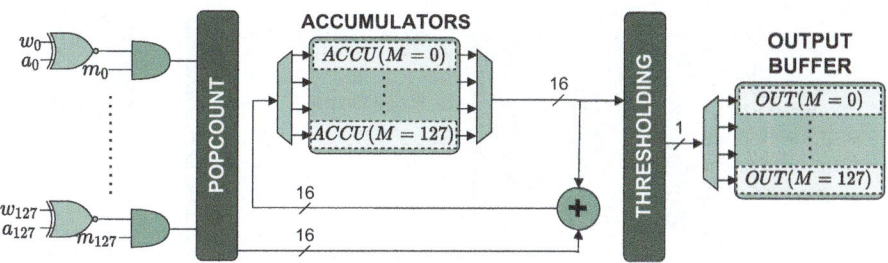

Fig. 7 Accelerator core of XNOR Neural Engine with TP = 128. The XNOR operation is performed on the activations a and weights w. Whenever the number of input operands is not a multiple of TP, the outputs can be masked by masking bits m to make sure that they do not contribute to the popcount output

In this case, the different pixels concurrently offered to the compute core belong to different input channels. Therefore, the choice of TP directly imposes a constraint on the C and M loops in order to run at full efficiency. Furthermore, the output of the popcount operation is directly fed through the binarization function; this means that partial (higher bit-width) results cannot be extracted, prohibiting their use for residual layers. For benchmarking the platform, a TP factor of 128 was chosen, which means that $v_C = 128$ and $v_M = 128$ for this design point.

3.2.2 ChewBaccaNN

ChewBaccaNN [1] is like XNE, an accelerator utilizing binary weights and binary activations. Contrary to XNE, this architecture does not implement a full SoC and is therefore purely based on the accelerator core. ChewBaccaNN is designed using GF22 technology and uses SCM to enable aggressive voltage scaling. A top-level view of the architecture is shown in Fig. 8. The components in this architecture are:

- *BPU Array* consists of seven Basic Processing Units (BPUs) and forms the computational heart of the accelerator; the BPU is detailed in Fig. 9 and is discussed in the next paragraph.
- *Feature Map Memory (FMM)* holds the input and output feature maps and also has the ability to store partial results (e.g., for residual layers). The FMM is implemented using SCM only. This enables aggressive voltage scaling for the whole chip at the expense of sacrificing memory capacity.
- *Row Banks* buffer the input feature map rows and kernel rows. The *crossbar* (x-bar) is utilized when the convolutional window slides down. Since each BPU processes one kernel row, the kernel weights can stay inside the BPUs, while the input feature map needs to move one row down. This is done by loading one new row and shifting the other rows by one BPU (using the crossbar).

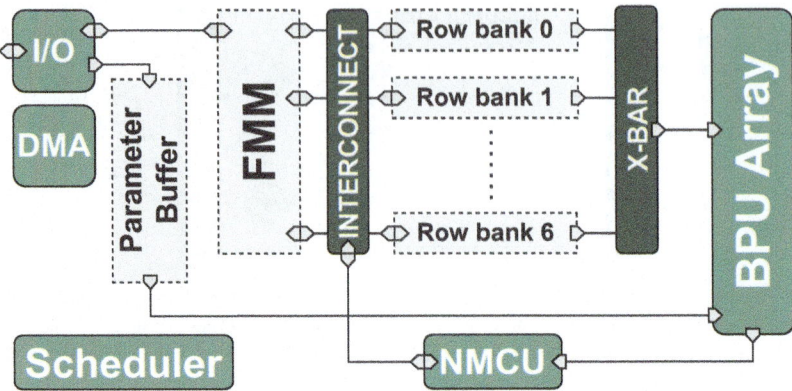

Fig. 8 The top-level architectural overview of the ChewBaccaNN accelerator. All the memories are implemented using latch-based standard-cell memory. The control signals are not shown in this overview

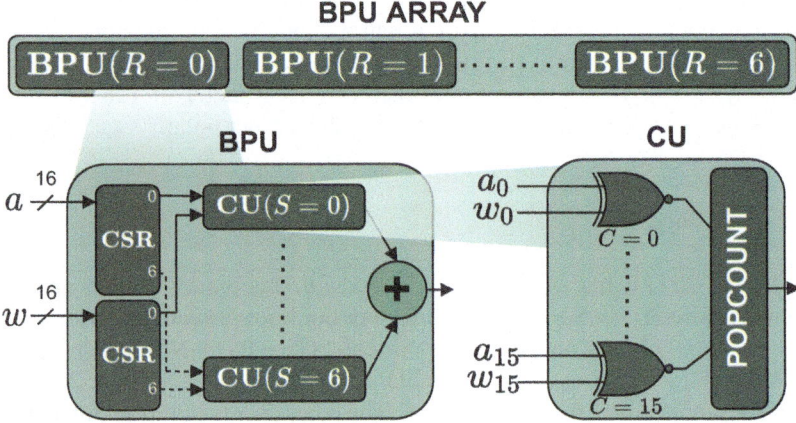

Fig. 9 ChewBaccaNN compute core. Hardware parallelization is performed over the kernel height (R) in the Basic Processing Unit (BPU) array, over the kernel width (S) inside a single BPU and over the input channel dimension (C) inside the compute unit (CU). The Controlled Shift Register (CSR) enables data reuse in a sliding window fashion. The architecture contains a total of $16 \times 7 \times 7 = 784$ ($v_C \times v_R \times v_S$) binary multipliers

- *Scheduler*, used to control the crossbar behavior and make sure that the row banks are timely rotated to the next BPU and the correct weights and IFM pixels are loaded.
- *Near Memory Compute Unit (NMCU)*, which writes output data from the BPU array to the correct location in the FMM, accumulates residual paths, rebinarizes results, and is used for bit-packing (rebinarized) outputs into 16-bit packets.

In Fig. 9, the compute core of ChewBaccaNN is depicted. It can be seen that several of the parameters listed in Fig. 2 are hard-wired into the design. The kernel height

(R) and width (S) are completely unrolled (in this case with a factor of 7), while the channel dimension (C) should be a multiple of 16 (the number of XNOR gates) to achieve full utilization; in other words, the vectorization factors are $v_C = 16$, $v_R = 7$, and $v_S = 7$.

The Controlled Shift Register (CSR) allows using the sliding window principle to get data reuse; for each IFM image row, initially, the full kernel width (in this case 7) is transferred, while the iterations thereafter only need one new column ($v_R \times 1 \times v_C$) of activations.

3.2.3 Completely Unrolled Ternary Inference Engine (CUTIE)

Completely Unrolled Ternary Inference Engine (CUTIE) [22] is, as the name suggests, an inference accelerator using Ternary operands. The main design philosophy behind CUTIE is to avoid iteration by spatially unrolling most of the convolutional loops found in Listing 1, namely the loops over the R, S, C, and M dimensions. Furthermore, ternary operands allow the representation of zero, therefore making the accelerator capable of exploiting neural network sparsity by silencing compute units. The top-level design of CUTIE is depicted in Fig. 10. The main components within the CUTIE architecture are:

- *Output Channel Compute Unit (OCU)*, the basic compute building block of this architecture, computing the output pixels belonging to one single output channel. A detailed view of the OCU is given in Fig. 11.
- *Feature Map Memory (FMM)*, used to store the inputs coming either from previous computations (OCUs) or from an external interface. The FMM is double-buffered such that the latency for loading new input feature maps can be hidden.

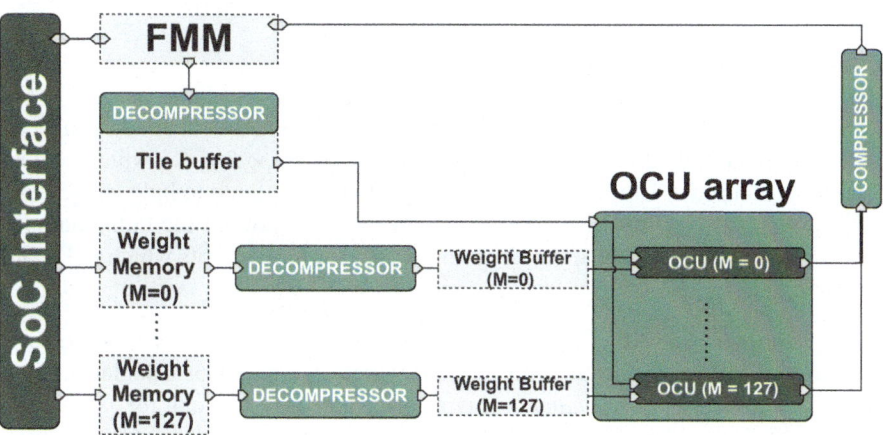

Fig. 10 CUTIE top-level architecture. The OCU array contains one output channel compute unit for each output channel in the neural network design

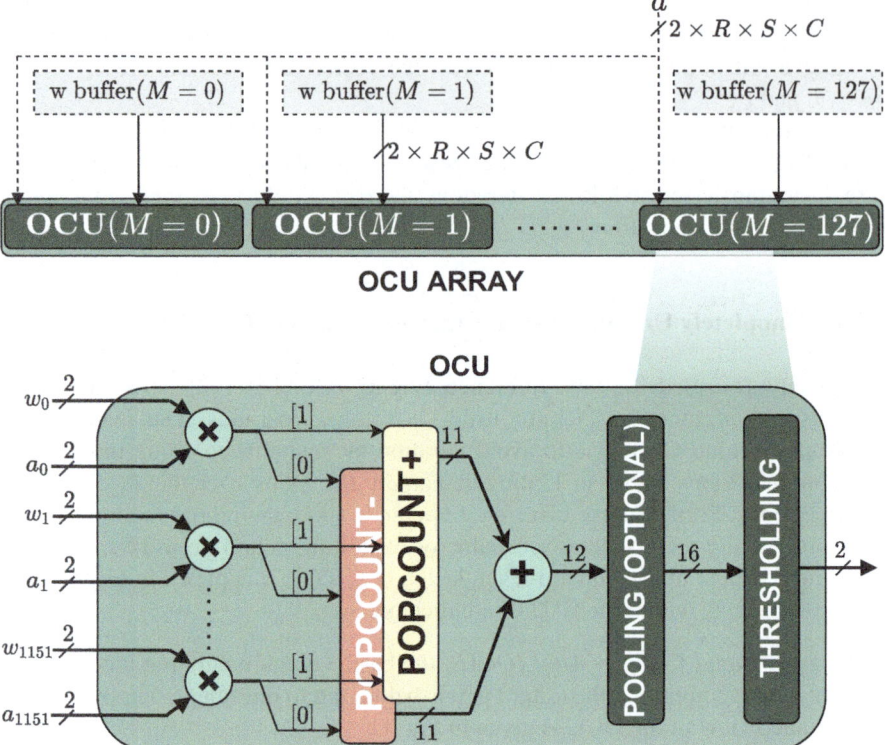

Fig. 11 CUTIE compute core, consisting of several Output Channel Compute Units (OCUs) and one weight buffer per OCU. For brevity, decompression and pipelining are omitted in this figure. The ternary multipliers are unrolled over the R, S, and C dimensions, which in this case gives $3 \times 3 \times 128 = 1152$ ternary multipliers. In total, the architecture can process $3 \times 3 \times 128 \times 128 = 147{,}456$ ($v_R \times v_S \times v_M \times v_C$) inputs each compute cycle

- *Tile buffer*, used to buffer IFM pixels in a sliding window fashion.
- *Weight buffer*, one is attached to each OCU: it is designed with enough capacity to contain the full kernel for a single output channel ($R \times S \times C$), which enables great weight reuse. The weight buffer is also double-buffered to hide latency.
- *Compression/decompression units* are used to shift between the computational form of the trits, i.e., 2 bits, and the compressed form of the trits which is 1.6 bits wide.

The compute core of CUTIE is depicted in Fig. 11. Its main workhorse is the Output Channel Compute Unit (OCU), which is a unit that calculates pixels *exclusive to a single output channel*. Having a separate compute unit for each output channel brings the advantage that the weight kernel can stay inside the weight buffer (w buffer) while moving the convolutional window over the IFM giving maximum weight data reuse. Alongside the weight reuse, there is also IFM reuse being utilized in two different ways: (1) the IFM is broadcasted to each of the OCUs and (2) just like ChewBaccaNN, when sliding the convolutional window over the IFM image, only R new IFM pixels are needed (i.e., only one new column of the IFM needs to be loaded, assuming a stride of 1).

Each Output Channel Compute Unit (OCU) processes $128 \times 3 \times 3$ ($v_C \times v_R \times v_S$) input pixels each cycle. By hard-wiring many of the convolutional layer parameters, CUTIE sacrifices area in favor of avoiding temporal iteration. This also means that this architecture sacrifices most flexibility by constraining C, M, R, and S. Therefore, the only dimensions that are freely schedulable are W and H. By constraining many of the dimensions into the hardware, flexibility crumbles, but the temporal mapping is greatly simplified. The fully spatially unrolled structure also minimizes the movement of (large) partial sums. Since each OCU directly computes an output pixel, there is no need, in contrast to the other architectures, to move around partial results. This is beneficial since the partial results have a higher bit-width than the final (requantized) results.

3.2.4 Binary Neural Network Accelerator in 10-nm FinFet

In [16], Knag et al. show a fully digital accelerator with `binary` operands which is implemented using 10nm FinFet technology. The SoC designed intersperses arithmetic with memory according to the Compute Near Memory (CNM) paradigm. Contrary to the other architectures discussed, this work focuses more on the physical implementation and circuit-level design choices rather than the architectural design aspects. The design of this accelerator is shown in Fig. 12. The main components of this accelerator are:

- *Control Unit*, which consists of four 256-bit wide SRAM memory banks used as main storage and a Finite-State Machine (FSM) that controls the flow of data between memory banks and the MEUs.
- *Memory Execution Unit (MEU)*. Each MEU can compute two output pixels in a time-interleaved manner (see Fig. 12, each MEU contains two output registers). The MEUs are interleaved with latch-based memories to utilize the compute near memory advantages. In total, there is an array of 16×8 MEUs. Having 8 weight SCM banks was found to be the right trade-off between energy consumed by the computational elements and energy consumed by the transportation of data to the compute units. The SRAM memory banks are connected to the MEUs by means of a crossbar network. Since the input feature map pixels are stored in an interleaved manner, the crossbar network allows any (2×2) combination of the input feature map to be read. The weights are also loaded from this memory.

The authors of the work do not discuss the external interfacing required on this chip.

Binary arithmetic is relatively cheap, compared to the cost of accessing memory (e.g., for loading weights). To amortize the costs of memory reads and data movement, the computational intensity should be sufficient to balance the energy

Control Unit

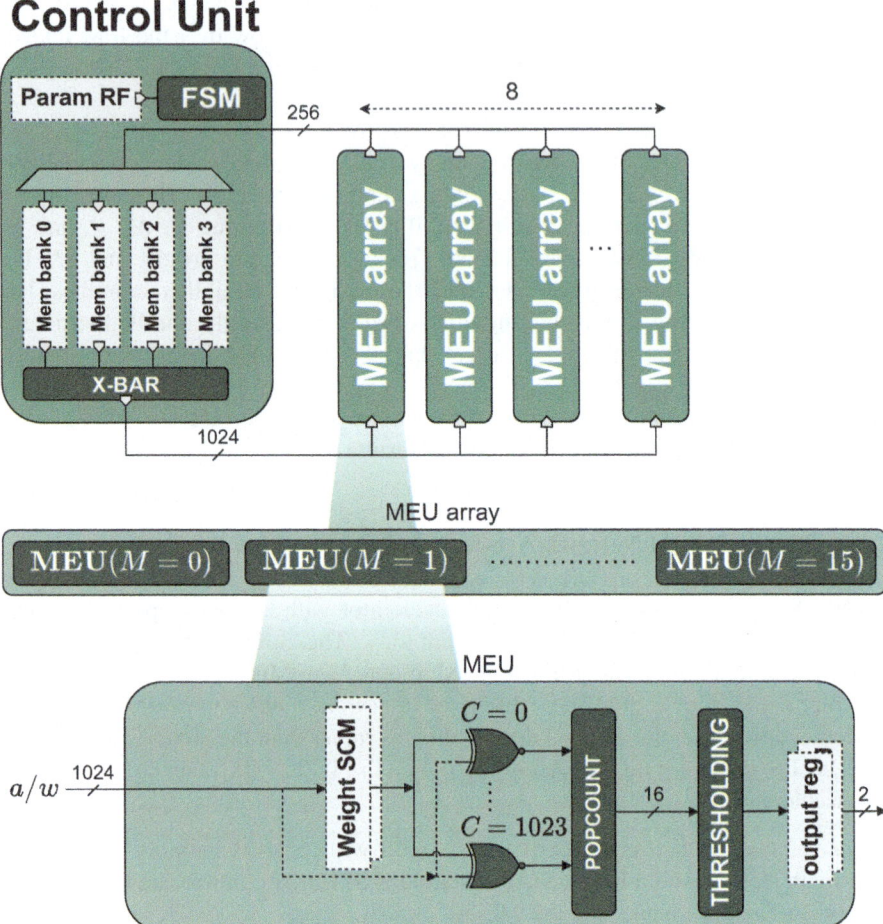

Fig. 12 Top-level view of the 10nm FinFet BNN accelerator. The central memory inside the control unit consists of 4× 256-bit wide SRAM banks to enable 2 × 2 convolutional window access in a single cycle and a finite-state machine (FSM). The MEUs are placed in an 8 × 16 array to exploit the compute near memory principle. In total, $1024 \times 16 \times 8 = 131{,}072$ ($v_C \times v_M \times 8$) binary operations can be performed each cycle

consumption. Parallelism of the MAC unit (as shown in Fig. 12) is used to balance the power mismatch of the (expensive, high bit-width) accumulator, present in the popcount module, and the (cheap) XNOR gates. By enlarging the number of inputs of the popcount module, the fixed accumulator cost is amortized by many XNOR gates. Like the other architectures, this accelerator parallelizes the MAC operation over the input channel (C) dimension. The parallelization should be high enough to offset the accumulator cost while being low enough to not impose unreasonable constraints on the number of input channels (C) required for full utilization. Therefore, a trade-off study was performed to see which level of parallelism was needed to offset the accumulator cost. A design with an input feature map parallelization factor of 1024 ($v_C = 1024$) was chosen as the sweet-spot. Negligible energy improvements were shown when going for more parallelism.

Furthermore, the idea of pipelining the `popcount-adder` tree was explored. When pipelining the design, the voltage can be lowered at iso-performance (i.e., iso-frequency). However, due to the sequential logic and clock-power dissipated while adding more pipeline stages, the final design choice was to not pipeline the `popcount-adder` tree.

3.2.5 BrainTTA

BrainTTA is a fully compiler-programmable mixed-precision flexible-datapath architecture. Contrary to the fixed-path accelerators, BrainTTA is based on the Transport-Triggered Architecture (TTA) [7] that provides a fully programmable datapath (via a compiler) directly to the user. Before diving into the BrainTTA architecture, a proper introduction to the Transport-Triggered Architecture is given.

Transport-Triggered Architectures are programmed by data movements instead of arithmetic operations typically found in Very Long Instruction Word (VLIW) architectures. This means that the movement of data between function units (FUs) and register files (RFs) is exposed to the programmer; the TTA is an explicit datapath architecture. This is in stark contrast to VLIW architectures, where the data movement is implicit and performed in hardware (i.e., not exposed to the programmer). With the control of the datapath given to the compiler, several optimizations can be performed like operand sharing and register file bypass.

An example instance of a TTA is displayed in Fig. 13. The TTA consists of a Control Unit (CU) used for instruction fetching and decoding, Register Files (RFs) for temporary storage, and Load-Store Units (LSUs) to access the memories. The gray circles inside the busses denote that this bus is connected to the corresponding input- or output-port of some function unit. This connectivity is design time configurable, visible to the compiler, and can be made as generic or specific for certain applications as desired; more connectivity is at the expense of larger instruction size and more switching activity in the interconnect. In [19], Multanen presented several ways to alleviate this effect by applying techniques that reduce the instruction overhead such as instruction compression. An example instruction is shown in Fig. 13, which shows that the instruction can be broken down into *move operations* for each bus.

BrainTTA is based on the TTA, built specifically for inference with precisions `integer8`, `binary`, and `ternary`. A top-level view of the BrainTTA SoC is shown in Fig. 14. BrainTTA is designed using the open-source toolchain TTA-based Co-design Environment (TCE) [9, 13]. The SoC consists of:

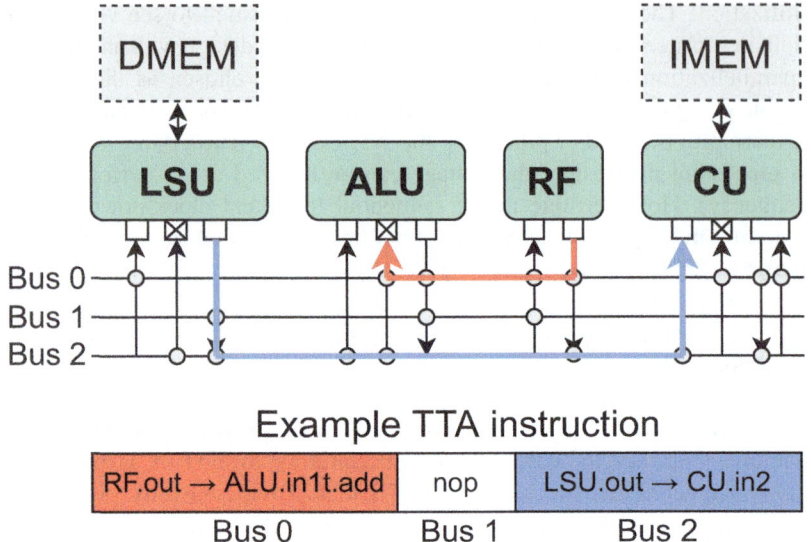

Fig. 13 An example TTA instance and instruction, the square blocks denote *input-* and *output-ports*. A cross denotes a *trigger-port*. The colored arrows drawn on the architecture illustrate the *move operations* inside the example instruction

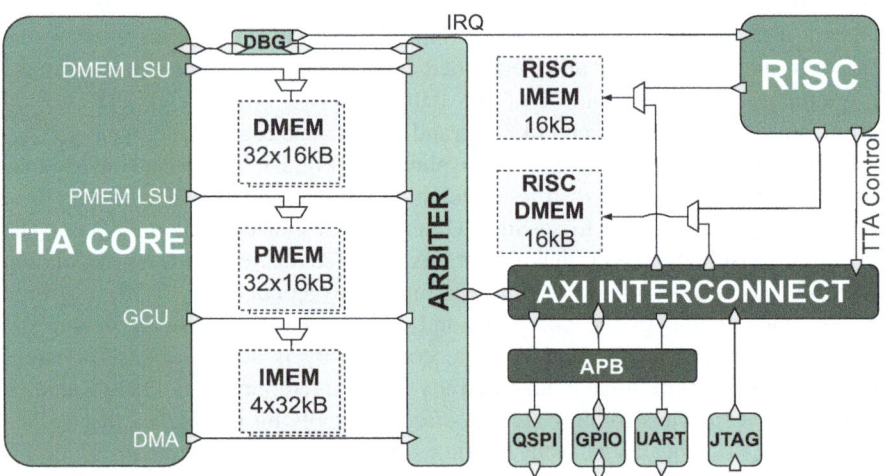

Fig. 14 Top-level view of the BrainTTA SoC, the arbiter forms the border between the RISC and TTA part of the SoC

- *RISC-V host processor*, which is taken from an open-source repository [24], the host processor starts and halts execution of the TTA core and takes care of the external communication (e.g., loading the on-chip memories).
- *TTA core*, the workhorse of the architecture, supports mixed-precision inference.

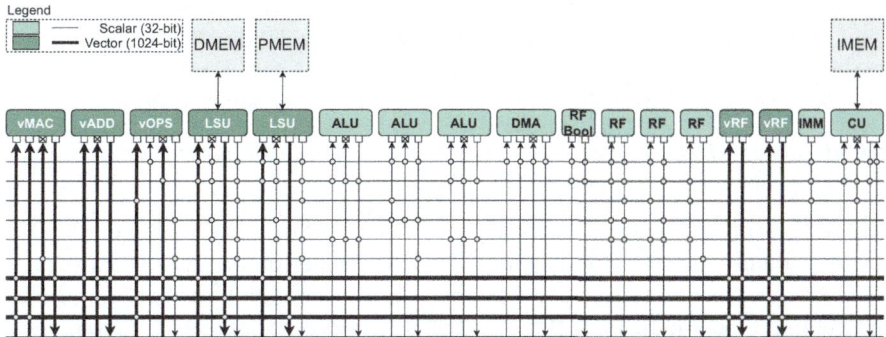

Fig. 15 BrainTTA core instance, thicker lines denote vector busses, thinner lines scalar busses

- *SRAM Memories*, separate memories for the RISC and TTA core, the TTA core memories are highly banked to allow efficient access of smaller bit-widths, while also supporting wide vector accesses. The TTA core is connected to three memories, the DMEM, used for storing input and output feature maps, the PMEM, used to store the weights, and the IMEM used for instructions to program the behavior of the TTA core.
- *Debugger (DBG)*, used to control the execution of the TTA core, can signal task completion to the RISC-V.
- *AXI interconnect*, used for on- and off-chip communication between the RISC, TTA core, and peripherals.

The workhorse of this architecture is the TTA core, where the actual inference happens. The details of the TTA core instantiation used in BrainTTA can be found in Fig. 15. The core contains different Function Units (FUs), divided into scalar and vector FUs. The FUs are interconnected via the busses, with 32-bit scalar busses (bus 0–5) and 1024-bit vector busses (bus 6–9). The core consists of the following units:

Control Unit (CU) it contains the logic to fetch and decode instructions and steers the other units to execute the correct operations. Furthermore, the CU contains a hardware loop buffer to save energy on the instruction memory accesses. This can be very beneficial since all network layers are essentially described by multiple nested loops (see Listing 1).

Vector Multiply–Accumulate (vMAC), the actual number cruncher. This unit supports the following operations: `integer8` MAC (scalar–vector product and vector–vector product), `binary` MAC, and `ternary` MAC. Its vector size is 1024-bit, with 32 entries of 32-bits each. The scalar–vector MAC multiplies a scalar by a vector by broadcasting the (32-bit) scalar value to all vector entries. This is beneficial when multiple inputs share the same weights (as in convolution).

For each precision MAC operation, the vectorization factor is different. All arithmetic circuitry contains 32 accumulators for the (intermediate) output channel result, i.e., $v_M = 32$. The number of concurrent input channels is the vector size (1024) divided by 32 (the number of output channels) divided by the operand size (i.e., 1, 2, or 8 bits). Therefore, the input channel vectorization is $v_C = 32$, $v_C = 16$, and $v_C = 4$ for binary, ternary, and integer8, respectively.

Vector Add (vADD) is used to add two (either 512- or 1024-bit) vectors. This can for example be used to support residual layers.

Vector Operations (vOPS), auxiliary (vector) operations that are required in the network, alongside to the computations. This FU can perform requantization, binarization, ternarization, as well as activation functions, e.g., ReLU and pooling functions such as MaxPool. Furthermore, various other operations to extract and insert scalar elements into a vector are also supported by this unit.

Register Files (RFs) come in different bit-widths, namely binary, 32-bit scalar, and 1024-bit vector. These registers can be used to facilitate data reuse and store intermediate results without performing (more costly) access to the SRAM.

Load-Store Units (LSUs) form the interface between the TTA core and the SRAM memory. For each memory, there is a separate LSU to facilitate concurrent weight and input loading. The units support loads and stores for different bit-widths ranging from 8 bits all the way up to 1024 bits. Since the memory is banked, a strobe signal can be used to selectively turn on banks when data with smaller bit-widths are loaded/stored, in order to save energy.

Scalar ALUs are mostly used for address calculations needed as inputs to the LSUs. These units support basic arithmetic on values up to 32 bit.

4   Comparison and Discussion

All architectures discussed in Sect. 3 are evaluated on flexibility and energy efficiency. These results are given in Table 1. This table is split into three sections: the implementation characteristics, performance characteristics as discussed in Sect. 3.1.2, and the flexibility aspects as discussed in Sect. 3.1.1.

The energy efficiency of the accelerators ranges from 1.6 to 115 fJ per operation for binary precision, a large range. It should be noted, however, that the two architectures that have the highest energy usage (XNE and BrainTTA) are the only architectures that show a full autonomous SoC including peripherals. Furthermore, all architectures except BrainTTA utilize voltage–frequency scaling to run the accelerator at lower than nominal supply voltage, trading off throughput for better energy efficiency.

Next to the energy efficiency, the table also lists the neural network layer requirements that these architectures impose in order to fully utilize the arithmetic hardware. It is seen that the most energy-efficient architectures, CUTIE [22] and the BNN accelerator in 10-nm FinFet from Knag et al. [16], are also the most constrained architectures, in terms of neural network layer requirements. Therefore, the question arises, does hard-wiring the neural network layer parameters directly improve the energy efficiency of an architecture, for different models, also when layer variety is high?

Interestingly, the XNE and BrainTTA share very similar layer constraints. Both are only constrained in the input channel (C) and output channel (M) dimensions. The energy consumption of BrainTTA is somewhat lower at an older technology node while using a higher supply voltage. The reason for this is that BrainTTA better exploits data reuse. The execution schedule for BrainTTA was tuned to maximize data reuse, while XNOR neural engine only reuses a set of input feature maps for TP (in this case 128) cycles while reloading the weights for each MAC operation.

The inefficient schedule of XNE is confirmed by the energy numbers of the implementation that only uses SCM. XNE was benchmarked using SCM only, severely cutting the very high energy cost associated with these redundant memory fetches, at the cost of losing memory capacity. Some architectures report energy numbers for an SCM as well as an SRAM implementation. The memory capacity of the SCM versions is very low compared to the SRAM versions, hindering the ability to run full-size networks on it without adding expensive off-chip memory accesses. For the sake of comparison, for all the architectures with an SRAM version available, the SRAM version is chosen for further analysis.

Support for residual layers can only be found in ChewBaccaNN and BrainTTA. Other architectures are not able to support this due to their fixed datapath. The dataflow through these accelerators is very static, and the accumulated value will directly be binarized or ternarized after all inputs are accumulated. This prohibits the use of residual layers since residual layers need the intermediate (larger bit-width) results that were obtained before requantization.

It is clear that parallelism and data reuse (either in the form of locally buffering or by broadcasting) are the keys to amortizing the memory access cost, which is so much larger than the low-precision arithmetic cost. Techniques to mitigate these costs are to replace SRAM with low-voltage SCM, hard-wire network parameters to enable broadcasting, and use the sliding window principle (like the FMM banks in combination with the crossbar in ChewBaccaNN [1]). In essence, all these solutions boil down to designing the architecture around the data movements in a less-flexible manner. These architectures solve the mapping problem by fixing most parameters using **spatial mapping**, greatly simplifying the task of **temporal mapping** at the cost of losing flexibility. XNE and BrainTTA fix the least number of parameters using **spatial mapping**, therefore leaving a larger **temporal mapping** space to be explored.

5 Summary and Conclusions

Neural networks are all around and are making an advance into the embedded domain. With the increasing popularity of edge computing, new methods are needed to port the typically power- and memory-hungry neural networks to devices that have limited storage and are subject to severe energy constraints. Quantization is a fundamental ingredient in overcoming these challenges. Very low precisions, down to 1 bit, have shown to achieve great energy efficiency while drastically reducing the model size and computational cost involved in neural network inference. To fully exploit the reduced computational complexity and memory requirements of these networks, neural network accelerators aimed specifically at these heavily quantized networks have been developed.

In this chapter, state-of-the-art low- and mixed-precision architectures are reviewed. Taking into account the variety present in network layers of CNNs, the architectures are compared against each other in terms of flexibility and energy efficiency. It was found that spatially mapping more dimensions of the neural network layer increases the energy efficiency as it allows minimization of data movement by tailoring the memory hierarchy design, which is a big contributor to energy cost in inference accelerators. Contrary to the group of accelerators that maps most layer dimensions spatially, there is a group of accelerators that minimizes the layer dimension requirements by less heavily relying on spatial mapping, retaining more freedom in the temporal mapping domain. They are more flexible and can handle a larger part of the neural architecture design space. In addition, they may have support for multiple bit precisions.

With new attempts to streamline the process of finding the best combination of temporal and spatial mappings [17], while co-designing the memory hierarchy, the question arises if an optimized temporal mapping in combination with memory hierarchy co-design can close the energy efficiency gap with the more constrained, heavily spatially mapped accelerators, giving better energy efficiency at a wider range of neural network layers.

References

1. Andri, R., Karunaratne, G., Cavigelli, L., Benini, L.: ChewBaccaNN: A flexible 223 TOPS/W BNN accelerator. arXiv (May), 23–26 (2020)
2. Bankman, D., Yang, L., Moons, B., Verhelst, M., Murmann, B.: An always-on 3.8 μ J/86% CIFAR-10 mixed-signal binary CNN processor with all memory on chip in 28-nm CMOS. IEEE J. Solid-State Circuits **54**(1), 158–172 (2019). https://doi.org/10.1109/JSSC.2018. 2869150. https://ieeexplore.ieee.org/document/8480105/
3. Bengio, Y., Léonard, N., Courville, A.: Estimating or Propagating Gradients Through Stochastic Neurons for Conditional Computation pp. 1–12 (2013). http://arxiv.org/abs/1308.3432
4. Blalock, D., Ortiz, J.J.G., Frankle, J., Guttag, J.: What is the State of Neural Network Pruning? (2020). http://arxiv.org/abs/2003.03033
5. Bulat, A., Tzimiropoulos, G.: XNOR-Net++: Improved binary neural networks. In: 30th British Machine Vision Conference 2019, BMVC 2019 pp. 1–12 (2020)
6. Conti, F., Schiavone, P.D., Benini, L.: XNOR neural engine: a hardware accelerator IP for 21.6-fJ/op binary neural network inference. IEEE Trans. Comput.-Aided Design Integr. Circuits Syst. **37**(11), 2940–2951 (2018). https://doi.org/10.1109/TCAD.2018.2857019
7. Corporaal, H.: Microprocessor Architectures: From VLIW to TTA. Wiley, Hoboken (1997)

8. Deng, L., Jiao, P., Pei, J., Wu, Z., Li, G.: GXNOR-Net: training deep neural networks with ternary weights and activations without full-precision memory under a unified discretization framework. Neural Netw. **100**, 49–58 (2018). https://doi.org/10.1016/j.neunet.2018.01.010

9. Esko, O., Jääskeläinen, P., Huerta, P., De La Lama, C.S., Takala, J., Martinez, J.I.: Customized exposed datapath soft-core design flow with compiler support. In: Proceedings - 2010 International Conference on Field Programmable Logic and Applications, FPL 2010, pp. 217–222 (2010). https://doi.org/10.1109/FPL.2010.51

10. Gholami, A., Kim, S., Dong, Z., Yao, Z., Mahoney, M.W., Keutzer, K.: A Survey of Quantization Methods for Efficient Neural Network Inference (2021). http://arxiv.org/abs/2103.13630

11. Gluska, S., Grobman, M.: Exploring Neural Networks Quantization via Layer-Wise Quantization Analysis (2020). http://arxiv.org/abs/2012.08420

12. Huang, S., Waeijen, L., Corporaal, H.: How flexible is your computing system? ACM Trans. Embedd. Comput. Syst. (2022). https://doi.org/10.1145/3524861. https://dl.acm.org/doi/10.1145/3524861

13. Jääskeläinen, P., Viitanen, T., Takala, J., Berg, H.: HW/SW co-design toolset for customization of exposed datapath processors. In: Computing Platforms for Software-Defined Radio, pp. 147–164. Springer International Publishing, Cham (2016). https://doi.org/10.1007/978-3-319-49679-5_8

14. Jacob, B., Kligys, S., Chen, B., Zhu, M., Tang, M., Howard, A., Adam, H., Kalenichenko, D.: Quantization and training of neural networks for efficient integer-arithmetic-only inference. Proceedings of the IEEE Computer Society Conference on Computer Vision and Pattern Recognition, pp. 2704–2713 (2018). https://doi.org/10.1109/CVPR.2018.00286

15. Kharya, P.: TensorFloat-32 in the A100 GPU Accelerates AI Training, HPC up to 20x (2020). https://blogs.nvidia.com/blog/2020/05/14/tensorfloat-32-precision-format/

16. Knag, P.C., Chen, G.K., Sumbul, H.E., Kumar, R., Hsu, S.K., Agarwal, A., Kar, M., Kim, S., Anders, M.A., Kaul, H., Krishnamurthy, R.K.: A 617-TOPS/W all-digital binary neural network accelerator in 10-nm FinFET CMOS. IEEE J. Solid-State Circuits **56**(4), 1082–1092 (2021). https://doi.org/10.1109/JSSC.2020.3038616

17. Mei, L., Houshmand, P., Jain, V., Giraldo, S., Verhelst, M.: ZigZag: enlarging joint architecture-mapping design space exploration for DNN accelerators. IEEE Trans. Comput. **70**(8), 1160–1174 (2021). https://doi.org/10.1109/TC.2021.3059962

18. Muller, O., Prost-Boucle, A., Bourge, A., Petrot, F.: Efficient decompression of binary encoded balanced ternary sequences. IEEE Trans. Very Large Scale Integr. Syst. **27**(8), 1962–1966 (2019). https://doi.org/10.1109/TVLSI.2019.2906678

19. Multanen, J.: Energy-Efficient Instruction Streams for Embedded Processors. Ph.D. Thesis, Tampere University (2021)

20. Parashar, A., Raina, P., Shao, Y.S., Chen, Y.H., Ying, V.A., Mukkara, A., Venkatesan, R., Khailany, B., Keckler, S.W., Emer, J.: Timeloop: A systematic approach to DNN accelerator evaluation. In: Proceedings - 2019 IEEE International Symposium on Performance Analysis of Systems and Software, ISPASS 2019, pp. 304–315 (2019). https://doi.org/10.1109/ISPASS.2019.00042

21. Rastegari, M., Ordonez, V., Redmon, J., Farhadi, A.: XNOR-net: ImageNet classification using binary convolutional neural networks. In: Computer Vision—ECCV 2016. Lecture Notes in Computer Science (including subseries Lecture Notes in Artificial Intelligence and Lecture Notes in Bioinformatics), LNCS, vol. 9908, pp. 525–542 (2016). https://doi.org/10.1007/978-3-319-46493-0_32

22. Scherer, M., Rutishauser, G., Cavigelli, L., Benini, L.: CUTIE: Beyond PetaOp/s/W Ternary DNN Inference Acceleration with Better-than-Binary Energy Efficiency pp. 1–14 (2020). http://arxiv.org/abs/2011.01713

23. Tan, M., Chen, B., Pang, R., Vasudevan, V., Sandler, M., Howard, A., Le, Q.V.: MnasNet: Platform-aware neural architecture search for mobile. In: Proceedings of the IEEE Computer Society Conference on Computer Vision and Pattern Recognition 2019-June, pp. 2815–2823 (2019). https://doi.org/10.1109/CVPR.2019.00293

24. Traber, A., Gautschi, M.: PULPino: Datasheet. ETH Zurich, University of Bologna (2017)
25. Ueyoshi, K., Papistas, I.A., Houshmand, P., Sarda, G.M., Jain, V., Shi, M., Zheng, Q., Giraldo, S., Vrancx, P., Doevenspeck, J., Bhattacharjee, D., Cosemans, S., Mallik, A., Debacker, P., Verkest, D., Verhelst, M.: DIANA: An end-to-end energy-efficient digital and ANAlog hybrid neural network SoC. In: 2022 IEEE International Solid- State Circuits Conference (ISSCC), pp. 1–3. IEEE (2022). https://doi.org/10.1109/ISSCC42614.2022.9731716. https://ieeexplore. ieee.org/document/9731716/
26. Valavi, H., Ramadge, P.J., Nestler, E., Verma, N.: A 64-Tile 2.4-Mb in-memory-computing CNN accelerator employing charge-domain compute. IEEE J. Solid-State Circuits 54(6), 1789–1799 (2019). https://doi.org/10.1109/JSSC.2019.2899730. https://ieeexplore.ieee.org/ document/8660469/
27. Wang, S., Kanwar, P.: BFloat16: The secret to high performance on Cloud TPUs (2019). https://cloud.google.com/blog/products/ai-machine-learning/bfloat16-the-secret-to-high-performance-on-cloud-tpus
28. Wu, B., Dai, X., Zhang, P., Wang, Y., Sun, F., Wu, Y., Tian, Y., Vajda, P., Jia, Y., Keutzer, K.: FBNET: Hardware-aware efficient convnet design via differentiable neural architecture search. In: Proceedings of the IEEE Computer Society Conference on Computer Vision and Pattern Recognition 2019-June, pp. 10726–10734 (2019). https://doi.org/10.1109/CVPR.2019.01099
29. Wu, Y.N., Emer, J.S., Sze, V.: Accelergy: An architecture-level energy estimation methodology for accelerator designs. In: IEEE/ACM International Conference on Computer-Aided Design, Digest of Technical Papers, ICCAD 2019-Nov (2019). https://doi.org/10.1109/ICCAD45719. 2019.8942149

Designing Resource-Efficient Hardware Arithmetic for FPGA-Based Accelerators Leveraging Approximations and Mixed Quantizations

Salim Ullah, Siva Satyendra Sahoo, and Akash Kumar

1 Introduction

With the recent advancements in the field of Artificial Intelligence (AI), Machine Learning (ML) is becoming an imperative part of modern applications such as autonomous driving [1], personalized healthcare [2], precision agriculture [3], smart factories [4], and smart homes [5]. Machine learning algorithms perform various tasks for these applications, such as scene perception, object recognition and classification, voice recognition and decision-making, and natural language processing. However, machine learning algorithms, such as Artificial Neural Networks (ANNs), are computationally expensive and have very high energy requirements and memory footprints [6]. Therefore, high-performance parallel architectures, such as Graphic Processing Units (GPUs), and cloud-based computing are typically utilized for training the ML models. Nonetheless, the GPUs' high power consumption makes them an infeasible choice for deploying the trained ML models on embedded devices at the edge. Similarly, factors such as high power consumption of data transmission from device to cloud, network costs, throughput, and data security are the primary reasons to avoid cloud-based inference and thus motivate the need for executing trained ML algorithms at the edge.

Embedded machine learning refers to utilizing and executing machine learning models on embedded systems to perform the aforementioned AI/ML-related tasks. The ubiquitous deployment of embedded systems in almost every application—from space rockets to microwave ovens—further emphasizes the need for smart embedded systems by utilizing ML models. Toward this end, various state-of-the-art works, such as [7], have presented various techniques to reduce these models'

S. Ullah · S. S. Sahoo · A. Kumar (✉)
Technische Universität Dresden, Dresden, Germany
e-mail: salim.ullah@tu-dresden.de; siva_satyendra.sahoo@tu-dresden.de;
akash.kumar@tu-dresden.de

© The Author(s), under exclusive license to Springer Nature Switzerland AG 2024
S. Pasricha, M. Shafique (eds.), *Embedded Machine Learning for Cyber-Physical, IoT, and Edge Computing*, https://doi.org/10.1007/978-3-031-19568-6_4

overall computational complexity, memory footprint, and storage requirements for execution on embedded systems. Most of these techniques exploit the inherent error resilience of ML models to introduce various approximations in the implementation of a trained ML model. This inherent error resilience enables an application to produce acceptable quality results despite some of its operations and data being approximate/inaccurate [8]. It should be noted that error-tolerant applications may not produce a single golden answer and instead produce multiple feasible answers. For example, a search engine can return multiple feasible options instead of a single golden answer while searching for the best theater in the town. Similarly, an ML model trained to recognize cat images can produce outputs with 70%, 80%, and 90% confidence values, and all these results are acceptable depending on an application's output quality requirements. For machine learning models such as Deep Neural Networks (DNNs), network pruning [9], quantization of trained parameters [10], and utilization of approximate arithmetic modules [11, 12] are the commonly utilized techniques to trade the output accuracy with corresponding performance gains in the implementation. To this end, TensorFlow [13], one of the most commonly used frameworks for developing machine learning models, also provides the TensorFlow Lite tool for optimizing ML models for embedded systems. These optimizations have enabled the execution of ML models on single-board computers such as those provided by Raspberry Pi and Arduino [7].

The various optimizations performed by tools, such as TensorFlow Lite, however, mainly focus on reducing an ML model size and utilizing 16-bit and 8-bit integer number schemes, along with single-precision floating-point numbers, to represent the trained parameters of a model. Many recent works have demonstrated that ML models' inherent error resilience can be exploited further to utilize fewer bits (less than 8 bits) to represent a model's parameters and still achieve acceptable quality results. For example, the binary and ternary networks utilize 1-bit and 2-bit number representation schemes, respectively, to represent the parameters of DNNs [16, 17]. The utilization of fewer bits to represent parameters of a model significantly reduces the storage and memory requirements of its implementation. However, these number schemes underutilize the computational resources of a standard processor. Moreover, many recent works, such as [10], have defined new number representation (quantization) schemes to utilize the available bit widths efficiently. These number schemes also underutilize the available processing resources in a general-purpose processor. These challenges can be addressed by utilizing custom architectures designed according to the employed number representation scheme. For instance, Fig. 1 compares the impact of deploying different number representation schemes across multiple performance parameters—behavioral (quantization-induced errors in weights), computational (critical path delay (CPD) of a Multiply and Accumulate (MAC) unit), and memory requirements (weights' storage) in the Conv2_1 layer of a pre-trained VGG16 network [14]. For this experiment, we have used single-precision floating point (FP32), fixed point (Fxp), and the recently proposed number representation scheme Posit (Pos) [18]. The results compare the accuracy and performance of FP32-based MAC with 16-bit and 8-bit fixed-point and Posit representations-based MACs (Fxp16, Pos16, Fxp8, Pos8). For this experiment, each

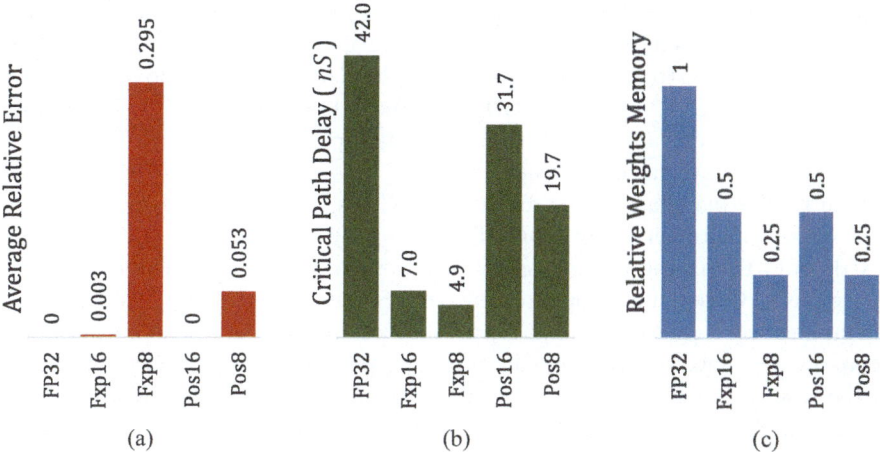

Fig. 1 Accuracy and performance comparison of various schemes of numbers representation for the Conv2_1 layer of pre-trained VGG16 [14]: (**a**) Average absolute relative error with respect to single-precision floating-point-based parameters, (**b**) CPD, (**c**) normalized memory footprints [15]. (**a**) Accuracy. (**b**) Computation. (**c**) Storage

technique has been implemented on the Xilinx UltraScale Field Programmable Gate Array (FPGA) to obtain the corresponding implementation results. The MAC units in this experiment have been implemented using 6-input Lookup Tables (LUTs) and with a latency constraint of a single cycle. The results show that the utilization of more bits for the representations of the parameters significantly reduces the errors induced by quantization schemes. However, the single-precision (FP32) implementation holds the highest memory footprint with the worst CPD of 42 ns. Similarly, compared to the Fxp-based number representation scheme, the Posit number representation schemes offer better coverage of the FP32-based pre-trained parameters. However, compared to the Posit-based arithmetic, the simplicity of the Fxp-based arithmetic results in significantly reducing the CPD of the MAC units. Therefore, there is a need to explore the various available number representation schemes and corresponding efficient arithmetic architectures to implement ML models on resource-constrained embedded systems.

FPGA vendors provide various MultiProcessor System On Chips (MPSoCs), such as Zynq UltraScale+ [19], to combine the power efficiency and programmability of general-purpose ARM processors with the reconfigurability and parallelism of FPGAs for embedded systems. The reconfigurable nature of FPGAs allows designing area-optimized, low-latency, and energy-efficient accelerators for various functions of an embedded application. Moreover, using custom architectures also facilitates achieving higher throughputs in embedded systems by exploiting the large parallelism supported by FPGAs. This chapter provides a comprehensive overview of some of the commonly utilized number representation schemes and their corresponding FPGA-optimized custom arithmetic architectures for the embedded machine learning. Specifically, it focuses on the architectures for multiplication operations as it is one of the most commonly used operations in ML models. For example, VGG16 network deploys $15.5G$ MAC operations to perform the inference on a single 224×224 RGB input image [14]. Therefore, the availability of resource-efficient and high-performance multiplier architectures can help in enabling embedded machine learning.

The rest of the chapter is organized as follows: Sect. 2 presents accurate and approximate multiplier architectures for fixed-point-based integer arithmetic. We first describe the fixed-point representation technique to represent fractions followed by various FPGA-optimized architectures for accurate signed multipliers. Considering the error-resilient nature of ML algorithms, we then present two architectures for approximate signed multipliers. Afterward, we present the implementation results of the discussed architectures and evaluate their efficacy by employing them in different high-level applications. Section 3 discusses the opportunities provided by both commonly used and novel number representation schemes. In particular, we discuss the Posit number representation scheme and the associated challenges for Posit-based arithmetic in embedded systems. To this end, we present a modified Posit representation for ML algorithms. Utilizing the modified scheme, we present a technique for deploying fixed-point arithmetic for the Posit numbers. Finally, we use DNNs as a benchmark application and present a detailed accuracy and performance analysis of utilizing Posit for an FPGA-based accelerator design for DNNs.

2 Integer Arithmetic for Embedded Machine Learning

As described in Sect. 1, high-performance Central Processing Units (CPUs) and GPUs are typically used to train ML models. These systems utilize IEEE single-precision and double-precision floating-point number schemes to provide high computational accuracy. However, due to the high computational cost of floating-point arithmetic, it is a common practice to exploit the inherent error resilience of ML models and represent the floating-point trained parameters of a model in the fixed-point number representation scheme. This process is commonly known as the quantization of a trained model. The quantized models are then executed using fixed-point arithmetic (integer arithmetic) on resource-constrained embedded systems.

2.1 Fixed-Point Representation

The commonly used technique to represent floating-point trained parameters in fixed-point representation is *linear quantization*. The linear quantization of a data tensor x from floating-point precision to N-bit fixed-point precision is illustrated by Eqs. (1)–(5). The step size Δ in Eq. (1) represents the minimum possible increment in the quantized value x_{quant}. Equation (2) transforms the step size Δ in a power of 2 so that it can be represented in the fixed-point representation. Depending upon the selected rounding function (round up and round down), the Δ and the resulting quantized values will change accordingly. Finally, Eq. (3) utilizes the calculated Δ to compute the fixed-point representation of a floating-point number. The clip function, defined in Eq. (5), ensures that a parameter does not violate the allowed range of values. The limited precision of the N-bit fixed-point representation can reduce the precision of the quantized numbers, as defined in Eq. (4); however, as defined previously, due to the inherent error resilience of ML models, the ML models can produce acceptable quality results in most situations.

$$\Delta = \frac{max\ (\mid x \mid)}{2^{N-1}} \tag{1}$$

$$\Delta = 2^{round(log_2(\Delta))} \tag{2}$$

$$x_{quant\_Rep} = clip\left(round\left(\frac{x}{\Delta}\right), -2^{N-1}, 2^{N-1} - 1\right) \tag{3}$$

$$x_{quant} = \Delta \bullet x_{quant\_Rep} \tag{4}$$

$$clip(x, Max, Min) = \begin{cases} x, & Min < x < Max \\ Max, & x \geq Max \\ Min, & \text{otherwise} \end{cases} . \tag{5}$$

2.2 Accurate Custom Signed Multipliers

FPGA vendors, such as Xilinx and Intel, provide Digital Signal Processing (DSP) blocks to achieve fast multipliers [20, 21]. However, as shown by the work presented in [22], it is necessary to have logic-based soft multipliers along with DSP blocks to obtain overall performance gains in different implementation scenarios. Consequently, Xilinx and Intel also provide logic-based soft multipliers [23, 24]. In this section, we describe some state-of-the-art FPGA-optimized accurate signed multipliers. These designs are based on the efficient utilization of the 6-input LUTs and associated fast carry chains of Xilinx FPGAs.

Baugh–Wooley's Multiplier (Mult-BW)
Baugh–Wooley's multiplication algorithm [25] eliminates the need for computing and communicating sign-extension bits by encoding the sign information in the

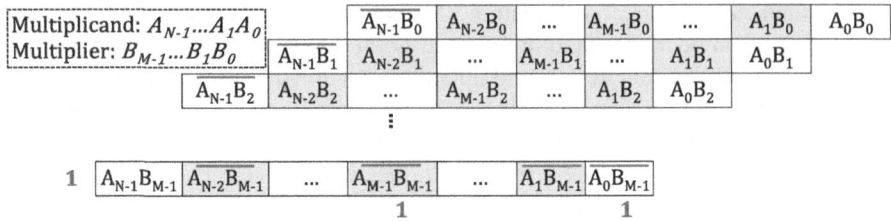

Fig. 2 Baugh–Wooley's N × M signed multiplier design [22]

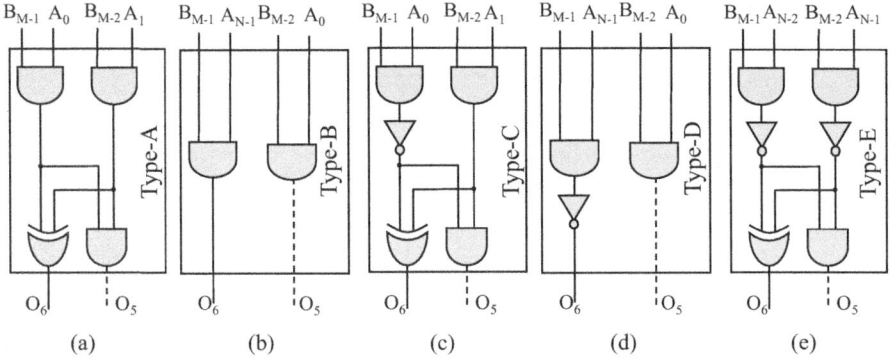

Fig. 3 The various functions implemented by LUTs to realize a Baugh–Wooley's multiplier: (**a**) LUT configuration-A, (**b**) LUT configuration-B, (**c**) LUT configuration-C, (**d**) LUT configuration-D, and (**e**) LUT configuration-E [22]

generated partial products. Figure 2 presents the graphical representation of Baugh–Wooley's algorithm. The authors of [22] have used this algorithm and defined different configurations of the 6-input LUTs to generate the signed partial products. These configurations are presented in Fig. 3. For an $N \times M$ signed multiplier, the proposed implementation generates only $\left\lceil \frac{M}{2} \right\rceil$ partial products by fusing every two consecutive partial products. Figure 4 presents the mapping of the LUTs configurations and carry chains to generate all partial products of an $N \times M$ signed multiplier. The proposed methodology utilizes LUTs- and carry chains-based binary and ternary adders to add all the generated partial products for computing the final product. Compared to a binary adder that can add two operands at a time, a ternary adder can add three operands simultaneously, as shown in Fig. 5. The proposed LUT-level design optimizations result in realizing resource-efficient implementations for various sizes of multiplications.

Booth's Multiplier (Mult-Booth)
The authors of [12, 26, 27] have used Booth's multiplication algorithm to present area-optimized, reduced-latency, and energy-efficient implementations of radix-4 Booth's multiplication algorithm [28]. These works utilize similar Booth's encoding techniques for the 6-input LUTs of the FPGAs that can be used to implement

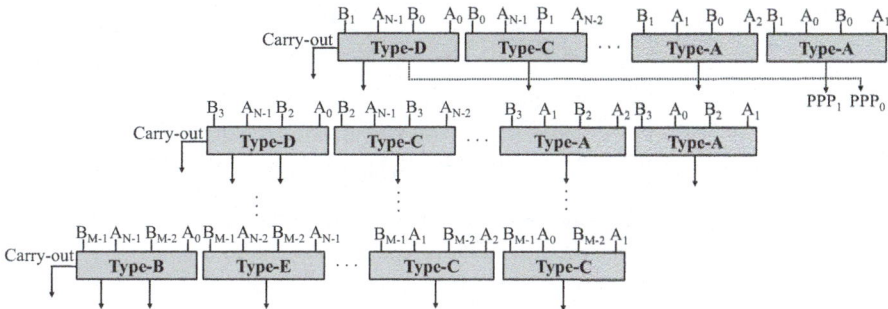

Fig. 4 Partial products generation for an N × M Mult-BW multiplier [22]

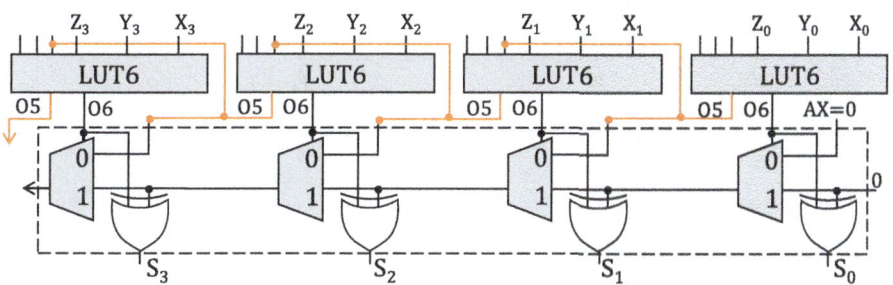

Fig. 5 LUTs- and carry chain-based ternary adder

multipliers of various sizes. The work presented in [12] generates the signed partial products sequentially, and each generated partial product is added with the previous partial product row in a single step, as shown in Fig. 6. Figure 7 describes the corresponding configurations of the LUTs. The sequential generation of the partial products significantly reduces the overall utilized resources of the multiplier. The work presented in [26] explores the parallel generation of all partial products and then utilizes 4:2 compressors and binary adders for the addition of the generated partial product rows to compute the final product. The parallel generation of partial product rows significantly reduces the overall latency of the implemented multipliers. The work presented in [27] has used Booth's algorithm to implement custom unsigned multipliers. The unsigned multipliers can be utilized to implement signed multipliers by employing dedicated signed–unsigned converters. These converters receive 2's complement numbers and generate corresponding numbers in sign-magnitude format. After multiplication in sign-magnitude format, the result is converted back to 2's complement scheme using a signed–unsigned converter. However, as described in Sect. 2.4, these converters result in increasing the total number of utilized resources, critical path delay, and power consumption of the whole circuit. Moreover, due to the limited dynamic range of sign-magnitude format, the -2^{N-1} number in 2's complement format cannot be represented in an N-bit sign-magnitude format. However, ML models, such as ANNs, process

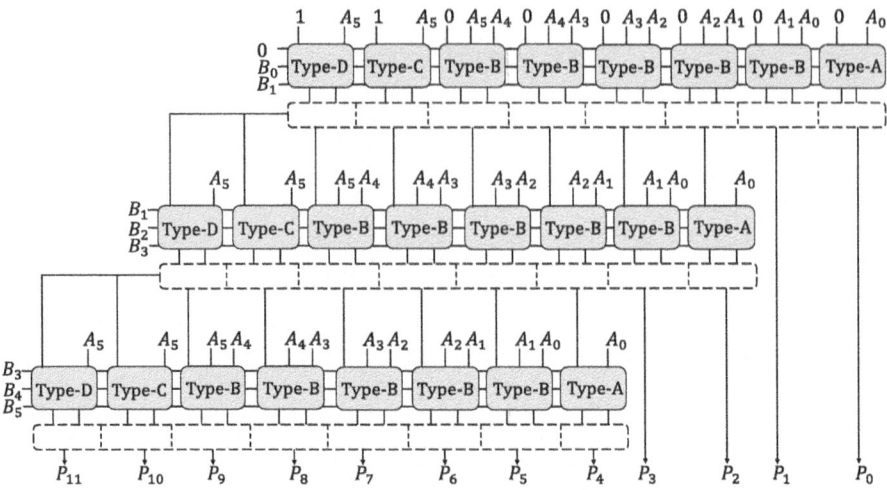

Fig. 6 A 6×6 area-optimized accurate Booth's multiplier [12]

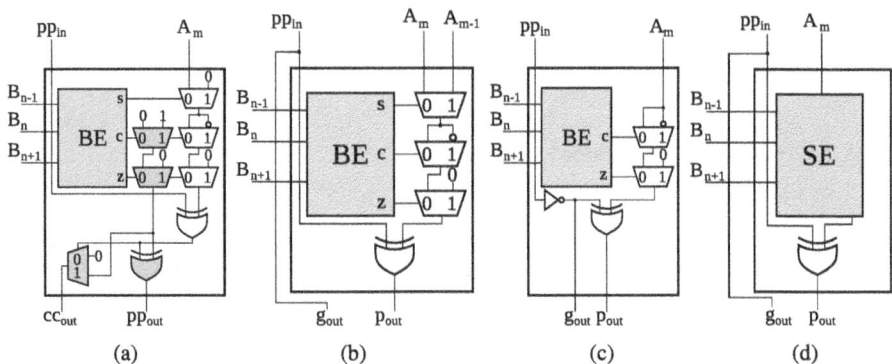

Fig. 7 Configuration of LUTs for implementing Booth's multiplier [12]. (**a**) Type-A. (**b**) Type-B. (**c**) Type-C. (**d**) Type-D

signed numbers; therefore, the employment of the unsigned multipliers and the dedicated signed–unsigned converters can degrade the overall performance of hardware accelerators for these applications.

2.3 Approximate Custom Signed Multipliers

As discussed in Sect. 1, the error resilience of ML models allows introducing different approximations at the various layers of the computation stack to trade the output accuracy with corresponding implementation performance gains. To this end, many approximate computing-related works have focused on the resources-optimized, high-performance, and energy-efficient architectures of approximate arithmetic operators. For example, the works in [11, 12] have proposed various strategies for designing approximate signed custom multipliers. The authors of [12] have analyzed the Booth's multiplier accurate implementation, presented in Fig. 6, for various sizes of multipliers and identified the logic elements that contribute the most to the CPD and the dynamic power consumption for all possible input combinations. For example, Fig. 8a presents this analysis for a 6×6 multiplier highlighting the five topmost power-consuming elements and the five worst critical paths. Based on the analysis, the authors have proposed various approximations to reduce the approximate implementation's overall dynamic power consumption and CPD. Figure 8b presents the approximate multiplier's dynamic power consumption and critical path analysis.

Further generalizing this idea, the authors of [11] have proposed a generic framework, "AppAxO," for implementing application-specific approximate operators optimized for FPGA-based systems. AppAxO employs the 6-input LUTs and the associated carry chains of FPGAs to implement approximate operators according to *binarized string* configurations. These configurations specify the LUTs, in an accurate operator implementation that should be disabled to realize a corresponding approximate operator. For example, for an $M \times N$ accurate multiplier, utilizing "T" LUTs, AppAxO explores the design space of 2^T approximate multipliers with different accuracy and performance parameters. To determine the feasible configurations for an application, AppAxO employs a Multi-objective Bayesian Optimization (MBO)-based exploration method to generate only those approximate operator configurations that fulfill an application's accuracy and performance constraints. The authors have shown that by considering application-specific accuracy–performance constraints, AppAxO provides novel approximate operators, providing better design points for an application than the traditional application-agnostic design methodology.

2.4   Comparison of Multiplier Designs

This section summarizes the performance (resources, CPD, and power consumption) and accuracy results of the discussed FPGA-optimized accurate and approximate multipliers. All presented multipliers have been implemented in VHDL and synthesized for Virtex-7 family FPGA using Xilinx Vivado. For Power-Delay Product (PDP) calculations, Vivado Simulator and Power Analyzer tools have been used.

Figure 9 compares the average performance of the Baugh–Wooley (Mult-BW) [22], Booth's multipliers Mult-Booth-1, Mult-Booth-2, and Mult-Booth-3 presented in [12, 26], and [27], respectively, with Vivado speed- and area-optimized multiplier IPs [23]. The Mult-Booth-3 design also employs signed–unsigned con-

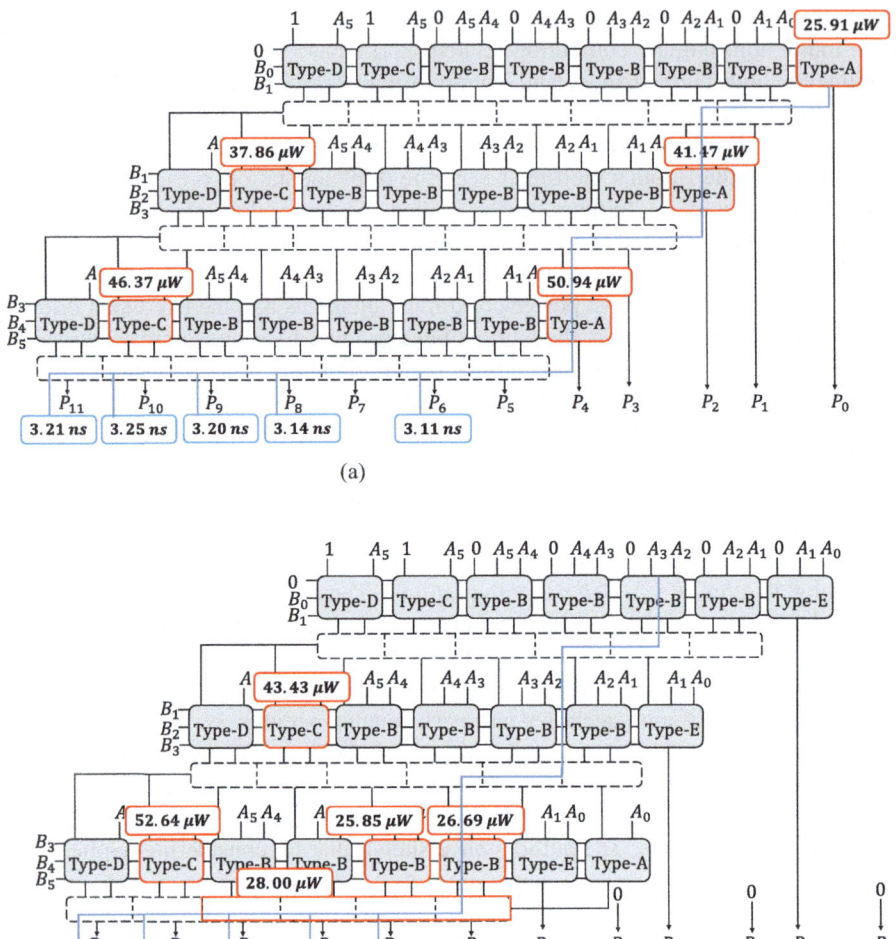

Fig. 8 CPD and dynamic power analysis to implement approximate circuits [12]. (**a**) 6 × 6 Booth multiplier showing top five critical paths per output (in blue) and top five most power-consuming elements. (**b**) 6 × 6 Booth approximate multiplier showing top five critical paths per output (in blue) and top five most power-consuming elements

verters to perform signed multiplications, as discussed in Sect. 2.2. We have utilized the *Average Performance* metric to compare the performance of the custom multipliers for 4×4, 8×8, 16×16, and 32×32 multipliers. The *Average Performance* metric is the average of the product of normalized values of LUTs utilization, CPD, and PDP, as shown in Eq. (6). All individual performance metrics of each multiplier have been normalized to the corresponding performance metrics of Vivado area-optimized multiplier IP [23]. It should be noted that a smaller average value of

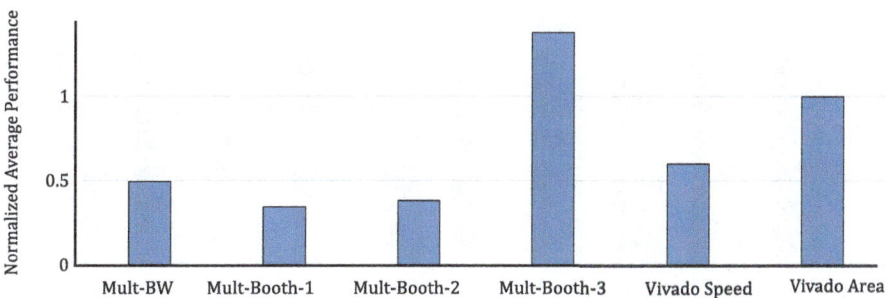

Fig. 9 Performance comparison of accurate signed multipliers across different sizes of multipliers: a smaller average value shows a design with a better performance

the metric presents an implementation with a better performance. As shown by the results in Fig. 9, the custom multipliers Mult-BW, Mult-Booth-1, and Mult-Booth-2 provide better overall performance than Vivado speed- and area-optimized IPs. Mult-Booth-1 is more resource- and energy-efficient than other designs due to its sequential generation and addition of partial products. Mult-Booth-2 offers significantly reduced critical path delay due to the parallel generation of the partial products. The signed–unsigned converters increase the total number of utilized LUTs, CPD, and dynamic power consumption of the Mult-Booth-3 design.

$$Average\ Performance = Average(Norm.\ \text{LUTs} \times Norm.\ \text{CPD} \times Norm.\ \text{PDP}). \quad (6)$$

Figure 10 presents the utilization of Mult-Booth-2 in the implementation of an accelerator of an ANN for the classification of the MNIST digits dataset [29]. The inference accuracy of the dataset using the single-precision floating-point number is 97%. The corresponding inference accuracy using 8-bit fixed-point quantization is 96.6%, resulting in an insignificant drop in output accuracy. First, the network is implemented using Vivado speed-optimized multiplier IP with as many neurons as possible using three different input sizes, 8×8, 16×16, and 32×32. The same setups are then used for the Vivado area-optimized multiplier and Mult-Booth-2 multipliers. The resulting LUT utilization, CPD, and PDP for each design are normalized to Vivado area-optimized IPs. Mult-Booth-2 produces the best results in the combined LUT×PDP averaged across all input sizes. Mult-Booth-2 outperforms Vivado speed- and area-optimized IPs by 8.4% and 29.4%, respectively. Mult-Booth-2 is comparable in PDP to Vivado's speed-optimized IP but requires an average of 8% fewer LUTs.

Figure 11 compares the implementation performance of various approximate multipliers with the Vivado speed- and area-optimized IPs [23]. The Booth-Approx multiplier is discussed in Sect. 2.3. *S1* [30] and *S2* [31] are two approximate unsigned multipliers originally designed for ASIC-based systems. These multi-

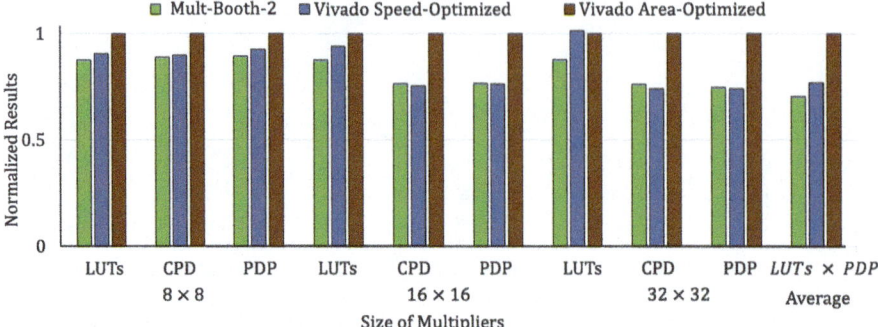

Fig. 10 Application-level performance comparison of accurate signed multipliers

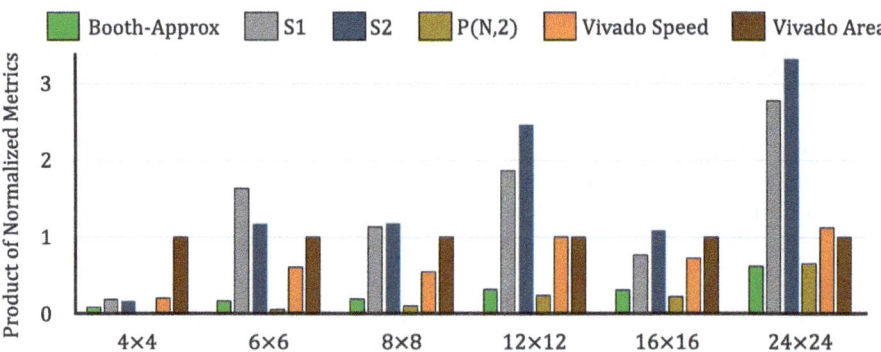

Fig. 11 Performance comparison of approximate signed multipliers

pliers are implemented again using signed–unsigned converters and synthesized for Virtex-7 FPGA for the performance comparison. The P(N,2) is a precision-reduced soft multiplier. P(N,2) truncates the two least significant bits (LSBs) of the input operands before multiplication and then utilizes a lower bit-width accurate multiplier, $(N − 2) × (N − 2)$, for multiplication. The computed product is shifted by 4 locations to calculate the final approximate result. Figure 11 depicts the product of normalized values of total utilized LUTs, CPD, and PDP for each design across different bit widths. All values have been normalized to corresponding Vivado area-optimized multiplier IP values. As previously stated, a lower product value (LUTs × CPD × PDP) indicates a better performing implementation. Although the P(N,2) multiplier outperforms the Booth-Approx multiplier for smaller designs, the performance gains do not scale proportionally for higher-order P(N,2) multipliers. For example, for 24 × 24 multipliers, Booth-Approx reduces the product of the normalized performance metrics by 5.2% when compared to the P(N,2) multiplier. Furthermore, a detailed error analysis of the approximate multipliers reveals that P(N,2) multipliers have lower accuracy across all error metrics.

Table 1 Error analysis of 8 × 8 approximate signed multipliers

Design	Error Occurrences %	Maximum Error	Average Abs. Error	Max. Abs. Relative Error	Avg. Abs. Relative Error
Booth-Approx	90.56	361	85.01	6	0.091
S1 [30]	86.46	7225	1842.44	1	0.362
S2 [31]	34.19	882	118.875	1	0.0223
P(8,2)	93	759	149.78	15	0.121

Gaussian Smoothed Images

Noise-induced Image	all-accurate	all-approximate	Difference Image
PSNR: 16.9	PSNR: 20.6, SSIM: 1.0	PSNR: 20.5, SSIM: 0.95	

Fig. 12 Comparison of Gaussian image smoothing application for 8 × 8 multiplier: Average PSNR=52.36, Average SSIM=0.99 for 15 images

Table 1 shows the error analysis for the Booth-Approx multiplier, as well as precision-reduced P(8,2) and other state-of-the-art signed approximate multipliers (using signed–unsigned converters). As the number −128 cannot be represented using the sign-magnitude format using 8-bit representation, the maximum error magnitude observed in *S1* and *S2* is 16,384. To provide a fair comparison, the range of 8-bit operands for computing the maximum error is limited to [−127, +127] for designs in *S1* and *S2*. The Booth-Approx multiplier has the lowest maximum error magnitude and average absolute error among all presented multipliers, as shown by the highlighted cells in the table. Furthermore, it can be seen that Booth-Approx outperforms the P(8,2) multiplier across all error parameters.

Figure 12 depicts the impact of using approximate multipliers on application-level accuracy. We used Gaussian Smoothing as a test case to determine the efficiency of the Booth-Approx multiplier. We processed 15 images from the USC-SIPI Database [32] for this experiment and reported their average output quality using the Peak Signal-to-noise Ratio (PSNR) and Structural Similarity Index (SSIM) metrics. In comparison to the accurate multipliers-based implementation, the approximate multipliers-based Gaussian smoothing produces insignificant output quality degradation. However, this slight reduction in output quality can be exchanged for considerable performance gains in the corresponding implementation.

3   Arithmetic for Novel Number Representation Schemes

A plethora of recent works has proposed different types of data representation techniques to reduce the memory and energy budgets of employing machine learning models. For example, the Google Tensor Processing Units (TPUs) utilize the Brain Floating-Point Format (bfloat16) for providing high-performance operations. The bfloat16, a subset of the IEEE 754 single-precision floating point, utilizes only 7 bits for storing the fraction (mantissa) [33], as shown in Fig. 13. Compared to the IEEE 754 half-precision, the bfloat16 provides the dynamic range of the single-precision format by committing 8 bits for storing the exponent value. However, compared to the single- and half-precision format, it utilizes only 7 bits for storing the mantissa. The bfloat16 is designed for reducing the storage requirements and accelerated computations of machine learning algorithms. It is currently used by different architectures such as Google TPUs [34], Intel FPGAs and Intel AI processors [35], ARM processors [36], and Nvidia's GPUs [37]. To accelerate the computation performance of ML models, Nvidia's GPUs also utilize a custom 19-bit floating-point representation, Tensor Float 32 (TF32). Figure 13d shows the structure of TF32. However, TF32 is used only for computation in the tensor cores, and the results of these computations are still stored in single-precision format.

Besides the commercially utilized number representation schemes, many recent works have also defined custom number representation schemes for ML algorithms. These schemes focus on the efficient mapping of the application-specific dynamic range of values of ML models to the available bit width. For example, the number representation scheme proposed in [38] focuses on computing the optimal quantization step sizes for features and parameters of DNNs. The proposed scheme

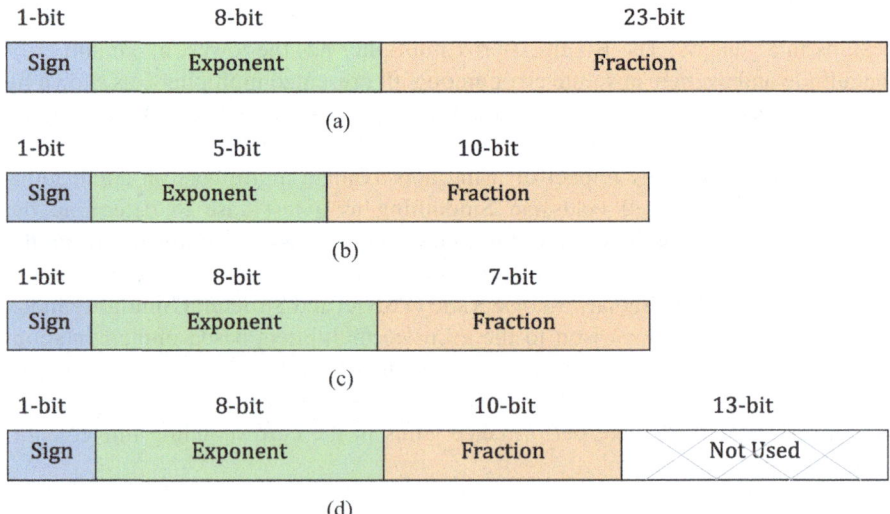

Fig. 13 Various commonly utilized number representation schemes. (**a**) IEEE 754 single-precision 32-bit float. (**b**) IEEE half-precision 16-bit float. (**c**) bfloat16. (**d**) 19-bit Tensor Float 32

iteratively adjusts the step size for each layer's data structure and records the generated errors in the layer under consideration. The individual computation of optimal step size for each layer helps reduce the quantization-induced errors by adjusting the step size according to the distribution of parameters in each layer. The authors of [10] have considered the *power of 2* quantization schemes to represent the floating-point trained parameters of ANN. The proposed scheme uses a custom template for storing the most significant fractional bits in the trained parameters of DNNs. In this technique, log_2 is used to find the location of the most significant 1 in the trained parameters. The location of the leading 1 fractional bit and the actual values of the following fractional bits are stored in the proposed template.[1] The experimental results of the proposed scheme in various networks and applications show that identifying and recording only the most significant fractional bits result in insignificant accuracy loss compared to the floating-point precision-based results. Moreover, the *power of 2* quantization allows the implementation of computationally complex multiplication operation using bit-shift and addition operations. Similarly, the authors of [39] have also utilized the *power of 2* quantization to use bit-shift and addition operations for implementing DNNs.

3.1 Posit Number Representation Scheme-Based Arithmetic

Compared to the IEEE 754 single-precision floating-point format, the recently developed Posit number representation scheme offers a larger dynamic range and greater precision for various applications [18]. The Posit number scheme's constituent fields, *sign*, *regime*, *exponent*, and *fraction*, are shown in Fig. 14. The number of bits sets aside for the exponent (*ES*), and the total number of bits (*N*) defines a Posit number configuration. Equation (7) describes the computation of a Posit value using the four fields of the Posit scheme. The value of k in Eq. (7) is determined using the regime field in Fig. 14. When an inverted bit ($\bar{r}$) is encountered, the regime field is terminated, and the associated value of k is decided by the number of identical bits (m); if the identical bits are a string of 0s, then $k = -m$; if they are a string of 1s, then $k = m - 1$. Next, the remaining bits are used to calculate the exponent (e) and fraction value (f). The Posit number scheme has a wider dynamic range thanks to the use of the regime field. For instance, according to the authors in [40], it is possible to obtain comparable output precision for some applications by substituting m-bit Posit-based numbers for n-bit floats (where $m < n$).

$$Posit\ value = s * (2^{2^{ES}})^k * 2^e * 1.f. \tag{7}$$

---

[1] The total number of recorded fractional bits depends on the deployed bit width of the quantization scheme.

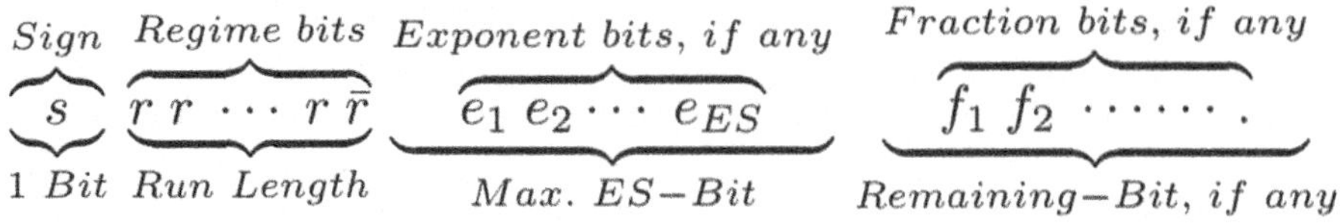

$$\underbrace{\overbrace{s}^{Sign}\ \overbrace{r\,r\ \cdots\ r\,\bar{r}}^{Regime\ bits}\ \overbrace{e_1\,e_2\cdots\,e_{ES}}^{Exponent\ bits,\ if\ any}\ \overbrace{f_1\,f_2\ \cdots\cdots\ .}^{Fraction\ bits,\ if\ any}}$$

Sign · Regime bits · Exponent bits, if any · Fraction bits, if any

1 Bit · Run Length · Max. ES−Bit · Remaining−Bit, if any

Fig. 14 Posit number representation

Table 2 Comparison of resource utilization of adders and multipliers for single-precision floating-point, fixed-point, and Posit [40] and [12]

		Adder	Multiplier	
Bit configuration		LUTs	LUTs	DSP Blocks
Single precision		1049	533	4
32-bit fixed point		32	167	0
Posit	(32,1)	934	576	4
	(32,2)	981	572	4
	(32,3)	951	582	4
	(31,3)	894	560	4
	(30,3)	873	655	3
	(29,3)	837	464	2
	(28,3)	821	459	2
Half-precision		356	212	1
16-bit fixed point		16	144	0
Posit	(16,1)	391	218	1
	(16,2)	404	223	1
	(16,3)	386	219	1
	(15,1)	382	207	1
	(14,1)	353	184	1
	(13,1)	290	181	1
	(12,1)	254	167	0

The associated arithmetic circuits, however, have larger critical path delays and resource utilization than the single-precision-based arithmetic units because of the dynamic nature of the various fields of the Posit scheme. For instance, Table 2 compares the resource usage of floating-point, fixed-point (integer), and Posit scheme-based adders and multipliers. These results include those Posit configurations that offer output accuracy comparable to the floating-point representation [40]. We also compare the resource usage of 32-bit and 16-bit fixed-point adders and multipliers [12]. The comparison demonstrates that fixed-point architectures use significantly fewer resources than the other number representation schemes. Depending on the utilized Posit configuration, the corresponding arithmetic circuit (adder or multiplier) may use more resources than the floating-point-based implementation. It is also noteworthy that for some Posit configuration, Posit-based adders can use more resources than multipliers. Such drawbacks of the Posit number scheme may prevent their utilization in resource-constrained embedded systems.

Designing resource-effective and performance-optimized hardware architectures for Posit-based arithmetic has received a lot of attention recently. For instance, the authors in [40] address the run-time variable field length by creating hardware arithmetic structures for conversion from Posit into floating point and vice versa. A tool to create pipelined Posit operators as a drop-in replacement in processing units is proposed in [41]. The architecture of a parameterized Posit arithmetic unit to implement Posit adders and multipliers of arbitrary bit width is presented in [42]. Similarly, PACoGen uses a three-stage procedure—Posit data extraction, core arithmetic processing, and Posit construction—to conduct parameterized Posit arithmetic, such as multiplication and division [43]. Additionally, Posit arithmetic has been incorporated into Clarinet [44], a RISC-V ISA-based processor that supports employing a Posit arithmetic core.

Some recent studies have also explored the improved dynamic range of Posit-based representation for the training and inference stages of various machine learning models. For instance, the work in [45] has presented vectorized extensions for the *cppPosit*, a C++ posit arithmetic library, using the ARM scalable vector extension Single Instruction Multiple Data (SIMD) engine. An Exact Multiply and Accumulate (EMAC) has been proposed in [46] to implement the MAC operations in ANN. This work demonstrates that the output accuracy of ANN is maintained more accurately by the Posit-based representation of network parameters than by the fixed-point-based representation. However, compared to fixed-point-based MAC operations, Posit-based EMACs operations have much higher resource utilization and Energy-Delay Product (EDP). In [47], the authors have also utilized the EDP metric to compare their presented Posit-based design with the floating-point- and fixed-point-based implementations. Their results show that the fixed-point-based implementations always have lower EDP values than the corresponding Posit-based designs. The research works in [48, 49] have considered the Posit techniques for storing the trained weights of ANNs and then employing floating-point-based computations to calculate output values.

3.2   Fixed-Point-Based Posit Arithmetic

ExPAN(N)D, proposed in [15], is a framework for investigating the joint usage of Posit and fixed-point representation for implementing ML models. By modifying the Posit number representation to store numbers (parameters of pre-trained ML models) within the sub-normal region and by implementing a Posit to Fixed-point (PoFx) converter, ExPAN(N)D aims to take advantage of Posit's useful storage capability and the compute efficiency of fixed-point-based arithmetic. ExPAN(N)D's top-level view is depicted in Fig. 15. The *hardware design* and characterization of the MAC units for various number representation (quantization) schemes form the central theme around which the other two methods—*behavioral analysis* and *accelerator design*—are implemented. The *behavioral analysis* allows the analysis of quantization-induced errors in a given ML model, such as ANNs, utilizing

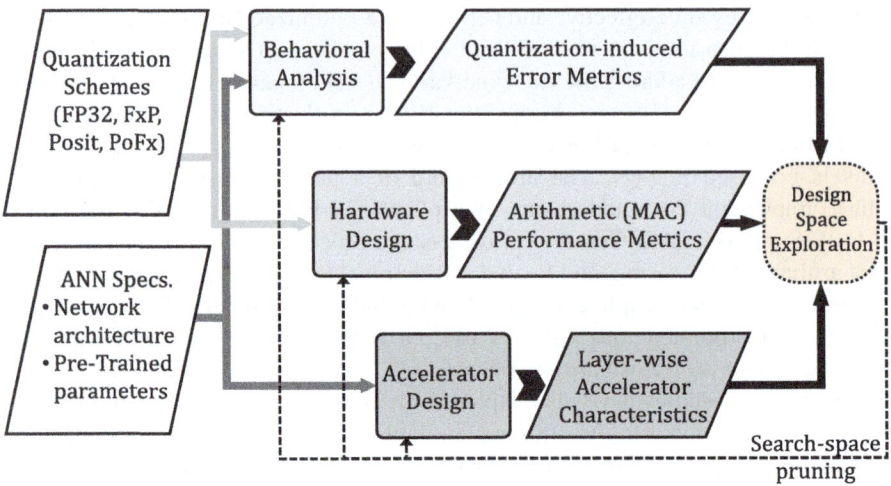

Fig. 15 ExPAN(N)D design methodology for various quantization schemes: FP32, FxP, Posit [15]

the proposed hardware designs. Using the *accelerator design* flow, a designer can also evaluate the performance–resource trade-offs resulting from the adoption of different quantization schemes in an accelerator for a specific layer of the ANN. The results from each of the three processes of ExPAN(N)D can be used to constrain the search space in the design of an efficient ML model using successive design space pruning. In this chapter, we discuss only the Posit-based representation, PoFx conversion, and the PoFx-based MAC design for implementing ML models.

Hardware Design

Normalized Posit Representation: Trained ML models, such as ANNs, have parameters with values between -1 and $+1$. The standard Posit-based representation of these values leads to partial utilization of the available dynamic range. The sub-optimal usage of the dynamic range can result in communication and storage overheads, as more than the required bits are utilized. Correspondingly, more bits than necessary for storing the information are processed during each computation. ExPAN(N)D [15] uses normalized Posit, a unique representation built on the Posit scheme that maintains its hardware implementation, tapered accuracy, and efficient encoding while doubling the number of usable bit-pattern values (x) inside the normalized range $(-1 \leq x < +1)$. This normalized Posit representation—a logical subset of Posits—is customized for quantizing and storing weight normalized FP32 values. For instance, Table 3 displays all possible bit patterns and their corresponding real values for a $N = 4$, $ES = 0$ Posit configuration. The highlighted rows in the table display the bit patterns that correspond to normalized numbers. An analysis of the normalized representation identifies that the two leading bits of the Posit representation are identical; ExPAN(N)D takes advantage of this observation to omit the leading Posit bit in the normalized Posit representation.

Table 3 Posit(N=4, ES=0) to normalized Posit representation

Posit	s	k	f	Value	ExPAN(N)D
0000	0	−3	0	0	000
0001	0	−2	0	0.25	001
0010	0	−1	0	0.5	010
0011	0	−1	0.5	0.75	011
0100	0	0	0	1	–
0101	0	0	0.5	1.5	–
0110	0	1	0	2	–
0111	0	2	0	4	–
1000	1	−3	0	NaR	–
1001	1	2	0	−4	–
1010	1	1	0	−2	–
1011	1	0	0.5	−1.5	–
1100	1	0	0	−1	100
1101	1	−1	0.5	−0.75	101
1110	1	−1	0	−0.5	110
1111	1	−2	0	−0.25	111

PoFx: Normalized Posit to Fixed-Point Converter: The PoFx conversion hardware can efficiently quantize and store weight normalized FP32 values in memory while also providing FxP converted values close to the processing components. The PoFx enables the efficient execution of ML models with very little conversion overhead. Posit representation, Posit(N, ES), is converted via the PoFx conversion method to fixed-point representation, FxP(M, F), where M is the overall length of the output and F is the length of the output's fraction. Based on the Posits numbers decoding scheme, this method successfully transforms a Posit into an FxP number. We demonstrate this conversion using the Posit($N = 4, ES = 0$) bit patterns in Table 3. The key to understanding this algorithm is to realize that the fraction field recovered from the Posit representation is the same as the one required in the FxP output. Once the data in the Posit bit pattern are extracted into its components s, k, e, and f; the Posit value can be computed by setting a bit and storing the extracted fraction bits to its right, followed by a final bit shift determined by the term $2^{ES} * k + e$. The term $2^{ES} * k + e$ can be computed by adding the e value to the bit sequence created by appending k to the ES number of zero bits. The Posit representation is given by the sign bit and the shifted bit sequence in sign-magnitude FxP format, which is easily convertible into a 2's complement format.

MAC Unit with PoFx Converter: Any application that can benefit from efficiently storing a large number of parameters can leverage the PoFx converter. As a particular case for ML models, the authors in [15] incorporate the normalized PoFx into MAC units to improve low-precision ANN inference. The schematic of their proposed parameterized PoFx converter-based MAC with a *ReLU* activation function is shown in Fig. 16. The figure shows that the weights/biases are assumed to be stored/communicated as Posit($N − 1, ES$) numbers. These values are then multiplied with the M-bit input activation values after being transformed to their equivalent M-

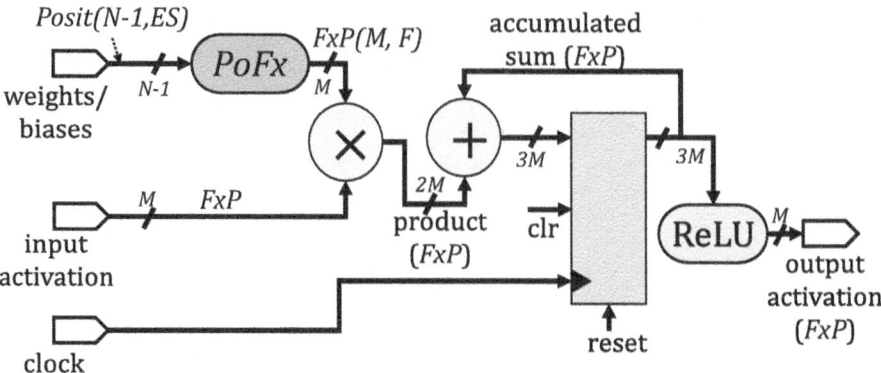

Fig. 16 MAC unit (with ReLU activation) using PoFx converter to convert numbers from Posit into fixed-point (FxP) representation

bit FxP representations. The authors have selected a $3M$-bit adder for accumulation across all configurations. This choice of adder size was made in order to account for the overflows caused by the accumulation of numerous $2M$-bit values and thus facilitate the evaluation of the proposed architecture. The $3M$-bit result for a single node in a layer of an ANN is supplied to the activation function after all values have been added up for that node. It should be noted that the PoFx-based MAC unit enables the designer to express the weights/biases with fewer bits while still being able to apply various FxP-based arithmetic optimizations, such as precision scaling, approximations, in the processing element.

3.3 Results

For ExPAN(N)D framework, the Posit-based arithmetic designs are produced using the SmallPosit HDL repository [50]. Verilog HDL is used to implement the PoFx converter and the related arithmetic blocks. Xilinx Vivado Design Suite is used to characterize the hardware designs. Every design has been implemented on the Xilinx Zynq UltraScale+ MPSoC (xczu3eg-sbva484-1-e device). Python and TensorFlow [13] are utilized to carry out the behavioral analysis of ANNs. All of the proposed methods can be applied to any arbitrary application. However, in the presented work, VGG16 network [14] is used as a test application.

MAC Design Analysis
The presented PoFx enables the use of high-performance, resource-efficient computation for Posit number systems. We compare 8-bit MAC units based on PoFx with the conventional FxP-based MAC units, in order to assess the effectiveness of the presented technique and estimate the associated overheads of the PoFx-based designs. In addition, we have created two types of designs for a deeper investigation

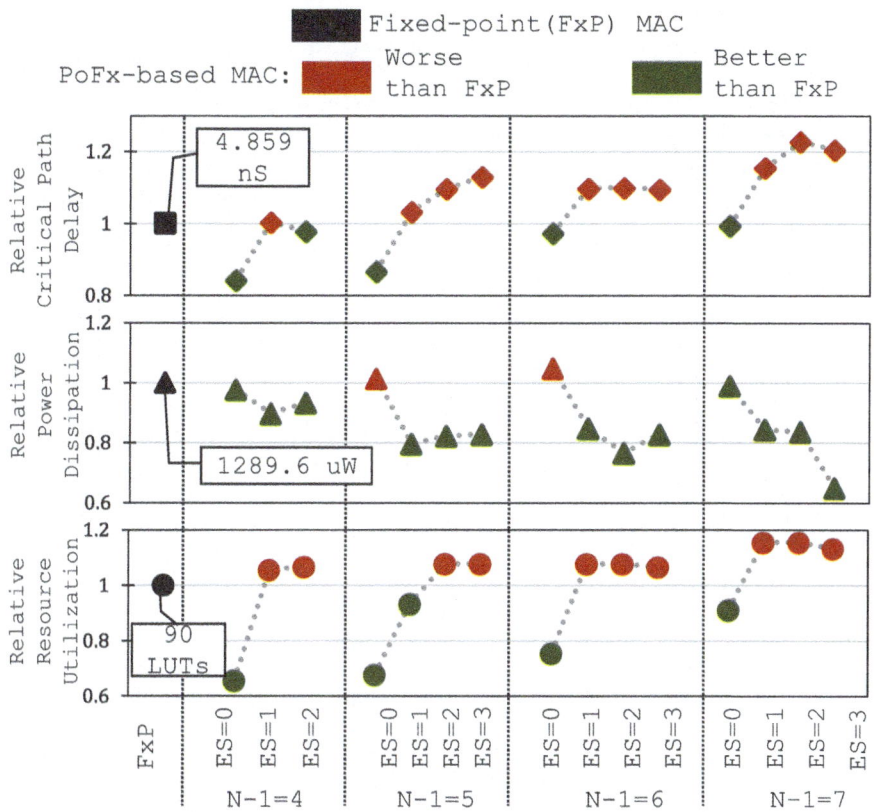

Fig. 17 Relative hardware performance metrics of PoFx-based MAC units with varying values of ES and $N - 1$ for Posit($N - 1$, ES) inputs to 8-bit FxP MAC. The PoFx-based design points with worse performance than the FxP-based design are highlighted in red

of the PoFx-based designs: *ToolOpt*, which enables the synthesis tool to optimize across the constituent blocks (decoder–encoder, multipliers, and adders), and *non-ToolOpt*, which performs optimization for the constituent blocks separately. Figure 17 presents the findings of comparisons made across multiple design metrics for different Posit configurations.[2] The MAC's critical path delay and resource usage exhibit a steadily increasing trend for both N and ES values. It should be observed that the PoFx-based MAC occasionally outperforms the FxP-only MAC in terms of critical path delay, power consumption, and LUT utilization. This behavior is especially true when $ES = 0$. The dynamic range of the Posit scheme is restricted for $ES = 0$, and the PoFx does not use the entire dynamic range of the FxP.

---

[2] The data in Fig. 17 refer to the design with the better metrics among the ToolOpt and non-ToolOpt versions.

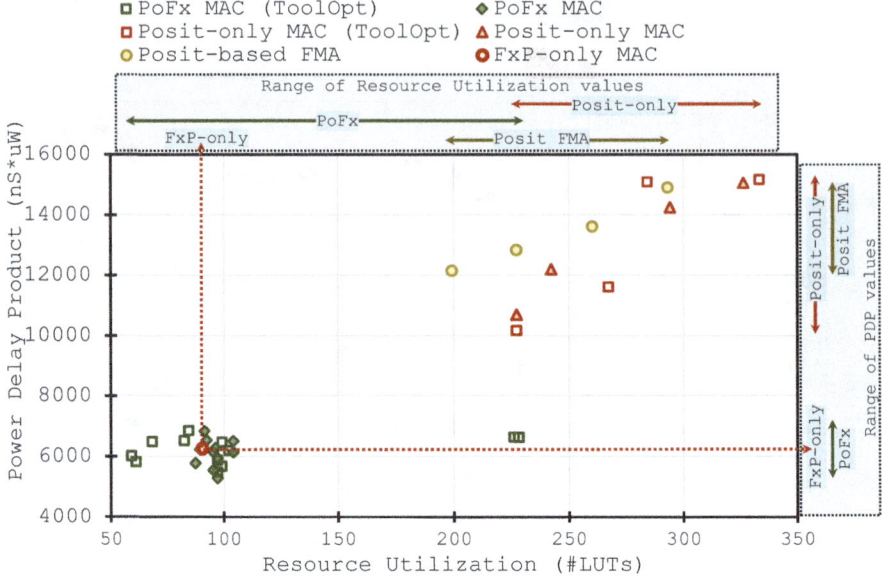

Fig. 18 Comparison of various 8-bit MAC implementations: for Posit($N - 1, ES$) $N - 1 \in$ [4 .. 7] *and* $ES \in \{0, 1, 2\}$. The PoFx MAC refers to the *non-ToolOpt* version. The range of performance values reported by each scheme is shown along the axes (top and right)

After conversion, the small pool of distinct FxP values allows the synthesis tool to improve the PoFx-based MACs' design and performance characteristics. Since power measurements are generated based on the bit switches (necessary to get the right bit sequence as the output), hence they do not follow a well-defined trend. We report worst-case overheads for critical path delay, power dissipation, and LUT utilization of 22.8%, 5.0%, and 15.5%, respectively, as compared to the FxP-only MAC.

We compare PoFx-based MAC designs with FxP-only MAC, Posit-only MAC, and Posit-based 3-input Fused Multiply Add (FMA) [51] to assess its efficacy further. Figure 18 compares the implementation results of these designs for 8-bit inputs in terms of PDP and utilized LUTs. Posit-only MAC, developed by employing a standalone N-bit Posit adder and N-bit Posit multiplier, has much higher PDP and LUT consumption due to the extraction and packaging of Posit numbers between stages. Despite being optimized to reduce the overheads of the encode–decode stages, the Posit-based FMA requires more hardware resources for its implementation. It can be observed that the results for PoFx-based MAC designs are very similar to those based on FxP while providing a wider range of designs with performance trade-offs. Furthermore, the PoFx-based designs generate a more exact $3N - bit$ output compared to the Posit-only MAC and Posit-based FMA designs, which both produce N-bit outputs. The higher output precision can enable reduced inter-layer quantization-induced errors in ANNs as one can analyze and determine

Table 4 Joint analysis of classification accuracy and MAC hardware characteristics of FxP, Posit, and PoFx-based designs

Configuration	N	ES	Top-1 [%]	Top-5 [%]	PDP [Maximum: 13616 uW*nS]	LUTs [Maximum: 319]
					Relative MAC Metrics	
FxP	16	–	69.66	89.02	0.763	1.000
	8	–	64.71	86.26	0.475	0.282
Posit (N,ES)	7	1	68.88	88.5	0.578	0.671
	8	1	69.59	**89**	**1.000**	0.815
	6	2	**66.32**	**86.99**	**0.441**	**0.555**
	7	2	68.77	88.54	0.550	0.618
	8	2	**69.65**	**89**	0.853	**0.837**
	7	3	68.02	87.97	0.469	0.567
	8	3	69.43	88.86	0.747	0.712
PoFx (N-1,ES)	6	1	64.38	85.94	0.432	**0.304**
	7	1	**64.48**	**86.15**	0.451	0.326
	5	2	58.27	81.99	0.417	0.310
	6	2	64.36	85.99	**0.388**	**0.304**
	7	2	64.4	86.08	**0.478**	0.326
	5	3	**57.13**	**81.13**	0.446	**0.304**
	6	3	62.67	84.62	0.418	**0.304**
	7	3	64.45	86.15	0.413	**0.361**

the type of rounding mechanism at the output to maintain as much precision as possible before communicating the values to the next stage in the network.

Application-Level Accuracy Analysis of PoFx-Based Arithmetic

To demonstrate the impact of different number representation schemes on the output accuracy of high-level applications, we used DNNs as a test case application. For this work, we classified the ImageNet dataset using a pre-trained VGG16 network [14]. VGG16 network—with thirteen convolutional and three fully connected layers—has 138 million trainable parameters, which makes it a good candidate for assessing the effectiveness of different number representation systems. For 50,000 validation images in the ImageNet dataset and utilizing single-precision FP32-based trained parameters, the network reports 69.72% and 89.09% as Top-1 and Top-5 percentage output classification accuracies, respectively.

Estimating the impact of the presented methods on the network's classification accuracy forms the primary component of the behavioral analysis. To this end, Table 4 presents the combined analysis of the ImageNet dataset classification accuracy and the related MAC designs for various number representation systems. In this experiment, the parameters (weights and biases) are encoded using multiple 8-bit representation schemes, and the activations have FP32 precision. The results also demonstrate the classification accuracy using a 16-bit FxP-based quantization approach for comparison. Compared to FP32-based results, the FxP-

16 and Posit($N = 8, ES = 2$) produce similar classification results by reducing the final output accuracy by only 0.06 and 0.07, respectively. The Top-1 and Top-5 classification accuracies are reduced by 5.01 and 2.83, respectively, using the FxP-8-based configuration. We take into account the normalized PoFx approach and make use of *Posit(N-1, ES)* configurations of N-bit Posit numbers for the PoFx-based schemes. It should be noted that Table 4 includes only Posit and PoFx variants with comparable accuracy and with feasible hardware designs. Moreover, the values for the PDP and LUT metrics in the table correspond to values relative to the maximum value displayed in the top row of the table.

The configurations of Posit($N = 8, ES = 1$) and FxP-16, respectively, show the maximum PDP and LUT utilization values. In Table 4, the bold text highlights the greatest and lowest values of the performance metrics for each of the two categories—Posit and PoFx. It is evident that Posit($N = 8, ES = 2$), the Posit configuration with the best Top-1 accuracy, corresponds to the MAC design with the maximum LUT utilization. Similarly, Posit($N = 8, ES = 1$) and Posit($N = 8, ES = 2$), the Posit configurations with the highest Top-5 accuracies, correspond to the highest PDP values. The design with the lowest PDP and LUT utilization among Posit-based MACs corresponds to the Posit configuration with the lowest accuracy, Posit($N = 6, ES = 2$).

PoFx-based designs showed comparable correlations as well. Typically, designs with higher PDP yield more precise results. Compared to FxP-8-based designs, PoFx($N - 1 = 7, ES = 1$) achieves comparable accuracy with reduced PDP ($\approx 5\%$) and slightly greater LUT overhead ($\approx 15.5\%$). The same is evident with PoFx ($N - 1 = 6, ES = 2$), which achieves equivalent accuracy with much smaller PDP ($\approx 18\%$) and LUT overheads ($\approx 8\%$). Further, the PoFx-based designs use fewer bits to express network characteristics. As a result, each layer of the network's accelerator design may result in lower communication and storage overheads.

Accelerator-Level Design Analysis

The benefits of employing PoFx-based arithmetic operators are evident in the design of accelerators. As observed in the experiment results, compared to Posit- and FxP-based accelerators, the suggested PoFx technique significantly reduces processing overheads at an insignificant cost to accuracy. We incorporated the potential solutions in the design of an accelerator for a fully connected layer of a DNN to calculate the system-level effects of employing the proposed PoFx approach. The accelerator was created in C++ and synthesized in Xilinx Vivado HLS. We used a matrix–vector multiplication to keep the design generic. The vector represents a single input activation, and the matrix represents the weights of a fully connected layer. We estimated the switching activity using thousand input activations to calculate the dynamic power dissipation of each design. The implemented accelerator employs the ReLU activation function. LUTRAMs were employed to store the local arrays, along with sufficient partitioning to facilitate parallel execution obtained by loop unrolling. In order to compare the effect of using Posit-based, PoFx-based, and FxP-based MAC units, the following variants of the accelerator were considered:

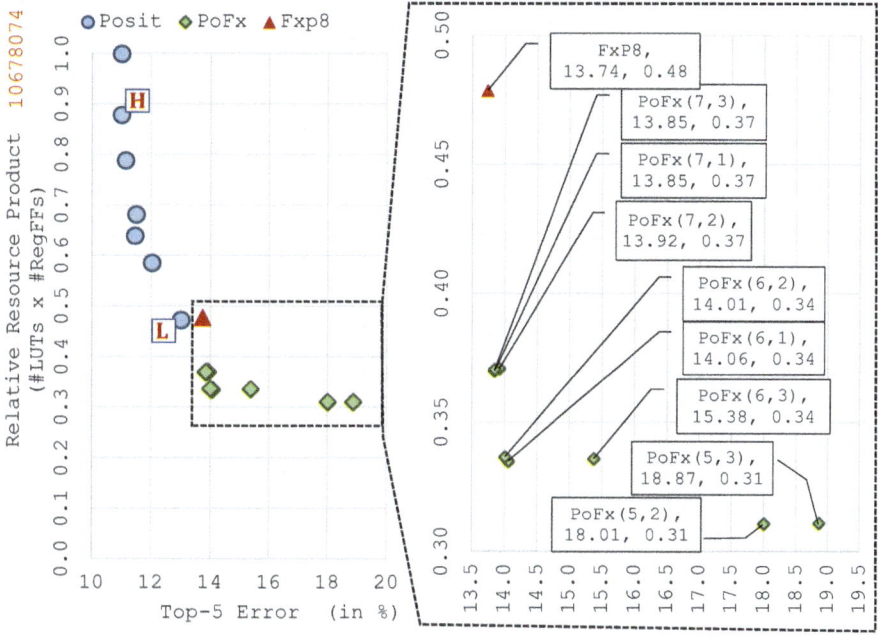

Fig. 19 Top-5 percentage errors in ImageNet dataset classification using VGG-16 vs. the resource utilization of a sample accelerator implementing a fully connected layer. The zoomed-in portion shows the detailed comparison of PoFx- and FxP8-based accelerator implementations

1. *Posit*: The accelerator stores and computes all operations in Posit(N, ES) format.
2. *PoFx*: The weights are moved from the main memory and stored in local memory in normalized PoFx($N - 1$, ES) format. During computation, the weights are fetched from local memory, converted into FxP($M = 8$), and used in the computation of the output activation values.
3. *FxP(8)*: The weights are moved from the main memory to the accelerator and stored in the local memory of the accelerator as FxP($M = 8$) numbers. The computation stage does not involve any conversions between number representations.

Figures 19, 20, and 21 plot the ImageNet dataset's classification accuracy using the VGG-16 network for FxP8, Posit, and PoFx, along with various performance metrics of an accelerator implementing multiply and accumulate operations with these number representation schemes. The accelerators created for generating the data in the figures correspond to a weight matrix of size 32 × 10. The plot with all the designs is shown on the left of each figure, along with a zoomed-in plot for comparison between FxP8-based and PoFx-based designs. The design points shown in the figures match the configurations displayed in Table 4 (except Fxp-16). The vertical axis of the graphs depicts the relative performance metric, and the horizontal axis displays the Top-5 *classification error* (in %) for the ImageNet dataset. Along the vertical axis, each performance metric's maximum values, which

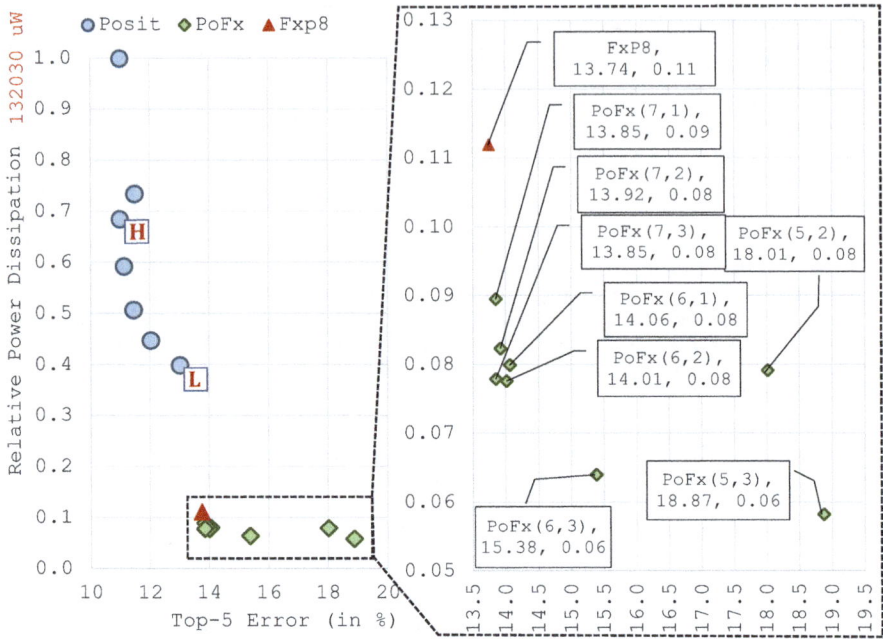

Fig. 20 Top-5 percentage errors in ImageNet dataset classification using VGG-16 vs. the power dissipation of a sample accelerator implementing a fully connected layer

correspond to 1.00, are displayed in red. The accelerator designs based on PoFx and FxP8 exhibit noticeably better performance (lower values on the vertical axis) than those based on Posit, as shown throughout Figs. 19, 20, and 21. The penalty of this enhanced performance is a slightly increased classification error.

The effect of using fixed-point operators, with inherently lower computational complexity than Posits, on the overall resource usage of the accelerators is depicted in Fig. 19. The dominant (Pareto) Posit-based designs in the figure are **H**(Posit(8,2)) and **L**(Posit(6,2)), with the highest and lowest resource utilization, respectively. The figure demonstrates that, in comparison to **H** and **L** designs, the FxP8-based design exhibits around 2.74% and 0.73% greater error, respectively. However, compared to **H** implementation, the FxP8-based design reduces the overall amount of utilized resources by nearly 45.5%. Resource usage for implementing the FxP8-based design is 1.2% higher than that for the **L** implementation. The increased use of RegFFs by the FxP8-based architecture to store weights is the primary cause for the rise in overall utilized resources. When comparing the PoFx(7,1)-based design to the FxP8-based designs, in the zoomed-in area of the figure, the PoFx(7,1)-based design has an additional 0.11% inaccuracy but uses 22.7% fewer resources. Similarly, PoFx(6,1)-based design utilizes 30% fewer resources than the FxP8-based version while adding 0.32% more error. The various PoFx-based design points offer different error–resource trade-offs. The PoFx-based design points' lower total resource usage, compared to the FxP8-based implementation, can be attributed to

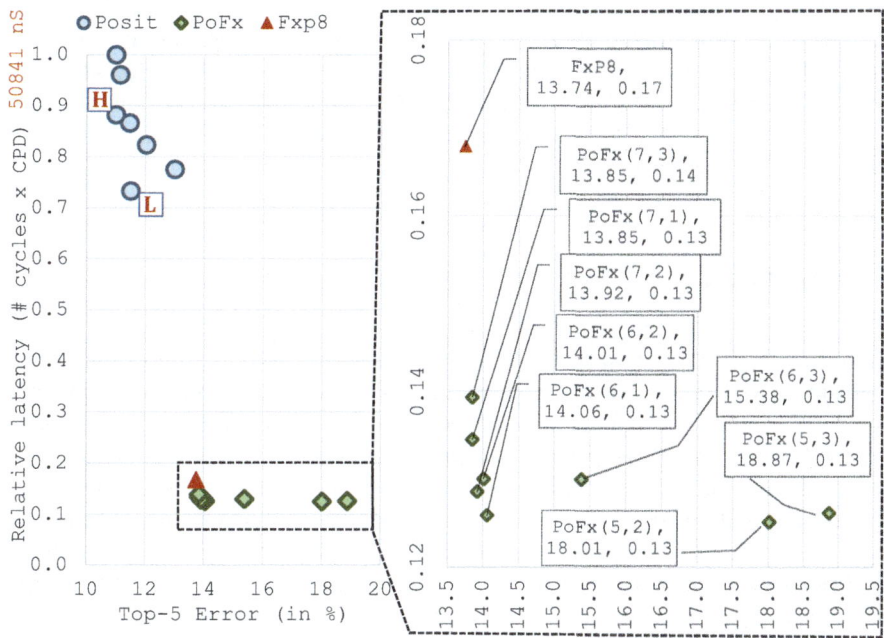

Fig. 21 Top-5 percentage errors in ImageNet dataset classification using VGG-16 vs. the *best-case latency* of a sample accelerator implementing a fully connected layer

their lower storage needs, which also improves resource efficiency by amortizing the conversion overheads of each PoFx-based MAC unit.

With regard to power dissipation, the advantages of employing PoFx-based designs are shown in Fig. 20. **H** (Posit(8,2)) and **L** (Posit(6,2)) are the dominant Posit-based design points with the highest and lowest power dissipations, respectively. The power consumption of the FxP8-based design is nearly 75.71 mW and 37.85 mW less than that of the **H** and **L** designs, respectively. This reduced power dissipation comes at the expense of additional classification errors of 2.74% and 0.73%, respectively. The PoFx-based designs exhibit even lower power dissipation. Compared to the FxP8-based implementation, designs utilizing PoFx(7,3), PoFx(6,3), and PoFx(5,3) report 4.49 mW, 6.33 mW, and 7.09 mW less power with 0.11%, 1.64%, and 5.13% more error, respectively. In the accelerator, the increased power dissipation of the Posit-based MAC units is made worse by routing power, which makes up a sizable portion of the overall power dissipation.

The *best-case*[3] latency of the accelerator for various Posit and PoFx-based designs is shown in Fig. 21. The Pareto-optimal Posit-based designs are shown as **H** and **L**, based on the highest and lowest accelerator performance metrics, respectively. In Fig. 21, Posit(8,1) and Posit(7,1) are the points denoted by the letters **H** and **L**,

---

[3] The best-case latency refers to the latency corresponding to the CPD of the design.

respectively. The FxP8-based design exhibits better performance than Posit-based designs with a minor decrease in classification accuracy (compared to **L**), similar to the results shown in Figs. 19 and 20. Additional design points that offer additional accuracy–performance trade-offs are provided by the PoFx-based designs. Because of having substantially lower CPD than Posit-based designs, FxP8- and PoFx-based systems have lower latency.

4 Conclusion

Embedded machine learning models are increasingly used to implement smart embedded systems for various application domains such as smart factories, personalized health care, and autonomous vehicles. However, the high computational cost and large memory footprints of the machine learning models are the challenges that hinder their ubiquitous deployment in resource-constrained embedded systems. To this end, state-of-the-art works exploit the inherent error resilience of machine learning algorithms to explore various techniques to optimize these models for embedded systems. This chapter presents selected novel approaches that address these challenges by focusing on the capabilities of various number representation schemes (quantization) and associated efficient arithmetic architectures for them. As multiplication is one of the most costly and commonly used operations in machine learning algorithms, the chapter mainly focuses on this operation, and it presents various architectures for accurate and approximate custom signed multipliers. The chapter also discusses the recently proposed number representation scheme Posit, which can provide the dynamic range and precision of the floating-point scheme. In particular, it introduces a modified Posit representation optimized for machine learning algorithms. Compared to the floating-point and fixed-point representations, Posit provides better storage efficiency by utilizing fewer bits to represent trained parameters of a machine learning model. However, Posit-based arithmetic has a high computational cost, and to this end, the chapter presents a Posit to the fixed-point converter to enable computationally efficient integer arithmetic for Posits.

The various architecture described in this chapter are available as open-source libraries at https://cfaed.tu-dresden.de/pd-downloads. Related future research may involve exploiting the inherent error tolerance of machine learning models across multiple degrees of freedom. For instance, in one of our recent works, we have proposed a framework to analyze the various approximation knobs available at different layers of the computation stack to implement high-performance accelerators for error-resilient applications. This framework can be extended to include and analyze the various degrees of freedom (approximation knobs) provided by the embedded machine learning model. Most state-of-the-art approximate computing-related works have focused on the basic arithmetic operators such as adders and multipliers. However, the approximation opportunities provided by larger operators, such as MAC, can result in more efficient architectures for machine learning models. The Posit to fixed-point converter discussed in this chapter combines the advantages of two different number representation schemes. Similar mixed quantization schemes can be further explored to enable efficient implementations of machine learning models on embedded systems.

References

1. Zablocki, É., Ben-Younes, H., Pérez, P., Cord, M.: Explainability of vision-based autonomous driving systems: Review and challenges. CoRR, vol. abs/2101.05307, 2021. https://arxiv.org/abs/2101.05307
2. Prabakaran, B.S., Akhtar, A., Rehman, S., Hasan, O., Shafique, M.: BioNetExplorer: Architecture-space exploration of biosignal processing deep neural networks for wearables. IEEE Internet Things J. **8**(17), 13251–13265 (2021)
3. Chlingaryan, A., Sukkarieh, S., Whelan, B.: Machine learning approaches for crop yield prediction and nitrogen status estimation in precision agriculture: A review. Comput. Electron. Agric. **151**, 61–69 (2018)
4. Kotsiopoulos, T., Sarigiannidis, P., Ioannidis, D., Tzovaras, D.: Machine learning and deep learning in smart manufacturing: The smart grid paradigm. Computer Science Review **40**, 100341 (2021). https://www.sciencedirect.com/science/article/pii/S157401372030441X
5. Control your smart home | google assistant. Accessed on 17 February, 2022. https://assistant.google.com/smart-home/
6. Lin, J., Chen, W.-M., Lin, Y., cohn, j., Gan, C., Han, S.: MCUNet: Tiny deep learning on IoT devices. In: Larochelle, H., Ranzato, M., Hadsell, R., Balcan, M.F., Lin, H. (eds.) Advances in Neural Information Processing Systems, vol. 33, pp. 11711–11722. Curran Associates, Inc., New York (2020). https://proceedings.neurips.cc/paper/2020/file/86c51678350f656dcc7f490a43946ee5-Paper.pdf
7. Warden, P., Situnayake, D.: TinyML: Machine learning with TensorFlow Lite on Arduino and ultra-low-power microcontrollers. O'Reilly Media (2019)
8. Chippa, V.K., Chakradhar, S.T., Roy, K., and Raghunathan, A.: Analysis and characterization of inherent application resilience for approximate computing. In: Proceedings of the 50th Annual Design Automation Conference (2013), pp. 1–9
9. Han, S., Pool, J., Tran, J., Dally, W.J.: Learning both weights and connections for efficient neural network. In: Proceedings of the 28th International Conference on Neural Information Processing Systems - Volume 1 (NIPS'15), pp. 1135–1143. MIT Press, Cambridge, MA (2015)
10. Gupta, S., Ullah, S., Ahuja, K., Tiwari, A., Kumar, A.: ALigN: A highly accurate adaptive layerwise log_2_lead quantization of pre-trained neural networks. IEEE Access **8**, 118899–118911 (2020)
11. Ullah, S., Sahoo, S.S., Ahmed, N., Chaudhury, D., Kumar, A.: AppAxO: Designing application-specific approximate operators for FPGA-based embedded systems. ACM Trans. Embed. Comput. Syst. (2022). https://doi.org/10.1145/3513262
12. Ullah, S., Schmidl, H., Sahoo, S.S., Rehman, S., Kumar, A.: Area-optimized accurate and approximate softcore signed multiplier architectures. IEEE Trans. Comput. **70**(3), 384–392 (2020)
13. Abadi, M., Agarwal, A., Barham, P., Brevdo, E., Chen, Z., Citro, C., Corrado, G.S., Davis, A., Dean, J., Devin, M., Ghemawat, S., Goodfellow, I., Harp, A., Irving, G., Isard, M., Jia, Y., Jozefowicz, R., Kaiser, L., Kudlur, M., Levenberg, J., Mané, D., Monga, R., Moore, S., Murray, D., Olah, C., Schuster, M., Shlens, J., Steiner, B., Sutskever, I., Talwar, K., Tucker, P., Vanhoucke, V., Vasudevan, V., Viégas, F., Vinyals, O., Warden, P., Wattenberg, M., Wicke, M., Yu, Y., Zheng, X.: TensorFlow: Large-scale machine learning on heterogeneous systems. In: 2015, Software available from tensorflow.org. https://www.tensorflow.org/

14. Simonyan, K., Zisserman, A.: Very deep convolutional networks for large-scale image recognition (2014)
15. Nambi, S., Ullah, S., Sahoo, S.S., Lohana, A., Merchant, F., Kumar, A.: ExPAN(N)D: Exploring posits for efficient artificial neural network design in FPGA-based systems. IEEE Access **9**, 103691–103708 (2021)
16. Courbariaux, M., Bengio, Y., David, J.-P.: BinaryConnect: training deep neural networks with binary weights during propagations. In: Proceedings of the 28th International Conference on Neural Information Processing Systems - Volume 2 (NIPS'15), pp. 3123–3131. MIT Press, Cambridge, MA, USA (2015)
17. Li, F., Zhang, B., Liu, B.: Ternary weight networks. arXiv preprint arXiv:1605.04711 (2016)
18. Gustafson, J.L., Yonemoto, I.T.: Beating floating point at its own game: Posit arithmetic. Supercomputing Frontiers and Innovations **4**(2), 71–86 (2017)
19. Xilinx: UltraScale Architecture Configuration: User Guide. https://www.xilinx.com/support/documentation/user_guides/ug570-ultrascale-configuration.pdf (2022)
20. Xilinx 7 Series DSP48E1 Slice. https://www.xilinx.com/support/documentation/user_guides/ug479_7Series_DSP48E1.pdf (2018)
21. Intel® Stratix® 10 Variable Precision DSP Blocks User Guide. https://www.intel.com/content/dam/www/programmable/us/en/pdfs/literature/hb/stratix-10/ug-s10-dsp.pdf (2020)
22. Ullah, S., Rehman, S., Shafique, M., Kumar, A.: High-performance accurate and approximate multipliers for FPGA-based hardware accelerators. In: IEEE Transactions on Computer-Aided Design of Integrated Circuits and Systems, pp. 1–1 (2021). https://doi.org/10.1109%2Ftcad.2021.3056337
23. Xilinx LogiCORE IP v12.0 . https://www.xilinx.com/support/documentation/ip_documentation/mult_gen/v12_0/pg108-mult-gen.pdf (2015)
24. Intel: Integer Arithmetic IP Cores User Guide. https://www.altera.com/en_US/pdfs/literature/ug/ug_lpm_alt_mfug.pdf (2020)
25. Baugh, C., Wooley, B.: A two's complement parallel array multiplication algorithm. IEEE Trans. Comput. **C-22**(12), 1045–1047 (1973)
26. Ullah, S., Nguyen, T.D.A., Kumar, A.: Energy-efficient low-latency signed multiplier for fpga-based hardware accelerators. IEEE Embed. Syst. Lett. **13**(2), 41–44 (2021)
27. Kumm, M., Abbas, S., Zipf, P.: An efficient softcore multiplier architecture for Xilinx FPGAs. In: 2015 IEEE 22nd Symposium on Computer Arithmetic, pp. 18–25. IEEE, New York (2015)
28. Booth, A.D.: A signed binary multiplication technique. Q. J. Mech. Appl. Math. **4**(2), 236–240 (1951)
29. MNIST-cnn. https://github.com/integeruser/MNIST-cnn (2016)
30. Rehman, S., El-Harouni, W., Shafique, M., Kumar, A., Henkel, J.: Architectural-space exploration of approximate multipliers. In: Proceedings of the 35th International Conference on Computer-Aided Design, ser. ICCAD '16. Association for Computing Machinery, New York (2016). https://doi.org/10.1145/2966986.2967005
31. Kulkarni, P., Gupta, P., Ercegovac, M.: Trading accuracy for power with an underdesigned multiplier architecture. In: 2011 24th International Conference on VLSI Design, pp. 346–351. IEEE, New York (2011)
32. SIPI Image Database. http://sipi.usc.edu/database/database.php?volume=misc (2019)
33. Kalamkar, D., Mudigere, D., Mellempudi, N., Das, D., Banerjee, K., Avancha, S., Vooturi, D.T., Jammalamadaka, N., Huang, J., Yuen, H., et al.: A study of BFLOAT16 for deep learning training. arXiv preprint arXiv:1905.12322 (2019)
34. Introduction to Cloud TPU. https://cloud.google.com/tpu/docs/intro-to-tpu
35. Intel® Deep Learning Boost New Deep Learning Instruction BFLOAT16 - Intrinsic Functions. https://www.intel.com/content/www/us/en/developer/articles/technical/intel-deep-learning-boost-new-instruction-bfloat16.html
36. Arm Armv9-A A64 Instruction Set Architecture. https://developer.arm.com/documentation/ddi0602/2021-12/?lang=en
37. TensorFloat-32 in the A100 GPU Accelerates AI Training, HPC up to 20x. https://blogs.nvidia.com/blog/2020/05/14/tensorfloat-32-precision-format/

38. Vogel, S., Springer, J., Guntoro, A., Ascheid, G.: Self-supervised quantization of pre-trained neural networks for multiplierless acceleration. In: 2019 Design, Automation and Test in Europe Conference and Exhibition (DATE), pp. 1094–1099. IEEE, New York (2019).
39. Sarwar, S.S., Venkataramani, S., Raghunathan, A., Roy, K.: Multiplier-less artificial neurons exploiting error resiliency for energy-efficient neural computing. In: 2016 Design, Automation and Test in Europe Conference and Exhibition (DATE), pp. 145–150. IEEE, New York (2016)
40. Chaurasiya, R., Gustafson, J., Shrestha, R., Neudorfer, J., Nambiar, S., Niyogi, K., Merchant, F., Leupers, R.: Parameterized posit arithmetic hardware generator. In: 2018 IEEE 36th International Conference on Computer Design (ICCD), pp. 334–341. IEEE, New York (2018)
41. Podobas, A., Matsuoka, S.: Hardware Implementation of POSITs and Their Application in FPGAs. In: 2018 IEEE International Parallel and Distributed Processing Symposium Workshops (IPDPSW), pp. 138–145 (2018)
42. Jaiswal, M.K., So, H.K.: Universal number posit arithmetic generator on FPGA. In: 2018 Design, Automation Test in Europe Conference Exhibition (DATE), pp. 1159–1162 (2018)
43. Jaiswal, M.K., So, H.K.: PACoGen: A Hardware Posit Arithmetic Core Generator. IEEE Access 7, 74586–74601 (2019)
44. Jain, R., Sharma, N., Merchant, F., Patkar, S., Leupers, R.: CLARINET: A RISC-V Based Framework for Posit Arithmetic Empiricism (2020)
45. Cococcioni, M., Rossi, F., Ruffaldi, E., Saponara, S.: Fast deep neural networks for image processing using posits and ARM scalable vector extension. J. Real-Time Image Proc. 17(3), 759–771 (2020)
46. Carmichael, Z., Langroudi, H.F., Khazanov, C., Lillie, J., Gustafson, J.L., Kudithipudi, D.: Deep Positron: A deep neural network using the posit number system. In: 2019 Design, Automation Test in Europe Conference Exhibition (DATE), pp. 1421–1426 (2019)
47. Langroudi, H.F., Carmichael, Z., Gustafson, J.L., Kudithipudi, D.: PositNN framework: Tapered precision deep learning inference for the edge. In: 2019 IEEE Space Computing Conference (SCC), pp. 53–59. IEEE, New York (2019)
48. Fatemi Langroudi, S.H., Pandit, T., Kudithipudi, D.: Deep learning inference on embedded devices: Fixed-point vs posit. In: 2018 1st Workshop on Energy Efficient Machine Learning and Cognitive Computing for Embedded Applications (EMC2), pp. 19–23 (2018)
49. Langroudi, H.F., Karia, V., Gustafson, J.L., Kudithipudi, D.: Adaptive posit: Parameter aware numerical format for deep learning inference on the edge. In: Proceedings of the IEEE/CVF Conference on Computer Vision and Pattern Recognition Workshops, pp. 726–727 (2020)
50. Wu, B.: SmallPositHDL. https://github.com/starbrilliance/SmallPositHDL (2020)
51. Xiao, F., Liang, F., Wu, B., Liang, J., Cheng, S., Zhang, G.: Posit arithmetic hardware implementations with the minimum cost divider and SquareRoot. Electronics 9(10), 1622 (2020). https://www.mdpi.com/2079-9292/9/10/1622

Efficient Hardware Acceleration of Emerging Neural Networks for Embedded Machine Learning: An Industry Perspective

Arnab Raha, Raymond Sung, Soumendu Ghosh, Praveen Kumar Gupta, Deepak A. Mathaikutty, Umer I. Cheema, Kevin Hyland, Cormac Brick, and Vijay Raghunathan

1 Introduction

The breakthroughs achieved by neural networks to solve challenging problems have ushered in a new era of demand for high-performance computing and domain-specific acceleration. The vast majority of existing AI-enabled applications still run on CPUs and GPUs due to the ease of programming and availability of high-level frameworks that make it easier to experiment with network parameters and deploy these solutions from the cloud to the edge. However, these solutions are not as energy-efficient nor do they match the throughput of domain-specific accelerators. As networks become more complex, the energy required for doing training and inference has resulted in a noticeable shift towards adopting specialized accelerators to meet strict latency and energy constraints that are prevalent in both edge and cloud deployments. These accelerators, which we call edge accelerators,

A. Raha (✉) · R. Sung · C. Brick
Intel Corporation, Santa Clara, CA, USA
e-mail: arnab.raha@intel.com; raymond.sung@intel.com; cormac.brick@intel.com

S. Ghosh · V. Raghunathan
Purdue University, West Lafayette, IN, USA
e-mail: ghosh37@purdue.edu; vr@purdue.edu

P. K. Gupta · U. I. Cheema
Intel Corporation, Hillsboro, OR, USA
e-mail: praveen.kumar.gupta@intel.com; umer.i.cheema@intel.com

D. A. Mathaikutty
Intel Corporation, Chandler, AZ, USA
e-mail: deepak.a.mathaikutty@intel.com

K. Hyland
Intel Corporation, Leixlip, Kildare, Ireland
e-mail: kevin.j.hyland@intel.com

© The Author(s), under exclusive license to Springer Nature Switzerland AG 2024
S. Pasricha, M. Shafique (eds.), *Embedded Machine Learning for Cyber-Physical, IoT, and Edge Computing*, https://doi.org/10.1007/978-3-031-19568-6_5

achieve high performance through parallelism in hundreds of processing elements and improve energy efficiency by minimizing data movement and maximizing resource utilization through data reuse [1, 2]. In this chapter, we will first provide a comprehensive summary of the problems that neural networks have been solving in the domains of Computer Vision, Natural Language Processing, Recommendation Systems, Networking, the Internet of Things (IoT), and Graph Processing (Fig. 1). The two phases of a neural network include training (or learning) and inference (or prediction). The number and type of layers are determined by the network architecture, whereas training determines the network weights. These weights are combined with new input activations to make predictions during inference. We will limit our focus to custom edge accelerators for inference and how individual layers of each of these different types of neural network can be accelerated in an energy-efficient way. In particular, we will focus on mapping neural networks on these architectures that attempt to minimize data movement by reusing input activations/weights during a particular compute round or across compute rounds. Work has to be distributed over multiple compute rounds when the layer cannot fit in its entirety onto the available processing elements. The bulk of our discussion will be on Convolutional Neural Networks (CNNs) due to their popularity on edge

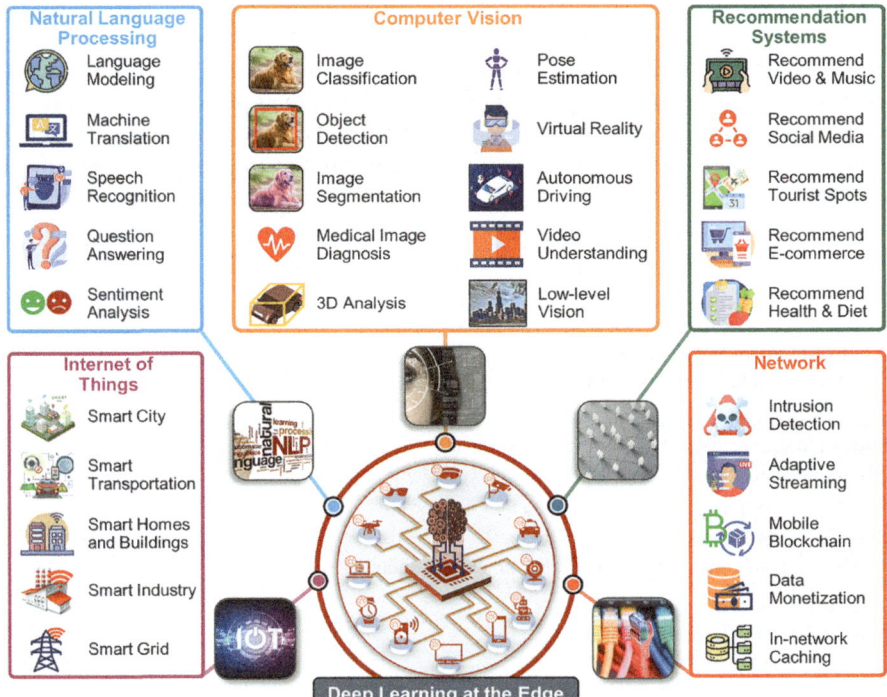

Fig. 1 Application domains of deep neural networks at the edge. *This image has been designed using some icons from Flaticon.com, Freepik.com, and Vecteezy.com*

accelerators. However, instead of discussing at length how to map the standard convolution operator itself, for which extensive prior research exists [3, 4], we will focus on the mapping of the special layers in a CNN. These layers can become the performance bottleneck after the standard convolutions have been accelerated, and there exists little insight on how these layers are accelerated in the literature. Furthermore, we will discuss the layers in newer lightweight CNNs that attempt to reduce the number of operations, such as in DepthWise Convolution, and how these might map to an accelerator originally optimized for standard convolutions. This is difficult because standard convolution lends itself to channel accumulation, while this is not the case for lightweight CNNs, resulting in reduced compute utilization. Newer neural networks such as Transformers and Graph Neural Networks present even more challenges for edge accelerators in the amount of data that must be processed and the shape and size of the input embeddings. The amount and types of computations in these emerging networks are non-optimal for edge accelerators, as they are currently being implemented, and may lead to newer design paradigms. It is not our intention to provide mappings for all the operators that one might encounter in different types of Deep Neural Networks (DNNs). In fact, exact mappings are specific to the resources found in a DNN accelerator. Instead, we seek to provide insights on the challenges that one might encounter during the design mapping process and the trade-offs to consider when designing an edge DNN accelerator for various types of layer. We will conclude by touching on future trends in neural network models and applications that can affect DNN accelerator design going forward.

2   Background

This section provides an overview of some of the key applications of DNNs in embedded machine learning and the various network models that are used in these applications. Figure 1 gives an overview of the DNN application domains in embedded machine learning. As demonstrated in the figure, the applications of DNNs are very diverse, but in this section we will cover some of the most popular ones in the domains of Computer Vision (CV), Natural Language Processing (NLP), Recommendation Systems (RSs), and Graph Neural Networks (GNNs). We also discuss the DNN models used in these domains and their progression over the years. Following the application domains and the corresponding models, we examine some common layer types used in DNNs today.

2.1   Computer Vision

Computer Vision is one of the most powerful and compelling types of Artificial Intelligence (AI) that enables computers and other execution platforms to analyze visual data from different sources, such as digital images, videos, and other visual systems, and extract valuable information, thereby providing recommendations or taking decisions based on the information. Unprecedented innovations in Deep Learning (DL) and DNNs in the past decade have enabled this field to progress leaps and bounds to such an extent that computers can now surpass human-level intelligence in multiple vision-related applications. Although the first CV research began in the 1950s, advances in DL powered by the exorbitant amount of visual data generated every day ($\sim$3 billion images shared online every day) coupled with significant improvements in hardware and computing resources (GPU, TPU [5], NPU [6]), have resulted in exponential growth in CV applications. Today, these applications in the real world are ubiquitous in health care, transportation, entertainment, business, and our daily lives. Given the multitude of these applications, it is important to understand the underlying DL architectures that power them. In the following sections, we discuss three of the most common types of DNN architecture used in CV, viz., CNNs, transformers, and multilayer perceptrons (MLPs).

2.1.1   Convolutional Neural Networks

Looking back at the 2010s, the renaissance of neural networks was mainly driven by CNNs. AlexNet [7] ushered in a new era in CV, and since then many popular CNNs have evolved with varying degrees of accuracy, efficiency, and scalability. In the past decade, the vision community has made tremendous efforts to improve the design of CNNs. Some representative networks that have revolutionized this field include VGGNet, InceptionV3, ResNet, MobileNet, EfficientNet, RegNet, U-Net, Faster RCNN, EfficientDet, and YOLOv(1-5) [8, 9]. Figure 2 shows the progression of the most popular CNNs since 2012 for different CV applications. These networks use the **convolution** layer as the core building block to extract features from images. Inherent group equivariance and spatial inductive bias in these models facilitate efficient learning of visual features. These features have made CNNs well adapted to various CV applications, such as image classification, object detection, and image segmentation, as indicated in Fig. 2. We can also infer that, compared to CNNs developed in the earlier part of the last decade, relatively newer models such as Mask RCNN, PANet, HRNet, EfficientDet, CondInst, and K-Net [9] are capable of handling multiple applications, as evident from the figure. In addition, this field has witnessed the continuous emergence of innovative applications powered by Generative Adversarial Networks (GANs) [10] that use CNNs as generator and discriminator models. Examples of some applications include image synthesis, image style transfer, image colorization, and image super-resolution, among others. In recent years, CNNs have also demonstrated their prowess in various video analytics applications such as video classification, continuous object tracking, and video prediction. In Table 1, we present a holistic view of traditional and emerging CV applications, together with the widely popular CNN architectures used for them.

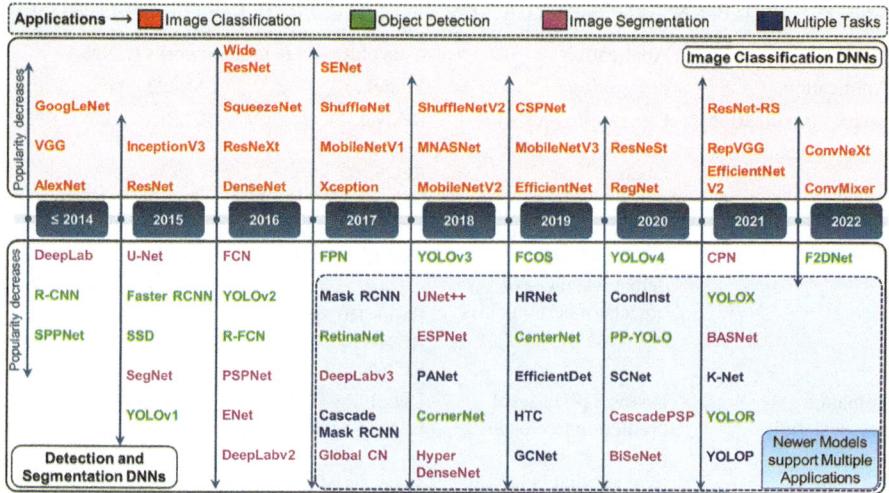

Fig. 2 Timeline of development of state-of-the-art convolutional neural networks. CNNs per year are arranged based on popularity and citation from publications

2.1.2 Emerging Deep Learning Architectures for Vision

Although CNNs have dominated as the mainstream DNN architecture in CV for many years, recently two types of architecture, namely Transformers and MLPs, have shown promising results similar to CNNs. This has ignited a spark in the research community to build better vision models by investing in these architectures.

Transformer-Based Vision Models As evident in Sect. 2.1.1, CNN models have been dominating the CV paradigm with tremendous success. Despite the overwhelming progress, CNNs lack a global understanding of the image itself and, therefore, are not able to model the dependencies between the extracted visual features. Any attempt to track long-range dependencies requires large receptive fields, which increases the model complexity many times. The attempt to overcome these limitations of inductive convolutional biases in CNNs led to the discovery of vision transformers. Before its grand debut in this domain, the first **Transformer** model based on a sequence-to-sequence architecture was proposed in [11] for machine translation. Since then, multiple pioneering breakthroughs using transformers have been made in NLP using state-of-the-art (SOTA) transformer-based models, such as BERT (Google), GPTv(1-3) (OpenAI), RoBERTa (Facebook) [9, 12]. These innovations sparked great interest in the CV community, leading to the design of transformers for visual and multi-modal learning tasks. Specifically, the advent of the Vision Transformer (ViT) [13] in 2021 initiated the design of multiple variants of transformer models for applications such as image classification, detection, and segmentation. Table 2 presents a holistic view of many emerging CV applications in addition to these, together with different variants of transformer architectures and

Table 1 Variants of CNNs for representative CV applications. Model details are available in [9]

Application	Application description	Convolutional neural networks (CNNs)	
		Model	Model type
Image classification	Classify images and assign class-specific labels	ResNet, InceptionV3, MobileNetV2, ResNext, EfficientNet, RegNet	Classification CNNs
Object detection	Combine classification and localization to detect instances of objects of certain class in image	SSD, RetinaNet, YOLOv5, FCOS	One-stage object detection models
		Faster RCNN, Mask RCNN, Cascade RCNN	Two-stage object detection models
Semantic segmentation	Perform pixel-level prediction to cluster parts of an image together that belong to same object class	DeepLabv3, FCN, U-Net++, HTC	Semantic segmentation models
		Mask RCNN, PaNet, HRNet, YOLOP	Instance segmentation models
Medical image diagnosis	Identify abnormalities in medical images and perform tissue-based detection and characterization	U-Net, SegNet, ESPNet, CPN	Medical image segmentation models
Image generation	Generate new images by learning unsupervised representations	DCGAN, StyleGan	Generative adversarial networks (GANs)
		SNGAN, SAGAN	
Low-level vision	Style transfer, super-resolution, denoising, colorization	Pix2Pix, CycleGAN, SRGAN, AgileGan	GANs
Video understanding	Video classification	Mask RCNN, DCGN	Video-based CNNs
	Object tracking	SORT, FairMOT, Re3	
	Video object segmentation	PreMVOS, AT-Net	
	Video prediction	PredNet, Vid2Vid	
Human pose estimation	Pose estimation	HRNet, VoxelPose	CNNs+GANs
	Face generation	DeblurGan, EDVR	
	Face deblurring	FaceShifter, Pose2Vid	
	Face swapping	DeepFaceLab	

the corresponding model categories/types. Examples of many popular transformer-based models are DETR, SwinT, DINO, T2T-ViT, TNT, and Twins [12]. In addition, interesting efforts to use the knowledge of CNNs led to a substantial improvement in ViT in terms of new designs, better accuracy, and training scalability. Consequently, transformers have emerged as a generic vision backbone and have demonstrated remarkable performance in a wide variety of new CV applications, as shown in Table 2. Deeper dives in these applications and models could be found in [12, 14].

Table 2 Variants of transformers for different CV applications. Model details are available in [12]

Application	Application description	Transformer-based vision architectures	
		Model	Model type
Image classification	Classify images and assign class-specific labels	ViT, DeiT	Uniform-scale vision transformer
		PVT, SwinT	Multi-scale vision transformer
Object detection	Combine classification and localization and detect instances of objects of certain class in image	DETR, Pix2Seq	CNN backbone + transformer detector
		YOLOS, PVT + DETR	Pure transformer
Semantic segmenta- tion	Perform pixel-level prediction to cluster parts of an image together that belong to the same object class	CMSA	CNN + cross-modal self-attention
		SETR, SegFormer	Pure transformer
Image generation	Generate new images by learning unsupervised representation	TransGAN DALL E	GAN-based transformer
Low-level vision	Super-resolution, denoising, colorization, image restoration	IPT, TTSR, ColTran	Transformer
Video under- standing	Video and language modeling	VideoBERT, VTN, MaskT	CNN backbone + transformer decoder
	Video action recognition Video instance segmentation	VisTR, VTN, MViT, TimeSFormer	ViT-based backone + transformer decorder
3D analysis	3D point cloud classification 3D segmentation	PT	Self-attention-based transformer
	3D pose reconstruction	PCT, METRO	Pure transformer
Multi-modal tasks	Visual question answering Visual commonsense reasoning	ViLBERT, PEMT, VLN, CLIP	Multi-stream transformer
	Cross-modal retrieval Image captioning	VisualBERT, UNITER, OSCAR, VLP	Single-stream transformer

Fundamentally, transformer models comprise the encoder and/or the decoder module, which again consist of multi-head attention layers and feedforward layers. Here, the attention layer only performs feature aggregation (Fig. 23), while the feedforward layer performs feature transformation (Fig. 24), that is, in contrast to simultaneous aggregation and transformation in CNNs. Despite exceptional performance, the high computational complexity and associated inference cost due to the enormous size and operations of these models hinder their widespread applicability to NLP and CV applications at the edge or mobile/end-user device. In fact, the $O(n^2)$ time and memory cost of self-attention operation is still a major challenge in applications such as detection, segmentation, super-resolution, etc. To address these challenges, researchers have come up with different approximation strategies in attention and lightweight transformer models, such as Hardware-aware-Transformer (MIT-IBM), MobileViT (Apple), and EdgeFormer (Microsoft) [9]. The optimization space in transformers is still untapped; therefore, emerging research will hopefully compress and accelerate SOTA transformer models and increase their applicability in edge computing devices. Some examples of how to map compute-intensive transformer layers on SOTA accelerators will be discussed in Sect. 5.1.

MLP-Based Vision Models Multilayer Perceptrons that originated in the 1980s have traditionally been used for simple regression, classification prediction, and, of late, machine translation and speech recognition. Although MLPs were considered insufficient since they mainly comprised fully connected layers and were inefficient due to inherent redundancy, MLP-Mixer (Google, 2021) [15] achieved competitive scores on image classification benchmarks, with ImageNet Top-1 accuracy of 87.8%. Interestingly, this model does not use the computation-hungry convolution layers and attention layers. On the contrary, the constituent layers are (1) per patch linear embedding, (2) mixer layers (channel-mixing MLP and token-mixing MLP), and (3) the classification head, which contributes to its simplicity. As a result, this model performs on-par with the SOTA vision transformer models and even outperforms them with respect to throughput and TPU training time. From the perspective of AI for resource-constrained embedded devices/IoT, these results are very promising and will definitely spark further research in MLP-based vision models for embedded CV. Furthermore, this breakthrough has led prominent research teams to propose other MLP-based models, viz., ResMLP, gMLP, RepMLPNet, and ConvMLP [9] in 2021.

2.2 Natural Language Processing

Natural Language Processing deals with the transformation of human language into a representation that computers can understand and manipulate. A wide range of neural network models, including Recurrent Neural Network (RNN), CNN, GAN, and Transformers, are used for NLP applications. Figure 3 shows a progression of the most popular neural networks since 2014 and shows the type of architectures to which they belong. As one can clearly comprehend, lately transformers have been the forerunner in this field. Here, we introduce some NLP subdomains where neural networks have found applications in embedded machine learning.

Language Modeling is an important application of NLP that deals with the prediction of the upcoming word sequence based on an earlier word sequence. Taking into account the importance of this application in various fields, numerous neural networks are proposed, as shown in Table 3. *Machine Translation* is another important application of NLP, which involves the computer-based automatic conversion of one natural language into another language while keeping the meaning of the input text intact. *Speech recognition* enables the conversion of natural language to text. Some examples of speech recognition at the edge are Amazon Alexa, Apple

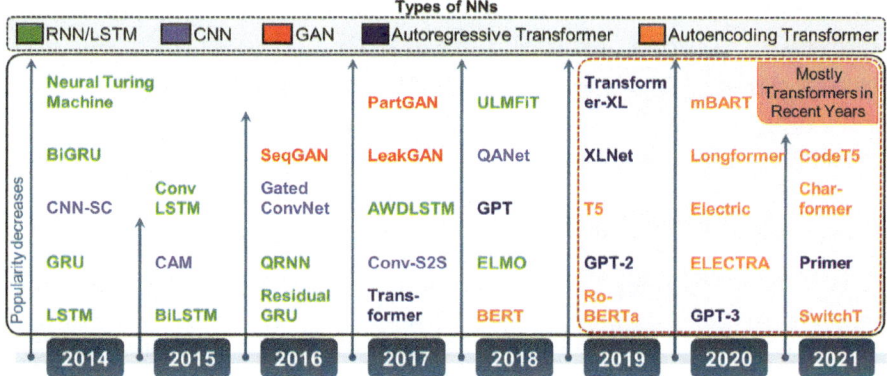

Fig. 3 Timeline of development of SOTA neural networks used for NLP with corresponding model type. Networks per year are ordered according to popularity and citation from publications

Siri, and Google Assistant. Although the processing of these assistants may be in the cloud, these devices utilize speech recognition to process the *wake* word at the edge terminal itself. *Text Generation*, generating text that is indistinguishable from human-written text, is another popular domain of NLP. Some of the most widely used text generator networks are based on autoencoders, GANs, and transformers. NLP-based *Question Answering (QA)* is about building systems that answer questions posed in natural language and abstaining if the question cannot be answered based on the given context. *Sentiment Analysis*, another important application in the age of social media, is a technique that is used to classify the polarity of a given language as positive, negative, or neutral. Table 3 shows the main types of NN models used for these applications and some of the latest high-performing models.

2.3   Deep Learning Based Recommendation Systems

The goal of recommendation systems (RSs) is to generate personalized recommendations for users based on collected user data. In other words, these systems predict the rating or ranking a user might give to a specific item and provide data to the user based on these predictions. Emerging innovations in this field have been continuously driven by improvements in Internet technology, smart edge devices, and e-Commerce. Traditionally, RSs have been based on techniques such as text mining, nearest neighbor, clustering, and matrix factorization [16]. However, in recent years, RSs have greatly benefited from the progress made in DNNs. Figure 4 highlights some of the SOTA and representative models from a wide range of neural networks such as MLP, Autoencoder, CNN, RNN, RBM, etc., and the most common data sources used by the models. As highlighted in the figure, GNNs

Table 3 Some major application of NLP and types and examples of models used [9]

Application	Common model types	Examples of some top models
Language modeling	Transformer	BERT, Megatron-LM, Transformer-XL, GPT3
	LSTM	LSTM (RMC), LSTM (Hebbian, Cache, MbPA), AWD-LSTM-MoS+ATOI, LSTM
	CNN	TaLK Convolutions, GCNN-8, TCN, Temporal CNN
Machine translation	Transformer	Temporal Cycle, Transformer+Rep, T5-11B, Transformer + R-Drop
	LSTM	GNMT+RL, MoE, RNN Enc-Dec Att, Deep-Att
	CNN	DynamicConv, TaLK Convolutions, ConvS2S, SliceNet
	Multiscale	MUSE
Speech recognition	Transformer	Conv+Transformer+Wav2vec2.0 + pseudo labeling, Wav2vec 2.0 with Libri-Light, Transformer+Time reduction+Self knowledge distillation
	Conformer	Conformer+Wav2vec 2.0 + Noisy training, SpeechStew, Conformer
	LSTM	ContextNet+SpecAugment-based Noisy training, LSTM transducer, tdnn+chain+rnnlm rescoring
	CNN	Multistream CNN with self-attentive SRU
Text generation	Autoencoders	Aggressive VAE, BART, CNN-VAE, SA-VAE
	GAN	LeakGAN, partGAN, RelGaN
	Transformer	GPT2, T5, UnitLM
Question answering	LSTM	SAN, FusionNet, BiDAF
	Transformer	LUKE, XLNet, SpanBERT
	DCN	DCN+,DCN+Char+CoVe
Sentiment analysis	Transformer	SMART-RoBERT, T5-3B, ALBERT
	LSTM	Block-sparse LSTM, bmLSTM, byte mLSTM7
	CNN	CNN Large, CNN, CNN+Logic rules
	MLP	gMLP-large

(covered in Sect. 2.4) have recently found increased usage in these systems due to the inherent graph-like structure of the input data [17]. Many of these are used by tech giants such as Amazon, Meta (Facebook), Google, Netflix, and Spotify to offer personalized AI-based ads to users. The success of DNNs in this field can be attributed to the inherent structure of RSs that these models could exploit [18]. In the context of user recommendation, this structure could correspond to a sequence of click logs or a sequence of specific words used in a sentence. The other property that makes DNNs a good fit for RSs is the composite nature that allows multiple neural

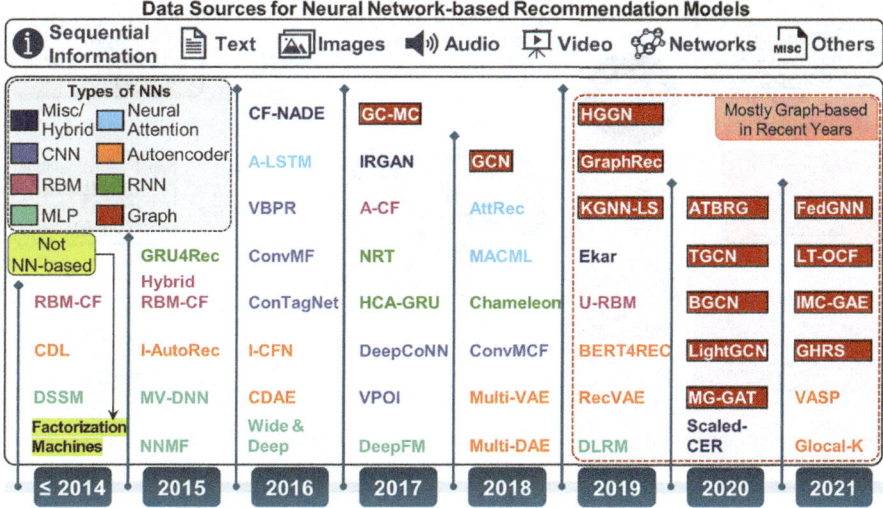

Fig. 4 Timeline of development of SOTA neural network-based Recommendation Systems along with common data sources used by these systems

building blocks to be composed into a large unit that could target the multi-modal nature of the data (data that include text, image, audio, video, among others).

Furthermore, compared to conventional recommendation models that are linear (e.g., factorization machines, sparse linear models, etc.), DNNs allow modeling of non-linearity in data using non-linear activation functions such as ReLU, sigmoid, tanh, etc., and with arbitrary precision by changing the activation combination and choices [18]. This property allows DNNs to model complicated patterns and behaviors with more precision. In addition, the inherent sequence modeling ability in modern DNNs allows them to mine the temporal context of user behavior and, hence, provide a better recommendation. Besides, the ability to learn new features automatically allows these networks to keep the recommendation up-to-date. These DNNs are also capable of processing heterogeneous content, such as video, audio, text, image, etc., allowing them to efficiently represent the underlying domain.

2.4 Graph Neural Networks

Graph Neural Networks are a set of connectivity-driven models that take advantage of the connectivity of structured graph data to learn and model relationships between graph nodes. Depending on the structure of the graph, these networks employ an iterative process to take input edges, vertex, and graph features (known attributes of the underlying application) and transform them into output features (e.g., target predictions). Figure 5 shows common GNN application domains, such as CV,

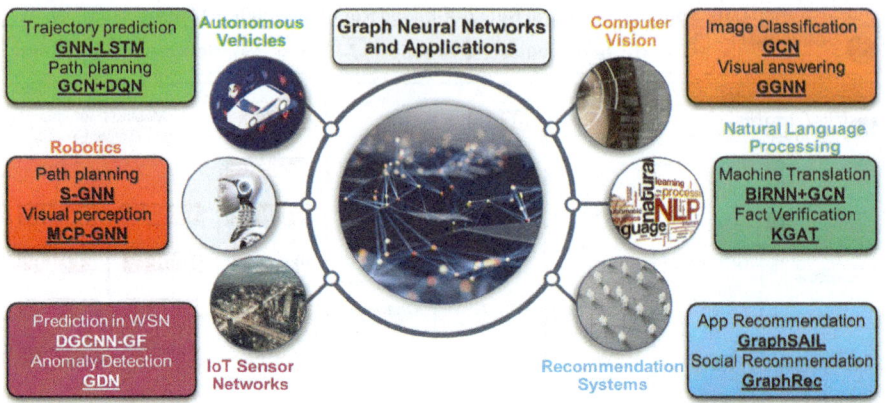

Fig. 5 Applications of Graph Neural Networks and corresponding GNN models. *Model names are underlined and in bold*

NLP, and RSs. In this figure, we have also denoted an example GNN model for each application in these domains. Furthermore, GNNs have been explored in the context of edge intelligence and IoT [19]. These networks can capture complex interactions within multi-modal sensory topology, enabling them to achieve SOTA results in application domains such as autonomous vehicles, IoT sensor networks, and robotics, as shown in Fig. 5. Other application areas include computer networks, science discovery (particle physics and chemistry), combinatorial optimization, and computer networks. We refer our readers to [20, 21] for more details on GNN models and applications.

Now that we have covered some key DL-based applications and the corresponding DNN models, we will briefly discuss some of the common layer types used in DNNs in Sect. 3. This description will be followed by a section on the efficient implementation of these layers in Sect. 4.

3 Common Layers Across Neural Networks

An overview of the common deep learning layers along with their operator-level description is given in Tables 4 and 5. For better readability, we have classified the layers into different categories, such as convolution layer, pooling layer, activation layer, normalization layer, combination layer, input/output layer, and fully connected layer. Other layers typically found in newer DNNs, such as Transformers and GNNs, have been categorized into miscellaneous layers. Note that we only list several representative layers rather than providing an exhaustive list. The following sections dive deep into the details of the most computationally complex layers among these and describe how they are mapped to a SOTA accelerator for inference on resource-constrained edge platforms.

Table 4 Common types of layers used in deep neural networks (1/2)

Layer	Subtypes	Layer description	Op level description
Convolution layers	n-D simple conv	Apply sliding convolution filters to n-D input	Standard convolution
	n-D grouped conv	Apply sliding convolutional filters on group of input channels	Group convolution
	1 × 1 conv	Used for dimensionality reduction, efficient low dimensional embeddings	1 × 1 convolution
	n-D dilated/atrous conv	Expands the input by inserting holes between consecutive kernel elements	Dilated convolution
	n-D transposed conv/deconvolution	Up-samples n-dimensional input feature maps	Standard convolution
	n-D depthwise conv	Apply a single convolutional filter for each input channel	Depthwise convolution
	Pointwise conv	Apply a 1 × 1 kernel to iterate through every single point of the input	Standard convolution
	n-D depthwise separable conv	Apply depthwise convolution and then pointwise convolution	Depthwise convolution
Pooling layers	n-D average pool	Average pooling operation for temporal/spatial/3D data	Pooling—combination/decode
	n-D max pool	Max pooling operation for temporal/spatial/3D data	Pooling
	n-D global avg pool	Global average pooling operation for temporal/spatial/3D data	Pooling
	n-D global max pool	Global max pooling operation for temporal/spatial/3D data	Pooling
Activation layers	ELU	Exponential linear unit	Activation functions
	Leaky ReLU	Leaky version of a rectified linear unit	Activation functions
	PReLU	Parametric rectified linear unit	Activation funtions
	ReLU	Rectified linear unit activation function	Activation functions
	Thresholded ReLU	Thresholded/clipped rectified linear unit	Activation functions
	Swish	Swish activation function	Activation functions
	Hyperbolic tangent	tanh activation function	Activation functions

(continued)

Table 4 (continued)

Layer	Subtypes	Layer description	Op level description
Normalization layers	Normalization	Preprocessing layer which normalizes continuous features	Normalization
	Batch normalization	Normalize mini-batch of data throughout all observations of each channel independently	Batch norm—combination
	Layer normalization	Normalize mini-batch of data for each observation independently throughout all channels	Elementwise other

Table 5 Common types of layers used in deep neural networks (2/2)

Layer	Subtypes	Layer description	Op level description
Combination layers	Add	Adds a list of inputs (from multiple layers) elementwise	Elementwise add-aggregation
	Average	Averages list of inputs elementwise	Math other—aggregation
	Multiply	Multiplies list of inputs elementwise	Elementwise
	Subtract	Subtracts two inputs elementwise	Elementwise
	Concatenate	Concatenates a list of inputs along a specified dimension	Other memory operation
Miscellaneous layers	Maximum	Computes the maximum (elementwise) from a list of inputs	Elementwise
	Minimum	Computes the minimum (elementwise) from a list of inputs	Elementwise
	Reshape	Reshapes inputs into the given shape	Memory other
	n-D ZeroPadding	Zero-pads layer for n-D data in specified dimension	Padding
	Text vectorization	Preprocessing layer that maps text features to integer sequences	Embedding

(continued)

Table 5 (continued)

Layer	Subtypes	Layer description	Op level description
	Embedding	Turns positive integers (indexes) into dense vectors of fixed size	Other memory operation
	Masking	Masks a sequence by using a mask value to skip timesteps	Elementwise
	Attention/self-attention	Dot-product attention layer, aka Luong-style attention	Matrix-matrix
	Multi head attention	Module for attention output from multiple self-attention layers	Matrix-matrix
Fully connected	Dense	Multiply the input by a weight matrix and then adds a bias vector	Vector to matrix/matrix— matrix
Input/output layers	Input layer	Used as an entry point into a network (or a graph of layers)	First layer
	Softmax layer	Applies softmax function to the input	Activation

4 Efficient Implementation of Emerging NN Operators

Since CNNs make up the majority of edge AI models, most embedded accelerators are primarily designed to process convolution layers. Convolution layers consist of seven nested loops where an output tensor, OFMAP, is produced from multiple kernel feature maps (FMAPs), on one or more input tensors, IFMAP, as shown in Fig. 6. The calculation of each point in the output volume is a multiply-and-accumulate (MAC) operation. As a result, DNN accelerators consist of one or more arrays of MAC units in their computation core. An example of a 1×1 convolution layer is the second convolution layer in ResNet50 where the IFMAP is represented by the dimensions $IX = 56$, $IY = 56$, $IC = 64$, and the filters are represented by the dimensions $FX = 1$, $FY = 1$, $IC = 64$, $OC = 256$. These are convolved together (with a batch size of $ON = 1$) to generate an OFMAP of dimensions $OX = 56$, $OY = 56$, $OC = 256$ with appropriate padding values.

To understand the basic principles of DNN acceleration, we first provide an IP level overview of a typical DNN accelerator, as shown in Fig. 7. There exist multiple types of core within this system: (i) a main scalar processor core that coordinates data movement between system memory (DRAM) and associated coprocessors, as well as issuing the required instructions, (ii) an associated vector DSP processor (VDP), and (iii) a neural network (NNP)/DNN accelerator [5]. Custom decoded instructions are communicated from the scalar processor to the NNP/VDP via a Network-On-Chip (NOC), while separate NOCs communicate the input feature

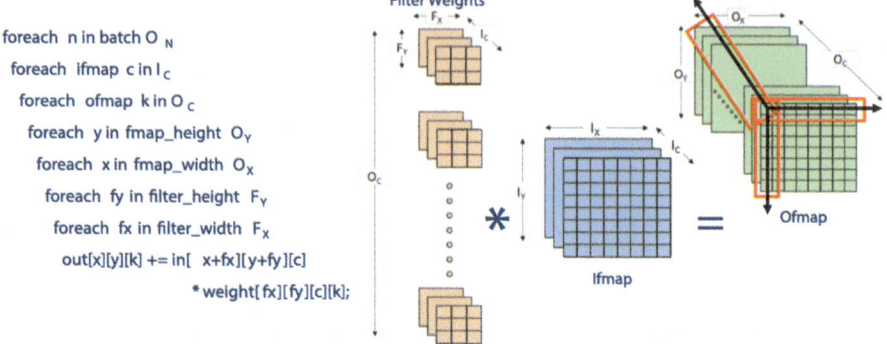

7-dimensional DNN tensor processing in convolution

Fig. 6 Convolution operation in DNN

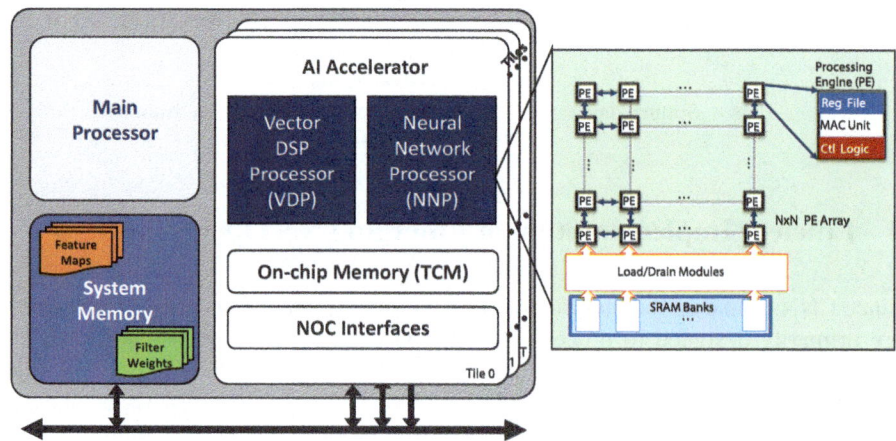

Fig. 7 High-level schematic of a typical DNN accelerator-based IP

maps, filter maps, and output feature maps from a tightly coupled on-chip SRAM memory that can be accessed by both the VDP and NNP.

The VDP is required to map DNN operators that either cannot be mapped efficiently to the NNP or in the worst case cannot be mapped at all to the DNN accelerator. Examples of such operators include various non-linear activation functions such as HardSwish, GeLU, HardTanh, etc. Although, considering the frequency of these operations, some of the recent DNN accelerators implement these non-linear activation functions using specialized programmable look-up tables [22]. It is also used to perform a host of pre-processing and post-processing steps on the input and output data before they can be executed on the NNP. Usually, the VDP has significantly lower throughput compared to the NNP because of a limited number of arithmetic units, a lower frequency of operation, and no local RF reuse compared

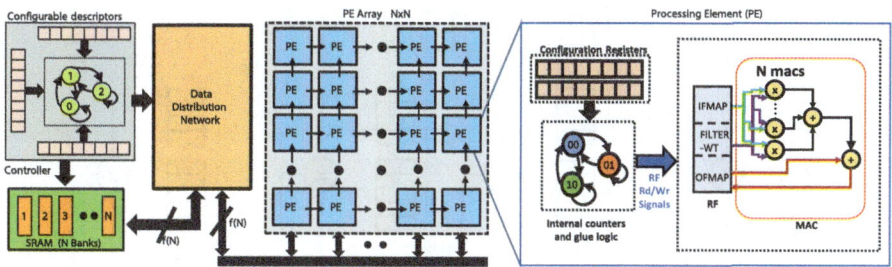

Fig. 8 Microarchitecture details of a generic DNN accelerator

to the NNP. Therefore, based on the exact arithmetic operator required, we always try to map any layer first on the NNP, and if that fails, the VDP is used.

In this chapter, we will focus entirely on the DNN accelerator (NNP), since this is the processing core that consumes the most area and power and is responsible for the bulk of the computations. Figure 8 shows a high-level diagram of a DNN accelerator. The main components of the NNP are the Processing Element Array (PEA) that is constructed using an NxM rectangular grid of Processing Elements (PE), a local SRAM memory to store and load activations and weights for each DNN layer, a tensor distribution network that consists of load and drain datapaths to and from the PE array, and finally the control logic that orchestrates the loading, computation, partial sum accumulation, and extraction of the output points to and from the PEA.

To understand how convolution layers are efficiently mapped to a DNN accelerator, it is useful to make some assumptions about the details of the underlying microarchitecture. Without any loss in generality, the processing element array (NxM) is assumed to be a square grid of NxN processing elements to simplify control logic both inside and outside the PE array. In DNNs, MAC operations are used to compute the dot product of many weights and hidden-layer activations to produce the output feature maps for the next layer. Each PE is capable of performing MAC operations using local datapaths consisting of register files, multipliers, and accumulators. Local register files (RF) are discrete or shared storage that contains input activations (IF), filter weights (FL), and output points (OF). There can be multiple MACs within a PE based on the performance requirements of the accelerator and limited by area and power constraints. The PEs have storage for multiple IF and FL operands mostly in the input channel (or IC) dimension, along which it can perform MAC operations over multiple consecutive clock cycles. All PEs work in parallel by sourcing IF and FL operands from their local RFs, as shown in Fig. 9 and high performance is achieved through parallelism over hundreds of processing elements. There exists a "Load Path" to retrieve the weights and activations from SRAM and distribute them to the register files within each PE. This type of architecture is efficient for data movement since it can take advantage of memory reuse, especially for convolutional neural networks, where a small kernel is multiplied by a large input matrix. Data movement has been shown to

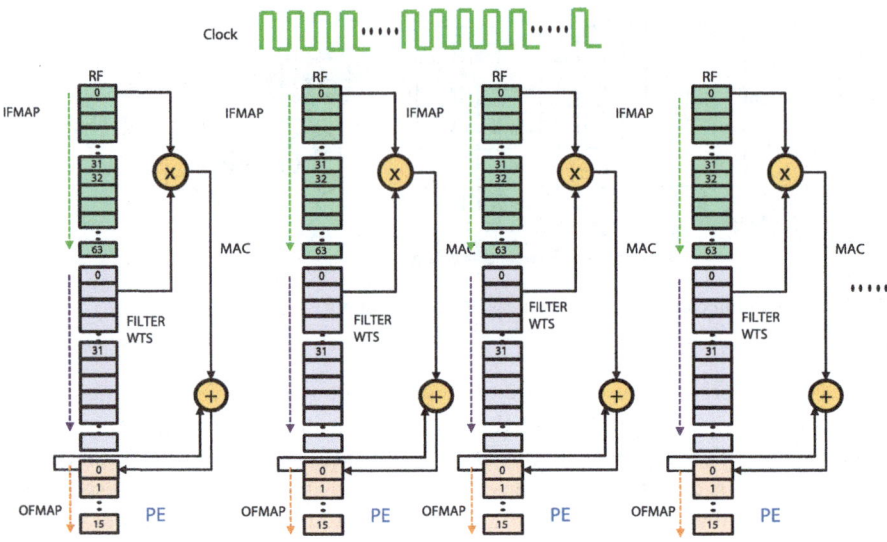

Fig. 9 Convolution operation in PE array

be a significant contributor to DNN accelerator energy cost. The Load is generally implemented as an NOC which allows it to broadcast, multicast, or unicast the input data to different processing elements with the goal of reusing as many of the inputs as possible depending on the neural network layer. The "Drain Path," also implemented as an NOC, is used to retrieve the output feature maps for each layer, running them through post-processing operations such as biasing and rounding, and eventually compressing the results before writing them back to memory. A popular example of such a spatial architecture in academia is Eyeriss [23, 24] and recent designs in the industry include Samsung and MediaTek NPUs [6, 25]. Another popular spatial architecture for performing General Matrix-Matrix Multiplication (GEMM) is systolic arrays. These provide the benefits of low area footprint and high frequency of operation, with a famous example being Google TPU [26]. However, systolic arrays are also known to suffer from low utilization issues due to limited programmability, inefficient mapping of odd dimensions [27] and are generally more power hungry due to limited reuse potential. Our high-level Fig. 8 diagram of a DNN accelerator can be easily extended to a systolic array with the modification that PEs within a systolic array usually lack local RF memories (or may contain just a staging buffer for IF and FL) and a single MAC unit that is responsible for the creation and forwarding of partial sums to adjacent PEs.

In some spatial microarchitectures, it may not be possible to map all the Input Channels (ICs) corresponding to an OC to the same PE. This may be the case for deeper layers within a DNN where the ICs are much larger compared to IX, IY, FX, or FY. In these cases, the partial sums (psums) generated across multiple PEs must be accumulated to generate the final OF point, as shown in Fig. 10.

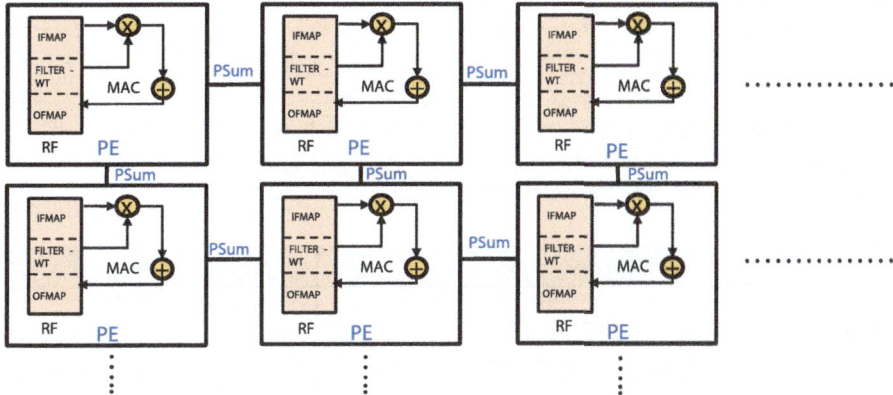

Fig. 10 Partial sum movement in PE array

We term this as internal psum accumulation. Due to memory limitations, most DNN architectures only allow accumulation over a maximum number of ICs while bringing the subsequent sets of ICs later. This requires the ability to spill and fill intermediate psums to and from the DNN accelerator. We term this as external psum accumulation. Note that most SOTA DNN accelerators support FP16/BF16 as well as INT8 precision arithmetic. Some accelerators even support lower precision and mixed precision, such as INT4, ternary, and binary, to increase overall throughput. However, we will concentrate only on the INT8 precision for DNN inference acceleration.

Apart from creating an efficient DNN accelerator, a well-defined software or compiler framework must exist to program the configuration registers, mapping different types of DNN operators, and preparing the input data and filter weights in the correct format. However, we do not go into the details of the compiler framework that is required for various hardware-software co-design mapping and optimization techniques, as described in this chapter.

A popular and efficient technique to improve the performance and reduce the energy consumption of DNN accelerators is to exploit the sparsity that is present in abundance in DNN networks [24, 28, 29]. Sparsity refers to the existence of zeros in the weights and activations in DNNs. Zero-valued activations occur from the processing of the IFMAP tensors through activation functions such as ReLU, which clamp negative values to zero. Zero-valued weights arise from the structured/unstructured pruning of weights and the quantization of weights from higher precision floating point numbers to narrow fixed point integers (converting FP32/FP16 to INT8/4/2/1, etc.). These zero-valued activations and weights do not contribute towards the result during multiply-and-accumulate operations, and hence, can be skipped during both computation and storage. Toward this end, machine learning accelerators can exploit the available sparsity to achieve significant speedup during compute, which leads to power savings because the same work is accomplished

Fig. 11 Zero-value compression (ZVC) of data

Load Round #	Sparsity Bitmap (binary)	Uncompressed Data reqd. in this round	Compressed Data Segment loaded every round
R0	1000 0010 0001 0100	0d000000 00000c00 0000000b 000a0000	1a0908070605040302010f0e0d0c0b0a Unused data Consumed data
R1	1010 1000 0100 1100	04000300 02000000 00010000 0f0e0000	2d2c2b2a1a0908070605040302010f0e
R2	0100 0000 0111 0000	00080000 00000000 00070605 00000000	242322212f2e2d2c2b2a1a0908070605
R3	0000 0000 0001 0001	00000000 00000000 0000001a 00000009	28272625242322212f2e2d2c2b2a1a09

in fewer cycles, as well as reducing the storage and bandwidth requirements for the weights and activations via efficient compression schemes. Reducing the total amount of data transferred through the memory hierarchy and decreasing the total computation time are critical to improving the energy efficiency of the NNP [24, 29]. Sparsity acceleration during computation is often bounded by the rate at which load data can keep the compute units busy. Previous works have addressed the load bottleneck by proposing techniques such as zero-value compression (ZVC) [30] to encode sparse weights and activations so that the loading of sparse data does not stall the compute. Figure 11 illustrates the ZVC scheme while Fig. 12 demonstrates how ZVC is leveraged during the loading of compressed data. Finally, Fig. 13 shows how sparse data are decompressed and the non-intersecting zeros between the weights and activations are extracted as input to the MAC within each PE to accelerate computation. Note that the OFMAPs will need to be compressed by the drain path, using the same ZVC, before they are written to the SRAM. Sparsity adds extra complexity to data loading, computation, and draining, and designers need to ensure that it can be enabled without adding too much overhead.

Before beginning our discussion of mapping different layers of the neural network, let us consider some characteristics of those layers that influence the utilization of the accelerator. If a layer is small, then only a subset of the compute units will be utilized, with no opportunity for further parallelism, since layers in a DNN need to be processed layer by layer. If the layer has large X, Y, and IC dimensions, then the layer must be split across multiple dimensions into smaller chunks of work both spatially across processing elements in the same accelerator (or even across multiple instances of the same accelerator) and over time. Irrespective of the layer dimensions, the goal is to maximize reuse by sharing weights/activations across processing elements in a single compute round and across compute rounds without refetching the same data from the memory hierarchy. The IC dimension

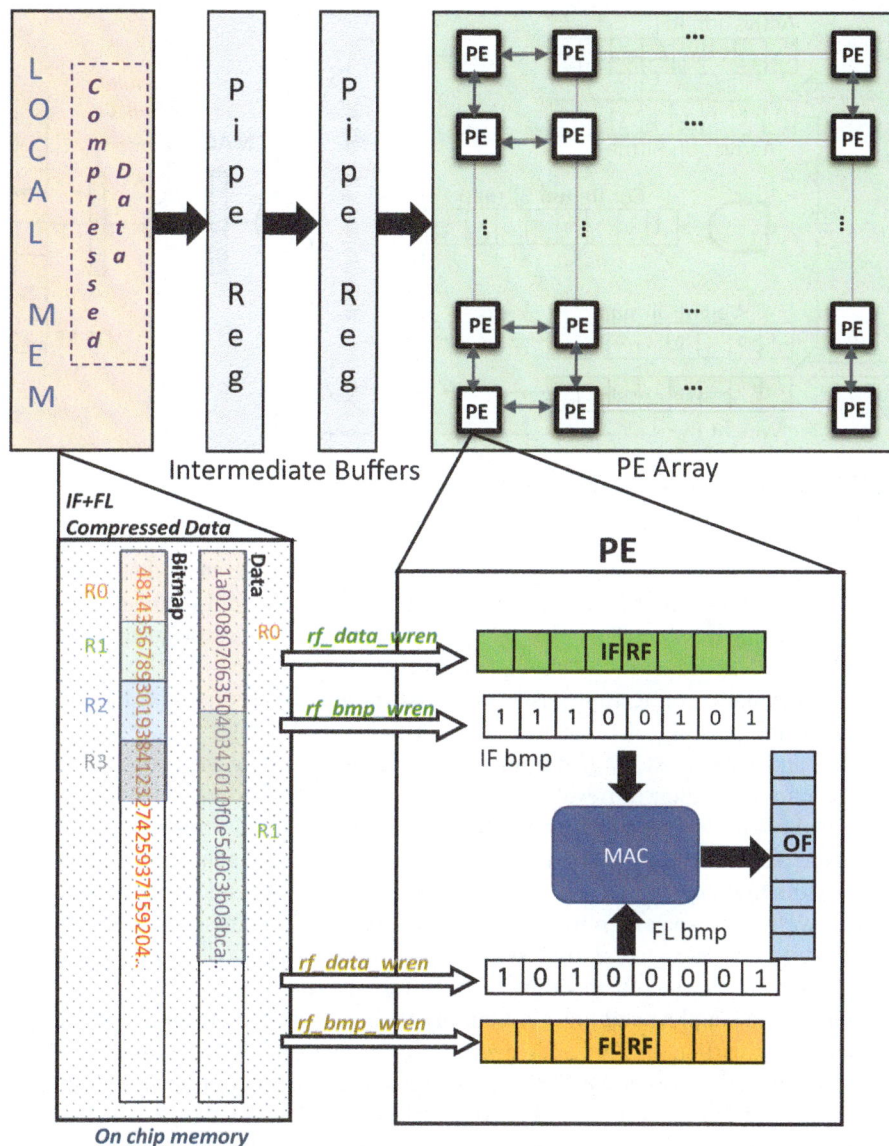

Fig. 12 Loading of compressed data from on-chip memory within each PE

requires special consideration since it determines various design trade-offs such as the size in bytes of the data that are distributed from the tensor distribution unit to and from the array, the amount and arrangement of the internal storage in each PE (# of ICs stored per X, Y) and the amount of sparsity decoding done during the compute round.

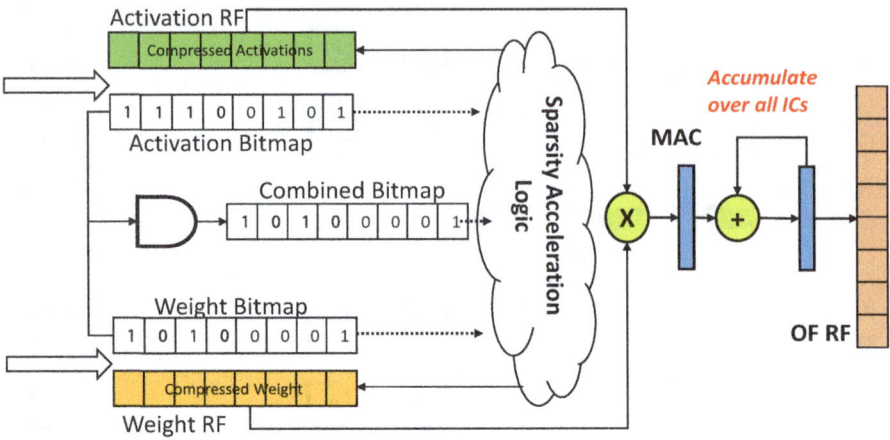

Fig. 13 Sparsity acceleration within each PE

It is reasonable to assume that many SOTA edge accelerators distribute ICs in multiples of 16, since most layers in modern DNNs contain channels that are divisible by this number. This also simplifies the storage and retrieval of ICs from memory, since each line in SRAM can also be a multiple of 16.

However, the complication arises when the number of ICs is not a multiple of 16, where the packing of these ICs across different X,Y points is not uniform in memory and requires storing additional information regarding the start and end of each X,Y to allow their retrieval during processing. Furthermore, it also introduces additional complexities in the exploitation of sparsity, where the amount of sparsity in a given X,Y is determined on the basis of the number of ICs that are zeros. Most accelerators that exploit sparsity need to manage the storage and retrieval of sparse data, as well as the sparsity bitmap from the memory and their distribution to the PE array. While there are several lossless compression techniques to leverage sparsity in storage, such as run-length encoding and compressed sparse row/column, we found that one of the simplest and most commonly used compression techniques is ZVC, as mentioned earlier. In ZVC, for every byte of data, one bit of sparsity bitmap is stored. Therefore, a typical SOTA accelerator needs to support sparsity bitmaps varying from 2B (16B of ICs) to the maximum size of ICs divided by 8. This implies the storage and retrieval of the sparse data and sparsity bitmap needs to be carefully considered to avoid over-designing the distribution network. A common solution is to always pad your ICs to be a multiple of 16, but this can have an adverse effect on the bandwidth of the distribution network and will result in a decrease in the overall compute efficiency for these padded layers.

Another aspect of ICs, specifically for layers that require accumulating over ICs, is that most SOTA accelerators implement an adder tree in some form or shape that allows them to generate output points by accumulating over ICs. To accomplish this, the accelerators have a structural layout of their PEs to allow the output of multiple

PEs to be accumulated through the adder tree. This design choice limits the mapping of layers that do not require IC accumulation (Separable Channels) and results in an inefficient mapping of these layers on the accelerator. In the following section, we will discuss how to overcome these design choices to perform efficient mapping of the different layer types.

Hyperparameters, such as padding, striding, and dilation, and their configuration for the kernel size of interest affect the shape of the output feature map. These hyperparameters help solve problems related to maintaining the original size or down-sampling or up-sampling the resolution of the output image, which provides the accelerator with a means to adjust the dimensionality of the data effectively. In order for accelerators to support these hyperparameters, the various stages of operation, such as load, compute, and output of an accelerator, need to incorporate them. We will cover the design implications of these hyperparameters for the individual layers in the following section.

4.1 Efficient Mapping and Acceleration of Special DNN Layers

Profiling a DNN shows that the total number of operations (and execution time) spent on the convolution layers of DNNs is significantly higher than all other layers combined. Therefore, a DNN accelerator is constructed primarily to accelerate the convolution layers of a DNN that consists of a series of nested loops of MAC operations, as demonstrated in Fig. 6. There exists a plethora of recent research and studies on best practices and design considerations for standard convolution layers on DNNs [1, 2] that have recently led to the design of several efficient DNN accelerators [23, 26, 31] recently. However, most of these accelerators (if not all) are inefficient for accelerating non-convolution layers such as the eltwise layer or even convolution layers with unique dimensions and characteristics such as the first layer of DNNs. We term these layers as special DNN layers. Once we have accelerated the standard convolution layers, these special layers become the performance (or energy) bottleneck for the entire DNN. Therefore, improving their efficiency can lead to significant improvement in network level performance. Here, we list some of these special layers and how they can be efficiently implemented on a DNN accelerator IP.

4.1.1 First Layer

The first layer is the visible/input layer of the network with three channels, where the channels correspond to the red, green, and blue components of the input image. The first layer is typically mapped to the NNP due to the large number of operations involved. Consider the first layer of the ResNet50 network; it has an activation layer (X, Y, IC) of dimension 224 × 224 × 3 and a weight layer (FX, FY, IC, OC) of dimension 7 × 7 × 3 × 64 and applies a stride parameter of 2, which results in the output layer of dimension $112 \times 112 \times 64$. The number of operations required to compute this layer is 118,013,952. Due to the large number of operations, if this layer is not efficiently mapped onto the accelerator, it will have an adverse effect on the overall network-level performance.

Design Consideration The two main characteristics of the first layer that affect its mapping on the NNP are the limited number of ICs and the channel-major layout of the data in memory due to the sources from which the data are captured. For NNPs that are typically built for IC accumulation, a smaller number of ICs will result in a drop in the number of MACs that are engaged for the first-layer processing. This aspect can be improved if the NNP has some flexibility in prioritizing either the larger X or Y dimensions of the activation or the higher FX or FY dimension of the filter over the ICs. The channel-major layout brings in additional complexity during the distribution of the input data, where there needs to be an agent that is aware of the first-layer layout in memory to perform an efficient load on the PE array. This load costs an additional area, and then the data needs to be repackaged in a way that it can be fed to a typical NNP. Performing these additional optimizations on the NNP to accelerate dimensions other than the ICs together with a specialized load can improve the efficient mapping of the first layer. The first layer is typically processed as dense due to the inability to introduce sparsity from the small number of ICs and therefore the load part of the distribution network needs to be provisioned for sufficient bandwidth so that it does not become the bottleneck for the network. Software / compiler can be used to transpose the tensor data with padded ICs to simplify the load considerations.

4.1.2   Eltwise Layer

Elementwise (eltwise) operations are deployed in various popular DNNs such as Residual Networks (ResNets), as well as Transformers and LSTMs. For example, in ResNets, the addition operation is the underlying eltwise operation that occurs between two tensors, which are output activations from two convolution layers, one earlier and another later in sequence. The elementwise summation between the two input tensors creates an output tensor with the same dimensions as that of the two input tensors. A second type of elementwise operation involves multiplying two tensors elementwise and this operator is mostly required for LSTM and transformer-based NNs.

Design Consideration The default option to map these eltwise operations is the VDP (Fig. 7), but this will result in significantly lower throughput compared to the VDP. On the other hand, the presence of MAC operators within the DNN PE array provides us with an inherent advantage of mapping these eltwise layers due to the existence of both addition and multiplication operations. However, we still need to ensure that we can bypass the multiplier and adder operators of the MAC during the add-eltwise and mult-eltwise, respectively, using multiplexers within the PE. For existing DNN accelerators that have dedicated data load paths for activations

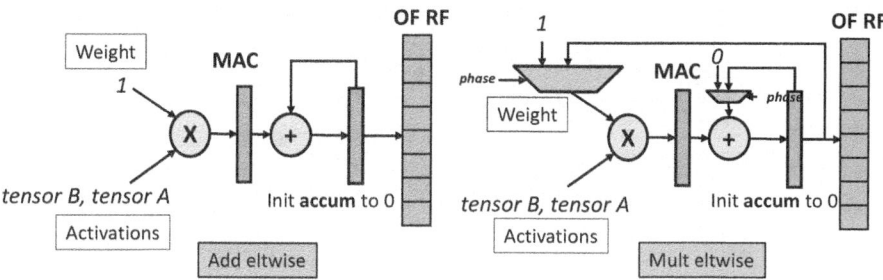

Fig. 14 Eltwise operation within a PE

and weights, an efficient way to perform eltwise is by loading the two tensors sequentially one after another within a PE using the activation load path. As part of this scheme, the weights are set to 1 so that we can bypass the multiplier and load the first tensor as it is in the accumulator. When the second tensor is loaded, we need to ensure that it uses either the MAC adder or the multiplier along with the loaded value to generate the final value. This can enable eltwise operation within an existing DNN accelerator with minimum overhead, as depicted in Fig. 14.

4.1.3 Fully Connected Layers

From an implementation point of view, fully connected (FC) layers can be assumed to be a special type of DNN convolution layer in which the IX, IY, FX, FY dimensions can be assumed to be set to 1. These densely connected linear layers usually form the last layer of a DNN and connect every input neuron to every output neuron. The output points are created by multiplying the input vector by a set of 2D weights. Due to the lack of IX, IY, FX, and FY dimensions, one cannot exploit any kind of data reuse for fully connected layers, resulting in bandwidth-limited execution of these layers in a DNN accelerator. Due to the streaming nature of input activations in FC layers, these layers can frequently be load-bound (from the weight side) or drain-bound (for output points). A secondary issue for fully connected layers can be the underutilization of available total MACs, which can be attributed to a fixed way of distributing the overall convolution work on the PE array. For example, some DNN accelerators split the total convolution work along the OX or OY dimensions, while the weights are multicast to these PEs. There will be considerable underutilization of MACs for such types of DNN accelerators due to non-existence of IX and IY dimensions for FC layers. A similar issue will arise if the fixed reuse pattern occurs along the FX, FY dimensions.

Design Consideration One way to improve the efficiency of FC layers is by splitting the convolution work along the IC and OC dimensions while allocating higher weight load and activation drain bandwidth. However, IC partitioning

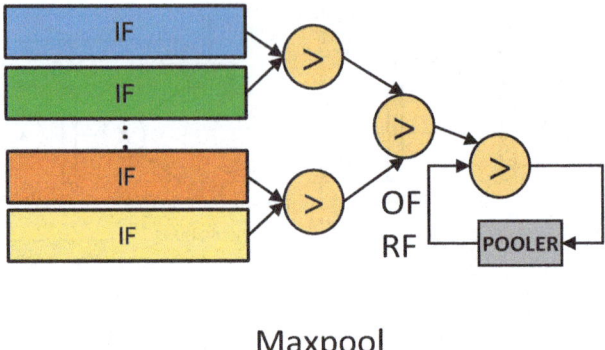

Maxpool

Fig. 15 Maxpool operation within a PE

requires additional logic like the presence of an adder tree to sum partial sums across the PEs, while the bandwidth increase also results in a significant area overhead. Note that the sparsity usually makes this bandwidth problem even worse. Since the FC layers usually occur at the end of a DNN, one alternative will be to execute them on the VDP which will free up the NNP for processing the next set of inputs. In general, VDP has fewer PEs that but that will usually be enough for allocated memory bandwidth.

4.1.4 Maxpool/Average Pool

The pooling operation reduces an NxN spatial window to generate a single point. A pool layer (maxpool or average pool) slides an $N \times N$ spatial pool window along the IX and IY dimensions of the input tensor to create a smaller output window (by a factor of N^2) per channel. This operation occurs only on the activations and does not involve any weights.

Design Consideration For implementation purposes, the pooling layer can be assumed to be almost similar to a depthwise layer, with the only difference being the reduction operation. For maxpool and average pool, the reduction operations are implemented using comparators and adders instead of a MAC based reduction for depthwise layers. We explain the inefficiencies of the implementation of a depthwise layer in Sect. 4.2.1 and also propose various architectural optimizations to improve its efficiency. All these optimizations are also valid for the pool layers. From a hardware perspective, the only additional component required to perform the maxpool operation is a comparator operator, as depicted in Fig. 15.

4.1.5 Activation Functions

Activation functions are used in DNNs to evaluate the output of a node. Depending on the choice of the activation function, the output value usually falls within a fixed range (e.g., [0, 1] or [−1, 1]). Hyperbolic Tangent (Tanh), Sigmoid, ReLU, and Swish are some of the activation functions that are popularly used. To implement an activation function in the hardware, various approximation methods are used, such as direct computation [32], piecewise linear approximation (PWL) [33], or look-up table (LUT) [34] or a combination of all these methods [35]. Usually, the activation function for a particular DNN is fixed. However, it may change from one DNN to another.

Design Consideration In order to support multiple activation functions in the hardware, we need to provision area and logic for each one of them in the hardware. We can potentially enable this in the VDP as it has a limited number of processing elements. Consequently, this requires the extraction of the output of the convolution to local shared memory, which incurs significant memory overhead. One way to eliminate this overhead is by fusing the activation layer with the convolution layer, where we can apply the activation function immediately after the convolution takes place before it is drained out to the shared memory. This can be implemented easily for hardware-amenable (linear) activation functions such as ReLU with minimal area overhead. The ReLU function can be integrated within the post-processing unit (PPE) of the accelerator before it is drained out. However, integrating non-linear activation functions that require a combination of LUTs and interpolation functions come with a significant area overhead. A possible efficient implementation will be to select a limited number of the most popular activation functions and integrate them within the PPE. The remaining ones can be mapped to the VDP.

4.2 Efficient Mapping and Acceleration of Layers in New Neural Networks

Similarly to the special layers of existing networks, current SOTA DNNs constitute multiple new types of layer with varying characteristics that may not be efficiently accelerated on a DNN accelerator without any optimization. The following subsections describe some of these layers and potential techniques for efficiently implementing them on existing DNN accelerators.

4.2.1 Channel Separable Depthwise Convolution Layers

In the case of standard convolution, a three-dimensional volume, representing a kernel or a filter, is slid across the input activation tensor. Subsequently, the activation and filter volumes (or tensors) are convolved to combine all the input

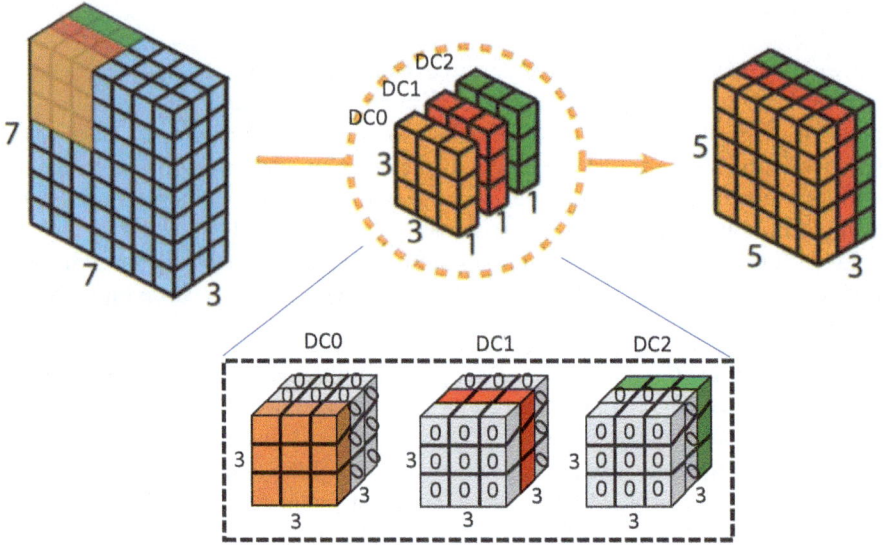

Fig. 16 Depthwise operation

channels to produce the output feature map. Multiple filters produce multiple output channels in the output feature map. In Depthwise Convolution (DWC), a three-dimensional volume representing a kernel or filter is still slid across the input tensor (Fig. 16). However, individual input channels are not combined, with multiplying and accumulations occurring only within a channel in the FX-FY directions, as shown in Fig. 17b (depthwise mode). This constraint substantially reduces the MAC utilization in almost all accelerator designs, since they handle DWC the same as standard convolution on the same hardware (Fig. 17a). The channel-combining step is done instead in the next layer using a 1×1 point-wise convolution to produce the output feature map. As before, multiple filters produce multiple output channels in the output feature map.

Design Consideration One obvious way to improve utilization is to use dedicated hardware for DWC but that has a prohibitively large silicon area and leakage power. Such a use of silicon real estate, extra leakage power, and the associated difficulties of distributing data to multiple processing arrays precludes such a solution for many Edge AI accelerators. Comparatively, a more desirable alternative would be to execute DWC on hardware optimized for standard convolution, but then it will only use a small fraction of the available MACs during DWC, and hence performance will suffer. All MACs that would normally operate on ICs in each cycle during standard convolution but cannot be used during DWC are marked with "0" in Figs. 16 and 17a. The percentage of underutilized MACs will continue to rise as the process nodes advance from one generation to the next due to memories and wires not scaling nearly as well as logic. It makes sense for future designs to

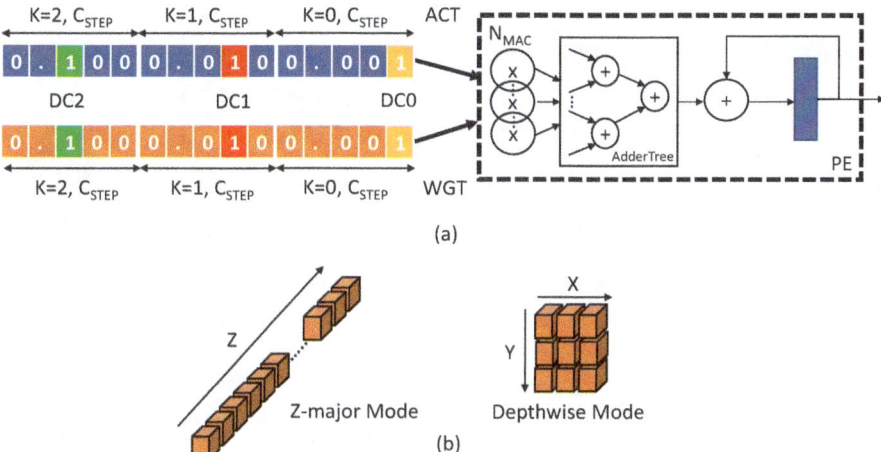

Fig. 17 Depthwise operation within a PE: (**a**) Depthwise operation performed using a conventional input channel (z-major) accumulating MAC; (**b**) Difference in accumulation pattern between conventional Z-major convolution mode and depthwise mode

increase the number of MACs since logic is "relatively free." However, if the MACs cannot be properly utilized for Edge AI workloads, then adding more of them does not make sense. Furthermore, many weights and activations need to be read from external memory, since it will not be possible to reuse the input activation data when performing the DWC. This will result in large amounts of data movement from external memory to local storage on PEs, causing high-power dissipation. In a standard AI accelerator, the number of activations that must be reloaded to process a typical 3×3-s1 DWC kernel is $\sim 5x$ in the activations if one were to look at the halo regions alone between successive rounds of DWC.

Significant improvement of DWC efficiency in DNN accelerators that are originally optimized for standard convolutions can be achieved via a combination of three techniques: (1) improved reuse of activations by multicast/broadcast of fetched data between PEs and sliding window multiplexing within a PE, while retaining the standard convolution mode of data delivery to the PEs, (2) ability to process multiple output points within a single PE, and (3) combination of internal and external adder tree hardware to perform reduction in the XY dimension (as opposed to the conventional Z dimension reduction in standard convolution). Each of the novel techniques can be controlled by means of software-programmed configuration registers, allowing our optimization to be effective for a wide range of layer dimensions.

A generic high-level architectural diagram of a typical AI accelerator running DWC is shown in Fig. 17a which basically accumulates in the IC (or Z) dimension as shown in Fig. 17b. In the figure, we use the notation DC to denote individual depthwise channels where the input channel is the same as the output channel.

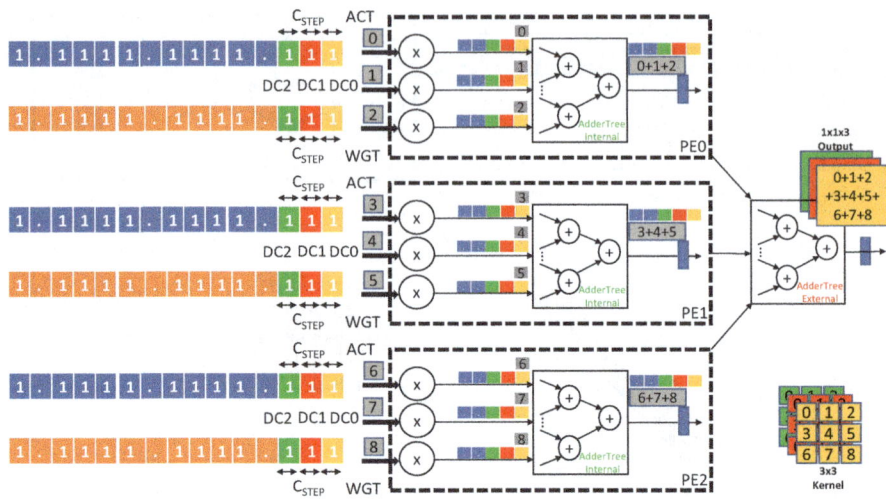

Fig. 18 Proposed efficient depthwise operation in a Z-major based PE array. The depthwise accumulation grid (columns and rows) is partitioned both within and across multiple PEs. This figure shows the use of both internal and external (to the PE) partial sum adder trees for adding the partial sums generated within and across PEs to generate the final output point

DC0, DC1, DC2 denote the depthwise channels 0, 1, 2, respectively. Putting the optimizations 1, 2, and 3 together, we arrive at the final proposed DNN architecture for improved DWC efficiency as shown in Fig. 18. For a 3×3 kernel, the filter rows are split between PE0, PE1, and PE2, while a single row is split between the subbanks within a single PE. The internal adder tree within the PE is used for performing intra-row-wise accumulation of the partial sums for each of the 16 DC, generating 16 output partial sum points. The external adder tree is used for accumulating the inter-row-wise partial sums of the 3 rows of the 3×3 kernel from PE0, PE1, and PE2 to generate the final output point. Yellow, red, and green are used to denote the input channel = 0, 1, 2, respectively, in the DC direction in Fig. 17 and 18. The gray boxes denoting 0, 1, 2, 3, 4, 5, 6, 7, 8 indicate the position in the 3×3 kernel that is contributing to the partial sum (row0 is 0, 1, 2, row1 is 3, 4, 5 and row2 is 6, 7, 8, respectively) within a subbank of the individual PE. Note that data reuse occurs across different IF RF subbanks (both within or across adjacent PEs) that can eliminate redundant loads of data already available inside using multiplexers within or across PE to reuse data for subsequent rounds of DWC operation with an adjacent window of activations. This reuse of IF data is demonstrated in Fig. 19. Note that this inherently requires the filter points to be stationary within the PEs, enhancing filter reuse.

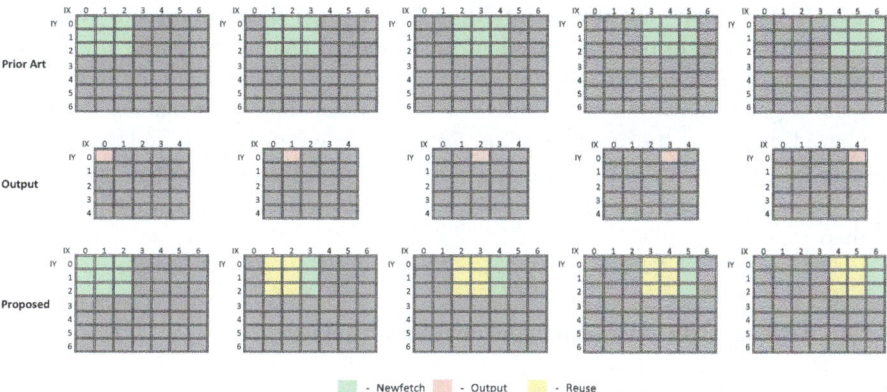

Fig. 19 Proposed reuse of data for depthwise operation. The spatial overlap of activation data in the sliding filter window (shown in the last (proposed) row) allows the elimination of the load of redundant activation data resulting in memory and compute savings

4.2.2 Group Convolution Layers

The concept of group convolutions was introduced in AlexNet [36, 37] to distribute the work over multiple GPUs as a means of introducing model parallelism. However, later with models such as ResNeXt, it was shown to improve classification accuracy. Consider a standard convolution with input channels and output channels; then the output feature map is produced when each of the filters is convolved with all the ICs to produce a single channel filter output and all outputs are concatenated. The total cost associated with a standard convolution is IC * OC * FX * FY. Now, let us consider that the filters are divided into two groups, where the first group of filters (g0) are convolved with the first IC/2 channels of the input and the other half of the filters (g1) are convolved with the second IC/2 channels of the input. The overall cost associated is now (g0 * IC/2 * FX * FY) + (g1 * IC/2 * FX * FY), which is half that of a standard convolution. This is a grouped convolution with a group size of 2. The total savings is typically a factor of the number of groups in a grouped convolution.

Design Considerations From the description, the group convolution can be parallelized, which allows splitting each group within a group convolution as a separate task and running them in parallel on multiple instances of the NNP on the AI accelerator. Since each group is just a smaller convolution, no specialized hardware is required to perform the convolution. The complexity arises when the output of the individual groups needs to be concatenated to produce the complete output for the next layer. If the different groups are run on the same NNP over time, then it simplifies the concatenation if the NNP has a continuous/back-2-back mode of operation, where it is told that is operating on tasks that are parts of a single tensor. If the groups are run on multiple NNPs, then their results are stored at different locations in the on-chip memory. Then an agent must perform a gather task to concatenate the outputs of each group to create the input for the next layer. Typically, group convolutions are processed as dense, since with sparsity the zeros are introduced along the input channels. Splitting the input channels and the sparsity bitmap into groups would require specialized hardware to decompress the sparse data to correctly identify the boundaries of each group and their corresponding bitmap.

4.2.3 Transposed Convolution/Deconvolution Layers

The intuition behind transposed convolution is to go in the opposite direction of a normal convolution, where we upsample the input feature map to a desired output feature map using some learnable parameters [38]. They are typically used in super-resolution networks to upscale the input image to a higher resolution and for semantic segmentation to decompress the abstract representation into a domain different from the input RGB input image. Consider the transposed convolution of a 2×2 input feature map with a 3×3 filter to generate a 5×5 output feature map, as shown in Fig. 20. The transposed convolution can be performed using normal convolution if the input feature map is expanded as shown in the figure. The expansion is done by padding the input tensor with zeros, and the number of zeros inserted would be equal to the stride of the normal convolution that was in the forward direction.

Design Considerations Most NNPs have the ability to pad the input tensor around the edges (left, right, top, and bottom). However, padding between the input coordinates requires additional support. The padding can be done during the output of the previous layer by packing the output with the padded zeros in the right locations, or it can be done as part of the input load, where the added specialized

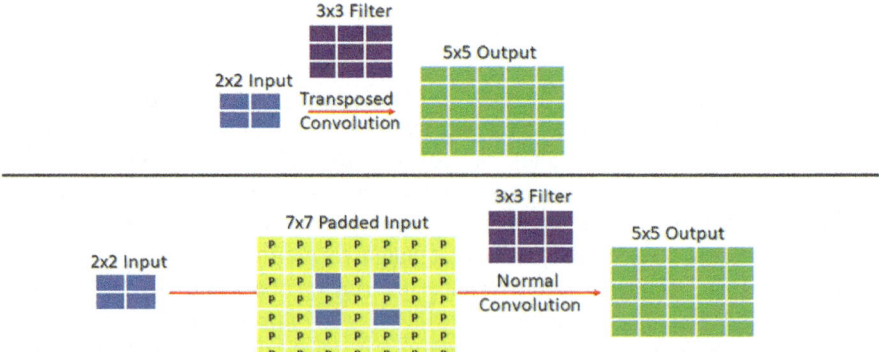

Fig. 20 Transposed convolution done using normal convolution

support inserts the zeros into the input data when it is on its way to the PE array from the memory as part of the distribution network. The second option will save memory storage and bandwidth, but will require additional hardware capable of inserting zeros into exact tensor coordinate locations to create an expanded tensor, as shown by the 7 × 7 padded tensor in Fig. 20 in yellow. Sparsity can be exploited to indicate which coordinates have valid data and which ones are padded with zeros. Note that in this scenario, sparsity is used to indicate which X,Y coordinates are sparse by marking all ICs for this coordinate as zeros.

4.2.4 Dilated Convolution/Atrous Convolution Layers

Dilated convolution is a technique to expand the filter input by inserting holes (dilation factor) between consecutive elements as a means to skip pixels and cover a larger area of the input. Alternate solutions such as pooling and/or strided convolutions reduce the resolution. Dilated convolution is used to exponentially expand the receptive view of the network without loss of resolution and maintaining the same computation and memory cost. It is a simple and effective way to detect fine details by processing inputs at higher resolution and utilizes the broader view of inputs to capture more contextual information with faster run times and fewer parameters [38]. It is typically used in semantic segmentation and to convert text to audio (WaveNet). Let us look at a 3 × 3 conv filter with dilation factors of 1, 2, and 3 as shown in Fig. 21.

Design Considerations These layers are typically mapped onto the NNP of the AI accelerator, but if the NNP tries to perform the convolution with the dilation, then there will be a drop in throughput due to the extra computation performed for the dilated coordinates. Furthermore, software intervention would be required to rearrange the filter layout in memory. Therefore, the goal should be to map this layer onto the NNP and perform a normal convolution without dilation. Another approach would be to break the tensor into a series of smaller 1 × 1 convolutions and compute the results through a continuous mode of operation on the NNP, but depending on the filter size, there can be a large number of tasks to complete for

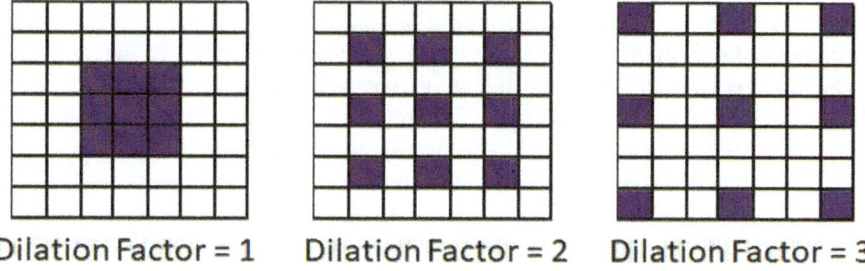

Dilation Factor = 1   Dilation Factor = 2   Dilation Factor = 3

Fig. 21 3 × 3 convolution filter with varying dilation factors

the entire tensor. A third approach is to enable a strided load from memory, which will skip the dilated coordinates during load and compute. Similarly to transposed convolution, sparsity can be exploited to indicate which coordinates have valid data and which ones are dilated. Note that the dilation factor value can vary from 1 to 32, which for larger filter sizes can cause a huge expansion of the filter, and therefore implementation constraints should be applied to NNP for the maximum supported dilation factor.

5 Efficient Mapping and Acceleration of Layers in Emerging Neural Networks

Now that we have looked at efficient mapping of some important NN operators, let us dive into efficient implementation of some important emerging neural networks such as Transformers and Graph Neural Networks (GNNs).

5.1 Transformers

Transformers were first introduced for sequence-to-sequence language translation [11] and have become one of the most important network architectures in Deep Learning. Within the context of NLP, traditionally LSTMs and RNNs were restricted to processing the tokens in sequential order, one word at a time, making them slow and unable to capture long-term dependencies between words that are spaced far apart. The self-attention mechanism in transformers, on the other hand, processes all words in the sentence at once, with each word attending to all the other words, as any of these can modify its meaning. A diagram of the vanilla transformer [11] is shown in Fig. 22. SOTA NLP networks extend the vanilla transformer by using i) stacks of encoders only such as BERT [39] for classification or sequence labeling problems, or ii) stacks of decoders only such as GPT-2 [40]/GPT-3 [41] for sequence generation, such as language modeling, or iii) multiple encoders/decoders such as T5 [42] for sequence-to-sequence modeling. Even more recently, Vision Transformers (as discussed in Sect. 2.1.2) have been applied to tasks normally associated with CNNs [13] through the idea that an image can be broken up into patches and embedding vectors are associated with each patch. These embedded patches are fed to the transformer model in the same way as words in a sentence. Great strides have also been made to reduce the number of operations in ViTs to that of larger CNNs through computing attention hierarchically and implementing shifted windows [43]. Irrespective of the transformer model, the two most computationally intensive building blocks are the "Multi-Headed Attention" and "Point-Wise Feed Forward." In these two blocks, the embeddings for each token require a large amount of memory, and compute in terms of matrix multiplications require the most Tera Operations per Second (TOPS).

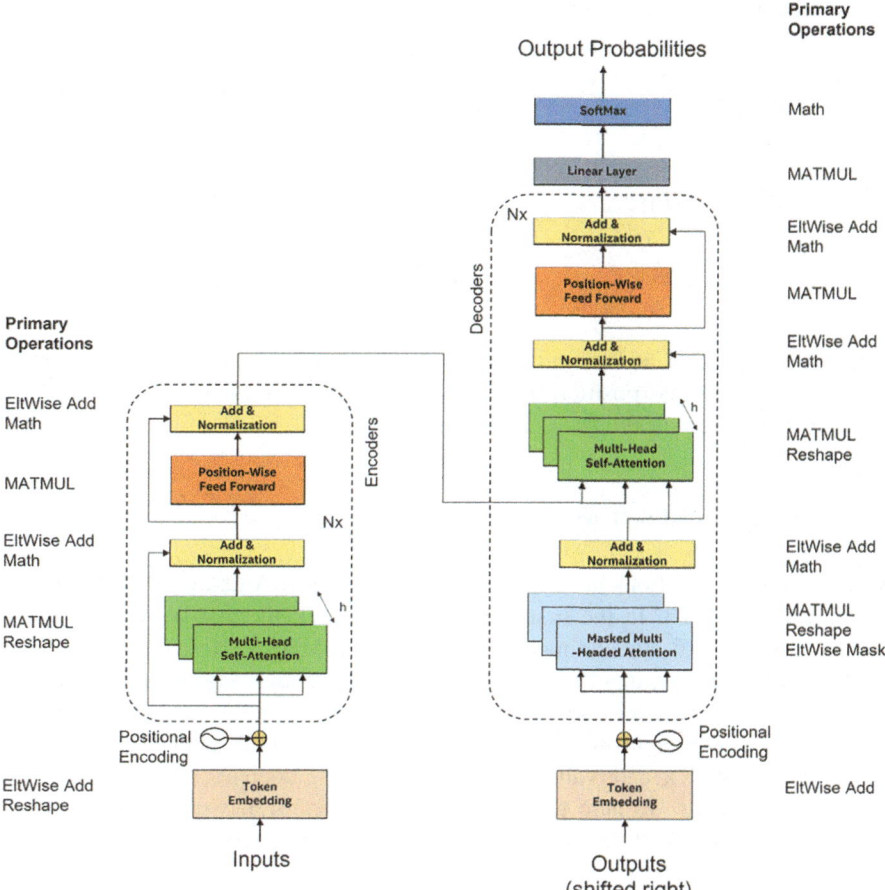

Fig. 22 Transformer network architecture

5.1.1 Input Embedding and Positional Encoding

Before the transformer encoders implement self-attention, each input word needs to index into the entire dictionary of words for the given task followed by every word being individually encoded with an embedding vector. During the training process, values of the learnable embedding vectors for words that often occur together will eventually be made more similar than those that do not occur together. Table 6 shows the size of the embedding vectors for different transformer networks d_{model}, as well as the dimensions of the feedforward vector d_{ff} and the number of multi-headed attention heads. Because, transformers process all the words in parallel, the position (relative or absolute) of each word is represented through positional embeddings of size d_{model}. The input embedding and positional encoding are merged using an

Table 6 Transformer model dimensions

Model	Embedding vector size (d_{model})	Feedforward vector dimension (d_{ff})	# multi-head attention heads (h)
Transformer-base	512	2048	8
Transformer-big	1024	4096	16
BERT$_{BASE}$	768	3072	12
BERT$_{LARGE}$	1024	4096	16
GPT-3$_{Small}$	768	3072	12
GPT-3$_{Large}$	1536	6144	16

elementwise addition and provided to the encoder and decoder blocks. For more details on how unique position embeddings are obtained using time signals, the reader is encouraged to refer to [11].

5.1.2 Multi-headed Self-Attention

All the relevant operations in a typical Multi-Headed Sell-Attention (MHSA) block are shown in Fig. 23. In the first transformer encoder, the matrix containing the position-aware word embeddings X is input to the MHSA. Note that all MHSA heads are fed the same input matrix X. Subsequent encoders take the output matrix from the previous encoder into the stack of encoders. The input matrix is multiplied by three different weight matrices W_Q, W_K, W_V in a linear layer to produce the Query (Q), Key (K), and Value (V) matrices whose contents are loosely based on retrieval systems. To help gain insight into the neural network, an example of retrieval might start with an online search using a broad query for a type of video. The search engine will map the Query against a set of Keys (title, genre, description, etc.) associated with candidates in the database and present the best matched Values (videos). Let us analyze how general matrix multiplications (GEMMs) might be mapped to the NNP of Fig. 8 to produce the Q , K , V matrices. As we proceed through the network, we will see that the transformers mainly compute large GEMMs of different dimensions. In our example, we limit the maximum sequence length of the sentence (s) to 64 and assume an embedding of $d_{model} = 1024$ and a dimension of $dim = 64$ for the linear layers. GEMMs are 2D convolutions and can be thought of as a 7D standard convolution, as shown in Fig. 6, with many of the loop parameters set to one. Using the notation of a standard convolution, let us look at how the linear layers of an MHSA are mapped to perform matrix multiplication (MATMUL).

$$IF\ (IX * IY * IC) = 1 * 64 * 1024$$

$$FL\ (FX * FY * IC * OC) = 1 * 1 * 1024 * 64$$

$$OF\ (OX * OY * OC) = 1 * 64 * 64$$

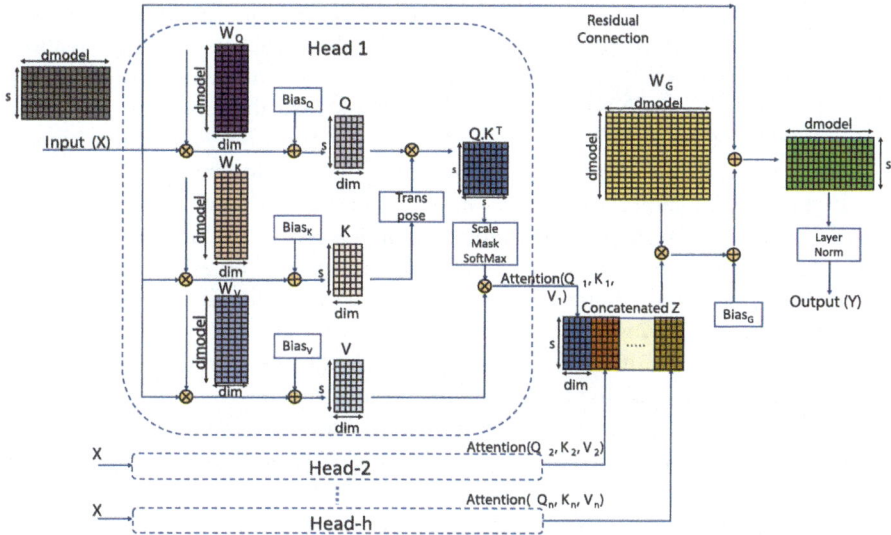

Fig. 23 Multi-headed self-attention

To demonstrate a potential mapping, assume an output stationary dataflow [4] and a 64 × 64 PE grid, where each PE contains a MAC and local storage for 32 IF and FL entries (IC = 32). Through this particular mapping, all 4096 MACs in the array can work together to produce the 4096 output points across 32 compute rounds. Note that there is no reuse across the different compute rounds since a different set of ICs need to be brought in from the on-chip memory. Each column of PEs can share a set of weights (OCs) and each row of PEs can share a set of words (OYs). The activation (X) or weight (Q, K, V) reuse is high since each set of weights and activations are reused 64 times to compute different output points. An important design consideration for transformers is that many MATMUL layers have one spatial dimension that is much smaller than the other. Because of this, one should consider designing an NNP where multiple PEs can work on different ICs simultaneously and pass accumulations between them to maximize the utilization of the array.

All of the (Q, K, V) matrices must be computed across all heads, and the intuition for employing multiple heads is that each will learn different features of the sentence in parallel. Let us assume that h = 16 for this example, as seen in the BERT_{LARGE}. Biasing operations to adjust the zero-point values of the matrices occur throughout the self-attention step, and these operations are mapped to elementwise operation of an Eltwise layer and are accelerated as such. Most SOTA neural networks, including transformer models, rely on a fixed layout of the tensor in memory to simplify load and drain design and build an efficient data distribution network. Typically matrices are stored in either row major or column major, and each element of the row/column is stored in the Z-major dimension to exploit sparsity savings seen in the ICs. To compute the similarity between the Query and the Key matrix, the Key matrix is transposed to align the dimensions with the Q matrix for multiplication. This operation requires a rearrangement in the memory or in the load path to move data efficiently into the NNP. In fact, it has been shown that on some platforms, such as CPU or GPU, MATMUL operation only accounts for 27% of the latency, with memory reshaping operations contributing more substantially to the total latency [44]. Many cloud accelerators have dedicated transpose/permute units [5] to handle this tensor reshaping operation, although these are prohibitive in area and power for edge inference. Instead, a power optimization alternative to consider is to perform the transpose on the drain path before writing the output results back to memory for the next layer.

Following the similarity matrix which calculated the attention score, there exists a scale operation which divides each value in the matrix by the square root of the dimensions of the query/key. This is similar to an Eltwise layer and can be accelerated as such. The next masking operation is only required when the MHSA is used in the transformer decoder to limit how much of the output sentence to attend to. This is also an elementwise operation that zeros out tokens that are supposed to be seen in the future. Softmax layers which produce output probabilities between 0 and 1, are difficult to efficiently implement within NNPs and will often be executed on the VDP. The final matrix multiplication in a head calculates the scaled dot product attention between the ($Q.K^T$ and V). The outputs from all the heads are then concatenated into a matrix Z of size 64×1024 and passed through a final linear layer. This linear layer is another learnable matrix (W_G) of size 1024×1024 which also has the effect of downsampling the output and reshaping it to the proper dimensions. Next, a residual connection combines the original input X and the output of the multi-headed attention through another elementwise addition. Residual connections were first introduced in [45] and have become extremely widespread across many DNNs including transformers. This is because they allow faster convergence during training through allowing the easier propagation of gradients during the backward pass and also allows DNNs to retain information from earlier in the network as deeper networks do not always perform better than shallower ones. Finally, before the point-wise feed-forward network, the matrix is layer normalized. Layer normalization normalizes the activations across features, whereas the more commonly known batch normalization normalizes the activations within a feature. In transformers, the feature vectors are each of the indices of d_{model} in the token embeddings. There are addition and subtraction operations in layer normalization, which can be accelerated on the NNP while the mean, standard deviation, and division operations can be executed on the VDP.

5.1.3 Point-Wise Feed-Forward

The Position-Wise Feed-Forward Network (PFFN), as shown in Fig. 24, is the other main contributor to the arithmetic intensity in transformers. The input Y from the MHSA is combined in a first linear layer with critical dimension d_{ff}. This is again a

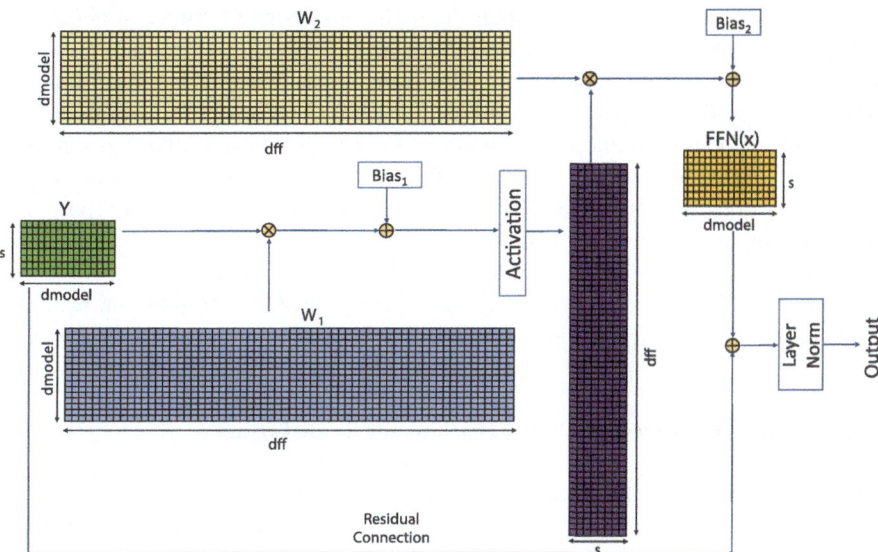

Fig. 24 Position-wise feed-forward

MATMUL operation with the same considerations as those that exist for MATMUL in the MHSA. This is then followed by an activation function, whose optimizations are discussed in Sect. 4.1.5. Newer transformers, such as BERT [39], use the GeLU activation function, while the original vanilla transformer [11] used the ReLU activation. This is important to consider when considering sparsity acceleration, as GeLU activations do not produce sparsity in the output matrix in the same way as ReLU. Another linear layer follows the activation function with critical dimension d_{ff}. We see that there are two large GEMMs, where $d_{ff} = 4d_{model}$, with individual MATMULs in the PFFN larger than the ones in the MHSA. Larger GEMMs will consume more of the total TOPS and runtime but have the added effect of keeping the PE array highly utilized. The high utilization is dependent of course, on the load path supplying the array with enough data. Smaller GEMMs can often underutilize the array because they are fully contained in only a subset of PEs. One simple optimization for increased efficiency would be to map MATMULs of different heads to different portions of the PE array, as there is no interaction between them. Finally, the different biases applied throughout this block and the final layer normalization are similar to what occurs in the MHSA.

5.1.4 Enabling Transformers on the Edge

Transformers have a huge number of learnable parameters/weights compared to most CNNs. For example, ResNet101 [45] has 1.7M parameters while BERT$_{BASE}$ [39] contains 110M parameters and GPT-3$_{Large}$ contains 760M parameters [41], which may limit their deployment in edge inference applications. In addition to energy concerns, the inference latency of a transformer can be prohibitive for real-time applications. However, much research has been done to reduce the number of computations and memory requirements for transformers [12]. These mainly focus on reducing the number of operations in self-attention since the complexity of that is $s^2 * d_{model}$. Some methods [14] include sparse attention, where every token does not attend to all other tokens, as well as query prototyping and memory compression, where the complexity of attention is reduced by eliminating the number of queries or key-value pairs. These approaches also improve the multi-head mechanism, where altering the behavior of different attention heads or allowing interaction across heads is allowed.

5.1.5 Summary of Design Considerations for Transformers

NNPs that accelerates Transformers can have slightly different design consider-ations than those that accelerate mainly convolutions. The computation of many GEMMs often have high reuse thus alleviating the throughput requirements of the load datapath. Furthermore, the matrices to be multiplied often have one spatial dimension which is smaller than the other. To maximize array utilization, there should be flexibility for mapping different portions of the larger matrix dimension across multiple PEs and also a means to accumulate partial sums across them. Some GEMMs are small enough in size that they can only fit entirely in a subset of the available PEs. It is important for the compiler to be able to map unrelated matrix multiplies, i.e., MATMULs of different heads, to different portions of the array to maximize utilization. Finally, newer transformer networks often use activation functions, such as GeLU, that do not produce sparsity in the output activations as much as older activations such as ReLU. This makes the case for adding the additional hardware for supporting activation sparsity much harder to justify.

5.2 Graph Neural Networks

GNNs are another class of emerging NNs that have found applications in multiple domains, as discussed in Sect. 2.4. These networks are employed for learning relationships in a graph-structured data. GNNs have become very popular in recent years because of their applicability in wide variety of real-world problems. From a network architecture perspective, GNNs can be viewed as an evolution of transformers. SOTA transformers can be applied to multi-modal application domains by adapting the input embedding layers and creating attention vectors which are capable of representing multiple modalities. Multi-headed attention layers can extract features from the embedded vectors based on the desired objective. In GNNs, the entire network architecture depends on the application or input graph, not just the input features. The neural network-based compute layers gather and extract

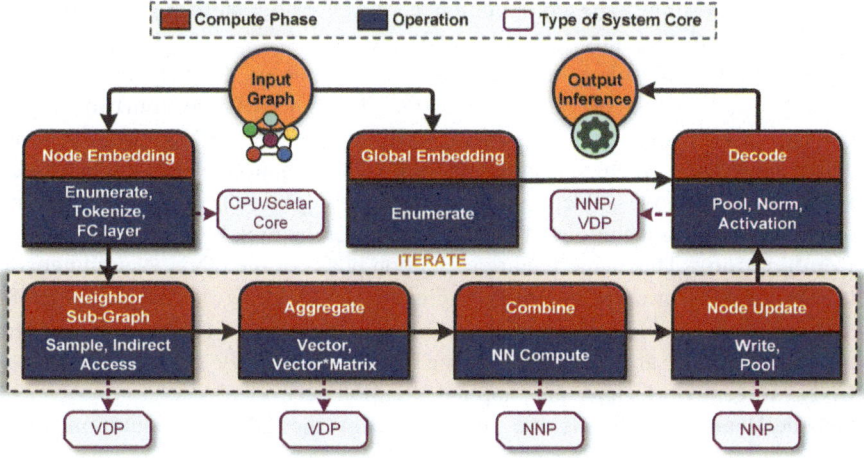

Fig. 25 Graph neural network—execution flow

information from graph vertex and edges. As we move to deeper layers, the network learns information/relations from more distant neighbors. Detailed execution flow of GNN layers is described in Sect. 5.2.1.

The computation pattern in GNNs stretches from sparse and irregular accesses during the Aggregation phase to dense and regular compute during the Combination phase. In the following sections, we provide detailed categorization of different compute phases. We will refer to a stripped-down structure of graph algorithm execution (as shown in Fig. 25) for the remaining part of this discussion.

5.2.1   Compute Phases of GNN

GNN computation (inference only) is a time evolving execution of the input graph, similar to RNNs. The execution is split into multiple layers, broadly classified into four categories, viz., Node Embedding, Aggregation, Combination, and Decode. Node Embedding and Decode stages are performed only at the start and end of the execution, respectively. Aggregation and Combination phases can be iterated multiple times depending on the choice of algorithm. Aggregation phase can be preceded by an optional sampling layer that creates a subgraph of the neighboring nodes. The vertices and edges are updated with the information gathered from their neighbors at the end of each layer. We now provide a detailed description of the computations and operators used in different GNN execution (inference only) layers.

- **Node Embedding**—This step is for transforming raw/real-world input data into feature vectors/matrices. It can be done either offline or online depending on the GNN algorithm. The computation kernels for this step are: vectorization, encoding transformations, layout changes, etc.

- **Aggregation**—This phase accumulates the features of the neighboring nodes (usually one-hop, but can vary depending on the choice of algorithm) and applies transformations, such as reduce mean, pool, fully connected NN, over the vertex and edge feature vectors. Vertex and Edge aggregation can follow different algorithms and node selection schemes. Some algorithms have also explored different sampling techniques for improved generalization [46].
- **Combination**—This is the feature extraction phase of the GNN. Neural network transformations are applied over aggregated Vertex/Edge feature vectors for deriving high-level feature representations/relations. SOTA GNNs have explored multiple network architectures ranging from simple fully connected layers to CNNs. Recent works have also applied attention networks [47].
- **Decode/Readout**—This is the final block of the GNN computation where the high-level features generated by multiple iterations of the Aggregation-Combination layers and the Graph global features are converted to output data or predictions. Compute kernels used in this phase is similar to other DNN algorithms and operators, e.g., MLP, Softmax.

In addition to the layers and operators mentioned above, Concatenation and Transpose operators are widely used at subblock boundaries in GNN architectures to improve inference accuracy/performance. Based on the layer-wise computations and operations described in the section, we have summarized the data access and compute patterns of the SOTA GNNs in Table 7.

5.2.2 Design Considerations

The Compute and Data access patterns in GNN layers are very diverse. In order to match the diversity of compute requirements, many SOTA hardware architectures built for GNN computation [48, 49] have taken a hybrid approach to efficiently process different GNN layers. Based on the compute characteristics of these layers

Table 7 GNN compute and data access patterns

Characteristic	Node embedding	Aggregation	Combination	Decode or readout
Compute kernels	Enumeration, FC	Vector, V*M	NN-MLP, conv, attention	Pool, norm, ReLU, ...
Access patterns	Regular	Indirect, irregular	Direct, regular	Direct, regular
Data reusability	Low	Mid	High	Low
Compute pattern	–	Dynamic, irregular	Static, regular	–
Compute intensity	Low	Low	High	Low
Execution bound	Memory	Memory	Compute	Memory

and the capabilities of the DNN accelerator-based IP considered in this work (Fig. 7), we believe that the following hardware design will be ideal.

- **Node Embedding**—This phase has low data reusability, is more control flow dependent and is usually done as part of offline processing/preparation. Therefore, this phase is ideal for processing in the CPU.
- **Aggregation**—This phase is dominated by sparse and random memory accesses, arithmetic and binary operations, followed by reduction or transformations. VDP block in our architecture (Fig. 7) is most suitable for the execution of these kernels. VDP hardware unit is a vector processor with additional capability to compute activation functions and matrix/vector transformations.
- **Combination**—As mentioned earlier, depending on the algorithm, different neural network topologies can be used in this phase, ranging from convolution to attention networks. Also, similarly to CNNs, the layer parameters are fully shared among the nodes in the combination phase. Thus, this phase is characterized by high compute intensity and high opportunity for data reuse. The NNP compute block in our architecture (Fig. 7) can provide the best hardware acceleration and efficiency for this computation. The hardware design requirements for computation of different Neural Network kernels is already covered in the previous sections.
- **Decode/Readout**—In the final phase, the high-level features extracted after multiple iterations of the Aggregation and Combination phase are transformed to obtain the final graph representation. Depending on the type of non-linear operators, this phase can be mapped to either the NNP or the VDP block.

5.2.3 GNN Data Flow

From an execution time perspective, most of the compute time in a GNN execution is spent on the Aggregation and Combination stage. Though some GNNs can also have a significant node embedding compute, this is not a common case. Figure 26 shows the computation flow for the Aggregation and Combination phases of a simple bidirectional Graph Network with seven nodes. Specific operators and network type are avoided to keep the generality. The node features are assumed to be a 1D vector of length equal to the feature size (assuming feature size = 1024 for the purpose of this discussion), henceforth referred to as h_i.

The Aggregation phase gathers the feature vectors of the immediate neighbors of node 1 (number of neighbors = 3). The one-hop neighbor feature map of node 1 after this gather stage is of dimension [4, 1024]. Note that the number of neighboring nodes on a practical GNN will be much higher. In addition, some GNNs [46] employ multi-hop neighbor aggregation strategies, which will lead to an exponential increase in the dimension of the feature map. Let us consider two aggregators for the gathered feature matrix, a mean and a pooling aggregator. Mean aggregator is a simple elementwise mean of the vectors representing the feature map of the neighboring nodes. On the other hand, the pooling aggregator independently multiplies each of

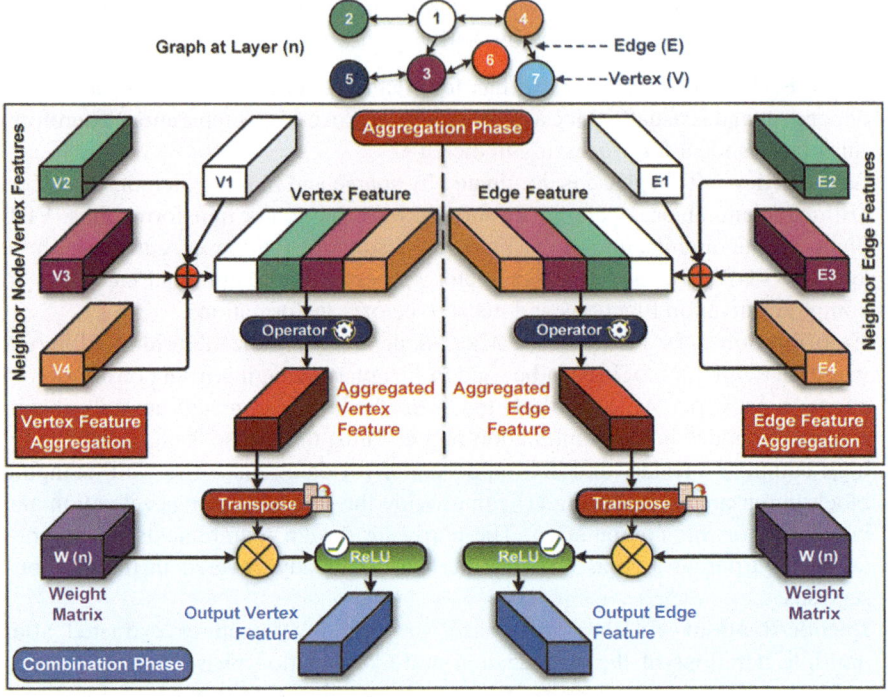

Fig. 26 Graph neural network—layer computation

the neighboring vectors with a fully connected network and applies ReLU, followed by an elementwise maxpool operation across neighbor sets. Mathematically, this can be expressed in the form: $Aggregate = EltwiseMax(ReLU([4, 1024]^T * [4, 4]))$. The output vector dimension will remain the same after elementwise operations i.e., [1, 1024]. Two different hardware mapping schemes for the aggregator phase is covered in the next subsection. In most of the research literature, this stage also includes a concatenation of updated feature vectors and history feature vectors (not shown in Fig. 26). Assuming concat layer, the feature vector size will be [1, 2048].

These operations have traditionally run on CPU with SIMD capabilities, and SOTA caching and prefetching mechanisms. But the additional requirements of network specific operators and fine-grained data/control sharing with rest of layer computes provides hardware acceleration opportunity by using VDP cores. As we observed, for a given layer, each vertex and edge performs an independent aggregation of its neighbors, with the possibility of shared neighbors, allowing data reuse. Note that the graph sizes are generally large compared to memory in a typical edge device. Another interesting characteristic is that the feature vectors in GNNs are usually very large and of different lengths. Different graph partitioning-based acceleration schemes that leverage the above characteristics are described in the next section. In the Combination phase, the aggregated node/edge feature (or concatenated feature) is multiplied by the layer weight matrix (W (n)), followed by the ReLU and Norm operation, as shown in Fig. 26. The Combination phase is optionally followed by the pooling operation. Further details of different NN compute (MLP, Conv, Attention) mappings to the NNP core are already covered in the previous sections.

As an illustration example, we will provide a description of the Attention Layer compute on a GNN, using the basic building blocks described in Sect. 5.1. Consider a graph feature map of dimension [number of nodes, number of features] (for the graph in Fig. 26, this will be [7, 1024]), where the dimension of the individual node feature (h_i) is [1, 1024]. The application of self-attention to the graph nodes is equivalent to determining the importance of node "m" for node "n." As the graph nodes are not fully connected, masked attention is performed. One of the masking approaches is to determine the self-attention for the immediate neighbors of a node. Therefore, the feature map of node 1 will be of dimension [4, 1024]. The self-attention of node "i" relative to node "j" is calculated using single-layer MLP followed by ReLU activation. Mathematically, this can be expressed in the form: $AttentionFactor = ReLU(a^T * concat(h_i, h_j))$, where a is the weight matrix of MLP. Softmax operation is applied over the output of the MLP layer. Attention factors are averaged to calculate the output feature. Single head attention described above can be easily extended to multi-headed attention. Each attention head independently executes the attention mechanisms, and the output feature is the concatenation of all. For the final layer decode, we can use the averaging function followed by Softmax to obtain the final output prediction. The readers are advised to refer to [47] for training and performance of Graph Attention Networks. From a hardware mapping perspective, the creation of node feature maps is performed as part of aggregation and mapped to the VDP block. As graph networks are very sparse, masked attention is adjacency list-based feature access (not a masking mechanism as done in language attention networks). The linear layers of MLP-based self-attention are mapped using the matrix multiplication (MATMUL) template of the NNP block.

Tensor Transformations and Memory Layout change operations (e.g., transpose, concat) within a compute phase or at phase boundaries account for significant execution latency. Both VDP and NNP blocks can perform these transformations as part of their write-back stage. Some commercial architectures also use dedicated hardware units for these operations. The ordering of aggregation and combination phase is an algorithm choice, either of the phases can be done first. Regardless of the ordering choice, both layers form a producer-consumer relationship and require hardware architecture support for synchronization. Vertex and edge features are updated at the end of a layer. Network weights within a layer are shared across nodes and can possibly be shared across layers.

5.2.4   Additional Opportunities for Hardware Acceleration

GNNs can leverage most of the hardware acceleration schemes described in the previous section. In addition, the following patterns can be exploited in the graphs to improve performance and efficiency.

Aggregation Phase Acceleration Section 5.2.3 has established the opportunities for the Aggregation phase in graph partitioning algorithms. On the basis of the observations, hardware mapping of Vertex/Edge aggregation across VDP cores can be done in two different ways, viz., (a) each vertex/edge is mapped to a single VDP core, and (b) the feature vector of a node is mapped across VDP cores. If we map our graph example from the previous section, mapping option (a) will map each of the seven nodes in the graph to a VDP core, whereas mapping option (b) will split the 1024 feature vectors between "N" VDP cores, where "N" can be an optimization parameter. Option (a) creates an imbalance between the execution time of the fast and slow vertices, which becomes more prominent in GNNs as the number of neighbors and the sparsity is dynamic, while option (b) solves the imbalance problem.

Sparsity Acceleration We can find the following two types of sparsity in a GNN. (a) Zero elements in the input activations and weights are seen in the Combination phase of the graph computation. Sparsity acceleration schemes are similar to those described in the previous sections. (b) Sparse connections/edges in Adjacency Matrix or sub-sampled graph are seen in the Aggregation phase of the graph computation. Sparsity can be due to either the input graph topology (static) or the sampling schemes employed in the algorithm (dynamic). Static sparsity can be handled by the graph partitioning algorithm executed by the compiler. The algorithm takes the graph topology as input and decides the grouping of vertex and edges for the best data reuse. Dynamic Sparsity, on the other hand, needs custom hardware acceleration support to eliminate redundant access of sparse edges. One such technique is covered in [48] where the vertex/edge partition boundaries are dynamically determined using window sliding and shrinking methods.

Data Flow In addition to the data reuse opportunities covered in the previous sections, further data flow optimizations are possible, as the GNN execution time shows a large amount of data copy and synchronization overheads. Different buffering and caching schemes are explored to accelerate iterations between layers and phase-by-phase computation within a layer. The control flow and data movement between NNP and VDP cores are described in the previous section.

Because of the dynamic nature and scale of GNNs, software acceleration schemes have also been abundantly explored in the research literature. However, software acceleration is not covered because it is beyond the scope of this chapter.

6   Future Trends: Networks and Applications

In summary, the previous sections provided a comprehensive survey of popular DNN architectures and corresponding applications in computer vision, natural language processing, recommendation systems, and graph neural networks. Subsequently, the mapping of common layers in these networks to SOTA hardware AI accelerators and insights derived on challenges and trade-offs are discussed. Now, we illuminate some of the hottest developments in DL-based applications and hardware that could very well be the foundation of the next frontier of AI.

1. **Emerging Applications and NNs.** The idea of combining CNNs and vision transformers has led to the emergence of new generation of *Hybrid Vision Models*, which essentially inherit the advantages of these complementary architectures [12]. These models, which either use depthwise convolutions or complete CNNs to tokenize an input, have led to better performance in low data regimes at reduced computational complexity. Some popular examples include CvT, LeViT, CoAtNets, NesT, TransCNN, AlterNet, and Conformer. Recent efforts such as ConvMLP have also explored combination of CNNs and MLPs for better accuracy *vs.* computation trade-offs. Interestingly, ConvNeXt (Facebook, 2022 [50]), built entirely of CNN modules have also demonstrated competitive results. On the other hand, *Multi-modal AI (MMA)* is another emerging field that consolidates heterogeneous data from multi-modal sensors and employs multiple AI algorithms such as conversational AI, image processing, natural language understanding, etc., to learn, reason, and synthesize information. Recent efforts in MMA have contributed state-of-the-art Transformer-based models catering to diverse workloads [9, 12], e.g., GLIDE, ALIGN, SwinBERT, Omnivore, FLAVA, Data2vec, and PerceiverIO among others. *Self-Supervised Learning (SSL)* is another emerging paradigm that enables learning from any random data as well as from any unbounded dataset. This makes SSL an ideal candidate for IoT applications. For example, Federated Learning that exploits decentralized computing power available in a network of edge devices could leverage SSL to learn from real-time unlabeled data. Popular examples such as SEER (Facebook), BYOL (Google), DINO, EsViT [12] uses transformers and exhibits exceptional performance in various CV applications. On the other hand, *Neuro-Symbolic AI (NeSy AI)* models [51] such as NCSCL, NSDR (MIT-IBM), and NLM (Google) integrate DL techniques with traditional rule-based symbolic AI approaches and generates error-resilient, explainable, and scalable models. Nevertheless, accelerating the low-operational-intensity computations and expensive data movement in these workloads are some of the challenges that needs to be addressed in the future. These discussions clearly indicate the need for future accelerator designs to cater to emerging NNs for compute and energy-efficient edge AI applications.
2. **Approximate Computing and Approximate Systems.** Approximate computing (AxC) is an emerging design paradigm that takes advantage of the inherent error resiliency of cutting-edge DNNs and increases the energy efficiency and

performance of the underlying edge computing systems, including custom accelerators, FPGA, GPU, and CPU [52]. In this paradigm, popular *algorithmic* and *software* approximation strategies include model scaling, early exit branch, channel pruning, weight sharing, sparsity prediction, and knowledge distillation [2]. *Hardware-software co-approximation* techniques such as quantization [53] have increased performance and energy efficiency of DNN inference in edge devices by adopting low-precision data types, such as FP16, BF16, 8-bit, 4-bit (NVIDIA [54]) and 2-bit (IBM [55]), while maintaining accuracy. Based on these efforts, we envision almost all DNN accelerators of the future to integrate lower precision MAC and have dedicated compiler support for the same. Apart from these compute approximations, recent efforts have also explored approximations in memory, sensor, and communication subsystems found in typical edge devices [2]. Finally, in their groundbreaking paper, authors [2, 56] have applied synergistic approximations across multiple subsystems to provide better system-level energy savings compared to individual subsystem approximations ushering the era of Approximate Systems (AxS). These studies suggest that future designs of DNN-based hardware accelerators should explore AxC and AxS to reap their benefits for energy-efficient edge AI applications.

3. **Processing in Memory.** The areas of In-memory Computing (IMC) and Near-memory Computing (NMC) are attracting growing attention among Non-Von Neumann computing architectures. This paradigm addresses the performance bottlenecks of modern DNN-based hardware accelerators which arises due to high data communication latency with off-chip memory (DRAM) and low memory bandwidth, popularly termed as the "memory wall" [57]. By implementing the processing units inside the memory chip, several IMC architectures have demonstrated significant performance optimization with orders of magnitude better throughput and energy efficiency than traditional Von Neumann architectures. Therefore, these are ideal for mobile and edge devices as well as real-time IoT applications. IMC designs developed for traditional memory platforms such as DRAM include AMBIT, DRISA, DrAcc, SCOPE, and LAcc [57]. From technology availability perspective, SRAM is suitable for IMC and pioneering works in SRAM-based IMCs include Neural Cache, IMAC, XNOR-SRAM, Conv-RAM, Twin-8T [57, 58]. Apart from these works, research efforts have also explored IMCs in novel non-volatile memory architectures such as ReRAM (ISSAC, PRIME), STT-MRAM (Binary CNN, MRIMA), and SOT-MRAM (IMCE, CMP-PIM) [57]. In fact, IMC technologies also leverage emerging approximate computing techniques like quantization of DNNs. However, future research should consider the major design challenges in IMC chips including analog-to-digital conversion (ADC) bottleneck, memory non-idealities, and analog compute variations to reflect similar gains in commercial settings.

4. **Dedicated Hardware For Emerging NNs.** To date, research on designing computing platforms dedicated for AI applications has mainly concentrated on accelerating CNNs. However, the growth of emerging neural network architectures such as Transformers and GNNs (Sects. 5.1 and 5.2), GANs, RNNs, sparse NNs, low-precision NNs, as well as new types of convolution operations

(Table 4) have very recently led to the design of domain-specific accelerators. For example, authors [59] have designed an FPGA-based CNN accelerator for depthwise separable convolution that provides optimal balance between speed and hardware resource. Another recent work [60] has developed a systolic array-based reconfigurable accelerator that employs matrix partitioning, dataflow optimizations, and non-linear function optimizations to accelerate transformers. Early research on GNN acceleration includes GRIP [61] that uses custom compute unit for arithmetic-intensive vertex-centric operations and memory-intensive edge-centric operations found in GNNs. GNNerator [49] exploits the inherent inter-stage parallelism in GNNs and developed a programmable accelerator composed of heterogeneous compute engines targeting sparse and dense computations in GNNs. These emerging trends lead us to expect further research in custom hardware designs specific to various emerging NNs, which could take advantage of AxC, AxS, and IMC and facilitate real-time and energy-efficient AI applications at the edge. Due to the area impact of additional hardware, this is still not seen as a viable solution for many edge devices, however, eventually an ensemble of different NN accelerators can soon become a reality with the emergence of smaller NN models and even smaller process technology.

References

1. Raha, A., Kim, S.K., Mathaikutty, D.A., Venkataramanan, G., Mohapatra, D., Sung, R., Brick, C., Chinya, G.N.: Design considerations for edge neural network accelerators: An industry perspective. In: 34th International Conference on VLSI Design and 20th International Conference on Embedded Systems, pp. 328–333 (2021)
2. Raha, A., Ghosh, S., Mohapatra, D., Mathaikutty, D.A., Sung, R., Brick, C., Raghunathan, V.: Special session: Approximate TinyML systems: Full system approximations for extreme energy-efficiency in intelligent edge devices. In: IEEE 39th International Conference on Computer Design (ICCD), pp. 13–16 (2021)
3. Sze, V., Chen, Y.H., Yang, T.-J., Emer, J.S.: Efficient processing of deep neural networks: A tutorial and survey. Proc. IEEE 105, 2295–2329 (2017)
4. Kwon, H., Chatarasi, P., Sarkar, V., Krishna, T., Pellauer, M., Parashar, A.: Maestro: A data-centric approach to understand reuse, performance, and hardware cost of DNN mappings. IEEE Micro 40, 20–29 (2020)
5. Norrie, T., Patil, N., Yoon, D.H., Kurian, G., Li, S., Laudon, J., Young, C., Jouppi, N.P., Patterson, D.A.: The design process for Google's training chips: Tpuv2 and tpuv3. IEEE Micro 41, 56–63 (2021)
6. Jang, J.-W., Lee, S., Kim, D., Park, H., Ardestani, A.S., Choi, Y., Kim, C., Kim, Y., Yu, H., et al.: Sparsity-aware and re-configurable NPU architecture for Samsung flagship mobile soc. In: ACM/IEEE 48th Annual International Symposium on Computer Architecture (ISCA), pp. 15–28, (2021)
7. Krizhevsky, A., Sutskever, I., Hinton, G.E.: ImageNet classification with deep convolutional neural networks. Commun. ACM 60, 6, 84–90 (2017). https://doi.org/10.1145/3065386
8. Zhao, Y., Wang, G., Tang, C., Luo, C., Zeng, W., Zha, Z.-J.: A battle of network structures: An empirical study of CNN, transformer, and MLP (2021). arXiv
9. Meta AI. The latest in machine learning | papers with code. https://paperswithcode.com/

10. Goodfellow, I., Pouget-Abadie, J., Mirza, M., Xu, B., Warde-Farley, D., Ozair, S., Courville, A., Bengio, Y.: Generative adversarial nets. In: Advances in Neural Information Processing Systems, vol. 27 (2014)

11. Vaswani, A., Shazeer, N.M., Parmar, N., Uszkoreit, J., Jones, L., Gomez, A.N., Kaiser, L., Polosukhin, I.: Attention is all you need (2017). arXiv

12. Khan, S., Naseer, M., Hayat, M., Zamir, S.W., Khan, F.S., Shah, M.: Transformers in vision: A survey. ACM Comput. Surv. **54**, 1–41 (2021)

13. Dosovitskiy, A., Beyer, L., Kolesnikov, A., Weissenborn, D., Zhai, X., Unterthiner, T., Dehghani, M., Minderer, M., Heigold, G., Gelly, S., Uszkoreit, J., Houlsby, N.: An image is worth 16x16 words: Transformers for image recognition at scale (2021). arXiv

14. Lin, T., Wang, Y., Liu, X., Qiu, X.: A survey of transformers (2021). arXiv

15. Tolstikhin, I.O., Houlsby, N., Kolesnikov, A., Beyer, L., Zhai, X., Unterthiner, T., Yung, J., Steiner, A., Keysers, D., Uszkoreit, J., et al.: MLP-mixer: An all-MLP architecture for vision. Adv. Neural Inf. Process. Syst. **34**, 24261–24272 (2021)

16. Ko, H., Lee, S., Park, Y., Choi, A.: A survey of recommendation systems: Recommendation models, techniques, and application fields. Electronics **11**(1), 141 (2022)

17. Wu, S., Sun, F., Zhang, W., Cui, B.: Graph neural networks in recommender systems: A survey (2020). arXiv

18. Zhang, S., Yao, L., Sun, A., Tay, Y.: Deep learning based recommender system: A survey and new perspectives. ACM Comput. Surv. **52**(1), 1–38 (2019)

19. Dong, G., Tang, M., Wang, Z., Gao, J., Guo, S., Cai, L., Gutierrez, R., Campbell, B., Barnes, L.E., Boukhechba, M.L: Graph neural networks in IoT: A survey. ACM Trans. Sensor Netw. (2022). http://nvdla.org/hw/v1/ias/lut-programming.html

20. Abadal, S., Jain, A., Guirado, R., López-Alonso, J., Alarcón, E.: Computing graph neural networks: A survey from algorithms to accelerators. ACM Comput. Surv. **54**(9), 1–38 (2021)

21. Zhou, J., Cui, G., Hu, S., Zhang, Z., Yang, C., Liu, Z., Wang, L., Li, C., Sun, M.: Graph neural networks: A review of methods and applications. AI Open **1**, 57–81 (2020)

22. NVDLA Open Source Project - LUT programming

23. Chen, Y.-H., Krishna, T., Emer, J.S., Sze, V.: Eyeriss: An energy-efficient reconfigurable accelerator for deep convolutional neural networks. IEEE J. Solid-State Circuits **52**(1), 127–138 (2017)

24. Chen, Y.-H., Yang, T.J., Emer, J., Sze, V.: Eyeriss v2: A flexible accelerator for emerging deep neural networks on mobile devices. IEEE J. Emerg. Sel. Topics Circuits Syst. **9**(2), 292–308 (2019)

25. Lin, C.-H., Cheng, C.-C., Tsai, Y.-M., Hung, S.-J., Kuo, Y.-T., Wang, P.H., Tsung, P.-K., Hsu, J.-Y., Lai, W.-C., et al.: 7.1 a 3.4-to-13.3tops/w 3.6tops dual-core deep-learning accelerator for versatile AI applications in 7nm 5g smartphone soc. In: IEEE International Solid-State Circuits Conference-(ISSCC), pp. 134–136 (2020)

26. Jouppi, N.P., Young, C., Patil, N., Patterson, D., Agrawal, G., Bajwa, R., Bates, S., Bhatia, S., Boden, N., Borchers, A., et al.: In-datacenter performance analysis of a tensor processing unit. SIGARCH Comput. Archit. News **45**(2), 1–12 (2017)

27. Qin, E., Samajdar, A., Kwon, H., Nadella, V., Srinivasan, S.M., Das, D., Kaul, B., Krishna, T.: Sigma: A sparse and irregular GEMM accelerator with flexible interconnects for DNN training. In: IEEE International Symposium on High Performance Computer Architecture (HPCA), pp. 58–70 (2020)

28. NVIDIA. Nvidia ampere architecture (2022). https://www.nvidia.com/en-us/data-center/ampere-architecture/

29. Parashar, A., Rhu, M., Mukkara, A., Puglielli, A., Venkatesan, R., Khailany, B., Emer, J., Keckler, S.W., Dally, W.J.: Scnn: An accelerator for compressed-sparse convolutional neural networks. In: Proceedings of the 44th Annual International Symposium on Computer Architecture, pp. 27–40 (2017)

30. Rhu, M., O'Connor, M., Chatterjee, N., Pool, J., Kwon, Y., Keckler, S.W.: Compressing DMA engine: Leveraging activation sparsity for training deep neural networks. In: IEEE International Symposium on High Performance Computer Architecture (HPCA), pp. 78–91 (2018)

31. Intel® Movidius™ Myriad™ X Vision Processing Unit (VPU). https://www.intel.com/content/www/us/en/products/details/processors/movidius-vpu/movidius-myriad-x.html
32. Lee, B., Burgess, N.: Some results on Taylor-series function approximation on FPGA. In: The Thirty-Seventh Asilomar Conference on Signals, Systems Computers, vol. 2, pp. 2198–2202 (2003)
33. Lin, C.-W., Wang, J.-S.: A digital circuit design of hyperbolic tangent sigmoid function for neural networks. In: 2008 IEEE International Symposium on Circuits and Systems (ISCAS), pp. 856–859 (2008)
34. Leboeuf, K., Namin, A.H., Muscedere, R., Wu, H., Ahmadi, M.: High speed VLSI implementation of the hyperbolic tangent sigmoid function. In: Third International Conference on Convergence and Hybrid Information Technology, vol. 1, pp. 1070–1073 (2008)
35. Zamanlooy, B., Mirhassani, M.: Efficient VLSI implementation of neural networks with hyperbolic tangent activation function. IEEE Trans. Very Large Scale Integr. Syst. **22**(1), 39–48 (2014)
36. Ioannou, Y.A., Robertson, D.P., Cipolla, R., Criminisi, A.: Deep roots: Improving CNN efficiency with hierarchical filter groups. In: 2017 IEEE Conference on Computer Vision and Pattern Recognition (CVPR), pp. 5977–5986 (2017)
37. Sun, K., Li, M., Liu, D., Wang, J.: Igcv3: Interleaved low-rank group convolutions for efficient deep neural networks. In: BMVC (2018)
38. Dumoulin, V., Visin, F.: A guide to convolution arithmetic for deep learning (2016). arXiv
39. Devlin, J., Chang, M.-W., Lee, K., Toutanova, K.: Bert: Pre-training of deep bidirectional transformers for language understanding. In: NAACL (2019)
40. Radford, A., Wu, J., Child, R., Luan, D., Amodei, D., Sutskever, I.: Language models are unsupervised multitask learners. OpenAI Blog **1**, 9 (2019)
41. Brown, T.B., Mann, B., Ryder, N., Subbiah, M., Kaplan, J., Dhariwal, P., Neelakantan, A., Shyam, P., et al.: Language models are few-shot learners (2020). arXiv
42. Raffel, C., Shazeer, N., Roberts, A., Lee, K., Narang, S., Matena, M., Zhou, Y., Li, W., Liu, P.J.: Exploring the limits of transfer learning with a unified text-to-text transformer (2019). arXiv
43. Liu, Z., Lin, Y., Cao, Y., Hu, H., Wei, Y., Zhang, Z., Lin, S., Guo, B.: Swin transformer: Hierarchical vision transformer using shifted windows. In: IEEE/CVF International Conference on Computer Vision (ICCV), pp. 9992–10002 (2021)
44. Wang, H., Zhang, Z., Han, S.: SpAtten: Efficient sparse attention architecture with cascade token and head pruning. In: IEEE International Symposium on High-Performance Computer Architecture (HPCA), pp. 97–110 (2021)
45. He, K., Zhang, X., Ren, S., Sun, J.: "Deep Residual Learning for Image Recognition," 2016 IEEE Conference on Computer Vision and Pattern Recognition (CVPR), pp. 770–778 (2016). https://doi.org/10.1109/CVPR.2016.90
46. Hamilton, W.L., Ying, R., Leskovec, J.: Inductive representation learning on large graphs (2017). arXiv
47. Veličković, P., Cucurull, G., Casanova, A., Romero, A., Liò, P., Bengio, Y.: Graph attention networks (2018). arXiv
48. Yan, M., Deng, L., Hu, X., Liang, L., Feng, Y., Ye, X., Zhang, Z., Fan, D., Xie, Y.: HyGCN: A GCN accelerator with hybrid architecture (2020). arXiv
49. Stevens, J.R., Das, D., Avancha, S., Kaul, B., Raghunathan, A.: GNNerator: A hardware/software framework for accelerating graph neural networks (2021). arXiv
50. Liu, Z., Mao, H., Wu, C.Y., Feichtenhofer, C., Darrell, T., Xie, S.: A convnet for the 2020s (2022). arXiv
51. Susskind, Z., Arden, B., John, L.K., Stockton, P., John, E.B.: Neuro-symbolic AI: An emerging class of AI workloads and their characterization (2021). arXiv
52. Wang, X., Han, Y., Leung, V.C., Niyato, D., Yan, X., Chen, X.: Convergence of edge computing and deep learning: A comprehensive survey. IEEE Commun. Surv. Tutor. **22**(2), 869–904 (2020)

53. Raha, A., Raghunathan, V.: qLUT: Input-aware quantized table lookup for energy-efficient approximate accelerators. ACM Trans. Embed. Comput. Syst. **16**(5s), 1–23 (2017)
54. Salvator, D., Wu, H., Kulkarni, M., Emmart, N.: Nvidia technical blog: Int4 precision for AI inference (2019). https://www.nvidia.com/en-us/data-center/ampere-architecture/
55. Choi, J., Venkataramani, S.: Highly accurate deep learning inference with 2-bit precision (2019). https://www.ibm.com/blogs/research/2019/04/2-bit-precision/
56. Ghosh, S.K., Raha, A., Raghunathan, V.: Approximate inference systems (axis) end-to-end approximations for energy-efficient inference at the edge. In: Proceedings of the ACM/IEEE International Symposium on Low Power Electronics and Design, pp. 7–12 (2020)
57. Bavikadi, S., Sutradhar, P.R., Khasawneh, K.N., Ganguly, A., Dinakarrao, S.M.P.: A review of in-memory computing architectures for machine learning applications. In: Proceedings of the Great Lakes Symposium on VLSI, pp. 89–94 (2020)
58. Yu, S., Jiang, H., Huang, S., Peng, X., Lu, A.: Compute-in-memory chips for deep learning: recent trends and prospects. IEEE Circuits Syst. Mag. **21**(3), 31–56 (2021)
59. Bai, L., Zhao, Y., Huang, X.: A CNN accelerator on FPGA using depthwise separable convolution. IEEE Trans. Circuits Syst. II: Express Briefs **65**(10), 1415–1419 (2018)
60. Lu, S., Wang, M., Liang, S., Lin, J., Wang, Z.: Hardware accelerator for multi-head attention and position-wise feed-forward in the transformer. In: IEEE 33rd International System-on-Chip Conference (SOCC), pp. 84–89. IEEE (2020)
61. Kiningham, K., Re, C., Levis, P.: Grip: A graph neural network accelerator architecture (2020). arXiv

Part II
Memory Design and Optimization for Embedded Machine Learning

An Off-Chip Memory Access Optimization for Embedded Deep Learning Systems

Rachmad Vidya Wicaksana Putra, Muhammad Abdullah Hanif, and Muhammad Shafique

1 Introduction

1.1 Overview

Artificial Intelligence (AI) is considered a prominent solution to process and analyze a continuous stream of data in the era of Internet-of-Things (IoT) and Big Data, where a large amount of data is generated every day by digital devices. Analyzing the generated data and inferring useful information are beneficial for improving the users' productivity and their quality of life [59]. In the last decade, the development and research in AI, specifically on Machine Learning (ML), have increased exponentially and spread across different fields, covering a wide range of applications [34]. The field of ML encompasses several algorithms, and the most influential ones in recent years are the brain-inspired ML algorithms, such as Artificial Neural Networks (ANNs) [7, 34, 59–61] and Spiking Neural Networks (SNNs) [39, 50, 53–55, 57, 58]. Among these algorithms, ANNs have achieved state-of-the-art performance/accuracy and even surpassed humans' accuracy through the Deep Learning (DL) or Deep Neural Network (DNN) algorithms [7, 61], as shown in Fig. 1. Consequently, nowadays, DL has become a *de facto* algorithm for solving many ML-based applications, such as computer vision [27, 42, 69], finance and business [15, 65, 66], healthcare [4, 5, 43], and autonomous driving systems (e.g., drones and cars) [13, 46].

R. V. W. Putra (✉)
Embedded Computing Systems, Institute of Computer Engineering, Technische Universität Wien, Vienna, Austria
e-mail: rachmad.putra@tuwien.ac.at

M. A. Hanif · M. Shafique
Division of Engineering, New York University Abu Dhabi, Abu Dhabi, UAE
e-mail: mh6117@nyu.edu; muhammad.shafique@nyu.edu

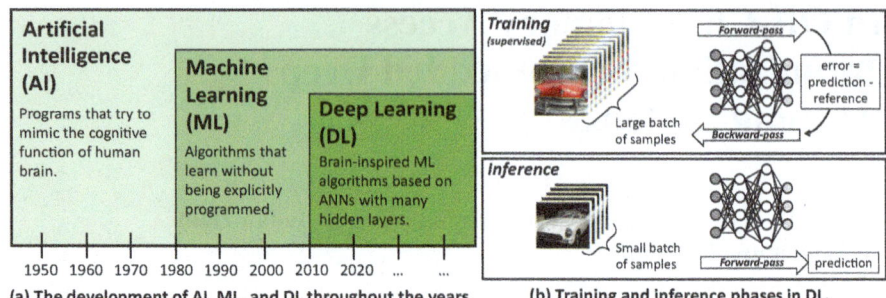

(a) The development of AI, ML, and DL throughout the years.   (b) Training and inference phases in DL.

Fig. 1 Illustrations of (**a**) Artificial Intelligence development throughout the years, including Machine Learning and Deep Learning, and (**b**) training and inference phases of Deep Learning (adapted from [7])

Although DL algorithms have achieved a cutting edge performance, they consume enormous memory and computing power/energy due to their memory- and compute-intensive nature during the training phase (at design time) and the inference phase (at run time) [1, 61]. In the training phase, a DNN experiences the forward-pass and the backward-pass, while in the inference phase, it only experiences a forward-pass. Therefore, the training phase needs a much more expensive cost of memory accesses and computational efforts than the inference. Recent trends show that many DL applications are moving towards mobile/embedded platforms, such as IoT-Edge and smart cyber-physical system (CPS) devices, mainly due to privacy and security reasons [7, 8, 59]. These embedded platforms usually employ the DNN models that have been trained at the cloud for performing inferences at the edge. However, performing such an inference is challenging since the embedded platforms are typically resource- and power/energy-constrained. For instance, the ResNet-50 requires more than 95MB of weight memory and more than 3.8 billion operations to process a single image input [21]. Such a high amount of processing is infeasible to be performed by embedded platforms for providing real-time results. Therefore, *it is necessary to provide specialized hardware accelerators for efficiently performing DL inference, hence fulfilling the memory and compute requirements of different types of DNN models for embedded DL systems.*

1.2 Design Constraints for Embedded DL Systems

In embedded applications, the resources of hardware platforms (e.g., accelerators) are tightly constrained. Embedded accelerators typically have *small on-chip memory* like 100KB–500KB [59, 61] and have to function properly using *low operational power* like within 5W [44, 52, 56, 59]. Moreover, some applications may even pose additional constraints such as *latency* and *throughput* [59], especially for safety-critical applications requiring correct real-time decision like autonomous driving. All these constraints make it even more challenging to efficiently run DL inference on embedded platforms. For instance, limited memory and power budgets lead to limited DNN operations that can be performed at the same time, and consequently, this leads to very long latency and very low throughput.

Previous studies have shown that many DL accelerators incur high energy consumption due to DL algorithms' memory- and compute-intensive nature [2, 36, 49, 51, 61, 62]. Recent works have identified that the energy consumption of DL accelerators is typically dominated by the off-chip memory (i.e., DRAM) [36, 51]. For instance, the DRAM operations in the Cambricon-X accelerator incur over 80% of the total energy consumption [71]. The reasons are the following:

- **High DRAM access energy:** The energy consumption for single DRAM access is significantly higher than other DNN operations. For instance, single DRAM access incurs about 200x energy consumption of a Multiply-And-Accumulate (MAC) operation [10, 61].
- **A large number of DRAM accesses:** The number of DRAM accesses required for a single inference is proportional to the number of data that need to be stored in and fetched from DRAM, including weights and feature maps (activations) [2, 49, 51]. Therefore, larger DNN models are likely to require more DRAM accesses than the smaller ones, thereby higher energy consumption.

Therefore, *optimizing the DRAM access energy is the key to minimizing the energy consumption of DL accelerators, hence enabling efficient embedded DL systems.*

In the following sections of this chapter, we discuss:

1. In Sect. 2, we provide *a brief overview of DL, hardware accelerators*, and *DRAM background.*
2. In Sect. 3, we discuss *our design methodology to optimize the DRAM access energy for DL accelerators.*
3. In Sect. 4, we discuss *the experimental evaluations* for our design methodology.
4. In Sect. 5, we conclude the chapter with a summary.

2 Preliminaries

2.1 Deep Learning

Deep Learning (DL) or Deep Neural Network (DNN) is a computational model inspired by biological neural networks and described as a network of interconnected neurons. Neurons are the fundamental units in a neural network, and each neuron performs a weighted sum of inputs (so-called *dot-product operation*) [18, 19]. These neurons are grouped into layers, encompassing *an input layer, multiple hidden layers*, and *an output layer* [6]. An input layer receives input signals, which are

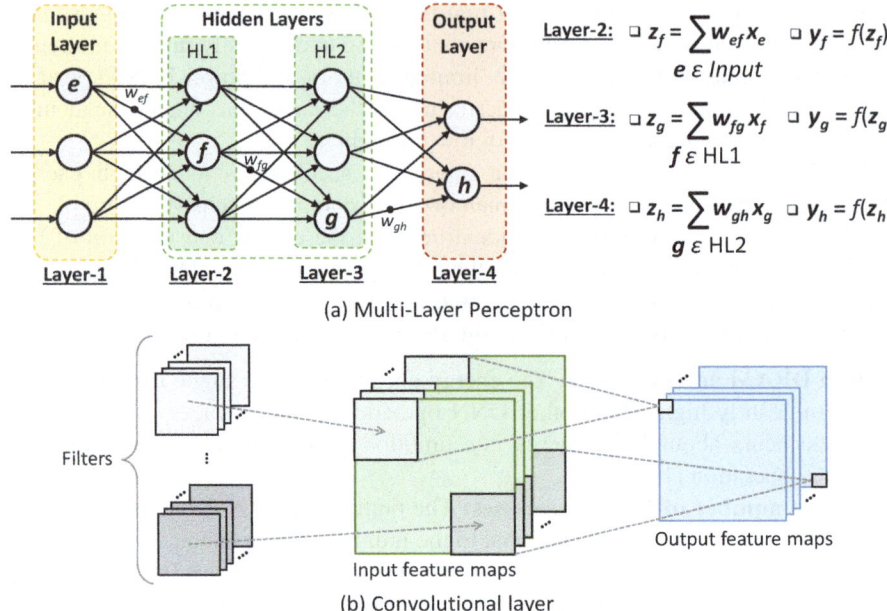

Fig. 2 Illustrations of (a) multi-layer perceptron, and (b) convolutional layer

then processed by hidden layers. Afterwards, the last layer obtains the result, i.e., the output layer. There are several types of DL, such as Multi-Layer Perceptrons (MLPs), Convolutional Neural Networks (CNNs), and Recurrent Neural Networks (RNNs) [34]. In this chapter, we focus on feed-forward neural network models, i.e., MLPs and CNNs, due to their widespread use in many ML applications.

An MLP is composed of multiple *fully connected layers*. In a fully connected layer, each neuron is connected to all neurons in the adjacent layers. An illustration of a simple three-layer MLP is shown in Fig. 2a. Meanwhile, a CNN is typically composed of *convolutional layers* and *fully connected layers*. An illustration of a convolutional layer is shown in Fig. 2b. In each convolutional layer, *multiple filters* are convolved with *input feature maps* to generate *output feature maps*. The input and output feature maps are also called *activation maps* or simply *activations*. The depth of filters (so-called *channel*) is the same as the depth of input feature maps. Convolution between input feature maps with one filter produces one output feature map, hence the total number of output feature maps equals the number of filters. A detailed discussion of the CNN operations can be found in [61].

2.2 Hardware Accelerators for Embedded DL Systems

To expedite DL inference for embedded applications in an energy-efficient manner, many specialized hardware accelerators have been proposed [9, 10, 16, 20, 26, 33, 38, 47, 63, 67, 71]. All these DL accelerators support specific dataflows while providing some unique advantages, and among them, *systolic array-based designs are considered among the most prominent ones* [20, 26, 38, 63].

A systolic array-based accelerator has a set of processing elements (PEs), which are tightly connected together in a homogeneous network, as shown in Fig. 3a. Each PE receives data from its nearest adjacent PEs, performs MAC operation, and passes the result and data to the adjacent PEs. In this manner, *data reuse* is exploited, reducing the need for expensive memory accesses and alleviating the memory bottleneck. Moreover, the systolic array is inherently suitable for performing matrix multiplications, which is the main operation in neural networks. For instance, the *Tensor Processing Unit (TPU)*, a DL accelerator developed by Google, has a systolic array architecture with 256x256 MAC units (i.e., PEs), and achieves 15x-30x faster performance and 30x-80x more efficiency as compared to the K80 GPU and the Haswell CPU [26]. The systolic array engine receives data from memories designed to meet the requirements of systolic array computations, such as latency and throughput, hence avoiding memory bottleneck. In this chapter, *the typical systolic array-based architecture of DL accelerators is considered*, as shown in Fig. 3b.

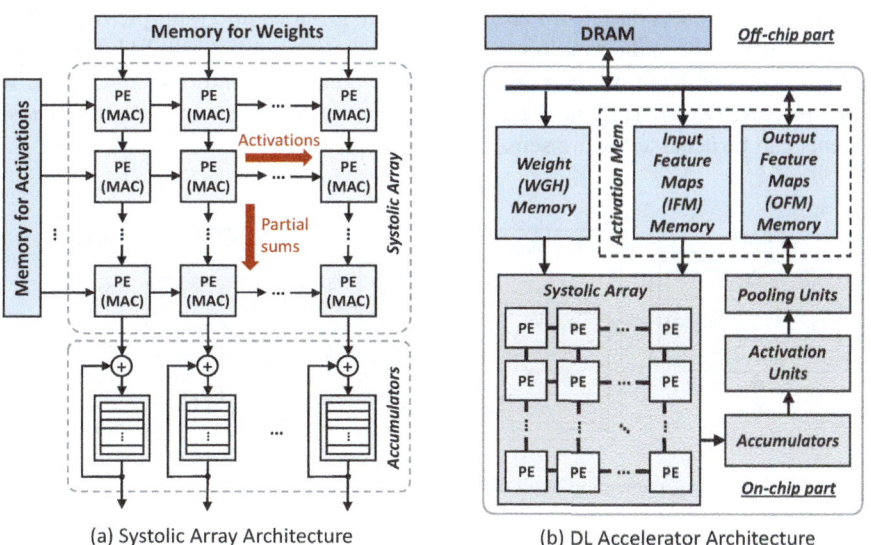

(a) Systolic Array Architecture   (b) DL Accelerator Architecture

Fig. 3 (**a**) Systolic array architecture (adapted from [20]). (**b**) Architecture of systolic array-based DL accelerators (adapted from [51])

Systolic Array Computations Before performing the neural network processing, weights are accessed from the weight memory, then loaded and held stationary in the PEs, in the manner that the same column of the array is loaded with weights from the same neuron or filter. During the processing, the activations are accessed from the activation memory and then streamed into the array. At each clock cycle, the activations are passed on adjacent PEs from left to right of the array, while *the partial sums* are moved downstream on adjacent PEs from top to bottom of the array. The activations across rows are aligned so that each activation reaches a particular PE at the same time when its corresponding partial sum also reaches the same PE, thereby generating a correct output partial sum. If the number of weights of a filter is larger than the number of rows in the array, output partial sums are divided into multiple portions. To support this, the accumulators hold the generated partial sums when the rest of the partial sums are computed by the array. A further detailed description of the architecture can be found in [20, 26, 70].

2.3 DRAM Fundamentals

2.3.1 Organization

DRAM is widely used as the main memory in modern computing systems, and it is organized hierarchically, as shown in Fig. 4. DRAM is accessed through *memory channel*. Each *channel* has specialized command, address, and data buses [45], and can be used for connecting multiple *memory modules*. Each memory module typically contains several *DRAM chips*, which are grouped into multiple *ranks*. The DRAM chips in the same rank operate in lockstep and parallel [28, 51]. Each chip typically contains several *banks*, and each bank consists of multiple *subarrays*. In each subarray, *DRAM cells* are organized into multiple *rows* and *columns*, whose contents can be accessed using sense amplifiers.

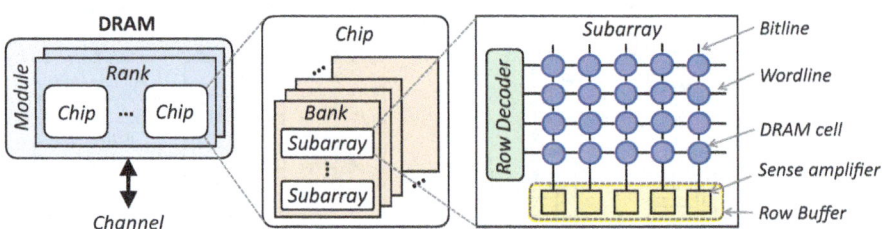

Fig. 4 A hierarchical organization of a DRAM (adapted from [28])

2.3.2 Operations

If there is a DRAM request, either read or write, a specific DRAM rank will respond. Afterwards, each DRAM request is decoded into a specific address of DRAM cells. Depending upon the request type (read/write), the contents of DRAM cells are accessed and read/written. To do this, the following DRAM operations are performed.

- *Activate (ACT)*: This operation activates a specific row of a DRAM bank, whose address is defined in the given command. Afterwards, data bits in the activated row are read into the *row buffer*.
- *Read (RD)* or *Write (WR)*: This operation accesses the data bits in the specified column of the row buffer. If the RD operation is executed, then the data bits are read and brought to DRAM I/O for on-chip computations. Meanwhile, if WR operation is executed, then the data bits are written by the given value.
- *Precharge (PRE)*: To activate a new row, the DRAM bank (with an activated row) should be brought back to the precharged state. To do this, PRE operation deactivates the row buffer and ensures that data bits from the row buffer are copied back to the corresponding activated row.

The above descriptions indicate that, accessing data that is already in the row buffer can be performed fast and efficiently. This condition is known as *a row buffer hit* [11]. Meanwhile, accessing data that is not in the row buffer, incurs higher latency and energy consumption than a row buffer hit. If such access happens when there is no activated row, then this is known as *a row buffer miss*, but if it happens when there is an activated row, then this is known as *a row buffer conflict* [11]. A row buffer conflict has a longer service time than a row buffer miss, as a row buffer conflict must wait to issue a PRE operation, and may also need to wait for an earlier request to complete. For a detailed explanation of DRAM fundamentals, we refer to DRAM papers [11, 28, 51].

3 DRAM Access Optimization for Embedded DL Systems

3.1 Overview

Several techniques can be employed for optimizing the DRAM access energy for embedded DL systems. These techniques can be categorized into run-time and design-time approaches, encompassing software- and hardware-level techniques, as shown in Fig. 5. At run time, *power management techniques*, such as *clock gating*, *power gating*, and *Dynamic Voltage and Frequency Scaling (DVFS)*, can be employed to reduce the dynamic power of DRAM [59]. However, these techniques need monitoring and decision units to manage the operational power properly, thereby requiring sophisticated designs to meet the constraints of the embedded applications (e.g., latency, throughput, and energy) [59]. At design time,

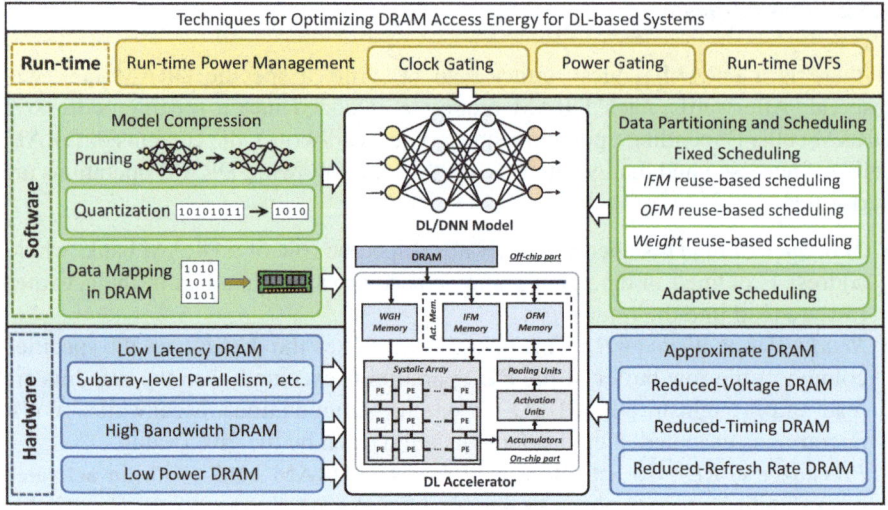

Fig. 5 Techniques for optimizing DRAM access energy for DL-based systems. Detailed descriptions for each technique can be found in [59]

software- and hardware-level techniques can be employed to reduce the DRAM access energy. The software-level techniques include *model compression* (such as *pruning* [3, 17, 22, 35, 41] and *quantization* [14, 17, 23, 40]), *data partitioning and scheduling* [36, 51, 62, 67], and *DRAM data mapping* [31, 49, 51]. Meanwhile, the hardware-level techniques include the employment of different DRAM designs, such as *low latency DRAM* [49], *high bandwidth DRAM* [24, 25, 37, 48], *low power DRAM* [30, 64], and *approximate DRAM* [31, 52].

In this chapter, we discuss our design methodology to optimize the DRAM access energy by ❶ *reducing the DRAM accesses through design space exploration (DSE)*, and *judiciously employing low latency DRAM through* ❷ *effective DRAM data mapping*, and ❸ *DSE that investigates the energy-delay-product (EDP) of DRAM accesses*; see Fig. 6. Our design methodology is a generic solution for different DNN models and different sizes of DL accelerators (e.g., sizes of memory and systolic array), thereby having high applicability for different embedded applications. In the following, we discuss the details for each proposed technique.

3.2 Reduction of DRAM Accesses

The size of a DNN model is usually larger than the size of on-chip memory in DL accelerators [61]. Therefore, to run the inference process on such accelerators, the data need to be partitioned into tiles.[1] Afterwards, a tile of *weights (WGH)* and a

---

[1] Tiling technique is widely used in the DL community to partition the DNN data, as it can exploit data reuse in convolutional processing [51].

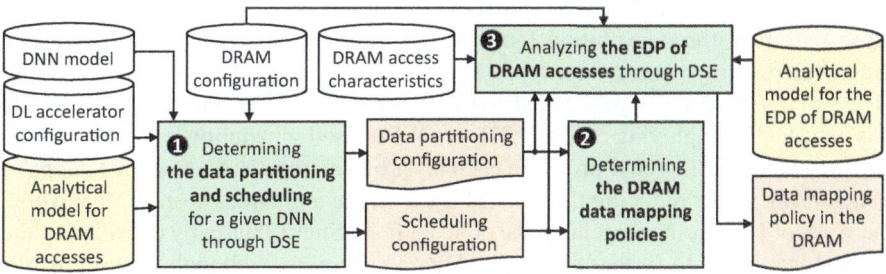

Fig. 6 Our design methodology for optimizing DRAM energy for embedded DL systems, showing the key proposed techniques (adapted from studies in [49, 51])

```
for ( b = 0; b < B; b++ ) {                                    Outer loops
  for ( x = 0; x < X; x += T_x ) {
    for ( y = 0; y < Y; y += T_y ) {
      for ( z = 0; z < Z; z += T_z ) {
        for ( c = 0; c < C; c += T_c ) {
          // load WGH, IFM, and OFM (partial sum)
          for ( k = 0; k < K; k++ ) {                          Inner loops
            for ( l = 0; l < L; l++ ) {
              for ( xi = x; xi < min(xi+T_x, X); xi++ ) {
                for ( yi = y; yi < min(yi+T_y, Y); yi++ ) {
                  for ( zi = z; zi < min(zi+T_z, Z); zi++ ) {
                    for ( ci = c; ci < min(ci+T_c, C); ci++ ) {
                      OFM [b][xi][yi][zi] += ...
                      WGH [k][l][ci][zi] * IFM [b][S*xi+k][S*yi+l][ci] }}}}}}
          // store OFM (partial sum) }}}}}
```

- B: number of samples in a batch.
- S: stride of convolution.
- OFM: height (X), width (Y), depth (Z).
- WGH: height (K), width (L), depth (C).
- T_x, T_y, T_z, and T_c: Tiling factor for X, Y, Z, and C, respectively.

Fig. 7 Pseudo-code of the tile-based convolution processing (adapted from studies in [36, 51])

tile of *input feature maps (IFM)* are transferred from DRAM to on-chip memory, and used in computations for producing a tile of *output feature maps (OFM)* at one time. This tile-based DNN processing can be represented as convolution loops, which can be divided into two parts, i.e., *outer and inner loops*, as shown in Fig. 7. The outer loops represent how the data transfer between off-chip and on-chip parts is scheduled, reflecting the DRAM accesses. Meanwhile, the inner loops represent how the on-chip computations are performed. However, *this tile-based processing usually requires redundant accesses for the same data to DRAM, thereby leading to high energy consumption.*

To address this issue, recent works have proposed different DSE techniques to find data partitioning and scheduling that offer minimum DRAM accesses for DNN workloads [36, 62, 67, 68]. These techniques aim at maximizing the on-chip data reuse based on the convolution loops in Fig. 7. Work of [67, 68] exploited data reuse based on a specific data type (i.e., either WGH, IFM, or OFM) across all layers of a network (i.e., so-called *fixed scheduling*). SmartShuttle [36] extended this concept by employing either WGH- or OFM-based scheduling based on the data type that has a higher reuse profile in each layer of a network (i.e., so-called *adaptive scheduling*), hence reducing the DRAM accesses than previous works. Meanwhile, the work of [62] further extended the concept from SmartShuttle by considering the DRAM bus-width to further reduce the DRAM accesses. Although all these works have provided some advantages, they do not optimize redundant accesses for the overlapping data (as shown in Fig. 8) and do not consider all possible scheduling schemes. Therefore, their optimization results are sub-optimal.

Towards this, we develop *a DSE technique to find the effective data partitioning and scheduling that lead to minimum DRAM accesses for each layer of a network, while considering overlapping data, all possible scheduling schemes, and DRAM organization* [49, 51], as shown by label ❶ in Fig. 6, through the following key steps.

- **Define the data partitioning:** We define different combinations of data partitioning for different data types (i.e., WGH, IFM, and OFM) in each layer of a network, while considering the available on-chip memory. We consider accessing a tile of IFM and a tile of WGH from DRAM, then storing them on-chip for computations that produce a tile of OFM. The generated OFM are then stored back to DRAM.
- **Define the adaptive scheduling:** For each combination of data partitioning for all data types, we consider different possible scheduling schemes (i.e., WGH-, IFM-, and OFM-based reuse scheduling) for evaluating the corresponding DRAM accesses. When defining the scheduling, we consider avoiding redundant DRAM accesses for the respective overlapping data, as shown in Fig. 8.
- **Evaluate the number of DRAM accesses:** To quickly explore different combinations of data partitioning and scheduling schemes, we employ the analytical model for estimating the number of DRAM accesses, while considering the DRAM organization and DRAM data alignment, as presented in Eqs. 1–6. *Further details on how we map the data in DRAM will be discussed in Sect. 3.3.1.*

Our Analytical Model for Estimating the Number of DRAM Accesses The total number of DRAM accesses of an inference ($\#DR_{access}$) is defined as the sum of the DRAM accesses from all layers of a network, as shown in Eq. 1. $\#DR^l_{access}$ denotes the total number of DRAM accesses in layer-l, and L denotes the number of layers in a given network.

$$\#DR_{access} = \sum_{l=1}^{L} \#DR^l_{access} \tag{1}$$

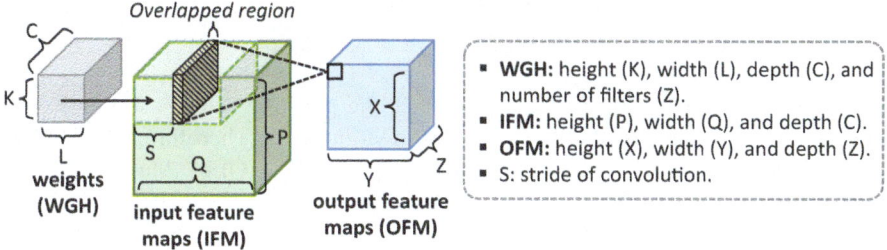

WGH: height (K), width (L), depth (C), and number of filters (Z).
IFM: height (P), width (Q), and depth (C).
OFM: height (X), width (Y), and depth (Z).
S: stride of convolution.

Fig. 8 Illustration of convolution processing showing the overlapped region (adapted from [51])

DRAM accesses-per-layer ($\#DR^l_{access}$) can be defined as the sum of DRAM accesses for all data types, as shown in Eq. 2. For each data type, we define its DRAM accesses-per-layer ($\#access^l_x$) as the sum of the DRAM accesses-per-tile ($\#access^t_x$), as shown in Eq. 3. We consider tile-based accesses since the size of on-chip memory typically limits the volume of data that can be stored at one time for on-chip computations.

$$\#DR^l_{access} = \#access^l_{WGH} + \#access^l_{IFM} + \#access^l_{OFM} \tag{2}$$

$$\#access^l_x = \sum_{t=1}^{T_x} \#access^t_x \quad \text{with } x \in \{WGH, IFM, OFM\} \tag{3}$$

We observe that the WGH and IFM only need DRAM *read*, hence we estimate their number of DRAM accesses-per-tile using Eqs. 4–5. Meanwhile, the OFM may have DRAM *read* and *write*. These two access types (*read* and *write*) happen when a tile of partial sums in on-chip memory still needs to be calculated with other partial sums to produce the final OFM, but they cannot be accumulated with the latest generated partial sums. Hence, these partial sums (that are stored in the on-chip memory), have to be transferred to DRAM so that the on-chip memory can store the latest generated data. Later, these partial sums (that are stored in the DRAM) will be transferred back to the on-chip memory for further computations generating a tile of final OFM. We estimate the number of DRAM accesses-per-tile for OFM using Eq. 6.

$$\#access^t_{WGH} = \left\lceil \frac{T_k \cdot T_l \cdot T_c \cdot T_z}{D_p} \right\rceil_{read} \tag{4}$$

$$\#access^t_{IFM} = \left\lceil \frac{T_p \cdot T_q \cdot T_c}{D_p} \right\rceil_{read} \tag{5}$$

$$\#access^t_{OFM} = \left\lceil \frac{T_x \cdot T_y \cdot T_z}{D_p} \right\rceil_{read} + \left\lceil \frac{T_x \cdot T_y \cdot T_z}{D_p} \right\rceil_{write} \tag{6}$$

Note, T_x denotes the tiling factor that defines the partition size for dimension-x, following the dimensions of data types, as shown in Fig. 8. Therefore, $x \in \{k, l, c, z\}$ for WGH, $x \in \{p, q, c\}$ for IFM, and $x \in \{x, y, z\}$ for OFM. Meanwhile, D_p denotes the number of DRAM chips-per-rank which operate in parallel.

3.3   Employment of Low Latency DRAM

DL accelerators can employ different types of DRAM based on the design requirements, e.g., low latency DRAM, low power DRAM, high bandwidth memory, etc. Since different DRAM types have the same internal organization, they have similar behavior of latency and energy consumption for every single access [11]. Therefore, the DRAM latency-per-access and energy-per-access always depend on whether a single access faces a row buffer hit, miss, or conflict [49]. Towards this, *we employ a low latency DRAM since it has lower latency-per-access and lower energy-per-access than conventional DRAM, which are beneficial for embedded DL systems.* Specifically, we employ a low latency DRAM design that exploits subarray-level parallelism (SALP) [28]. The reason is that any techniques that exploit SALP will be applicable for any DRAM types since a commodity DRAM bank is typically implemented as multiple subarrays with multiple local row buffers, as shown in Fig. 4. Work of [28] has proposed three variants of SALP architectures, i.e., SALP-1, SALP-2, and Multitude of Activated Subarrays (MASA), whose key ideas are explained in the following.

- **SALP-1:** It reduces the service time of commodity DRAM by overlapping ACT of one subarray with PRE of another subarray. To do this, a re-interpretation of the timing constraint for PRE operation is required.
- **SALP-2:** It reduces the service time more than SALP-1 by overlapping the ACT of one subarray with the latency of write-recovery for an active subarray. To do this, additional circuitry is required to activate two subarrays at the same time.
- **SALP-MASA:** It reduces the service time more than SALP-2 by activating multiple subarrays at the same time. To do this, additional circuitry is required to activate multiple subarrays at the same time, and it is more complex than the circuitry for SALP-2.

To understand the characteristics of SALP architectures, we perform experiments to observe their latency-per-access and energy-per-access for each row buffer hit, row buffer miss, row buffer conflict, subarray-level parallelism, and bank-level parallelism. The experimental results in Fig. 9 indicate that SALP architectures have the potential to reduce the latency-per-access and energy-per-access as compared to commodity DRAM. To judiciously employ such low latency DRAM architectures for embedded DL systems, *we define an effective DRAM data mapping policy and*

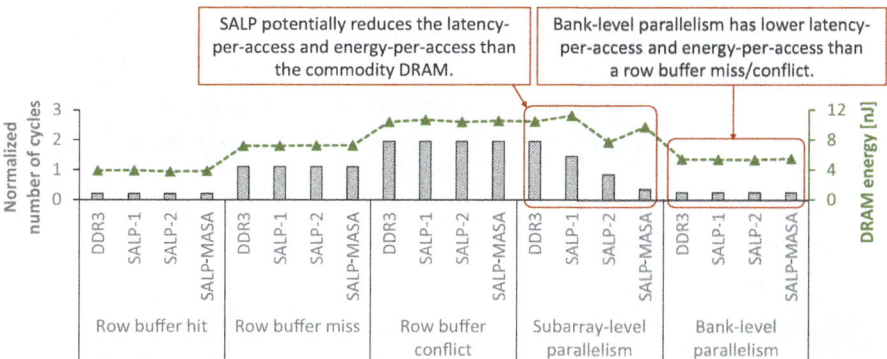

Fig. 9 Experimental results for the latency and energy consumption for a row buffer hit, a row buffer miss, a row buffer conflict, a subarray-level parallelism, and a bank-level parallelism on different DRAM architectures, i.e., DDR3, SALP-1, SALP-2, and SALP-MASA (adapted from [49]). For DDR3, we use DDR3-1600 2Gb x8 configuration, while for SALP, we use 2Gb x8 with 8 subarrays-per-bank. We generate data using state-of-the-art cycle-accurate DRAM simulators [12, 29]

conduct a DSE that quickly evaluates the EDP of DRAM accesses, which will be discussed in the Sects. 3.3.1 and 3.3.2, respectively.

3.3.1 Devising the Data Mapping Policy in DRAM

Figure 9 shows that different DRAM architectures (e.g., DDR3, SALP-1, SALP-2, and SALP-MASA) have similar patterns in terms of latency-per-access and energy-per-access. To exploit such patterns, we develop *a DRAM data mapping policy that incurs the lowest DRAM latency-per-access and energy-per-access for DL accelerators*, as shown by label ❷ in Fig. 6. Its idea is to *orderly prioritize the row buffer hit, bank-level parallelism, and subarray-level parallelism (if applicable)* for each given data. Following are the key steps for performing our data mapping.

1. We map data from a given data tile to different columns in the same row of a bank to maximize the row buffer hits. If multiple DRAM chips exist in the same rank, then this step is also performed in different chips for maximizing the chip-level parallelism.
2. If all columns in the same row are filled, then the remaining data are placed on different banks in the same DRAM chip to maximize bank-level parallelism. If multiple DRAM chips exist in the same rank, then this step is also performed in different chips.
3. If all columns in the same row' of all banks are filled, then the remaining data are placed on a different subarray in the same bank to maximize subarray-level parallelism. If multiple DRAM chips exist in the same rank, then this step is also performed in different chips.

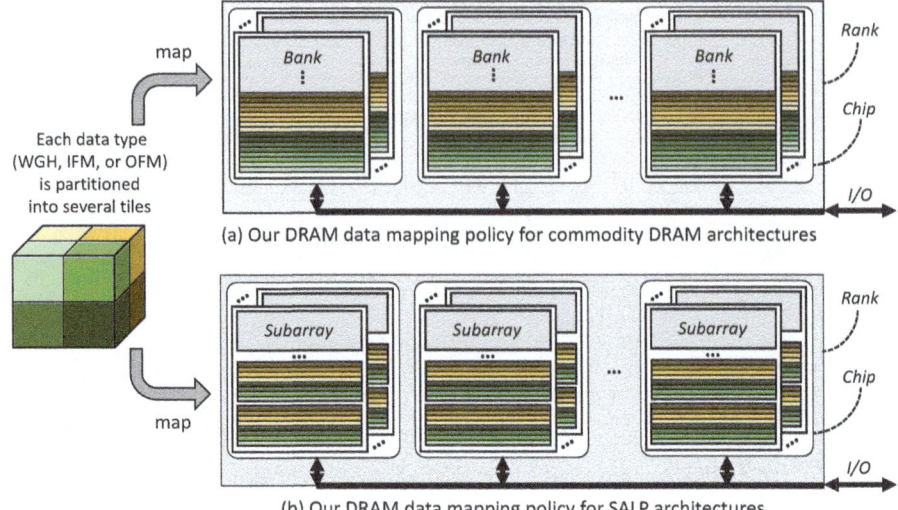

(a) Our DRAM data mapping policy for commodity DRAM architectures

(b) Our DRAM data mapping policy for SALP architectures

Fig. 10 Our data mapping policy (**a**) for commodity DRAM architectures, and (**b**) for SALP architectures, i.e., SALP-1, SALP-2, and SALP-MASA (adapted from studies in [49, 51])

4. If there are remaining data, then steps (1) to (3) are performed for different subarrays until all data are stored in the same rank.
5. If there are remaining data, then steps (1) to (4) are performed for different ranks, modules, and channels subsequently, if applicable.

For commodity DRAM architectures, our data mapping policy is shown in Fig. 10a, while for SALP architectures, our data mapping policy is shown in Fig. 10b.

3.3.2 Analysis for the EDP of DRAM Accesses

Our Analytical Model for Estimating the EDP of DRAM Accesses We leverage the experimental results in Fig. 9 to estimate the DRAM access latency and energy for a given network. Such estimation is beneficial for quickly investigating different possible configurations (e.g., DRAM organization, on-chip memory size, etc.) and determining the DRAM data mapping policy that can support data partitioning and scheduling, while incurring minimum DRAM access latency and energy. Towards this, we develop *an analytical model for estimating the energy-delay-product (EDP) of the DRAM accesses for a given network* [49], as shown by label ❸ in Fig. 6. Our analytical model employs the following key ideas.

- We define the EDP of DRAM accesses for an inference as the sum of the EDP-per-layer of a given network. Meanwhile, the EDP-per-layer is obtained by multiplying the DRAM latency-per-layer and the DRAM energy-per-layer.

- We calculate the latency-per-layer by accumulating all latency values required by the DRAM accesses for all data tiles in a layer. Then, we calculate the energy-per-layer by accumulating all energy values required by the DRAM accesses for all data tiles in a layer.
- For each data tile, we estimate the number of cycles required for DRAM accesses (N_{tile}) using Eq. 7 and estimate the energy consumption for DRAM accesses (E_{tile}) using Eq. 8.

$$N_{tile} = A_{dif\_column} \cdot N_{dif\_column} + A_{dif\_row} \cdot N_{dif\_row} +$$
$$A_{dif\_subarray} \cdot N_{dif\_subarray} + A_{dif\_bank} \cdot N_{dif\_bank} \tag{7}$$

$$E_{tile} = A_{dif\_column} \cdot E_{dif\_column} + A_{dif\_row} \cdot E_{dif\_row} +$$
$$A_{dif\_subarray} \cdot E_{dif\_subarray} + A_{dif\_bank} \cdot E_{dif\_bank} \tag{8}$$

Note, A_{dif_x} denotes the number of accesses to different DRAM x-location. N_{dif_x} denotes the number of cycles required for accessing different DRAM x-location. E_{dif_x} denotes the energy consumption required for accessing different DRAM x-location. For all terms, $x \in \{$column, row, subarray, bank$\}$.

Our DSE for Evaluating the EDP of DRAM Accesses To show that our DRAM data mapping policy always leads to minimum EDP in different possible conditions, we perform a DSE that leverages our analytical models on DRAM accesses and EDP estimation. The key idea of our DSE is to evaluate the EDP under different possible conditions, including different DRAM mapping policies (as presented in Table 1), different data partitioning and scheduling schemes (i.e., WGH-, IFM-, and OFM-based scheduling), and different DRAM architectures (e.g., DDR3, SALP-1, SALP-2, and SALP-MASA). *This DSE is important to show that the best solution that achieves the minimum EDP for each given condition is the same as our design methodology through (1) effective data partitioning and scheduling, and (2) effective DRAM data mapping policy.*

Table 1 Different (loop-based) DRAM mapping policies which are considered in the DSE. Note, our DRAM mapping policy is the same as the Mapping 3

Mapping	Inner-most loop to outer-most loop
1	column, subarray, bank, row
2	subarray, column, bank, row
3	column, bank, subarray, row
4	bank, column, subarray, row
5	subarray, bank, column, row
6	bank, subarray, column, row

4   Experimental Evaluations

To evaluate our design methodology, we build the following experimental setup.

DRAM Simulators We use a cycle-accurate DRAM simulator [29] to obtain the number of cycles for different DRAM access conditions, i.e., row buffer hit, row buffer miss, row buffer conflict, subarray-level parallelism, and bank-level parallelism. We also use a real experiments-based DRAM power/energy simulator [12] to obtain the energy consumption values for different DRAM access conditions.

DSE Simulator We develop our DSE simulator to find the data partitioning and scheduling that offer minimum EDP of DRAM accesses, while considering the network information (e.g., number of layers, data size, etc.), the DRAM access statistics (e.g., number of cycles and energy consumption), the configuration of a DL accelerator, and our analytical models on DRAM accesses and EDP estimation.

DL Accelerator and Workload We employ a TPU-like DL accelerator with reduced size of on-chip memories and systolic array. Details of the DL accelerator are provided in Table 2. To represent different DRAM architectures, we employ DDR3 and SALP (i.e., SALP-1, SALP-2, and SALP-MASA). For DRAM mapping policies, we use different mapping policies shown in Table 1. For the workload, we use AlexNet [32] with the ImageNet dataset.

4.1   Reduction of DRAM Accesses

Evaluation results for the number of DRAM accesses are shown in Fig. 11. Our design methodology decreases the number of DRAM accesses over other state-of-the-art works, e.g., by 12% over the BWA design. These improvements come

Table 2 Configuration of the systolic array-based DL accelerator

Module	Description
Systolic array	8 × 8 of PEs (MAC units)
On-chip memories	WGH: 64KB; IFM: 64KB; OFM: 64KB
Memory controller	Policy = open row; Scheduler = FCFS
DDR3-1600	Configuration: 2Gb x8
	1 channel, 1 rank-per-channel,
	1 chip-per-rank, 8 banks-per-chip
SALP	Configuration: 2Gb x8
	1 channel, 1 rank-per-channel,
	1 chip-per-rank, 8 banks-per-chip,
	8 subarrays-per-bank

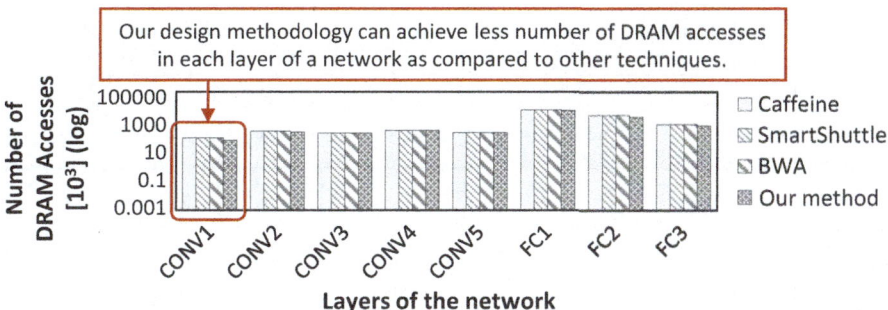

Fig. 11 Experimental results for the number of DRAM accesses on AlexNet under DDR3-1600 DRAM configuration, using Caffeine [68], SmartShuttle [36], Bus-Width Aware (BWA) [62], and our methodology (adapted from [51]). Note, CONV denotes the convolution processing, and FC denotes the fully connected processing

from the effective data partitioning and scheduling that are found using our DSE, which can be associated with several reasons. First, our analytical model for DRAM accesses reduces redundant accesses for the overlapping data. Second, our DSE considers more possible scheduling schemes than other works, i.e., by observing WGH-, IFM-, and OFM-based reuse scheduling. Therefore, our design methodology employs a wider search space than other works, as it considers a more detailed analytical model as well as more data partitioning and scheduling schemes, which lead to a higher possibility of finding less number of DRAM accesses. These results show that our methodology offers reductions of the DRAM accesses on a layer-wise basis, which is in-line with the defined analytical model. Furthermore, these results also emphasize that the adaptive scheduling scheme can achieve the lowest DRAM accesses compared to the fixed one.

4.2 Impact of Different DRAM Mapping Policies on EDP

Our DSE investigates different possible scheduling schemes, such as fixed scheduling (i.e., either WGH-, IFM-, or OFM-based reuse scheduling) and adaptive scheduling. The experimental results suggest that the adaptive one always offers the lowest DRAM accesses, thus in-line with the observation in Sect. 4.1. Evaluation results for the impact of different DRAM mapping policies on EDP under *adaptive scheduling* and different DRAM architectures, are shown in Fig. 12. From these results, we make the following key observations.

- **Observation ①:** Mapping-3 (our DRAM mapping policy) achieves the lowest EDP across different network layers and different DRAM architectures. It indicates that our mapping policy is the most effective DRAM data mapping

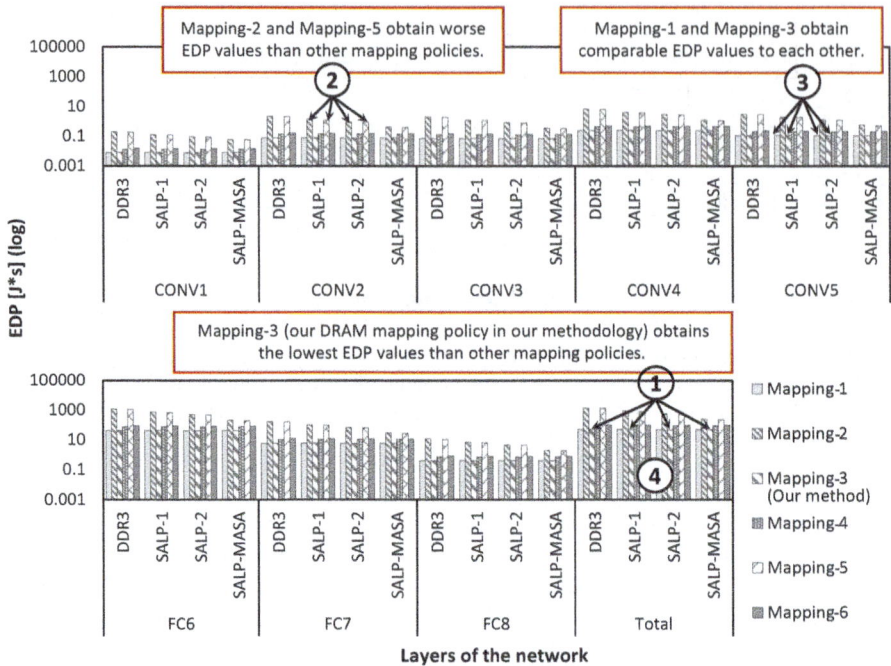

Fig. 12 Experimental results for the EDP of DRAM accesses on AlexNet under adaptive scheduling for different DRAM mapping policies and across different DRAM architectures, i.e., DDR3, SALP-1, SALP-2, and SALP-MASA (adapted from [49])

policy for different possible conditions that any DL accelerators may face. The reason is that, based on Table 1, Mapping-3 orderly prioritizes placing data on:

(1) different columns in the same row of a bank, hence maximizing row buffer hits in both DDR3 and SALP,

(2) different banks in the same chip, hence maximizing bank-level parallelism in both DDR3 and SALP,

(3) different subarrays in the same bank, hence maximizing subarray-level parallelism in SALP, but leading to row buffer conflicts in DDR3, and

(4) different rows in the same subarray, hence causing row buffer conflicts in both DDR3 and SALP.

Furthermore, any DL accelerators with any scheduling schemes can also benefit from our mapping policy to optimize their DRAM access latency and energy. Our mapping policy improves the EDP by up to 96% for DDR3 and 80%–94% for SALP, as compared to other mapping policies.

• **Observation ②:** Mapping-2 and Mapping-5 obtain worse EDP values across different layers of the network and different DRAM architectures, as compared to other mapping policies. The reason is that Mapping-2 and Mapping-5 prioritize placing data on different subarrays of the same bank. These mapping policies exploit subarray-level parallelism in SALP, but cause row buffer conflicts in DDR3, thereby consuming higher access latency and energy than row buffer hits and bank-level parallelism.

- **Observation ③:** Mapping-1 and Mapping-3 obtain comparable EDP values since they prioritize placing data on different columns of the same row, which leads to row buffer hits in both DDR3 and SALP. The difference between Mapping-1 and Mapping-3 is that Mapping-1 prioritizes exploiting subarray-level parallelism than bank-level parallelism, while Mapping-3 is the opposite. Therefore, Mapping-1 incurs higher EDP than Mapping-3.

- **Observation ④:** Generally, employing SALP architectures provides EDP improvements over DDR3 due to latency and energy reduction when exploiting subarray-level parallelism. The EDP improvements achieved by employing SALP as compared to DDR3 are 0.6%–3.9% (for Mapping-1), 19.9%–81% (for Mapping-2), 0.6%–3.9% (for Mapping-3), 0.5%–1.4% (for Mapping-4), 19.8%–81.8% (for Mapping-5), and 3.2%–7.6% (for Mapping-6).

4.3   Further Discussion

To further improve the energy efficiency of DL-based systems, a DNN model may go through a compression framework for achieving a compact model that can be deployed in tightly constrained embedded devices [17]. Recently, it has been observed that the structured pruning techniques are highly desirable due to their feasibility to be deployed on DL hardware accelerators [3, 22]. Towards this, our design methodology can be combined with the state-of-the-art structured pruning techniques, like the AutoML for Model Compression (AMC) [22], to efficiently expedite the sparse DNN model, thereby further reducing the DRAM access energy for efficient embedded DL systems [51].

5   Conclusion

In this chapter, we discuss our design methodology to optimize the DRAM access energy for embedded DL systems. It employs a DSE that incorporates our analytical model for DRAM accesses, to find the effective data partitioning and scheduling that offer the minimum DRAM accesses. It also employs low latency DRAM and effective DRAM data mapping policy to ensure that each DRAM request always incurs minimum latency-per-access and energy-per-access. We also employ a DSE that incorporates our analytical model for EDP estimation to corroborate that our design choice (data partitioning, scheduling, and DRAM mapping policy) always provides minimum EDP of DRAM accesses. In this manner, our design methodology can determine how the data partitioning, scheduling, and data mapping in DRAM should be performed for the given DL accelerator and network for meeting the design constraints of embedded DL systems.

Acknowledgments This work was partly supported by the Indonesia Endowment Fund for Education (IEFE/LPDP) Graduate Scholarship Program, Ministry of Finance, Republic of Indonesia, under Grant PRJ-1477/LPDP.3/2017.

References

1. Achararit, P., Hanif, M.A., Putra, R.V.W., Shafique, M., Hara-Azumi, Y.: APNAS: Accuracy-and-performance-aware neural architecture search for neural hardware accelerators. IEEE Access **8**, 165319–165334 (2020). https://doi.org/10.1109/ACCESS.2020.3022327
2. Ahmad, H., Arif, T., Hanif, M.A., Hafiz, R., Shafique, M.: SuperSlash: A unified design space exploration and model compression methodology for design of deep learning accelerators with reduced off-chip memory access volume. IEEE Trans. Comput. Aided Des. Integr. Circ. Syst. **39**(11), 4191–4204 (2020). https://doi.org/10.1109/TCAD.2020.3012865
3. Anwar, S., Hwang, K., Sung, W.: Structured pruning of deep convolutional neural networks. J. Emerg. Technol. Comput. Syst. **13**(3) (2017). https://doi.org/10.1145/3005348
4. Arslan, A.K., Yasar, S., Colak, C.: An intelligent system for the classification of lung cancer based on deep learning strategy. In: 2019 International Artificial Intelligence and Data Processing Symposium (IDAP), pp. 1–4 (2019). https://doi.org/10.1109/IDAP.2019.8875896
5. Barata, C., Marques, J.S.: Deep learning for skin cancer diagnosis with hierarchical architectures. In: 2019 IEEE 16th International Symposium on Biomedical Imaging (ISBI 2019), pp. 841–845 (2019). https://doi.org/10.1109/ISBI.2019.8759561
6. Bengio, Y.: Learning deep architectures for AI. Found. Trends Mach. Learn. **2**(1), 1–127 (2009). https://doi.org/10.1561/2200000006
7. Capra, M., Bussolino, B., Marchisio, A., Shafique, M., Masera, G., Martina, M.: An updated survey of efficient hardware architectures for accelerating deep convolutional neural networks. Future Internet **12**(7), 113 (2020)
8. Capra, M., Peloso, R., Masera, G., Ruo Roch, M., Martina, M.: Edge computing: A survey on the hardware requirements in the internet of things world. Future Internet **11**(4) (2019). https://doi.org/10.3390/fi11040100. https://www.mdpi.com/1999-5903/11/4/100
9. Chen, T., Du, Z., Sun, N., Wang, J., Wu, C., Chen, Y., Temam, O.: Diannao: A small-footprint high-throughput accelerator for ubiquitous machine-learning. In: 19th International Conference on Architectural Support for Programming Languages and Operating Systems, pp. 269–284 (2014). https://doi.org/10.1145/2541940.2541967
10. Chen, Y.H., Emer, J., Sze, V.: Eyeriss: A spatial architecture for energy-efficient dataflow for convolutional neural networks. In: 2016 ACM/IEEE 43rd Annual International Symposium on Computer Architecture, pp. 367–379 (2016). https://doi.org/10.1109/ISCA.2016.40
11. Ghose, S., Li, T., Hajinazar, N., Cali, D.S., Mutlu, O.: Demystifying complex workload-DRAM interactions: An experimental study. Proc. ACM Meas. Anal. Comput. Syst. **3**(3) (2019). https://doi.org/10.1145/3366708
12. Ghose, S., et al.: What your DRAM power models are not telling you: Lessons from a detailed experimental study. Proc. ACM Meas. Anal. Comput. Syst. **2**(3), 38:1–38:41 (2018). https://doi.org/10.1145/3224419
13. Grigorescu, S., Trasnea, B., Cocias, T., Macesanu, G.: A survey of deep learning techniques for autonomous driving. J. Field Rob. **37**(3), 362–386 (2020). https://doi.org/10.1002/rob.21918
14. Gupta, S., Agrawal, A., Gopalakrishnan, K., Narayanan, P.: Deep learning with limited numerical precision. In: Int. Conf. on Machine Learning (ICML), pp. 1737–1746 (2015)

15. Ha, V.S., Lu, D.N., Choi, G.S., Nguyen, H.N., Yoon, B.: Improving credit risk prediction in online peer-to-peer (p2p) lending using feature selection with deep learning. In: 2019 21st International Conference on Advanced Communication Technology (ICACT), pp. 511–515 (2019). https://doi.org/10.23919/ICACT.2019.8701943

16. Han, S., Liu, X., Mao, H., Pu, J., Pedram, A., Horowitz, M.A., Dally, W.J.: EIE: Efficient inference engine on compressed deep neural network. In: 2016 ACM/IEEE 43rd Annual International Symposium on Computer Architecture, pp. 243–254 (2016). https://doi.org/10.1109/ISCA.2016.30

17. Han, S., Mao, H., Dally, W.J.: Deep compression: Compressing deep neural networks with pruning, trained quantization and Huffman coding. Preprint (2015). arXiv:1510.00149

18. Hanif, M.A., Khalid, F., Putra, R.V.W., Rehman, S., Shafique, M.: Robust machine learning systems: Reliability and security for deep neural networks. In: 2018 IEEE 24th International Symposium on On-Line Testing And Robust System Design, pp. 257–260 (2018). https://doi.org/10.1109/IOLTS.2018.8474192

19. Hanif, M.A., Khalid, F., Putra, R.V.W., Teimoori, M.T., Kriebel, F., Zhang, J.J., Liu, K., Rehman, S., Theocharides, T., Artusi, A., et al.: Robust computing for machine learning-based systems. In: Dependable Embedded Systems, pp. 479–503. Springer, Cham (2021)

20. Hanif, M.A., Putra, R.V.W., Tanvir, M., Hafiz, R., Rehman, S., Shafique, M.: MPNA: A massively-parallel neural array accelerator with dataflow optimization for convolutional neural networks. Preprint (2018). arXiv:1810.12910

21. He, K., Zhang, X., Ren, S., Sun, J.: Deep residual learning for image recognition. In: Proceedings of the IEEE Conference on Computer Vision and Pattern Recognition, pp. 770–778 (2016)

22. He, Y., Lin, J., Liu, Z., Wang, H., Li, L.J., Han, S.: AMC: AutoML for model compression and acceleration on mobile devices. In: The European Conference on Computer Vision, pp. 784–800 (2018)

23. Jacob, B., Kligys, S., Chen, B., Zhu, M., Tang, M., Howard, A., Adam, H., Kalenichenko, D.: Quantization and training of neural networks for efficient integer-arithmetic-only inference. In: 2018 IEEE/CVF Conference on Computer Vision and Pattern Recognition, pp. 2704–2713 (2018). https://doi.org/10.1109/CVPR.2018.00286

24. Jain, A.K., Kumar, S., Tripathi, A., Gaitonde, D.: Sparse deep neural network acceleration on HBM-enabled FPGA platform. In: 2021 IEEE High Performance Extreme Computing Conference (HPEC), pp. 1–7 (2021). https://doi.org/10.1109/HPEC49654.2021.9622804

25. Jiang, W., He, Z., Zhang, S., Zeng, K., Feng, L., Zhang, J., Liu, T., Li, Y., Zhou, J., Zhang, C., et al.: FleetRec: Large-scale recommendation inference on hybrid GPU-FPGA clusters. In: Proceedings of the 27th ACM SIGKDD Conference on Knowledge Discovery & Data Mining, pp. 3097–3105 (2021)

26. Jouppi, N.P., Young, C., Patil, N., Patterson, D., Agrawal, G., Bajwa, R., Bates, S., Bhatia, S., Boden, N., Borchers, A., Boyle, R., Cantin, P., Chao, C., Clark, C., Coriell, J., Daley, M., Dau, M., Dean, J., Gelb, B., Ghaemmaghami, T.V., Gottipati, R., Gulland, W., Hagmann, R., Ho, C.R., Hogberg, D., Hu, J., Hundt, R., Hurt, D., Ibarz, J., Jaffey, A., Jaworski, A., Kaplan, A., Khaitan, H., Killebrew, D., Koch, A., Kumar, N., Lacy, S., Laudon, J., Law, J., Le, D., Leary, C., Liu, Z., Lucke, K., Lundin, A., MacKean, G., Maggiore, A., Mahony, M., Miller, K., Nagarajan, R., Narayanaswami, R., Ni, R., Nix, K., Norrie, T., Omernick, M., Penukonda, N., Phelps, A., Ross, J., Ross, M., Salek, A., Samadiani, E., Severn, C., Sizikov, G., Snelham, M., Souter, J., Steinberg, D., Swing, A., Tan, M., Thorson, G., Tian, B., Toma, H., Tuttle, E., Vasudevan, V., Walter, R., Wang, W., Wilcox, E., Yoon, D.H.: In-datacenter performance analysis of a tensor processing unit. In: 2017 ACM/IEEE 44th Annual Int. Symp. on Computer Architecture, pp. 1–12 (2017). https://doi.org/10.1145/3079856.3080246

27. Kaskavalci, H.C., Gören, S.: A deep learning based distributed smart surveillance architecture using edge and cloud computing. In: 2019 International Conference on Deep Learning and Machine Learning in Emerging Applications (Deep-ML), pp. 1–6 (2019). https://doi.org/10.1109/Deep-ML.2019.00009

28. Kim, Y., Seshadri, V., Lee, D., Liu, J., Mutlu, O.: A case for exploiting subarray-level parallelism (SALP) in DRAM. In: 2012 39th Annual International Symposium on Computer Architecture, pp. 368–379 (2012). https://doi.org/10.1109/ISCA.2012.6237032
29. Kim, Y., et al.: Ramulator: A fast and extensible DRAM simulator. IEEE Comput. Archit. Lett. 15(1), 45–49 (2016). https://doi.org/10.1109/LCA.2015.2414456
30. Ko, J.H., Na, T., Amir, M.F., Mukhopadhyay, S.: Edge-host partitioning of deep neural networks with feature space encoding for resource-constrained internet-of-things platforms. In: 2018 15th IEEE International Conference on Advanced Video and Signal Based Surveillance, pp. 1–6 (2018). https://doi.org/10.1109/AVSS.2018.8639121
31. Koppula, S., Orosa, L., Yağlıkçı, A.G., Azizi, R., Shahroodi, T., Kanellopoulos, K., Mutlu, O.: Eden: Enabling energy-efficient, high-performance deep neural network inference using approximate DRAM. In: 52nd Annual IEEE/ACM Int. Symp. on Microarchitecture, pp. 166–181 (2019). https://doi.org/10.1145/3352460.3358280
32. Krizhevsky, A., Sutskever, I., Hinton, G.E.: ImageNet classification with deep convolutional neural networks. In: Advances in Neural Information Processing Systems, pp. 1097–1105 (2012)
33. Kwon, H., Samajdar, A., Krishna, T.: Maeri: Enabling flexible dataflow mapping over DNN accelerators via reconfigurable interconnects. In: 23th International Conference on Architectural Support for Programming Languages and Operating Systems, pp. 461–475 (2018). https://doi.org/10.1145/3173162.3173176
34. LeCun, Y., Bengio, Y., Hinton, G.: Deep learning. Nature 521(7553), 436–444 (2015)
35. Li, H., Kadav, A., Durdanovic, I., Samet, H., Graf, H.P.: Pruning filters for efficient convnets. Preprint (2016). arXiv:1608.08710
36. Li, J., Yan, G., Lu, W., Jiang, S., Gong, S., Wu, J., Li, X.: SmartShuttle: Optimizing off-chip memory accesses for deep learning accelerators. In: 2018 Design, Automation Test in Europe Conference Exhibition, pp. 343–348 (2018). https://doi.org/10.23919/DATE.2018.8342033
37. Li, Z., Zhang, Y., Wang, J., Lai, J.: A survey of FPGA design for AI era. J. Semicond. 41(2), 021402 (2020). https://doi.org/10.1088/1674-4926/41/2/021402
38. Lu, W., Yan, G., Li, J., Gong, S., Han, Y., Li, X.: FlexFlow: A flexible dataflow accelerator architecture for convolutional neural networks. In: 2017 IEEE International Symposium on High Performance Computer Architecture, pp. 553–564 (2017). https://doi.org/10.1109/HPCA.2017.29
39. Maass, W.: Networks of spiking neurons: The third generation of neural network models. Neural Networks 10(9), 1659–1671 (1997). https://doi.org/10.1016/S0893-6080(97)00011-7
40. Marchisio, A., Bussolino, B., Colucci, A., Martina, M., Masera, G., Shafique, M.: Q-CapsNets: A specialized framework for quantizing capsule networks. In: 2020 57th ACM/IEEE Design Automation Conference
41. Marchisio, A., Hanif, M.A., Martina, M., Shafique, M.: Prunet: Class-blind pruning method for deep neural networks. In: 2018 Int. Joint Conf. on Neural Networks, pp. 1–8 (2018). https://doi.org/10.1109/IJCNN.2018.8489764
42. Minaee, S., Boykov, Y.Y., Porikli, F., Plaza, A.J., Kehtarnavaz, N., Terzopoulos, D.: Image segmentation using deep learning: A survey. IEEE Trans. Pattern Anal. Mach. Intell. (TPAMI), 1–1 (2021). https://doi.org/10.1109/TPAMI.2021.3059968
43. Mohsen, H., El-Dahshan, E.S.A., El-Horbaty, E.S.M., Salem, A.B.M.: Classification using deep learning neural networks for brain tumors. Future Comput. Inf. J. 3(1), 68–71 (2018). https://doi.org/10.1016/j.fcij.2017.12.001. https://www.sciencedirect.com/science/article/pii/S2314728817300636
44. Nvidia: Nvidia jetson nano. https://developer.nvidia.com/embedded/jetson-nano-developer-kit
45. Olgun, A., Luna, J.G., Kanellopoulos, K., Salami, B., Hassan, H., Ergin, O., Mutlu, O.: Pidram: A holistic end-to-end FPGA-based framework for processing-in-DRAM. Preprint (2021). arXiv:2111.00082

46. Palossi, D., Loquercio, A., Conti, F., Flamand, E., Scaramuzza, D., Benini, L.: Ultra low power deep-learning-powered autonomous nano drones. CoRR **abs/1805.01831** (2018). http://arxiv.org/abs/1805.01831

47. Parashar, A., Rhu, M., Mukkara, A., Puglielli, A., Venkatesan, R., Khailany, B., Emer, J., Keckler, S.W., Dally, W.J.: SCNN: An accelerator for compressed-sparse convolutional neural networks. In: 2017 ACM/IEEE 44th Annual International Symposium on Computer Architecture, pp. 27–40 (2017). https://doi.org/10.1145/3079856.3080254

48. Park, K., Han, Y., Kim, L.S.: Deferred dropout: An algorithm-hardware co-design DNN training method provisioning consistent high activation sparsity. In: 2021 IEEE/ACM International Conference On Computer Aided Design, pp. 1–9 (2021). https://doi.org/10.1109/ICCAD51958.2021.9643433

49. Putra, R.V.W., Hanif, M.A., Shafique, M.: DRMap: A generic DRAM data mapping policy for energy-efficient processing of convolutional neural networks. In: 2020 57th ACM/IEEE Design Automation Conference, pp. 1–6 (2020). https://doi.org/10.1109/DAC18072.2020.9218672

50. Putra, R.V.W., Hanif, M.A., Shafique, M.: Respawn: Energy-efficient fault-tolerance for spiking neural networks considering unreliable memories. In: 2021 IEEE/ACM International Conference On Computer Aided Design, pp. 1–9 (2021). https://doi.org/10.1109/ICCAD51958.2021.9643524

51. Putra, R.V.W., Hanif, M.A., Shafique, M.: ROMANet: Fine-grained reuse-driven off-chip memory access management and data organization for deep neural network accelerators. IEEE Trans. Very Large Scale Integr. (VLSI) Syst. **29**(4), 702–715 (2021). https://doi.org/10.1109/TVLSI.2021.3060509

52. Putra, R.V.W., Hanif, M.A., Shafique, M.: SparkXD: A framework for resilient and energy-efficient spiking neural network inference using approximate DRAM. In: 2021 58th ACM/IEEE Design Automation Conference, pp. 379–384 (2021). https://doi.org/10.1109/DAC18074.2021.9586332

53. Putra, R.V.W., Hanif, M.A., Shafique, M.: SoftSNN: Low-cost fault tolerance for spiking neural network accelerators under soft errors. Preprint (2022). arXiv:2203.05523

54. Putra, R.V.W., Shafique, M.: FSpiNN: An optimization framework for memory-and energy-efficient spiking neural networks. IEEE Trans. Comput. Aided Des. Integr. Circ. Syst. **39**(11), 3601–3613 (2020). https://doi.org/10.1109/TCAD.2020.3013049

55. Putra, R.V.W., Shafique, M.: Q-SpiNN: A framework for quantizing spiking neural networks. In: 2021 International Joint Conference on Neural Networks, pp. 1–8 (2021). https://doi.org/10.1109/IJCNN52387.2021.9534087

56. Putra, R.V.W., Shafique, M.: SpikeDyn: A framework for energy-efficient spiking neural networks with continual and unsupervised learning capabilities in dynamic environments. In: 2021 58th ACM/IEEE Design Automation Conference, pp. 1057–1062 (2021). https://doi.org/10.1109/DAC18074.2021.9586281

57. Putra, R.V.W., Shafique, M.: lpSpikeCon: Enabling low-precision spiking neural network processing for efficient unsupervised continual learning on autonomous agents. Preprint (2022). arXiv:2205.12295

58. Putra, R.V.W., Shafique, M.: tinySNN: Towards memory-and energy-efficient spiking neural networks. Preprint (2022). arXiv:2206.08656

59. Shafique, M., Marchisio, A., Putra, R.V.W., Hanif, M.A.: Towards energy-efficient and secure edge ai: A cross-layer framework ICCAD special session paper. In: 2021 IEEE/ACM International Conference On Computer Aided Design, pp. 1–9 (2021). https://doi.org/10.1109/ICCAD51958.2021.9643539

60. Shafique, M., Naseer, M., Theocharides, T., Kyrkou, C., Mutlu, O., Orosa, L., Choi, J.: Robust machine learning systems: Challenges, current trends, perspectives, and the road ahead. IEEE Des. Test **37**(2), 30–57 (2020)

61. Sze, V., Chen, Y., Yang, T., Emer, J.S.: Efficient processing of deep neural networks: A tutorial and survey. Proc. IEEE **105**(12), 2295–2329 (2017). https://doi.org/10.1109/JPROC.2017.2761740

62. Tewari, S., Kumar, A., Paul, K.: Bus width aware off-chip memory access minimization for CNN accelerators. In: 2020 IEEE Computer Society Annual Symposium on VLSI, pp. 240–245 (2020). https://doi.org/10.1109/ISVLSI49217.2020.00051
63. Wei, X., Yu, C.H., Zhang, P., Chen, Y., Wang, Y., Hu, H., Liang, Y., Cong, J.: Automated systolic array architecture synthesis for high throughput CNN inference on FPGAs. In: 2017 54th ACM/EDAC/IEEE Design Automation Conference, pp. 1–6 (2017). https://doi.org/10.1145/3061639.3062207
64. Yamada, Y., Sano, T., Tanabe, Y., Ishigaki, Y., Hosoda, S., Hyuga, F., Moriya, A., Hada, R., Masuda, A., Uchiyama, M., Jobashi, M., Koizumi, T., Tamai, T., Sato, N., Tanabe, J., Kimura, K., Ojima, Y., Murakami, R., Yoshikawa, T.: A 20.5 tops multicore soc with DNN accelerator and image signal processor for automotive applications. IEEE J. Solid State Circ. **55**(1), 120–132 (2020). https://doi.org/10.1109/JSSC.2019.2951391
65. Ying, J.J.C., Huang, P.Y., Chang, C.K., Yang, D.L.: A preliminary study on deep learning for predicting social insurance payment behavior. In: 2017 IEEE International Conference on Big Data, pp. 1866–1875 (2017). https://doi.org/10.1109/BigData.2017.8258131
66. Zanc, R., Cioara, T., Anghel, I.: Forecasting financial markets using deep learning. In: 2019 IEEE 15th International Conference on Intelligent Computer Communication and Processing, pp. 459–466 (2019). https://doi.org/10.1109/ICCP48234.2019.8959715
67. Zhang, C., Li, P., Sun, G., Guan, Y., Xiao, B., Cong, J.: Optimizing FPGA-based accelerator design for deep convolutional neural networks. In: ACM/SIGDA International Symposium on Field-Programmable Gate Arrays, pp. 161–170 (2015). https://doi.org/10.1145/2684746.2689060
68. Zhang, C., Sun, G., Fang, Z., Zhou, P., Pan, P., Cong, J.: Caffeine: Toward uniformed representation and acceleration for deep convolutional neural networks. IEEE Trans. Comput. Aided Des. Integr. Circ. Syst. **38**(11), 2072–2085 (2019). https://doi.org/10.1109/TCAD.2017.2785257
69. Zhang, D., Liu, S.E.: Top-down saliency object localization based on deep-learned features. In: 2018 11th International Congress on Image and Signal Processing, BioMedical Engineering and Informatics, pp. 1–9 (2018). https://doi.org/10.1109/CISP-BMEI.2018.8633218
70. Zhang, J., Rangineni, K., Ghodsi, Z., Garg, S.: ThUnderVolt: Enabling aggressive voltage underscaling and timing error resilience for energy efficient deep learning accelerators. In: Proceedings of the 55th Annual Design Automation Conference, DAC '18. Association for Computing Machinery, New York, NY, USA (2018). https://doi.org/10.1145/3195970.3196129
71. Zhang, S., Du, Z., Zhang, L., Lan, H., Liu, S., Li, L., Guo, Q., Chen, T., Chen, Y.: Cambricon-x: An accelerator for sparse neural networks. In: 2016 49th Annual IEEE/ACM International Symposium on Microarchitecture, pp. 1–12 (2016). https://doi.org/10.1109/MICRO.2016.7783723

In-Memory Computing for AI Accelerators: Challenges and Solutions

Gokul Krishnan, Sumit K. Mandal, Chaitali Chakrabarti, Jae-sun Seo, Umit Y. Ogras, and Yu Cao

1 Introduction

1.1 Machine Learning in Modern Times

Machine learning (ML) or artificial intelligence (AI) has made an enormous impact on the society. ML algorithms, such as deep neural networks (DNNs), achieve accuracy that exceeds human-level perception for a variety of applications, including computer vision, natural language processing, and medical imaging [19, 59, 63]. The popularity of ML algorithms has been driven by two main sources. First, the availability of big datasets for various applications, such as image classification, object detection, and segmentation [28, 59, 61]. Second, the increased computation power, provided by the next generation machine learning hardware accelerators and general purpose server platforms, has made both training and inference of large ML models more accessible.

Figure 1 shows the taxonomy of ML algorithms, which can be broadly classified into supervised and unsupervised learning. Unsupervised learning refers to the process of extracting features from a distribution without any annotation for the data. Applications of unsupervised learning include selecting samples from a distribution,

G. Krishnan
Arizona State University, Tempe, AZ, USA
e-mail: gkrish19@asu.edu

S. K. Mandal · U. Y. Ogras
University of Wisconsin-Madison, Madison, WI, USA
e-mail: skmandal@wisc.edu; uogras@wisc.edu

C. Chakrabarti · J.-s. Seo · Y. Cao (✉)
School of ECEE, Arizona State University, Tempe, AZ, USA
e-mail: chaitali@asu.edu; jseo28@asu.edu; Yu.Cao@asu.edu

© The Author(s), under exclusive license to Springer Nature Switzerland AG 2024
S. Pasricha, M. Shafique (eds.), *Embedded Machine Learning for Cyber-Physical, IoT, and Edge Computing*, https://doi.org/10.1007/978-3-031-19568-6_7

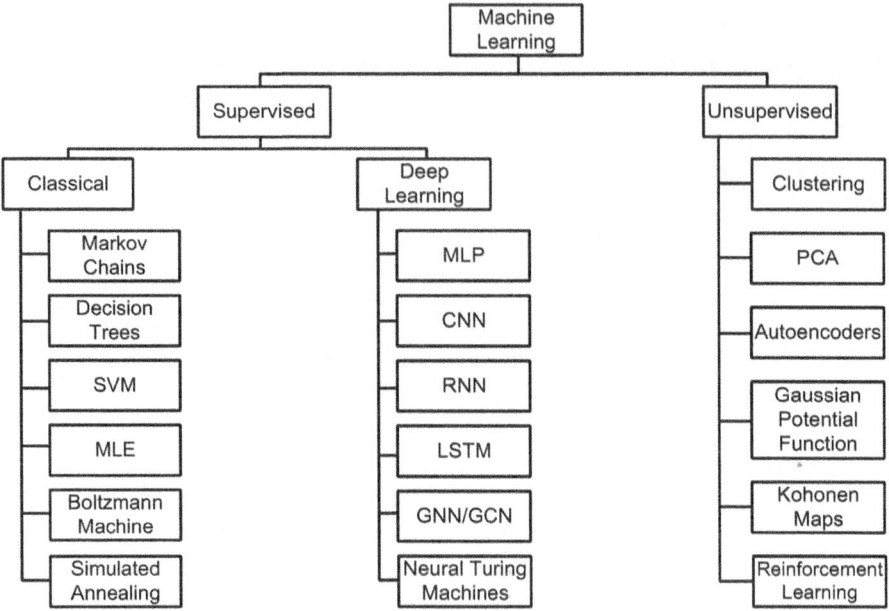

Fig. 1 Taxonomy of machine learning showing different types of learning and the associated techniques

learning to denoise data, and clustering data into different groups. The unsupervised learning algorithm aims to find the most optimal representation of the data. The optimal representation preserves maximum information about the input data x, while utilizing constraints to ensure the representation is simpler than the data itself. The three main ways of defining the simpler representation are lower dimensional representation, sparse representation, and independent representation [64, 86]. Popular unsupervised learning techniques include clustering, principal component analysis (PCA), autoencoders, Gaussian potential functions, etc.

Supervised learning deals with the ML model being trained with a set of labelled training set and testing it with a labelled testing set. Supervised learning can be classified into two types, classical approaches and deep learning. Classical approaches focus on conventional techniques that utilize a probabilistic model to determine the next state based on a set of parameters. Some of the popular classical techniques include Markov chains, decision trees, support vector machines (SVM), and maximum likelihood estimation (MLE), among others [23, 26, 48, 84]. But classical techniques suffered from several drawbacks including lack of generalization, difficulty in scaling, and the need for significant data engineering for each algorithm.

Deep learning algorithms are built on top of the classical techniques and resolve the drawbacks within them. In this chapter, we focus on the deep learning techniques for supervised learning. Convolutional neural networks (CNNs) are the most popular deep learning algorithm due to their ability to perform exceedingly well

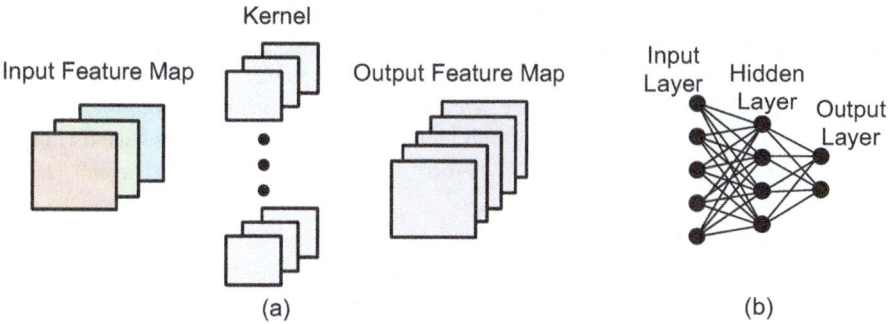

Fig. 2 (**a**) Convolution operation within a CNN consisting of the input feature map (IFM), kernel, and the output feature map (OFM). The kernel window slides over the IFM to generate the OFM and (**b**) fully connected (FC) layer in a CNN. Each neuron within the FC layer is connected to a neuron in the subsequent layer. The edges represent weights of the FC layer

for a variety of machine learning tasks, such as computer vision, object detection, and object segmentation. In addition, recurrent neural networks (RNNs) have been very effective in processing temporal data, while graph convolutional networks (GCNs) have combined both graphs and neural networks for a wide variety of applications. We will discuss the recent advancements in CNNs, RNNs, and GCNs with focus on structures, training methods, and execution efficiency for both training and inference operations. Conventional CNNs consist of a set of layers connected in a sequential manner or with skip connections. In addition to convolutional layers, ReLU, pooling, and batch-normalization are utilized for better performance.

Figure 2 shows the typical structure of a convolution and fully connected layer. The sequential layers usually consists of a stack of convolution (Conv) layers that perform feature extraction from the input. Examples of Conv layer kernels include 7×7, 5×5, 3×3, and 1×1. In addition, depth-wise convolutions proposed in MobileNet [32] break down a given $N\times N$ convolution into two parts. First, an $N\times1$ is performed, and the result is then run through a $1\times N$ convolution. Depth-wise convolution results in better accuracy and lower hardware complexity. Pooling layers are utilized periodically to reduce the feature map size and in turn truncate noisy input. Finally, a set of classifier layers or fully connected (FC) layers are utilized to perform classification on the extracted features. The Conv and FC layers have a set of weights that are trained to achieve the best accuracy. Popular CNN structures include AlexNet [59], GoogleNet [101], ResNet [29], DenseNet [34], MobileNet [32], and SqueezeNet [35]. CNNs such as DenseNet and ResNet feature skip connections from prior layers that result in a highly branched structure. The skip connections aim to improve the feature extraction process and are present within the Conv layers only.

On the other hand, conventional CNNs suffer from a wide range of drawbacks, including over-parameterization [21, 49, 51], higher hardware cost in training and inference, and vanishing gradient problem, among others. Network architecture search (NAS) was introduced to automatically search the optimal neural network architecture based on the target design point. The design point is determined by the target application. For example, higher accuracy, better generalization, higher hardware efficiency, lower memory access, etc. are some of the popular design points used in NAS. Some of the popular techniques proposed include NasNet [117], FBNet [105], AmoebaNet [85], PNAS [66], ECONas [115], and MNasNet [102], among others.

RNN is also a popular deep learning technique, which provides an efficient solution to model data with temporal or sequential structure and varying length inputs and outputs, across different applications [43, 62, 100]. RNNs process sequential data one element at a time utilizing a connectionist model with the ability to selectively pass information. Through this, RNNs model input and/or output data consisting of a sequence of elements that are dependent. Furthermore, RNNs can simultaneously model sequential and time dependencies at different scales. RNNs utilize a feedforward network that utilizes the edges that span adjacent time steps, introducing time to the model. RNNs do not have cycles among conventional edges, while edges that connect adjacent time steps, called recurrent edges, can form cycles. Modern RNN architectures can be classified into two main categories. Long short-term memory (LSTM) introduces the memory cell, a unit of computation that replaces traditional nodes in the hidden layer of a network [30]. The other variant of RNNs includes bi-directional RNNs (BRNNs) proposed in [90].

Finally, while CNNs and RNNs effectively capture the hidden patterns within Euclidean data, an increasing number of applications utilize graphs to represent data. For example, for e-commerce, a graph-based learning system can exploit the interactions between users and products to make highly accurate recommendations. However, the complexity of graphs and the underlying irregularity pose significant challenges to existing DNNs. Hence to address this, graph neural networks (GNNs) were introduced. GNNs can be categorized into three types: recurrent GNNs (RecGNNs) [24, 27, 89], convolutional GNNs (CGNNs) [16, 67, 106], and graph autoecnoders (GAEs) [18, 70, 95]. RecGNNs aim to learn node representations with recurrent neural architectures. RecGNNs assume that a node within the graph constantly exchanges information with the neighboring nodes until a stable equilibrium is reached. Convolutional GNNs (CGNNs) were introduced to generalize the operation of convolution to graph data. CGNNs utilize an aggregation of a given node's features and the features from neighboring nodes. Furthermore, CGNNs stack multiple graph convolution layers to extract high-level node representation. Finally, GAEs map nodes into a latent feature space and decode the graph information from latent representations. GAEs are used to learn network embeddings or generate new graphs. A low-dimensional vector is used to represent the node that preserves the node's topological information. GAEs learn network embeddings using an encoder to extract network embeddings. A decoder is used to enforce network embeddings to preserve the graph topological information (PPMI matrix and the adjacency matrix).

1.2   Hardware Implications of DNNs

The diverse structures of state-of-the-art DNNs spanning across CNNs, RNNs, and GCNs result in significant compute and memory requirements. The higher accuracy achieved by these ML models requires increased computational complexity and model size, which in turn requires more memory to store both the weights and activations. In addition to memory and computation, the total volume of on-chip data movement is increased due to the increased model size and complexity. For example, ResNet-50 [29] for the ImageNet dataset [59] requires 50MB of memory and needs to perform 4 GFLOPs for each inference. Simultaneously, DenseNet-121 [34] for ImageNet requires 110MB of memory and required 8GFLOPs for each inference. Furthermore, due to limited on-chip memory capacity, conventional architectures that separate memory and computation result in a significant number of external memory access. The increased off-chip memory access leads to reduced energy efficiency and performance. The average cost of an external memory access is 1000× higher than the energy required to perform computations [31]. To further understand the impact on the hardware platform, we analyze the total energy spent in performing the inference for both VGG-16 and ResNet-50 using conventional von-Neumann architectures. A floating-point 32-bit (FP-32) multiplication results in 3.2pJ, and an FP-32 add requires 0.9pJ in the 45-nm technology node [25]. Therefore, only accounting for computations, to perform inference for one image, 65mJ of energy is consumed using the VGG-16 CNN, while ResNet-50 takes 16mJ. Scaling the computation energy up for 1000 inference performs, *VGG-16 takes 65J while ResNet-50 consumes 16J of energy*. Through this, we conclude that the increased accuracy achieved in DNNs results in higher computation complexity, increased memory requirements, higher off-chip memory access, and lower energy efficiency.

In this chapter, we discuss an alternative to conventional von-Neumann architecture, in-memory computing (IMC), that provides higher energy efficiency, better performance, and reduced off-chip memory access. In-memory computing (IMC) has emerged as a promising method to address the memory access, energy efficiency, and performance bottleneck introduced by DNN applications. Both SRAM and nanoscale non-volatile memory (e.g., resistive RAM or RRAM)-based IMC hardware architectures provide a dense and parallel structure to achieve high performance and energy efficiency [17, 22, 39, 47, 52, 56, 92, 96, 97, 103, 108–111]. However, IMC-based AI accelerators also require on-chip communication calling for an energy-efficient on-chip interconnect. Hence, we detail the different choices of on-chip interconnect and the impact on the overall performance of the accelerator. Finally, to perform efficient design space exploration, a quick and efficient simulator suite is necessary. To this end, this chapter explores the different benchmarking simulators for IMC architectures.

2   In-Memory Computing Architectures

In the earlier section, we discussed the hardware implications of modern DNNs, specifically the memory and computation complexity in von-Neumann architectures. For example, dense structures like DenseNet perform approximately 2.7×10^7 off-chip memory accesses to process a frame of an image [34]. The increased number of off-chip memory access degrades energy efficiency of the overall system. In-memory computing (IMC) architectures offer a promising alternative to conventional von-Neumann architectures. Figure 3 shows the generic block diagram of an IMC architecture with RRAM/SRAM memory cells. IMC utilizes either analog- or digital-domain computation to perform the multiply-and-accumulate (MAC) operations. Specifically, the crossbar-based IMC structure efficiently combines both memory access and analog-domain computation into a single unit for the acceleration of DNN workloads. Overall, the enhanced energy efficiency is attributed to a full-custom design, higher density, and higher memory bandwidth [52, 78, 92]. Therefore, IMC-based systems are becoming more popular for implementing compute- and memory-intensive AI applications. In this section, we will discuss different IMC architectures in detail using both SRAM and RRAM memory cells.

2.1   RRAM/SRAM-Based IMC Architectures

2.1.1   RRAM Device

RRAM-Based IMC architectures consist of an RRAM memory cell at each cross point within the IMC crossbar array. RRAM is a two-terminal device with programmable resistance representing the weights of the neural network and has high integration density, fast read speed, high memory accessing bandwidth, and good

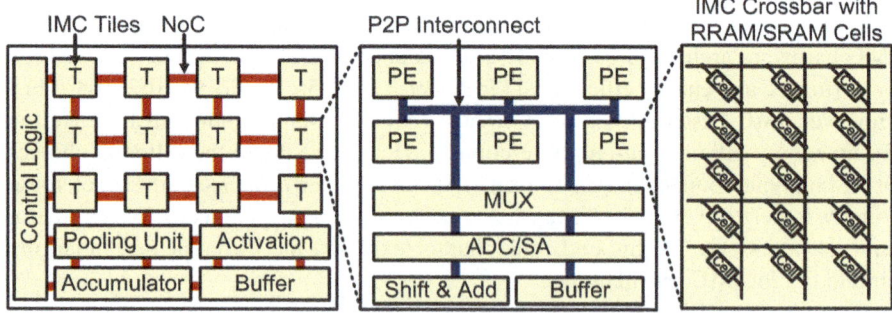

Fig. 3 Generic block diagram of an IMC architecture for DNN acceleration. It consists of an array of IMC tiles connected by an NoC with each tile consisting of a number of IMC arrays

compatibility with CMOS fabrication technology. For example, the RRAM device stack can include a TiN bottom electrode, HfO_2 mem-resistive switching layer, a PVD Ti oxygen exchange layer (OEL), and ~40 nm TiN top electrode [57, 60]. This specific stack is implemented between the M1 and M2 metallization layers, using a FEOL-compatible process flow.

Each RRAM cell can be characterized by the number of resistance levels that can be accessed within them. Broadly, RRAM can be classified into single-level cells (SLCs) and multi-level cells (MLCs). SLC only has two resistance levels, i.e., they can store only binary data. On the other hand, MLC cells have multiple resistance levels that represent higher precision data. The number of available resistance levels is governed by the ratio of the off resistance (R_{off}) to the on resistance (R_{on}). The ratio provides the range of resistances that are accessible for the given RRAM device. The overall resistance range can be divided into two main states, a low resistance state (LRS) and a high resistance state (HRS). LRS deals with the lower spectrum within the resistance band, while the HRS deals with the upper band of resistance of the RRAM device.

To program the RRAM device, a series of steps need to be followed. First, the RRAM device is formed by applying a large voltage across the two terminals. This process breaks the barrier and then allows for the flow of electrons across the terminals. Next, the RRAM is programmed to the required resistance by passing a specific current (compliance current) through the two electrodes. Depending on the compliance current, the RRAM can be programmed at different resistances. Furthermore, depending on the RRAM device (SLC or MLC), different levels of resistance can be achieved. Finally, once the RRAM device is programmed, we can perform a read by applying a voltage across the device electrodes. For the RRAM device proposed in [57, 60], a read voltage of up to 0.4V can be sustained by the RRAM device. The application of a higher voltage results in the damage of the device or goes into the write state, resulting in the change of the programmed resistance level.

2.1.2 IMC Architecture

Studies involving crossbar architectures have demonstrated that a $100\times$ to $1000\times$ improvement in energy efficiency is achieved as compared to traditional CPU and GPU architectures [39, 52, 75, 92, 96, 108, 108, 110]. Figure 3 shows the block diagram of an IMC architecture with an RRAM/SRAM memory cell. The architecture consists of an array of IMC tiles connected by a network on chip (NoC). The architecture also consists of a global pooling unit, nonlinear activation unit, accumulator, and input/output buffers. A global control logic performs the overall handling of the blocks within the architecture.

Each tile consists of an array of processing elements (PEs), where each PE is an IMC crossbar array with either an SRAM or an RRAM cell. Each IMC crossbar array consists of a set of peripheral circuits that enable the MAC computations.

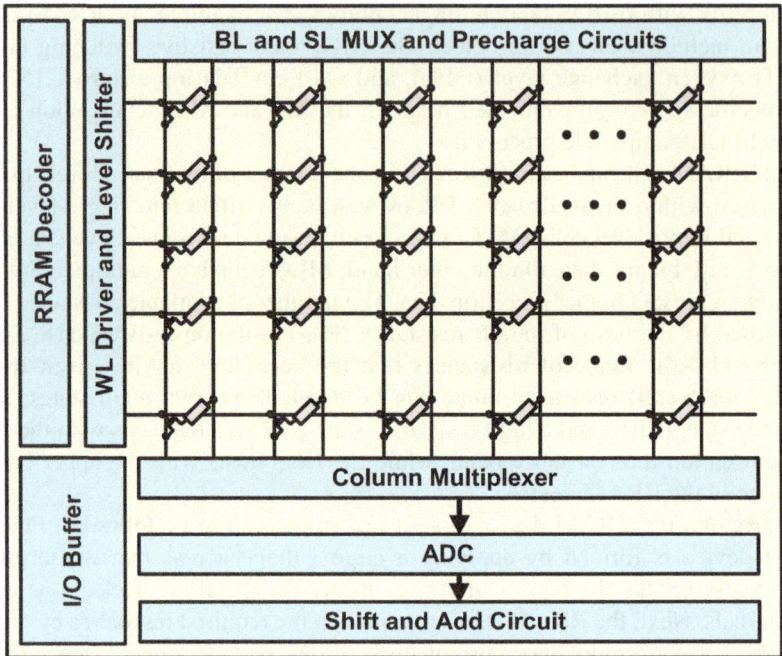

Fig. 4 Block diagram of an RRAM-based IMC crossbar array. An array of RRAM cells form the IMC crossbar array. Peripheral circuits such as bitline (BL)/select-line (SL)/column multiplexer (MUX), precharge circuit, wordline (WL) decoder and driver, buffers, level shifters, ADC, and shift and add circuit complete the RRAM-based IMC

Figure 4 shows the generic block diagram for a single RRAM-based IMC crossbar array. In the case of RRAM IMC, a transistor connects the gate to the wordline (WL) of the IMC crossbar array [57]. For the SRAM-based IMC with a conventional 6T structure, the WL connects to the access transistors. The IMC crossbar arrays consist of a wordline (WL) decoder, WL driver, a column multiplexer, analog-to-digital converter (ADC) or a sense amplifier, shifter and add circuit, control logic, and input/output buffers. The WL decoder turns on and off the WL for the IMC crossbar array. Meanwhile, the WL driver and level shifter are used to ensure that the driver can turn on the memory cell. Next, for an N×N IMC crossbar array, M columns are shared across the read-out circuit. The read-out circuit consists of the ADC, shift and add circuit, and the precharge circuit for the read operation. To enable the sharing of M columns, a column multiplexer is used. Finally, a custom control logic is utilized to drive the control signals during the operation of the IMC crossbar array. We will now go over the operation for both the SRAM- and RRAM-based IMC architectures. First, we will detail the working of the RRAM-based IMC architecture. Figure 4 shows the generic block diagram for a single RRAM-based IMC crossbar array. The RRAM devices are programmed by connecting the two terminals to a given voltage. To facilitate this, the terminals are connected to the bitline (BL) and the select-line (SL). By applying a voltage across the BL and the SL, forming, programming, and read operations are performed in a cell-by-cell fashion. During the write state of the IMC, each cell is chosen, and then the write is performed. During the compute state, the RRAM undergoes the read operation. Two kinds of read-out are performed, parallel and serial. During the parallel read-out, all/multiple WLs are turned on simultaneously, and the output is accumulated across the BL. Two kinds of input schemes are employed for single- and multi-bit inputs. The first method uses a digital-to-analog converter (DAC) to convert the input vector to an analog voltage and performs the computation in the charge domain [92]. The second method is to perform bit-serial computing, where each bit in the input vector is computed one at a time. The bit significance for each input vector is handled by using a shift and add circuit [52, 75, 110].

Depending on the resistance stored in the RRAM, an output current/charge is generated by the product of the voltage and resistance (conductance). This operation is analogous to multiply with the MAC. This current/charge is then accumulated across all rows for a given column to perform addition in the MAC. In the case of the serial read-out, a row-wise access of the IMC array is performed for MAC computations. Overall, the final MAC output is generated by accumulating across all rows of the IMC crossbar array.

Figure 5 shows the generic block diagram for a single SRAM-based IMC crossbar array. Next, we will discuss the operation for an SRAM-based IMC architecture [20, 39, 94, 98, 103, 109, 110]. Depending on the SRAM bitcell type and the degree of parallelism, the IMC design can be largely divided into three categories [91]: 6T bitcell with parallel compute, 6T bitcell with local compute, and (6T+extra-T) bitcell with parallel compute. Originally, SRAM-based IMC architecture employed the 6T bitcell with a parallel computation [45, 112]. The parallel computation was achieved by turning on all the WLs together to perform the MAC operations. The WLs are driven by the input vector where a 1 means it turns on that cell, while a 0 means the cell is turned off. Next, a 6T bitcell with a local compute structure is utilized where a special compute engine is designed to perform the MAC operation [98]. Here, the MAC operation is performed in a row-by-row fashion, similar to the serial read-out in RRAM-based IMC. Finally, in addition to the 6T cell, extra transistors are added in each bitcell to perform parallel compute [20, 94, 109]. In addition to the bitcell structure, peripheral circuits such as precharge circuit, ADCs, write driver, column multiplexer, row decoder, and row drivers are used.

2.1.3 Challenges with IMC Architectures

IMC architectures provide improved energy efficiency and throughput but suffer from certain drawbacks. The limited precision with the IMC crossbar array, specifically the memory cell, and the ADC impact the inference accuracy for DNNs [15, 50]. In addition, the impact of noise within the analog computation also adversely impacts the inference accuracy of DNNs.

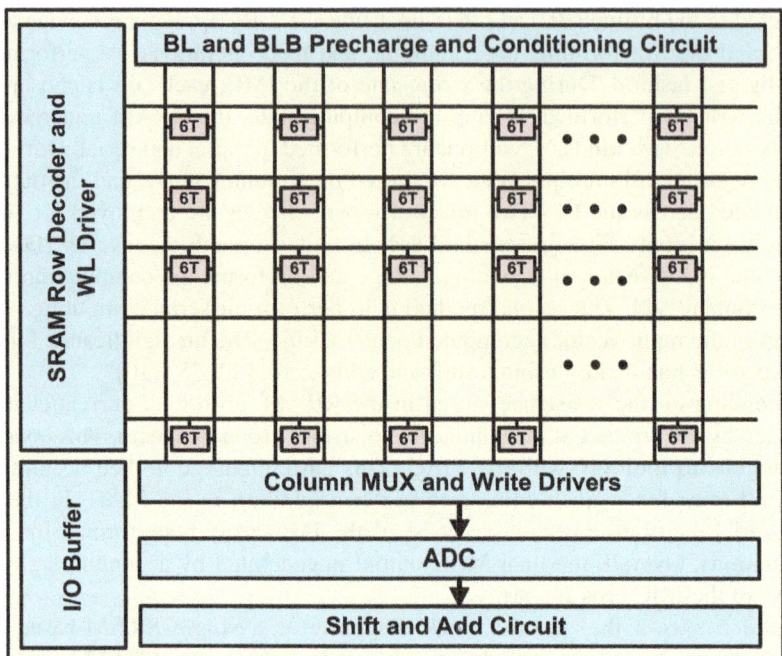

Fig. 5 Block diagram of an SRAM-based IMC crossbar array. An array of SRAM cells (6T or 6T+additional circuit) form the IMC crossbar array. Peripheral circuits such as bitline (BL) and bitline bar (BLB) precharge and conditioning circuits, row decoder and WL driver, column multiplexers, buffers, write drivers, ADC, and shift and add circuit complete the SRAM-based IMC

First, we will discuss the challenges with an RRAM-based IMC architecture. RRAM device suffers from several non-idealities such as limited resistance levels, device-to-device write variations, stuck at faults, and limited R_{off}/R_{on} ratio, posing a significant challenge to designing reliable RRAM-based IMC architectures [7–9, 44, 57, 58, 68, 69, 99, 107, 114]. The non-idealities within the RRAM device result in a deviation of the programmed weights values (resistance value), causing a significant reduction in post-mapping accuracy for DNNs. Furthermore, the crossbar structure of the IMC, with its limited array size, requires splitting of the large convolution (conv) or fully connected (FC) layers into partial operations. Such partial operation (conv/FC) results in further error due to the limited precision of the peripheral circuits (ADC and shift and add) of the RRAM-based IMC crossbar.

Several methods have been proposed in prior works to mitigate the post-mapping accuracy loss for RRAM-based in-memory acceleration of DNNs. Closed-Loop-on-Device (CLD) and Open-Loop-off-Device (OLD) perform iterative read–verify–write (R–V–W) operations at the RRAM device till the resistance converges to the desired value [33, 65]. References [7, 10] utilize variation-aware training (VAT) based on known device variation (σ) characterized from RRAM devices, while [68] combines VAT with dynamic precision quantization to mitigate the post-mapping accuracy loss. In [14], RRAM macro-measurement results that include variability and noise have been injected during the DNN training process to improve the DNN accuracy of the RRAM IMC hardware. Reference [79] utilizes a post mapping training by selecting a random subset of weights and mapping them to an on-chip memory to recover the accuracy. Meanwhile, [9] utilizes knowledge distillation and online adaptation for accuracy mitigation. The authors in [9] utilize an SRAM-based IMC as the parallel network, while the authors in [79] propose to use a register file and a randomization circuit. At the same time, [69, 99] propose to use a custom unary mapping scheme by mapping the MSB and LSB of the weights to RRAM devices based on the individual cell variations and bit significance.

Next, we discuss the challenges associated with SRAM-based IMC architectures. A compromise between parallelism and reliability is employed for best performance. In a conventional 6T SRAM IMC architecture, the parallel computation is achieved by turning on all or multiple rows. The higher parallelism raises the critical issue of read disturbance, resulting in the WL voltage to be driven with a lower voltage [45, 87, 112]. To mitigate this, a reduced parallelism is employed by exploiting the local compute engine [98]. The reduced parallelism results in reduced throughput for DNN inference. References [20, 94, 109] propose to utilize additional transistors that isolate the bitcell and employ parallel computation. Such a solution comes at the cost of increased area overhead, thus limiting the density of the SRAM-based IMC architecture. The additional transistor solution is typically implemented using a resistance or a capacitance. The resistive IMC method implements a multi-bit MAC operation by utilizing a resistive pull-up/-down by using transistors [20, 94, 109]. The pull-up/-down characteristics of the transistors exhibit a nonlinear behavior for the read bitline (RBL) transfer curve across different voltage ranges, thus having reduced reliability. At the same time, the capacitive SRAM-based IMC utilized a capacitor per bitcell and utilizes charge sharing and capacitive coupling to perform the MAC operations [39]. The capacitive SRAM IMC exhibits a more linear transfer characteristic on the RBL, but at the cost of a capacitor per bitcell. Finally, the limited precision of the ADC and the noise on the bitline (BL) requires careful design of the algorithm to achieve best inference accuracy [87].

3 Interconnect Challenges and Solutions

In the earlier sections, we have discussed that the IMC technique reduces the need for notoriously power-hungry off-chip memory (e.g., DRAM) accesses. Therefore, these techniques have a great potential to deliver energy-efficient AI accelerators. Deep neural networks consist of layer-by-layer operation, i.e., the output to the kth layer is the input of the $(k + 1)$th layer. In IMC-based accelerators, the data movement (i.e., communication) between the DNN layers is enabled by on-chip interconnects. Since the number of parameters of the neural networks has grown

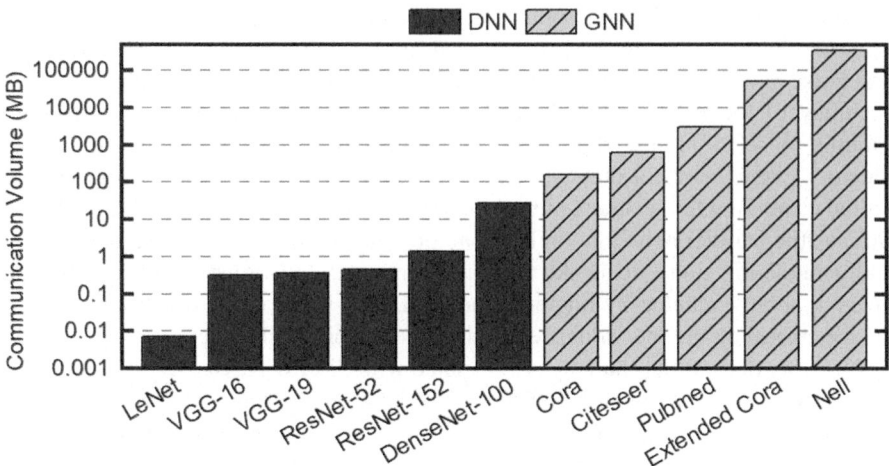

Fig. 6 Comparison of communication volume across different AI workloads

over past ten years as shown in [74], the on-chip communication volume also increases. We also observe this trend in other emerging AI algorithms, such as graph convolutional networks (GCNs). Figure 6 shows the total communication volume (in MB) for different AI algorithms. Increasing communication volume, in turn, increases energy consumption due to communication which can mask the energy benefit of IMC technology itself. Therefore, an energy-efficient communication strategy is required for AI accelerator to compliment the energy benefits of IMC technology. Indeed, a single technique may not be suitable for all kinds of AI algorithms as discussed in [53].

Emergence of newer SoC paradigms (beyond monolithic integration) necessitates novel interconnect approaches. Therefore, the rest of this section discusses various interconnect technologies proposed for different AI algorithms as well as different types of SoCs. First, we highlight various interconnect techniques proposed for planar IMC-based SoCs targeting AI acceleration in Sect. 3.1. Then, we explore different interconnect techniques for monolithic 3D ICs with IMC technology in Sect. 3.3. After that, we discuss multiple research efforts that target on-package communication for chiplet-based AI accelerator in Sect. 3.2. A more comprehensive survey on an efficient on-chip network for DNN accelerator in general can be found in [80].

3.1 Interconnect for IMC-Based Planar AI Accelerators

There exist several NoC architectures for DNN accelerators. A recent study aims to maximize local data reuse and reduce data access from DRAM [12]. To this end, a row-stationary data flow is proposed, where filter weights and input feature maps (ifmap) are reused to minimize movement of ifmaps and filter weights. The architecture has later been extended to incorporate compact and sparse neural networks in [13]. In the extended version [13], a hierarchical mesh NoC is incorporated in the architecture. It consists of 16 PE clusters and 16 global buffer clusters distributed in an 8×2 array. Each PE cluster consists of 12 PEs arranged in a 3×4 array. However, both architectures consider a system with off-chip memory where frequent data transfer from off-chip to on-chip is required. Therefore, the NoC optimizations incorporated in these architectures are not applicable for IMC-based accelerators. ISAAC is one of the first IMC-based DNN accelerators proposed in the literature [92]. It uses an NoC with concentrated mesh (c-mesh) topology. However, only one NoC router is connected to each IMC tile, and no special interconnect optimization is considered in this architecture. Since larger DNNs (e.g., DenseNet with 100 layers) may contain 100s of IMC tiles, this architecture may require 100s of NoC routers, which may not be practical at all.

The growing number of NoC routers increases area as well as on-chip interconnect power consumption. To that end, a recent study proposes an optimization technique to determine the number of NoC routers for a given DNN [52]. The authors first construct an objective function for communication energy, which considers the number of activations between two consecutive layers for each layer as input. Then the objective function is minimized to obtain the number of routers needed for all layers of the DNN. A scheduling technique is also proposed in this work to minimize the congestion in the on-chip network. The optimized number of routers along with the scheduling technique provides up to 78% improvement in energy-delay product with respect to another DNN accelerator [92]. Although this work minimizes congestion in the on-chip network with a scheduling technique, it does not guarantee minimum latency for a given DNN. The authors in [75] proposed an NoC architecture that guarantees minimum possible communication latency for a given DNN. However, the proposed NoC architecture is customized for a single DNN. Therefore, a reconfigurable NoC is also proposed in [75], where a certain number of routers (determined with handful DNNs known at the design time) are allocated for each DNN layer. At runtime, if a new (not considered in the design time) DNN appears, then first the number of routers required for each DNN layer is computed first. If the number of routers required for a particular layer is more than the number of available routers on-chip, then the DNN layer will occupy the maximum number of routers available for that layer. The reconfigurable NoC shows 60%–80% improvement in communication latency over state-of-the-art 2D-mesh NoC. A communication-aware IMC accelerator for graph convolutional network (GCN), COIN, is proposed in [76]. In this work, a number of GCN nodes are implemented in a compute element, and multiple CEs are connected with an NoC router. The number of NoC routers is obtained by minimizing inter-CE and intra-CE communication energy. COIN shows up to 105× improvement in energy consumption with respect to state-of-the-art GCN accelerator.

3.2   On-Package Communication for Chiplet-Based AI Accelerators

The area of monolithic hardware accelerators increases with an increasing number of parameters of AI algorithms. Higher silicon area of a single monolithic system reduces the yield, which in turn increases fabrication cost [56]. Chiplet-based system solves the issue of higher fabrication cost by integrating multiple small chips (known as chiplets) on a single die. Since the area of each chiplet in the system is considerably lower than a monolithic chip (for the same AI algorithm), the yield of the chiplet-based system increases, which reduces the fabrication cost. The communication between chiplets is performed through network on package (NoP), as shown in Fig. 7. There are several works in the literature which propose NoP for chiplet-based system considering different performance objectives (e.g., latency, energy) [5, 56, 93, 104].

Kite is a family of NoP proposed in [5], which is mainly targeted for general purpose processors. In this work, three topologies are proposed—Kite-Small, Kite-Medium, and Kite-Large. First, an objective function is constructed with combination of the average delay between source and destination, diameter, and bisection bandwidth of the NoP. Experimental evaluations on synthetic traffic show that the proposed Kite topologies reduce latency by 7% and improve the peak throughput by 17% with respect to other well-known interconnect topologies. A chiplet-based system with 96-core processor, INTACT, is proposed in [104]. The chiplets are connected through a generic chiplet-interposer interfaces (called as 3D-plugs in the chapter). 3D-plugs consist of micro-bump arrays. However, both Kite and INTACT are not specific to AI workloads.

Shao et al. designed and fabricated a 36-chiplet system called SIMBA for deep learning inference [93]. The chiplets in the system are connected through a mesh NoP. Ground-referenced signaling (GRS) is used for intra-package communication. The NoP follows a hybrid wormhole/cut-through flow control. The NoP bandwidth is 100 GBps/chiplet, and the latency for one hop is 20 ns. Extensive evaluation on the fabricated chip shows up to 16% speed up compared to baseline layer mapping

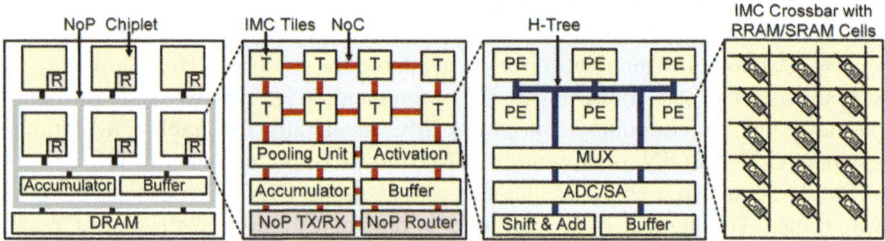

Fig. 7 Architecture utilized within SIAM [56] that includes an NoP for on-package communication, NoC for on-chip communication within each chiplet, and a point-to-point network like H-Tree for within tile communication

for ResNet-50. A simulator for chiplet-based system, SIAM, is proposed in [56], targeting AI workloads. In this simulator, a mesh topology is considered for NoP. It is shown that up to 85% of the total system area is contributed by NoP. In this work, multiple studies were performed by varying NoP parameters. For example, it is shown that increasing NoP channel width increases energy-delay product of the NoP for ResNet-110. This phenomenon is demonstrated for systems with 25 and 36 chiplets. However, none of the prior works considered any workload-aware optimization for the NoP. Therefore, there is ample opportunity of future research which considers NoP optimization for AI accelerators.

3.3 Interconnect for Monolithic 3D (M3D)-Based AI Accelerators

With increasing complexity of AI algorithms, the computing resource needed to execute the algorithms also increases. Therefore, complex AI algorithms require a large number of processing elements on-chip. For example, DenseNet-110 with 28.1M parameters requires 2184 ReRAM tiles on a single system [56]. Increasing the number of on-chip tiles results in long-range inter-tile communication. Too many long-range communications hurt energy efficiency of the system. Therefore, monolithic 3D (M3D)-based AI accelerators have emerged to facilitate energy-efficient communication between multiple processing elements. In M3D-based accelerators, multiple processing elements are placed in each plane. The processors across different planes are connected using through silicon vias (TSV).

REGENT is such an approach which integrates ReRAM-based IMC tiles as well as GPU cores on an M3D IC [42]. The processors in the IC are connected through a 3D-NoC. REGENT is optimized to perform energy-efficient CNN training. Specifically, a bin-package-based framework is adopted to map CNN layers on processing cores as well as physically place the cores in such a way that the overall system meets certain performance objectives. However, REGENT does not consider hardware implementation of normalization layers. To address this drawback, the authors in [41] propose a 3D-NoC-enabled IMC-based system considering normalization layers. Apart from considering hardware implementation of normalization layers, a performance-thermal aware mapping of CNN layers is also proposed in this work. The mapping helps to reduce thermal noise which can degrade the quality of CNN training. As a result, the proposed architecture is able to perform CNN training which achieves accuracy similar to GPU. The accelerator proposed in the aforementioned work is further extended in [40] by considering fewer normalization layers for CNNs. In this work, the authors show that considering few normalization layers actually improves CNN classification accuracy, since normalization helps to reduce bias occurring from a weight with high absolute value. Then, Bayesian optimization is utilized to construct an M3D system. The communication between multiple processing elements is facilitated by a mesh-NoC with XYZ routing. The accelerator proposed in this work reduces the latency by $15\times$ compared to conventional GPU-based system. However, all these works only consider CNN training on IMC-based M3D system.

Recently, several other works proposed IMC-based M3D systems which are capable of training graph neural networks (GNNs). ReGraphX is a 3D-NoC-enabled heterogeneous IMC-based system which performs energy-efficient GNN training [2]. In this system, there are two types of processing elements: V-PEs, which perform vertex computations, and E-PEs, which perform edge computation pertaining to a GNN. V-PEs consist of 128×128 crossbar arrays, whereas E-PEs consist of 8×8 crossbar arrays. Experimental evaluations show that ReGraphX reduces energy consumption by $11\times$ with respect to conventional GPUs. The authors in [4] show performance and accuracy trade-offs in 3D-NoC-enabled IMC-based GNN accelerator. In this work, a stochastic rounding technique is proposed to reduce the precision of ReRAM crossbar arrays. The reduced precision helps to improve energy efficiency of the accelerator. A DropLayer-aware M3D-based manycore ReRAM architecture for training GNNs, DARe, is proposed in [3]. The DropLayer-based technique reduces on-chip communication volume in the system, which, in turn, prevents communication hotspot. Reduced communication hotspot improves the energy efficiency of the overall system. The proposed architecture demonstrates $1.9\times$ reduction in execution time with respect to ReGraphX [2]. Thus, M3D-based systems with 3D NoC provide energy-efficient platform for CNN as well as GNN training.

4   Evaluation Frameworks for IMC-Based AI Accelerator

Since in-memory computing (IMC)-based AI accelerators are recently drawing more attention due to their energy efficiency, extensive evaluations of power, performance and area for the accelerators are required. The pre-silicon evaluations help to identify the bottleneck of the systems as well as compare the performance with other systems. There exist multiple simulators that evaluate the performance of systems with general purpose processing elements. Gem5 is the most popular cycle accurate simulator which considers various architectural parameters of a system and evaluates its performance [6]. Furthermore, a detailed on-chip interconnect simulator, GARNET, is integrated with Gem5 [1]. However, cycle accurate simulators incur significant simulation time which is prohibitive for fast design space exploration. To accelerate the design space exploration process, several prior works proposed analytical model-based performance evaluation of the underlying system [46, 71–73, 77, 81]. However, none of these evaluation techniques specifically target AI applications. AI applications (e.g., DNNs, GNNs) mainly consist of multiply and accumulate operations which can be implemented as systolic arrays. The authors in [88] propose a systolic array-based simulator, SCALE-sim, which is able to evaluate system performance executing DNN workloads. Nonetheless, none of the aforementioned approaches consider performance evaluation of IMC-based

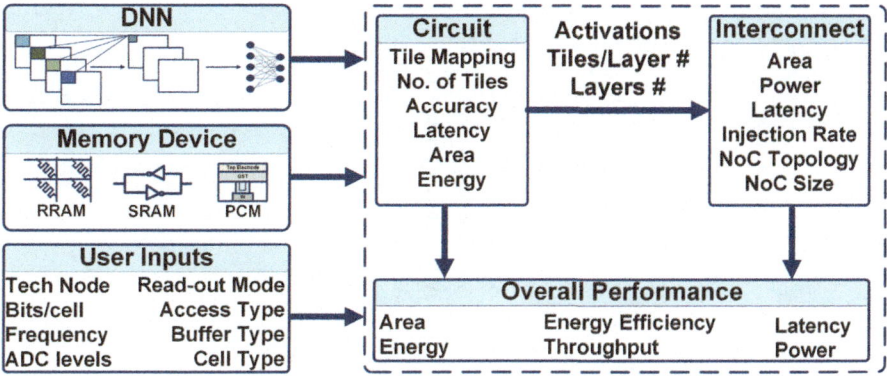

Fig. 8 Block diagram on an IMC benchmarking simulator proposed in [54]. The simulator consists of a circuit part and an interconnect part that perform system-level benchmarking of IMC architectures

accelerators (Fig. 8). In the next sections, we will discuss multiple performance evaluation technique for IMC-based AI accelerators. Table 1 provides a summary of evaluation frameworks proposed for IMC-based DNN accelerators.

4.1 Evaluation Frameworks for Monolithic AI Accelerators

Multiple researchers have proposed evaluation frameworks for IMC-based AI accelerators. NeuroSim is the first simulator which evaluates performance IMC-based AI accelerators [11]. The performance evaluation metrics in NeuroSim include area, latency, and power consumption of an IMC system under a given DNN workload. NeuroSim provides excellent flexibility to users to evaluate the performance of IMC-based AI accelerators under different system specifications. For example, it considers conventional CMOS-based memory technology (e.g., SRAM) as well as emerging non-volatile memory technologies (e.g., ReRAM, STT-MRAM) for the in-memory compute elements. NeuroSim assumes a tile-based architecture as proposed in [92]. Specifically, the architecture under consideration in NeuroSim consists of multiple *tiles*. The *tiles* consist of *PE's* (processing elements), and *PEs* consist of IMC-based crossbar arrays. The lower level components (e.g., buffers, ADC, multiplexers) in NeuroSim are simulated using the Predictive Technology Model (PTM) [113], and verified against circuit simulation (e.g., SPICE), reaching more than 90% accuracy. Furthermore, Peng et al. [83] created an interface between NeuroSim and popular machine learning frameworks (PyTorch and TensorFlow), which make NeuroSim more user friendly. One important drawback of NeuroSim is that it assumes H-Tree-based bus interconnect for inter-tile communication. H-Tree-based bus interconnect is not practical since it can consume up to 90% total energy consumption of DNN inference [53]. Network on chip (NoC) is a promising

Table 1 Summary of different evaluation frameworks for IMC-based architectures

Simulator	System	Operation	Memory elements	Interconnect	Performance metric
Chen et al. [11]	Monolithic	Inference	SPICE model	Bus-based	Area, Latency, Energy
Zhu et al. [116]	Monolithic	Inference	Behavioral model	NoC	Area, Latency, Energy
Krishnan et al. [54]	Monolithic	Inference	SPICE model	NoC	Area, Latency, Energy
Jain et al. [36]	Monolithic	Inference	SPICE model	Not supported	Accuracy
Peng et al. [82]	Monolithic	Training	SPICE model	Bus-based	Area, Latency, Energy
Krishnan et al. [56]	Chiplet(2.5D)	Inference	SPICE model	NoC+NoP	Area, Latency, Energy

alternative for inter-tile communication. Therefore, Krishnan et al. [54] proposed an evaluation framework for IMC-based AI accelerator which considers cycle accurate NoC simulation. Specifically, a customized version of BookSim [38] is integrated with NeuroSim to provide more realistic performance evaluation of AI accelerators. MNSIM [116] also considers performance evaluation of IMC-based system to execute AI applications similar to NeuroSim. Apart from evaluating the system with a baseline architecture, MNSIM also integrates software–hardware co-design technique in the evaluation framework. Chakraborty et al. proposed GeneiX, an evaluation framework for crossbar-based IMC accelerator considering the non-idealities in the memory elements [7]. While hardware performance evaluation under AI workload is crucial, evaluation of accuracy of the AI workload while implemented on-chip is also important. Non-idealities in the memory elements can reduce the accuracy of DNNs. RxNN is a framework where accuracy of a given DNN workload is evaluated in the presence of memory non-idealities [36]. All these techniques consider performance evaluation of IMC systems executing DNN inference. However, emerging edge devices perform online learning which require training the DNN. Therefore, performance evaluation of AI accelerators while executing DNN inference is not enough.

An evaluation framework for IMC-based AI accelerators with on-chip training is presented in [82]. In this work, the authors introduce "non-linearity, asymmetry, device-to-device and cycle-to-cycle variation of weight update into the python wrapper, and peripheral circuits for error/weight gradient computation in NeuroSim core" for a given AI workload. The training framework is based on authors' prior work [37], where SRAM-based transposable function is proposed. SRAM-based arrays are able to perform write operations fast while consuming low energy. Therefore, the weight-gradient computation function is implemented through SRAM-based arrays as opposed to other non-volatile memory technology.

4.2 Evaluation Framework for Chiplet-Based AI Accelerators

Chiplet-based systems are becoming popular for large-scale integration due to its yield and fabrication cost benefit. Apart from general purpose workloads, chiplet-based systems have shown superior energy efficiency for AI workloads too [93]. Therefore, it is important to have an evaluation framework for chiplet-based systems executing AI workloads. SIAM, as shown in Fig. 9, is such a simulator where the performance of chiplet-based systems with IMC is evaluated for a given DNN workload. Specifically, SIAM integrates evaluation of IMC-based compute elements, on-chip interconnect within a chiplet and on-package communication between chiplets. This simulator utilizes model-based as well as cycle-accurate simulation components to evaluate system performance for a wide range of DNNs. SIAM automatically maps DNN workloads into multiple chiplets with a given mapping algorithm. The simulation time taken by SIAM is low compared to cycle-accurate simulators. For example, performance evaluation of a chiplet-based system

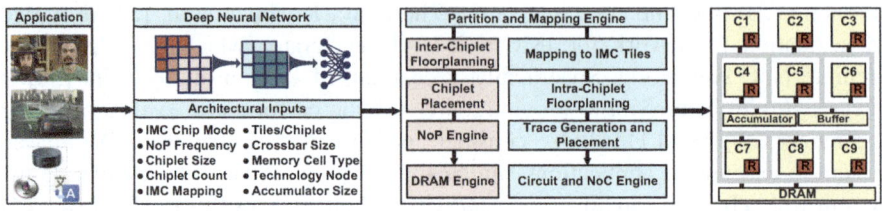

Fig. 9 Block diagram of the chiplet-based IMC architecture simulator SIAM [56]

for ResNet-110 with 1.7M parameters takes only 12 min enabling fast design space exploration. SIAM also provides chiplet-level as well as DNN layer-level performance evaluation which enables fine-grained analysis of the system as well as the AI workload. A summary of SIAM can be found in [55].

5 Conclusion

In this chapter, we discussed about in-memory computing-based AI accelerators. In-memory computing technique reduces on-chip energy consumption of AI accelerators. We discuss various in-memory computing architectures proposed in the literature. Both CMOS (e.g., SRAM)- and memristor (e.g., RRAM)-based IMC architectures are discussed. We also introduce the challenges associated with IMC architectures and introduce some of the solutions proposed in the literature. Although in-memory computing improves energy efficiency of computing elements, it increases on-chip communication volume. Increasing on-chip communication volume may mask the benefits of in-memory computing technique. We discuss multiple research which aim to construct energy-efficient interconnect for IMC-based AI accelerators. Finally, we discussed few frameworks to evaluate the performance of IMC-based AI accelerators.

References

1. Agarwal, N., Krishna, T., Peh, L.S., Jha, N.K.: GARNET: A Detailed on-chip Network Model inside a Full-system Simulator. In: 2009 IEEE International Symposium on Performance Analysis of Sand Software, pp. 33–42 (2009)
2. Arka, A.I., Doppa, J.R., Pande, P.P., Joardar, B.K., Chakrabarty, K.: ReGraphX: NoC-enabled 3D heterogeneous ReRAM architecture for training graph neural networks. In: 2021 Design, Automation & Test in Europe Conference & Exhibition (DATE), pp. 1667–1672. IEEE (2021)
3. Arka, A.I., Joardar, B.K., Doppa, J.R., Pande, P.P., Chakrabarty, K.: DARe: DropLayer-aware manycore ReRAM architecture for training graph neural networks. In: 2021 IEEE/ACM International Conference On Computer Aided Design (ICCAD), pp. 1–9 (2021)

4. Arka, A.I., Joardar, B.K., Doppa, J.R., Pande, P.P., Chakrabarty, K.: Performance and accuracy tradeoffs for training graph neural networks on ReRAM-based architectures. IEEE Trans. Very Large Scale Integr. (VLSI) Syst. **29**(10), 1743–1756 (2021)

5. Bharadwaj, S., Yin, J., Beckmann, B., Krishna, T.: Kite: A family of heterogeneous interposer topologies enabled via accurate interconnect modeling. In: 2020 57th ACM/IEEE Design Automation Conference (DAC), pp. 1–6. IEEE (2020)

6. Binkert, N., Beckmann, B., Black, G., Reinhardt, S.K., Saidi, A., Basu, A., Hestness, J., Hower, D.R., Krishna, T., Sardashti, S., et al.: The gem5 simulator. ACM SIGARCH Comput. Archit. News **39**(2), 1–7 (2011)

7. Chakraborty, I., Ali, M.F., Kim, D.E., Ankit, A., Roy, K.: Geniex: A generalized approach to emulating non-ideality in memristive Xbars using neural networks. In: 2020 57th ACM/IEEE Design Automation Conference (DAC), pp. 1–6 (2020)

8. Charan, G., Mohanty, A., Du, X., Krishnan, G., Joshi, R.V., Cao, Y.: Accurate inference with inaccurate rram devices: A joint algorithm-design solution. IEEE J. Explor. Solid State Comput. Dev. Circuits **6**(1), 27–35 (2020a)

9. Charan, G., et al.: Accurate inference with inaccurate RRAM devices: statistical data, model transfer, and on-line adaptation. In: DAC. IEEE (2020b)

10. Chen, L., et al.: Accelerator-friendly neural-network training: learning variations and defects in RRAM crossbar. In: DATE. IEEE (2017)

11. Chen, P.Y., Peng, X., Yu, S.: Neurosim: A circuit-level macro model for benchmarking neuro-inspired architectures in online learning. IEEE Trans. Comput. Aided Des. Integr. Circ. Syst. **37**(12), 3067–3080 (2018)

12. Chen, Y.H., Krishna, T., Emer, J.S., Sze, V.: Eyeriss: An energy-efficient reconfigurable accelerator for deep convolutional neural networks. IEEE J. Solid State Circ. **52**(1), 127–138 (2016)

13. Chen, Y.H., Yang, T.J., Emer, J., Sze, V.: Eyeriss v2: A flexible accelerator for emerging deep neural networks on mobile devices. IEEE J. Emerg. Sel. Top. Circ. Syst. **9**(2), 292–308 (2019)

14. Cherupally, S.K., Meng, J., Rakin, A.S., Yin, S., Yeo, I., Yu, S., Fan, D., Seo, J.: Improving the accuracy and robustness of RRAM-based in-memory computing against RRAM hardware noise and adversarial attacks. Semicond. Sci. Technol. **37**(3), 034001 (2022). https://doi.org/10.1088/1361-6641/ac461f

15. Cherupally, S.K., Meng, J., Rakin, A.S., Yin, S., Yeo, I., Yu, S., Fan, D., Seo, J.S.: Improving the accuracy and robustness of rram-based in-memory computing against rram hardware noise and adversarial attacks. Semicond. Sci. Technol. **37**(3), 034001 (2022)

16. Chiang, W.L., Liu, X., Si, S., Li, Y., Bengio, S., Hsieh, C.J.: Cluster-gcn: An efficient algorithm for training deep and large graph convolutional networks. In: Proceedings of the 25th ACM SIGKDD International Conference on Knowledge Discovery & Data Mining, pp. 257–266 (2019)

17. Chih, Y.D., Lee, P.H., Fujiwara, H., Shih, Y.C., Lee, C.F., Naous, R., Chen, Y.L., Lo, C.P., Lu, C.H., Mori, H., et al.: An 89tops/w and 16.3 tops/mm 2 all-digital sram-based full-precision compute-in memory macro in 22nm for machine-learning edge applications. In: 2021 IEEE International Solid-State Circuits Conference (ISSCC), vol. 64, pp. 252–254. IEEE (2021)

18. De Cao, N., Kipf, T.: Molgan: An implicit generative model for small molecular graphs. Preprint (2018). arXiv:1805.11973

19. Deng, L., Hinton, G., Kingsbury, B.: New types of deep neural network learning for speech recognition and related applications: an overview. In: 2013 IEEE International Conference on Acoustics, Speech and Signal Processing, pp. 8599–8603. IEEE (2013)

20. Dong, Q., Sinangil, M.E., Erbagci, B., Sun, D., Khwa, W.S., Liao, H.J., Wang, Y., Chang, J.: 15.3 a 351tops/w and 372.4 gops compute-in-memory sram macro in 7nm finfet cmos for machine-learning applications. In: 2020 IEEE International Solid-State Circuits Conference-(ISSCC), pp. 242–244. IEEE (2020)

21. Du, X., Krishnan, G., Mohanty, A., Li, Z., Charan, G., Cao, Y.: Towards efficient neural networks on-a-chip: Joint hardware-algorithm approaches. In: 2019 China Semiconductor Technology International Conference (CSTIC), pp. 1–5. IEEE (2019)

22. Fujiwara, H., Mori, H., Zhao, W.C., Chuang, M.C., Naous, R., Chuang, C.K., Hashizume, T., Sun, D., Lee, C.F., Akarvardar, K., et al.: A 5-nm 254-tops/w 221-tops/mm 2 fully-digital computing-in-memory macro supporting wide-range dynamic-voltage-frequency scaling and simultaneous mac and write operations. In: 2022 IEEE International Solid-State Circuits Conference (ISSCC), vol. 65, pp. 1–3. IEEE (2022)
23. Gagniuc, P.A.: Markov Chains: From Theory to Implementation and Experimentation. Wiley (2017)
24. Gallicchio, C., Micheli, A.: Graph echo state networks. In: The 2010 International Joint Conference on Neural Networks (IJCNN), pp. 1–8. IEEE (2010)
25. Gholami, A., Kim, S., Dong, Z., Yao, Z., Mahoney, M.W., Keutzer, K.: A survey of quantization methods for efficient neural network inference. Preprint (2021). arXiv:2103.13630
26. Goodfellow, I., Bengio, Y., Courville, A.: Deep Learning. MIT Press (2016)
27. Gori, M., Monfardini, G., Scarselli, F.: A new model for learning in graph domains. In: Proceedings. 2005 IEEE International Joint Conference on Neural Networks, vol. 2, pp. 729–734. IEEE (2005)
28. Hamilton, W., Ying, Z., Leskovec, J.: Inductive representation learning on large graphs. Adv. Neural Inf. Proces. Syst. **30**, (2017). arXiv:1706.02216
29. He, K., Zhang, X., Ren, S., Sun, J.: Deep residual learning for image recognition. In: Proceedings of the IEEE Conference on Computer Vision and Pattern Recognition, pp. 770–778 (2016)
30. Hochreiter, S., Schmidhuber, J.: Long short-term memory. Neural Computation **9**(8), 1735–1780 (1997)
31. Horowitz, M.: Computing's energy problem (and What We Can Do About It). In: IEEE ISSCC, pp. 10–14 (2014)
32. Howard, A.G., Zhu, M., Chen, B., Kalenichenko, D., Wang, W., Weyand, T., Andreetto, M., Adam, H.: Mobilenets: Efficient convolutional neural networks for mobile vision applications. Preprint (2017). arXiv:1704.04861
33. Hu, M., Li, H., Chen, Y., Wu, Q., Rose, G.S.: BSB training scheme implementation on memristor-based circuit. In: IEEE CISDA. IEEE (2013)
34. Huang, G., Liu, Z., Van Der Maaten, L., Weinberger, K.Q.: Densely connected convolutional networks. In: Proceedings of the IEEE Conference on Computer Vision and Pattern Recognition, pp. 4700–4708 (2017)
35. Iandola, F.N., Han, S., Moskewicz, M.W., Ashraf, K., Dally, W.J., Keutzer, K.: Squeezenet: Alexnet-level accuracy with 50x fewer parameters and <0.5 mb model size. Preprint (2016). arXiv:1602.07360
36. Jain, S., Sengupta, A., Roy, K., Raghunathan, A.: RxNN: A framework for evaluating deep neural networks on resistive crossbars. IEEE Trans. Comput. Aided Des. Integr. Circ. Syst. **40**(2), 326–338 (2020)
37. Jiang, H., Huang, S., Peng, X., Su, J.W., Chou, Y.C., Huang, W.H., Liu, T.W., Liu, R., Chang, M.F., Yu, S.: A two-way SRAM array based accelerator for deep neural network on-chip training. In: 2020 57th ACM/IEEE Design Automation Conference (DAC), pp. 1–6 (2020)
38. Jiang, N., et al.: A detailed and flexible cycle-accurate network-on-chip simulator. In: 2013 IEEE International Symposium on Performance Analysis of Systems and Software (ISPASS), pp. 86–96. IEEE (2013)
39. Jiang, Z., Yin, S., Seo, J.S., Seok, M.: C3SRAM: An in-memory-computing SRAM macro based on robust capacitive coupling computing mechanism. IEEE J. Solid State Circ. **55**(7), 1888–1897 (2020). https://doi.org/10.1109/JSSC.2020.2992886
40. Joardar, B.K., Deshwal, A., Doppa, J.R., Pande, P.P., Chakrabarty, K.: High-throughput training of deep CNNs on ReRAM-based heterogeneous architectures via optimized normalization layers. IEEE Trans. Comput. Aided Des. Integr. Circ. Syst. **41**(5), 1537–1549 (2021)
41. Joardar, B.K., Doppa, J.R., Pande, P.P., Li, H., Chakrabarty, K.: AccuReD: high accuracy training of CNNs on ReRAM/GPU heterogeneous 3-D architecture. IEEE Trans. Comput. Aided Des. Integr. Circ. Syst. **40**(5), 971–984 (2020)

42. Joardar, B.K., Li, B., Doppa, J.R., Li, H., Pande, P.P., Chakrabarty, K.: REGENT: A heterogeneous ReRAM/GPU-based architecture enabled by NoC for training CNNs. In: 2019 Design, Automation & Test in Europe Conference & Exhibition (DATE), pp. 522–527. IEEE (2019)

43. Jordan, M.I.: Serial order: A parallel distributed processing approach. In: Advances in Psychology, vol. 121, pp. 471–495. Elsevier (1997)

44. Joshi, V., et al.: Accurate deep neural network inference using computational phase-change memory. Nature Communications (2020)

45. Kang, M., Kim, Y., Patil, A.D., Shanbhag, N.R.: Deep in-memory architectures for machine learning–accuracy versus efficiency trade-offs. IEEE Trans. Circ. Syst. I Regul. Pap. **67**(5), 1627–1639 (2020)

46. Kiasari, A.E., Lu, Z., Jantsch, A.: An analytical latency model for networks-on-chip. IEEE Trans. Very Large Scale Integr. (VLSI) Syst. **21**(1), 113–123 (2012)

47. Kim, H., Yoo, T., Kim, T.T.H., Kim, B.: Colonnade: A reconfigurable sram-based digital bit-serial compute-in-memory macro for processing neural networks. IEEE J. Solid State Circ. **56**(7), 2221–2233 (2021)

48. Kotsiantis, S.B.: Decision trees: a recent overview. Artif. Intell. Rev. **39**(4), 261–283 (2013)

49. Krishnan, G., Du, X., Cao, Y.: Structural pruning in deep neural networks: A small-world approach. Preprint (2019). arXiv:1911.04453

50. Krishnan, G., Hazra, J., Liehr, M., Du, X., Beckmann, K., Joshi, R.V., Cady, N.C., Cao, Y.: Design limits of in-memory computing: Beyond the crossbar. In: 2021 5th IEEE Electron Devices Technology & Manufacturing Conference (EDTM), pp. 1–3. IEEE (2021)

51. Krishnan, G., Ma, Y., Cao, Y.: Small-world-based structural pruning for efficient fpga inference of deep neural networks. In: 2020 IEEE 15th International Conference on Solid-State & Integrated Circuit Technology (ICSICT), pp. 1–5. IEEE (2020)

52. Krishnan, G., Mandal, S.K., Chakrabarti, C., Seo, J.s., Ogras, U.Y., Cao, Y.: Interconnect-aware area and energy optimization for in-memory acceleration of DNNs. IEEE Des. Test **37**(6), 79–87 (2020)

53. Krishnan, G., Mandal, S.K., Chakrabarti, C., Seo, J.S., Ogras, U.Y., Cao, Y.: Impact of on-chip interconnect on in-memory acceleration of deep neural networks. ACM J. Emerg. Technol. Comput. Syst. (JETC) **18**(2), 1–22 (2021)

54. Krishnan, G., Mandal, S.K., Chakrabarti, C., Seo, J.s., Ogras, U.Y., Cao, Y.: Interconnect-centric benchmarking of in-memory acceleration for DNNs. In: 2021 China Semiconductor Technology International Conference (CSTIC), pp. 1–4. IEEE (2021)

55. Krishnan, G., Mandal, S.K., Chakrabarti, C., Seo, J.S., Ogras, U.Y., Cao, Y.: System-level benchmarking of chiplet-based IMC architectures for deep neural network acceleration. In: 2021 IEEE 14th International Conference on ASIC (ASICON), pp. 1–4 (2021)

56. Krishnan, G., Mandal, S.K., Pannala, M., Chakrabarti, C., Seo, J.S., Ogras, U.Y., Cao, Y.: SIAM: Chiplet-based scalable in-memory acceleration with mesh for deep neural networks. ACM Trans. Embed. Comput. Syst. (TECS) **20**(5s), 1–24 (2021)

57. Krishnan, G., Sun, J., Hazra, J., Du, X., Liehr, M., Li, Z., Beckmann, K., Joshi, R.V., Cady, N.C., Cao, Y.: Robust RRAM-based in-memory computing in light of model stability. In: IRPS. IEEE (2021)

58. Krishnan, G., Yang, L., Sun, J., Hazra, J., Du, X., Liehr, M., Li, Z., Beckmann, K., Joshi, R., Cady, N.C., et al.: Exploring model stability of deep neural networks for reliable RRAM-based in-memory acceleration. IEEE Trans. Comput. **71**(11), 2740–2752 (2022)

59. Krizhevsky, A., Sutskever, I., Hinton, G.E.: Imagenet classification with deep convolutional neural networks. In: Advances in Neural Information Processing Systems, pp. 1097–1105 (2012)

60. Liehr, M., Hazra, J., Beckmann, K., Rafiq, S., Cady, N.: Impact of switching variability of 65nm CMOS integrated hafnium dioxide-based ReRAM devices on distinct level operations. In: IIRW. IEEE (2020)

61. Lin, T.Y., Maire, M., Belongie, S., Hays, J., Perona, P., Ramanan, D., Dollár, P., Zitnick, C.L.: Microsoft coco: Common objects in context. In: European Conference on Computer Vision, pp. 740–755. Springer (2014)

62. Lipton, Z.C., Berkowitz, J., Elkan, C.: A critical review of recurrent neural networks for sequence learning. Preprint (2015). arXiv:1506.00019

63. Litjens, G., Kooi, T., Bejnordi, B.E., Setio, A.A.A., Ciompi, F., Ghafoorian, M., Van Der Laak, J.A., Van Ginneken, B., Sánchez, C.I.: A survey on deep learning in medical image analysis. Med. Image Anal. **42**, 60–88 (2017)

64. Liu, B., Chen, Y., Liu, S., Kim, H.S.: Deep learning in latent space for video prediction and compression. In: Proceedings of the IEEE/CVF Conference on Computer Vision and Pattern Recognition, pp. 701–710 (2021)

65. Liu, B., et al.: Reduction and IR-drop compensations techniques for reliable neuromorphic computing systems. In: ICCAD. IEEE (2014)

66. Liu, C., Zoph, B., Neumann, M., Shlens, J., Hua, W., Li, L.J., Fei-Fei, L., Yuille, A., Huang, J., Murphy, K.: Progressive neural architecture search. In: Proceedings of the European Conference on Computer Vision (ECCV), pp. 19–34 (2018)

67. Liu, Z., Chen, C., Li, L., Zhou, J., Li, X., Song, L., Qi, Y.: Geniepath: Graph neural networks with adaptive receptive paths. In: Proceedings of the AAAI Conference on Artificial Intelligence, vol. 33, pp. 4424–4431 (2019)

68. Long, Y., She, X., Mukhopadhyay, S.: Design of reliable DNN accelerator with un-reliable ReRAM. In: DATE. IEEE (2019)

69. Ma, C., et al.: Go unary: A novel synapse coding and mapping scheme for reliable ReRAM-based neuromorphic computing. In: DATE. IEEE (2020)

70. Ma, T., Chen, J., Xiao, C.: Constrained generation of semantically valid graphs via regularizing variational autoencoders. Preprint (2018). arXiv:1809.02630

71. Mandal, S.K., Ayoub, R., Kishinevsky, M., Islam, M.M., Ogras, U.Y.: Analytical performance modeling of NoCs under priority arbitration and bursty traffic. IEEE Embed. Syst. Lett. **13**(3), 98–101 (2020)

72. Mandal, S.K., Ayoub, R., Kishinevsky, M., Ogras, U.Y.: Analytical performance models for NoCs with multiple priority traffic classes. ACM Trans. Embed. Comput. Syst. (TECS) **18**(5s), 1–21 (2019)

73. Mandal, S.K., Krishnakumar, A., Ayoub, R., Kishinevsky, M., Ogras, U.Y.: Performance analysis of priority-aware NoCs with deflection routing under traffic congestion. In: Proceedings of the 39th International Conference on Computer-Aided Design, pp. 1–9 (2020)

74. Mandal, S.K., Krishnakumar, A., Ogras, U.Y.: Energy-efficient networks-on-chip architectures: design and run-time optimization. In: Network-on-Chip Security and Privacy, p. 55 (2021)

75. Mandal, S.K., Krishnan, G., Chakrabarti, C., Seo, J.S., Cao, Y., Ogras, U.Y.: A latency-optimized reconfigurable NoC for in-memory acceleration of DNNs. IEEE J. Emerg. Sel. Top. Circ. Syst. **10**(3), 362–375 (2020)

76. Mandal, S.K., Krishnan, G., Goksoy, A.A., Nair, G.R., Cao, Y., Ogras, U.Y.: COIN: Communication-aware in-memory acceleration for graph convolutional networks. IEEE J. Emerg. Sel. Top. Circ. Syst. **2**(2), 472–485 (2022)

77. Mandal, S.K., Tong, J., Ayoub, R., Kishinevsky, M., Abousamra, A., Ogras, U.Y.: Theoretical analysis and evaluation of NoCs with weighted round-robin arbitration. In: 2021 IEEE/ACM International Conference On Computer Aided Design (ICCAD), pp. 1–9 (2021)

78. Mao, M., et al.: MAX2: An ReRAM-based neural network accelerator that maximizes data reuse and area utilization. IEEE J. Emerg. Sel. Top. Circ. Syst. **9**(2), 398–410 (2019)

79. Mohanty, A., et al.: Random sparse adaptation for accurate inference with inaccurate multi-level RRAM arrays. In: IEDM. IEEE (2017)

80. Nabavinejad, S.M., Baharloo, M., Chen, K.C., Palesi, M., Kogel, T., Ebrahimi, M.: An overview of efficient interconnection networks for deep neural network accelerators. IEEE J. Emerg. Sel. Top. Circ. Syst. **10**(3), 268–282 (2020)

81. Ogras, U.Y., Bogdan, P., Marculescu, R.: An analytical approach for network-on-chip performance analysis. IEEE Trans. Comput. Aided Des. Integr. Circ. Syst. **29**(12), 2001–2013 (2010)

82. Peng, X., Huang, S., Jiang, H., Lu, A., Yu, S.: DNN+ NeuroSim V2. 0: An end-to-end benchmarking framework for compute-in-memory accelerators for on-chip training. IEEE Trans. Comput. Aided Des. Integr. Circ. Syst. **40**(11), 2306–2319 (2020)
83. Peng, X., Huang, S., Luo, Y., Sun, X., Yu, S.: DNN+ NeuroSim: An end-to-end benchmarking framework for compute-in-memory accelerators with versatile device technologies. In: 2019 IEEE International Electron Devices Meeting (IEDM), pp. 32–35 (2019)
84. Pisner, D.A., Schnyer, D.M.: Support vector machine. In: Machine Learning, pp. 101–121. Elsevier (2020)
85. Real, E., Aggarwal, A., Huang, Y., Le, Q.V.: Regularized evolution for image classifier architecture search. In: Proceedings of the AAAI Conference on Artificial Intelligence, vol. 33, pp. 4780–4789 (2019)
86. Rubinstein, R., Bruckstein, A.M., Elad, M.: Dictionaries for sparse representation modeling. Proc. IEEE **98**(6), 1045–1057 (2010)
87. Saikia, J., Yin, S., Cherupally, S.K., Zhang, B., Meng, J., Seok, M., Seo, J.S.: Modeling and optimization of SRAM-based in-memory computing hardware design. In: 2021 Design, Automation & Test in Europe Conference & Exhibition (DATE), pp. 942–947. IEEE (2021)
88. Samajdar, A., Zhu, Y., Whatmough, P., Mattina, M., Krishna, T.: Scale-sim: systolic CNN accelerator simulator. Preprint (2018). arXiv:1811.02883
89. Scarselli, F., Gori, M., Tsoi, A.C., Hagenbuchner, M., Monfardini, G.: The graph neural network model. IEEE Trans. Neural Networks **20**(1), 61–80 (2008)
90. Schuster, M., Paliwal, K.K.: Bidirectional recurrent neural networks. IEEE Trans. Signal Process. **45**(11), 2673–2681 (1997)
91. Seo, J.: Advances in digital vs. analog AI accelerators (2022). In: Tutorial at IEEE International Solid-State Circuits Conference (ISSCC)
92. Shafiee, A., et al.: ISAAC: A convolutional neural network accelerator with in-situ analog arithmetic in crossbars. ACM SIGARCH Comput. Archit. News **44**(3), 14–26 (2016)
93. Shao, Y.S., Clemons, J., Venkatesan, R., Zimmer, B., Fojtik, M., Jiang, N., Keller, B., Klinefelter, A., Pinckney, N., Raina, P., et al.: Simba: Scaling deep-learning inference with multi-chip-module-based architecture. In: Proceedings of the 52nd Annual IEEE/ACM International Symposium on Microarchitecture, pp. 14–27 (2019)
94. Si, X., Chen, J.J., Tu, Y.N., Huang, W.H., Wang, J.H., Chiu, Y.C., Wei, W.C., Wu, S.Y., Sun, X., Liu, R., et al.: 24.5 a twin-8t SRAM computation-in-memory macro for multiple-bit CNN-based machine learning. In: 2019 IEEE International Solid-State Circuits Conference-(ISSCC), pp. 396–398. IEEE (2019)
95. Simonovsky, M., Komodakis, N.: Graphvae: Towards generation of small graphs using variational autoencoders. In: International Conference on Artificial Neural Networks, pp. 412–422. Springer (2018)
96. Song, L., Qian, X., Li, H., Chen, Y.: Pipelayer: A pipelined ReRAM-based accelerator for deep learning. In: 2017 IEEE International Symposium on High Performance Computer Architecture (HPCA), pp. 541–552 (2017)
97. Spetalnick, S.D., Chang, M., Crafton, B., Khwa, W.S., Chih, Y.D., Chang, M.F., Raychowdhury, A.: A 40nm 64kb 26.56 tops/w 2.37 mb/mm 2 rram binary/compute-in-memory macro with 4.23 x improvement in density and >75% use of sensing dynamic range. In: 2022 IEEE International Solid-State Circuits Conference (ISSCC), vol. 65, pp. 1–3. IEEE (2022)
98. Su, J.W., Si, X., Chou, Y.C., Chang, T.W., Huang, W.H., Tu, Y.N., Liu, R., Lu, P.J., Liu, T.W., Wang, J.H., et al.: 15.2 a 28nm 64kb inference-training two-way transpose multibit 6t SRAM compute-in-memory macro for AI edge chips. In: 2020 IEEE International Solid-State Circuits Conference-(ISSCC), pp. 240–242. IEEE (2020)
99. Sun, Y., et al.: Unary coding and variation-aware optimal mapping scheme for reliable ReRAM-based neuromorphic computing. TCAD (2021)
100. Sutskever, I., Vinyals, O., Le, Q.V.: Sequence to sequence learning with neural networks. In: Advances in Neural Information Processing Systems, pp. 3104–3112 (2014)
101. Szegedy, C., Liu, W., Jia, Y., Sermanet, P., Reed, S., Anguelov, D., Erhan, D., Vanhoucke, V., Rabinovich, A.: Going deeper with convolutions. In: Proceedings of the IEEE Conference on Computer Vision and Pattern Recognition, pp. 1–9 (2015)

102. Tan, M., Chen, B., Pang, R., Vasudevan, V., Sandler, M., Howard, A., Le, Q.V.: Mnasnet: Platform-aware neural architecture search for mobile. In: Proceedings of the IEEE/CVF Conference on Computer Vision and Pattern Recognition, pp. 2820–2828 (2019)
103. Valavi, H., Ramadge, P.J., Nestler, E., Verma, N.: A 64-tile 2.4-mb in-memory-computing CNN accelerator employing charge-domain compute. IEEE J. Solid State Circ. **54**(6), 1789–1799 (2019)
104. Vivet, P., Guthmuller, E., Thonnart, Y., Pillonnet, G., Fuguet, C., Miro-Panades, I., Moritz, G., Durupt, J., Bernard, C., Varreau, D., et al.: IntAct: A 96-core processor with six chiplets 3D-stacked on an active interposer with distributed interconnects and integrated power management. IEEE J. Solid State Circ. **56**(1), 79–97 (2020)
105. Wu, B., Dai, X., Zhang, P., Wang, Y., Sun, F., Wu, Y., Tian, Y., Vajda, P., Jia, Y., Keutzer, K.: Fbnet: Hardware-aware efficient convnet design via differentiable neural architecture search. In: Proceedings of the IEEE/CVF Conference on Computer Vision and Pattern Recognition, pp. 10734–10742 (2019)
106. Xu, K., Hu, W., Leskovec, J., Jegelka, S.: How powerful are graph neural networks? Preprint (2018). arXiv:1810.00826
107. Yang, X., et al.: Multi-objective optimization of ReRAM crossbars for robust DNN inferencing under stochastic noise. In: ICCAD. IEEE/ACM (2021)
108. Yin, S., Jiang, Z., Kim, M., Gupta, T., Seok, M., Seo, J.s.: Vesti: energy-efficient in-memory computing accelerator for deep neural networks. IEEE Trans. Very Large Scale Integr. (VLSI) Syst. **28**(1), 48–61 (2019)
109. Yin, S., Jiang, Z., Seo, J.S., Seok, M.: XNOR-SRAM: In-memory computing sram macro for binary/ternary deep neural networks. IEEE J. Solid State Circ. **55**(6), 1733–1743 (2020)
110. Yin, S., Zhang, B., Kim, M., Saikia, J., Kwon, S., Myung, S., Kim, H., Kim, S.J., Seok, M., Seo, J.s.: Pimca: A 3.4-mb programmable in-memory computing accelerator in 28nm for on-chip DNN inference. In: 2021 Symposium on VLSI Technology, pp. 1–2. IEEE (2021)
111. Yue, J., Liu, Y., Yuan, Z., Feng, X., He, Y., Sun, W., Zhang, Z., Si, X., Liu, R., Wang, Z., et al.: Sticker-im: A 65 nm computing-in-memory NN processor using block-wise sparsity optimization and inter/intra-macro data reuse. IEEE J. Solid State Circ. **57**(8), 2560–2573 (2022)
112. Zhang, J., Wang, Z., Verma, N.: In-memory computation of a machine-learning classifier in a standard 6t SRAM array. IEEE J. Solid State Circ. **52**(4), 915–924 (2017)
113. Zhao, W., Cao, Y.: New generation of predictive technology model for Sub-45 nm early design exploration. IEEE Trans. Electron Dev. **53**(11), 2816–2823 (2006)
114. Zhou, C., Kadambi, P., Mattina, M., Whatmough, P.N.: Noisy machines: understanding noisy neural networks and enhancing robustness to analog hardware errors using distillation. Preprint (2020). arXiv:2001.04974
115. Zhou, D., Zhou, X., Zhang, W., Loy, C.C., Yi, S., Zhang, X., Ouyang, W.: Econas: Finding proxies for economical neural architecture search. In: Proceedings of the IEEE/CVF Conference on Computer Vision and Pattern Recognition, pp. 11396–11404 (2020)
116. Zhu, Z., Sun, H., Qiu, K., Xia, L., Krishnan, G., Dai, G., Niu, D., Chen, X., Hu, X.S., Cao, Y., et al.: MNSIM 2.0: A behavior-level modeling tool for memristor-based neuromorphic computing systems. In: Proceedings of the 2020 on Great Lakes Symposium on VLSI, pp. 83–88 (2020)
117. Zoph, B., Vasudevan, V., Shlens, J., Le, Q.V.: Learning transferable architectures for scalable image recognition. In: Proceedings of the IEEE Conference on Computer Vision and Pattern Recognition, pp. 8697–8710 (2018)

Efficient Deep Learning Using Non-volatile Memory Technology in GPU Architectures

Ahmet Inci, Mehmet Meric Isgenc, and Diana Marculescu

1 Introduction

Over the last decade, the performance boost achieved through CMOS scaling has plateaued, necessitating sophisticated computer architecture solutions to gain higher performance in computing systems while maintaining a feasible power density. These objectives, however, are concurrently challenged by the limitations of the performance of memory resources [1]. In contrast to the initial insight of Dennard on power density [2], deep CMOS scaling has exacerbated static power consumption, causing the heat density of ICs to reach catastrophic levels unless properly addressed [3–5].

As computers suffer from memory- and power-related limitations, the demand for data-intensive applications has been on the rise. With the increasing data deluge and recent improvements in GPU architectures, deep neural networks (DNNs) have achieved remarkable success in various tasks such as image recognition [6, 7], object detection [8], and chip placement [9] by utilizing inherent massive parallelism of GPU platforms. However, DNN workloads continue to have large memory footprints and significant computational requirements to achieve higher accuracy. Thus, DNN workloads exacerbate the memory bottleneck that degrades the overall performance of the system. To this end, while deep learning (DL) practitioners focus on model compression techniques [10–12], system architects investigate hardware

A. Inci · M. M. Isgenc
Carnegie Mellon University, Pittsburgh, PA, USA
e-mail: ainci@andrew.cmu.edu; misgenc@andrew.cmu.edu

D. Marculescu (✉)
Carnegie Mellon University, Pittsburgh, PA, USA
The University of Texas at Austin, Austin, TX, USA
e-mail: dianam@utexas.edu; dianam@cmu.edu

© The Author(s), under exclusive license to Springer Nature Switzerland AG 2024
S. Pasricha, M. Shafique (eds.), *Embedded Machine Learning for Cyber-Physical, IoT, and Edge Computing*, https://doi.org/10.1007/978-3-031-19568-6_8

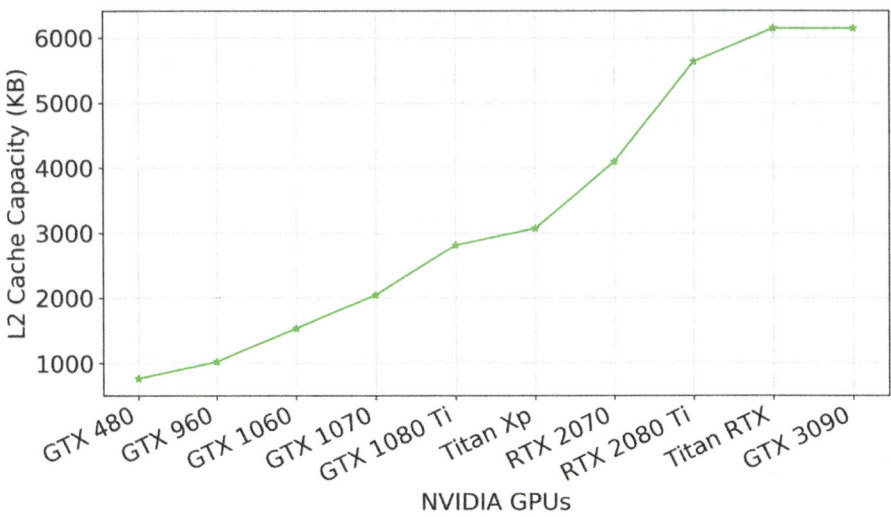

Fig. 1 L2 cache capacity in recent NVIDIA GPUs [29]

architectures to overcome the memory bottleneck problem and improve the overall system performance [13–24].

We note the current trend of GPU architectures is toward increasing last-level cache capacity as shown in Fig. 1. Our analysis shows that conventional SRAM technology incurs scalability problems as far as power, performance, and area (PPA) are concerned [21, 25–27]. Non-volatile memory (NVM) technology is one of the most promising solutions to tackle memory bottleneck problem for data-intensive applications [28]. However, because much of emerging NVM technology is not available for commercial use, there is an obvious need for a framework to perform design space exploration for these emerging NVM technologies for DL workloads.

In this chapter, we present *DeepNVM++* [19], an extended and improved framework [18] to characterize, model, and optimize NVM-based caches in GPU architectures for deep learning workloads. Without loss of generality, we demonstrate our framework for spin-transfer torque magnetic random access memory (STT-MRAM) and spin-orbit torque magnetic random access memory (SOT-MRAM), keeping in mind that it can be used for any NVM technology, GPU platform, or deep learning workload. Our cross-layer analysis framework incorporates both circuit-level characterization aspects and the memory behavior of various DL workloads running on an actual GPU platform. *DeepNVM++* enables the evaluation of *power, performance, and area* of NVMs when used for last-level (L2) caches in GPUs and seeks to exploit the benefits of this emerging technology to improve the performance ·of deep learning applications.

To perform *iso-capacity* analysis, we carry out extensive memory profiling of various deep learning workloads for both training and inference on the existing GPU platforms. For the *iso-area* analysis, the existing platforms cannot be used for varying cache sizes, so we rely on architecture-level simulation of GPUs to quantify and better understand last-level cache capacity and off-chip memory accesses. In both cases, our framework automatically combines resulting memory statistics with circuit and microarchitecture-level characterization and analysis of emerging NVM technologies to gauge their impact on DL workloads running on future GPU-based platforms.

We make the following contributions:

1. **Circuit-level bitcell characterization.** We perform detailed circuit-level characterization combining a commercial 16nm CMOS technology and prominent STT [30] and SOT [31] models from the literature to iterate through our framework in an end-to-end manner to demonstrate the flexibility of *DeepNVM++* [19] for future studies.

2. **Microarchitecture-level cache design exploration.** We use *NVSim* [32] to perform a fair comparison between SRAM, STT-MRAM, and SOT-MRAM by incorporating the circuit-level models developed in 1) using 16nm technology and choosing the best cache configuration for each of them.

3. **Iso-capacity analysis.** To compare the efficacy of magnetic random access memory (MRAM) caches to conventional SRAM caches, we perform our novel iso-capacity analysis based on *actual platform profiling* results for the memory behavior of various DNNs by using the *Caffe* framework [33] on a high-end NVIDIA 1080 Ti GPU (implemented in 16nm technology) for the ImageNet dataset [34].

4. **Iso-area analysis.** Because of their different densities, we compare SRAM and NVM caches in an iso-area analysis to quantify the benefits of higher density of NVM technologies on DL workloads running on GPU platforms. Since the existing platforms do not support resulting iso-area cache sizes, we extend the GPGPU-Sim [35] simulator to run DL workloads and support larger cache capacities for STT-MRAM and SOT-MRAM.

5. **Scalability analysis.** Finally, we perform a thorough scalability analysis and compare SRAM, STT-MRAM, and SOT-MRAM in terms of power, performance, and area to project and gauge the efficacy of NVM- and SRAM-based caches for DL workloads as cache capacity increases.

To the best of our knowledge, putting everything together, *DeepNVM++* [19] is the *first comprehensive framework* for cross-layer characterization, modeling, and analysis of emerging NVM technologies for deep learning workloads running on GPU platforms. Our results show that in the iso-capacity case, STT-MRAM and SOT-MRAM achieve up to $3.8\times$ *and* $4.7\times$ *energy-delay product reduction* and $2.4\times$ *and* $2.8\times$ *area reduction* compared to SRAM baseline, respectively. In the iso-area case, STT-MRAM and SOT-MRAM achieve up to $2.2\times$ *and* $2.4\times$ *energy-delay product reduction* and accommodate $2.3\times$ *and* $3.3\times$ *larger cache capacity* compared to SRAM, respectively.

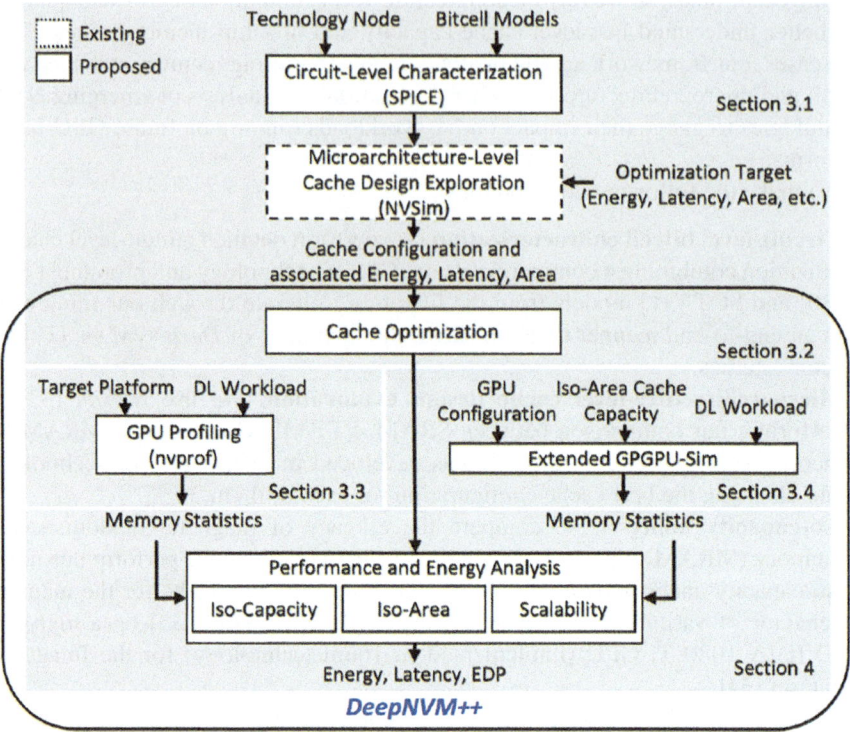

Fig. 2 Overview of the *DeepNVM++* [19] cross-layer analysis flow

Next, we present our cross-layer analysis framework, as shown in Fig. 2. First, we present the background and related work on non-volatile memory technologies (Sect. 2). Next, we show our detailed circuit-level characterization analysis using CMOS, STT, and SOT device models (Sect. 3.1). After developing bitcell models, we present our microarchitecture-level cache design methodology to obtain cache area, latency, and energy results (Sect. 3.2). Next, we describe our iso-capacity analysis flow in which we gather actual memory statistics through GPU profiling (Sect. 3.3). Furthermore, we detail our iso-area analysis in which we extend GPGPU-Sim to run deep learning workloads and support larger cache capacities for STT-MRAM and SOT-MRAM (Sect. 3.4). Next, we present experimental results demonstrating the efficiency of STT-MRAM and SOT-MRAM over the conventional SRAM for iso-capacity and iso-area cases (Sect. 4). We then discuss the implications of the results shown in this chapter (Sect. 5). Finally, we conclude this chapter by summarizing the results (Sect. 6).

2 Related Work

Although 16nm has become a commonplace technology for high-end customers of foundries, an intriguing inflection point awaits the electronics community as we approach the end of the traditional density, power, and performance benefits of CMOS scaling [36, 37]. To move beyond the computing limitations imposed by staggering CMOS scaling trends, MRAM has emerged as a promising candidate [28].

The enabling technology of MRAM consists of magnetic tunnel junction (MTJ) pillars that can store data as a resistive state [38]. An MTJ pillar consists of a thin oxide film sandwiched by two ferromagnetic layers. One of these ferromagnetic layers has a fixed magnetization that serves as a reference layer. The magnetization of the other layer can be altered by changing the direction of the current that flows through the pillar. If the magnetization of the free layer and the reference layer are in parallel, the device is in the low-resistance state. If the magnetization of layers is in opposite directions, the device is in the high-resistance state [39].

STT bitcells [40] use an MTJ pillar as their core storage element and an additional access transistor to enable read and write operations. Although STT bitcells offer non-volatility, low read latency, and high endurance [41], the write current is also high [42–44], which increases power consumption. To this end, SOT bitcells have been proposed to overcome the write current challenges by isolating the read and write paths [45]. Because the read disturbance errors are much less likely in SOT bitcells, both read and write access devices can be tuned in accordance with the lower current requirements [46, 47]. The read and write current requirements of STT and SOT bitcells can have a crucial impact on the eventual MRAM characteristics because they affect the CMOS access transistors, bitcell area, and peripheral logic. Thus, a comparison of these bitcells and the traditional SRAM merits a meticulous analysis that takes these factors into account.

Prior work has proposed effective approaches to overcome the shortcomings of emerging NVM technologies such as using hybrid SRAM- and NVM-based caches that utilize the complementary features of different memory technologies [48–51], relaxing non-volatility properties to reduce the high write latency and energy [52–55], and implementing cache replacement policies [56–58] for higher level caches such as L1 caches and register files. However, NVM technology appears to be a better choice for lower level caches such as L2 or L3 caches due to its long write latency and high cell density. Higher level L1 caches are latency-sensitive and optimized for performance, whereas last-level caches are capacity-sensitive and optimized for a high hit rate to reduce off-chip memory accesses. Therefore, NVM-based caches provide a better use case for replacing SRAM in last-level caches due to their high cell density when compared to SRAM-based caches. To this end, we evaluate power, performance, and area of NVM technology when used for last-level caches in GPU platforms.

While prior work has shown the potential of NVM technologies for generic applications to some extent, there is a need for a cross-layer analysis framework to explore the potential of NVM technologies in GPU platforms, particularly for DL workloads. The most commonly used modeling tool for emerging NVM technologies is *NVSim* [32], a circuit-level model for performance, energy, and area estimation. However, *NVSim* is not sufficient to perform a detailed cross-layer analysis for NVM technologies for DL workloads since it does not take architecture-level analysis and application-specific memory behavior into account. To this end, prior work has proposed cross-layer evaluation frameworks for non-traditional architectures such as processing-in-memory-based analog and digital architectures [59–61]. However, there is still a need for a cross-layer analysis framework to perform design space exploration of NVM technologies for GPU architectures for DL workloads. In this chapter, we incorporate *NVSim* with our cross-layer modeling and optimization flow including novel architecture-level iso-capacity and iso-area analysis flow to perform design space exploration for conventional SRAM and emerging NVM caches for DL workloads running on GPU architectures.

3   Methodology

3.1   Circuit-Level NVM Characterization

A vast majority of work in the literature use simple bitcell models [46] to assess the PPA of corresponding cache designs. Because bitcells are the core components of the memory, the methodology to calculate the bitcell latency, energy, and area is crucial for accurate comparisons. To this end, we use a commercial 16nm bitcell design as a baseline as we model the STT and SOT bitcells. This technology node also matches the fabrication technology of the GPU platform that we use to gather actual memory statistics in Sect. 3.3.

The key bitcell parameters needed for cache modeling are read and write currents and latency values for high-to-low and low-to-high resistive transitions. These parameters can be optimized by tuning the size of the access transistors. While larger access transistors enable faster reads and writes, they increase the energy consumption and the bitcell layout size. The optimal sizing of the access transistor and the array architecture varies based on the bitcell type. The access transistor sizing optimization is crucial since it impacts the eventual PPA characteristics of the bitcell and the cache. To address the array architecture differences between STT and SOT MRAM for a fair comparison, we performed transient simulations.

For our simulations, we used perpendicular to the plane STT [30] and SOT [31] models and a commercial 16nm FinFET model that takes post-layout effects into account. To find the latency and energy parameters, we used parameterized SPICE netlists wherein the read/write pulse widths were modulated to the point of failure. Furthermore, we swept a range of fin counts for the access devices to find the optimal balance between the latency, energy, and area. For the transient SPICE simulations, we picked the FinFET models corresponding to the worst delay and

Table 1 STT-MRAM and SOT-MRAM bitcell parameters after device-level characterization

	STT-MRAM	SOT-MRAM
Sense Latency (ps)	650	650
Sense Energy (pJ)	0.076	0.020
Write Latency (ps)	8400 (set)/7780 (reset)	313 (set)/243 (reset)
Write Energy (pJ)	1.1 (set)/2.2 (reset)	0.08 (set)/0.08 (reset)
Fin Counts	4 (read/write)	3 (write) + 1 (read)
Area (normalized)	0.34[a]	0.29[a]

[a] Area is normalized with respect to the foundry SRAM bitcell

power scenarios. To calculate the bitcell area for the 16nm layout design rules, we used the bitcell area formulations provided in prior work [62].

We summarize the obtained bitcell parameters in Table 1. The sensing delay is measured from wordline activation to the point where the bitline voltage difference reaches 25mV. The sense energy is the integration of the power consumed over the sensing time window. For both magnetic flavors, the sense delay is similar; however, SOT-MRAM is more energy-efficient in terms of read operation owing to the separation of the read/write terminals. The write latency in this context refers to the time between the arrival of the write-enable signal to the access transistor and a complete magnetization change for the MTJ. The write latencies for STT and SOT bitcells are significantly different, as expected. This difference can be seen in the energy values as well. The access device is more than double the width of the technology minimum device in order to enable a larger current flow to the STT bitcell, causing the 1T1R STT bitcell to occupy a larger area than the 2T1R SOT bitcell. The isolation of the read and write terminals in the SOT bitcell allows for a smaller write access device. The area values are normalized by the foundry bitcell area. We highlight the significant area difference and demonstrate its impact on the cache characteristics in Sect. 3.2. We use these bitcell parameters for energy-delay-area product (EDAP)-optimized cache design exploration as discussed in the next section.

3.2   Microarchitecture-Level Cache Design Exploration

In order to demonstrate the impact of using STT and SOT bitcells in L2 caches, we use *NVSim* [32], a circuit-level analysis framework that delivers energy, latency, and area results. After developing *NVSim*-compatible bitcell models as described in Sect. 3.1, we analyzed a range of cache capacities (1MB to 32MB) for all possible configurations and cache access types to demonstrate the potential of STT-MRAM and SOT-MRAM as the cache capacity tends to grow. Such a scalability study will help in determining the benefits of switching from conventional SRAM- to NVM-based caches in future GPU platforms as depicted by the trend in Fig. 1.

Algorithm 1: EDAP-optimal cache tuning algorithm

Input: Memory type *mem*, Cache capacity *cap*, Optimization target *opt*, ...
... Access type *acc*
Output: EDAP-tuned cache configuration
1 $mem \in M = \{SRAM, STT, SOT\}$;
2 $cap \in C = \{1, 2, 4, 8, 16, 32\}$;
3 $opt \in O = \{Read_{Latency}, Write_{Latency}, Read_{Energy}, Write_{Energy}, Read_{EDP}, ...$
4 ...$Write_{EDP}, Area, Leakage\}$;
5 $acc \in \mathcal{A} = \{Normal, Fast, Sequential\}$;
6 **for** *each* $mem \in M$ **do**
7   **for** *each* $cap \in C$ **do**
8         $Q' \leftarrow \infty$;
9   **for** *each* $opt \in O$ **do**
10   **for** *each* $acc \in \mathcal{A}$ **do**
11              $Q \leftarrow calculate(EDAP)$;
12   **if** $Q < Q'$ **then**
13                 $Q' \leftarrow Q$;
14   **end**
15   **end**
16   **end**
17        $TunedConfig.append(argv(Q))$;
18   **end**
19 **end**
20 **return** $TunedConfig$;

Algorithm 1 depicts the EDAP-optimal cache tuning algorithm. Based on the optimization target used in *NVSim*, the cache PPA values vary substantially. Therefore, we independently choose the best configuration for each type of memory technology in terms of EDAP metric to perform a fair comparison that encompasses all and not just one of the design constraint dimensions.

As described in Sect. 3.1, we use a commercial 16nm bitcell design. To ensure a correct analysis, we modified the internal technology file of *NVSim* to the corresponding 16nm technology parameters. Next, we compare SRAM, STT-MRAM, and SOT-MRAM for various cache capacities in terms of area, latency, and energy results. Based on these, we determine the EDAP for the cache (as denoted by *calculate(EDAP)* in Algorithm 1).

Table 2 shows the latency, energy, and area results that correspond to the cache capacity of NVIDIA GTX 1080 Ti GPU (3MB) and to the larger MRAM caches that fit into the same area of SRAM baseline. We convert read and write latencies to clock cycles based on 1080 Ti GPU's clock frequency for our calculations. For STT-MRAM and SOT-MRAM, we show parameters for both iso-capacity and iso-area when compared to SRAM. We use these parameters to evaluate the workload-dependent impact of memory choices using DL workloads with diverse structures and multiply–accumulate operation (MAC) configurations.

The energy and latency benefits of STT-MRAM and SOT-MRAM depend on the data characteristics of a given workload. To account for differences in the data-

Table 2 Latency, energy, and area results for SRAM, STT-MRAM, and SOT-MRAM caches for iso-capacity and iso-area

	SRAM	STT-MRAM		SOT-MRAM	
		Iso-capacity	Iso-area	Iso-capacity	Iso-area
Capacity (MB)	3	3	7	3	10
Read Latency (ns)	2.91	2.98	4.58	3.71	6.69
Write Latency (ns)	1.53	9.31	10.06	1.38	2.47
Read Energy (nJ)	0.35	0.81	0.93	0.49	0.51
Write Energy (nJ)	0.32	0.31	0.43	0.22	0.40
Leakage Power (mW)	6442	748	1706	527	1434
Area (mm^2)	5.53	2.34	5.12	1.95	5.64

Table 3 Configurations for DNNs under consideration

	AlexNet [63]	GoogLeNet [64]	VGG-16 [65]	ResNet-18 [66]	SqueezeNet [67]
Top-5 Error (%)	16.4	6.7	7.3	10.71	16.4
CONV Layers	5	57	13	17	26
FC Layers	3	1	3	1	0
Total Weights	61M	7M	138M	11.8M	1.2M
Total MACs	724M	1.43G	15.5G	2G	837M

related read/write characteristics, we used a simple model where we multiply the number of read and write transactions by the corresponding latency and energy values for those operations.

Implications in Architecture-Level Analysis To gauge the benefits of using MRAM technology, we consider two scenarios: (i) First, one could replace the SRAM cache in a GPU with the same capacity MRAM with a smaller area. (ii) Alternatively, by using the same area dedicated to the cache, one can increase the on-chip cache capacity, thereby reducing costly DRAM traffic. We analyze and discuss both approaches through platform profiling results for iso-capacity scenario and a set of architecture-level simulations for iso-area scenario.

3.3 Architecture-Level Iso-Capacity Analysis

As the target platform to demonstrate our work, we use a high-end NVIDIA GTX 1080 Ti GPU that is fabricated in a commercial 16nm technology node that also matches our bitcell and cache models. We use the *Caffe* [33] framework to run various DNNs such as AlexNet [63], GoogLeNet [64], VGG-16 [65], ResNet-18 [66], and SqueezeNet [67] for the ImageNet [34] dataset as shown in Table 3. Our analysis is generalizable to other types of neural network architectures since we cover a wide range of DNN configurations with various workload characteristics. Furthermore, we also use the high-performance conjugate gradients (HPCG)

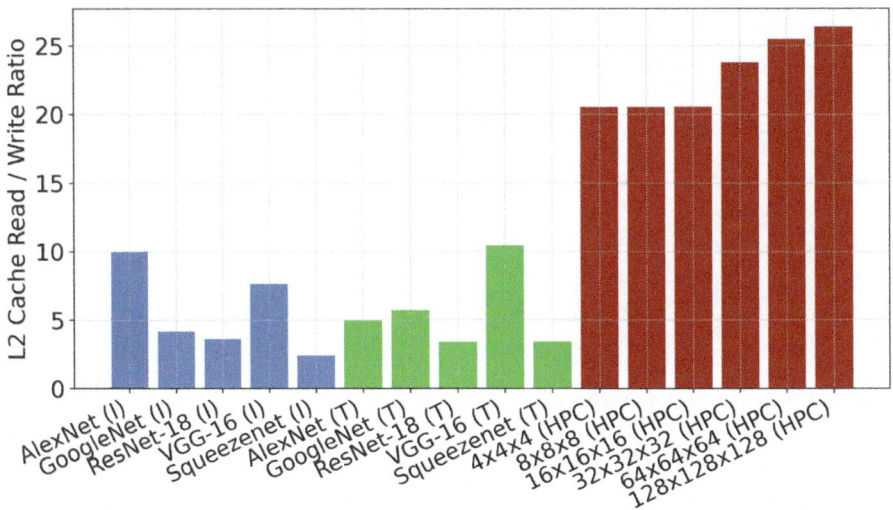

Fig. 3 Profiling results for L2 cache read/write ratio for various workloads

[68] benchmark, a widely used high-performance computing (HPC) workload, to demonstrate the generalizability of our analysis to different workloads besides deep learning applications.

We use the NVIDIA profiler [69] to obtain the device memory and L2 cache read and write transactions to better understand both on-chip and off-chip memory behavior of various deep learning and HPC workloads. To this end, Fig. 3 shows the profiling results for L2 cache read/write ratio for various deep learning and HPC workloads. In particular, we run the HPCG benchmark with different input local subgrid dimensions such as $4\times4\times4$, $8\times8\times8$, $16\times16\times16$, $32\times32\times32$, $64\times64\times64$, and $128\times128\times128$. We show that the ratio of the total number of read transactions to the total number of write transactions in L2 cache varies significantly from 2 to 26. Therefore, these profiling results also show that we cover a wide range of workloads with different workload characteristics in our analysis. To this end, we use $128\times128\times128$, $32\times32\times32$, and $8\times8\times8$ workload configurations for our analysis in the rest of the chapter that we refer to as HPCG-L, HPCG-M, and HPCG-S, respectively.

3.4 Architecture-Level Iso-Area Analysis

Since the iso-area larger capacities enabled by higher density NVM implementations do not exist in the existing platforms, we use *GPGPU-Sim* [35] to explore power and performance implications of having these larger L2 caches in GPU architectures for DNN workloads. For comparison, we model the high-end NVIDIA

Table 4 GPGPU-Sim
configurations

	NVIDIA GTX 1080 Ti
Number of Cores	28
Number of Threads/Core	2048
Number of Registers/Core	65,536
L1 Data Cache	48 KB, 128 B line, 6-way LRU
L2 Data Cache	128 KB/channel, 128 B line, 16-way LRU
Instruction Cache	8 KB, 128 B line, 16-way LRU
Number of Schedulers/Core	4
Core Frequency	1481 MHz
Interconnect Frequency	2962 MHz
L2 Cache Frequency	1481 MHz
Memory Frequency	2750 MHz

GTX 1080 Ti GPU. The configurations for NVIDIA GTX 1080 Ti GPU are shown
in Table 4. We extend the *GPGPU-Sim* simulator to support the cache capacity
of NVIDIA GTX 1080 Ti GPU. This GPU is built using a commercial 16nm
technology node that matches our bitcell and cache models. In particular, for
GPGPU-Sim compatibility, we set L2 cache capacity to 3MB. We use this capacity
for our analysis in the rest of the chapter. We measure the number of DRAM
transactions to quantify and better understand the relationship between larger L2
caches and the overall system power and performance. As a DNN benchmark, we
use AlexNet [63] with the ImageNet [34] dataset that is provided by the *DarkNet*
[70] framework. We extend *DarkNet* source code to enable deep learning workloads
on *GPGPU-Sim*.

4   Experimental Results

We analyze STT-MRAM and SOT-MRAM in terms of energy, performance, and
area results by using GPU profiling results for both iso-capacity and iso-area cases
in Sects. 4.1 and 4.2, respectively. In Sect. 4.2, we use iso-area cache parameters as
shown in Table 2, and we use *GPGPU-Sim* to quantify the DRAM access reduction
in the iso-area case at larger cache capacities. We include DRAM accesses in our
performance and energy calculations for iso-area case. In Sect. 4.3, we perform a
scalability analysis to project the implications of the current GPU trend shown in
Fig. 1 on performance and energy results.

4.1 Performance and Energy Results for Iso-Capacity

By combining the actual technology-dependent latency and energy metrics from Table 2, we can perform a performance and energy analysis for replacing conventional SRAM caches with MRAM caches. We choose batch size 4 for inference and 64 for training for our workloads as it is typically used in related work [71].

Figure 4 shows normalized dynamic energy and leakage energy breakdown results for NVIDIA GTX 1080 Ti GPU based on actual platform memory statistics and our MRAM cache models at the same cache capacity. We use our cache parameters and profiling results to calculate results for various DNNs for both inference and training workloads as well as HPCG workloads with different input sizes.

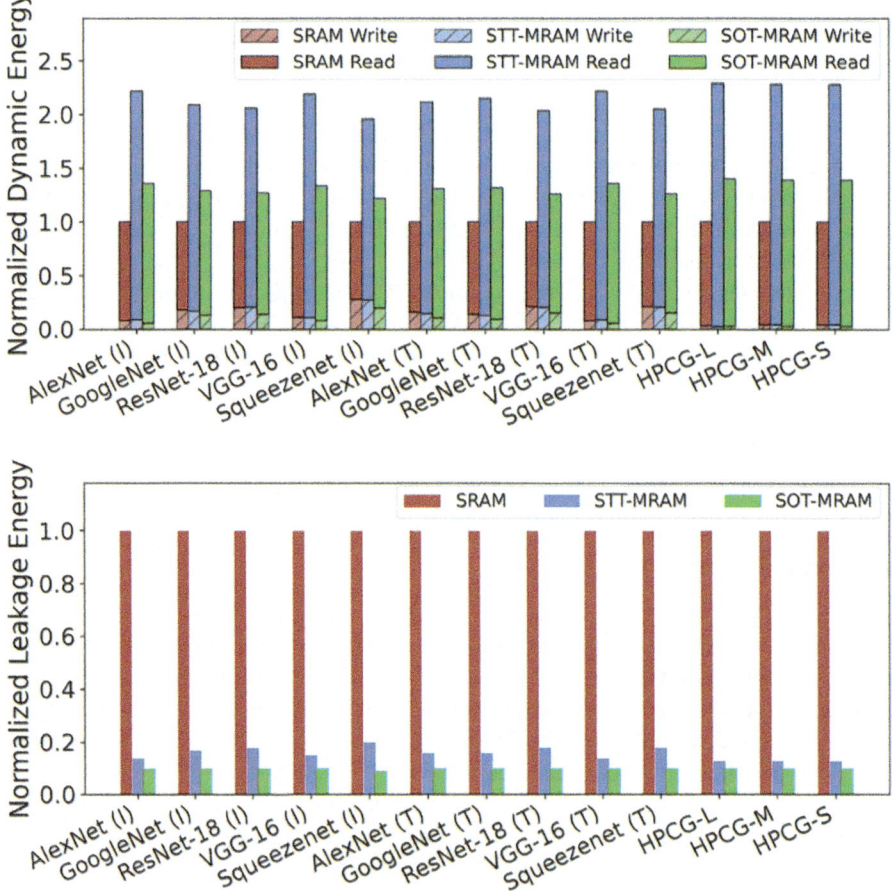

Fig. 4 Dynamic energy (top chart) and leakage energy (bottom chart) (lower is better) normalized with respect to SRAM by using NVMs with iso-capacity (3MB) for inference (I) and training (T) stages

In Fig. 4, we observe that STT-MRAM consumes 2.2× more dynamic energy, whereas SOT-MRAM has 1.3× more dynamic energy on average when compared to the SRAM baseline. Furthermore, our results show that 83% of the total dynamic energy of SRAM comes from read operations, whereas write operations only make for 17% of all transactions on average across deep learning workloads. For HPCG workloads, read operations take 96% of the total dynamic energy of SRAM, and write operations only make for 4% of the total energy. Our profiling results also support these findings as read operations dominate write operations in these DL and HPCG workloads.

On the other hand, Fig. 4 also shows that STT-MRAM and SOT-MRAM provide 6.3× and 10× lower leakage energy on average when compared to SRAM, respectively. Based on this result, Fig. 5 shows significant total normalized energy reduction of STT-MRAM and SOT-MRAM when compared to SRAM given that leakage energy dominates the total energy. In more detail, STT-MRAM and SOT-MRAM achieve 5.3× *and* 8.6× *energy reduction* on average across all workloads compared to SRAM baseline, respectively, due to their significantly low leakage energy. Moreover, Fig. 5 shows that STT-MRAM and SOT-MRAM provide up to 3.8× *and* 4.7× *EDP reduction and* 2.4× *and* 2.8× *area reduction*, respectively.

The Impact of Batch Size on EDP We perform this study to better understand the relationship between batch size and its implications for performance and energy results of SRAM, STT-MRAM, and SOT-MRAM. Figure 6 shows the impact of batch size on EDP results for AlexNet during training and inference stages based on NVIDIA GTX 1080 Ti memory profiling statistics. We show that batch size significantly affects the improvement of STT-MRAM and SOT-MRAM for training. For training, STT-MRAM provides 2.3× to 4.6× EDP reduction as batch size increases. On the other hand, SOT-MRAM provides 7.2× to 7.6× EDP reduction when compared to SRAM baseline. For inference, STT-MRAM and SOT-MRAM achieve 4.1× to 5.4× and 7.1× to 7.3× EDP reduction, respectively. These results also confirm the different workload characteristics of training and inference. STT-MRAM provides higher EDP reduction for training workloads as batch size increases. On the other hand, SOT-MRAM follows the same pattern for inference workloads due to their different access characteristics as shown in Table 2. We observe that training workloads become more read dominant, whereas inference workloads have lower read/write ratio as batch size increases.

4.2   Performance and Energy Results for Iso-Area

As in the iso-capacity study, for iso-area analysis, we use a batch size 4 for inference and 64 for training. Figure 7 shows the reduction in the total number of DRAM accesses as L2 cache capacity increases. We use *GPGPU-Sim* and start with the baseline configuration that is 3MB for NVIDIA GTX 1080 Ti and double its cache capacity up to 24MB to quantify the percentage of DRAM access reduction for STT-

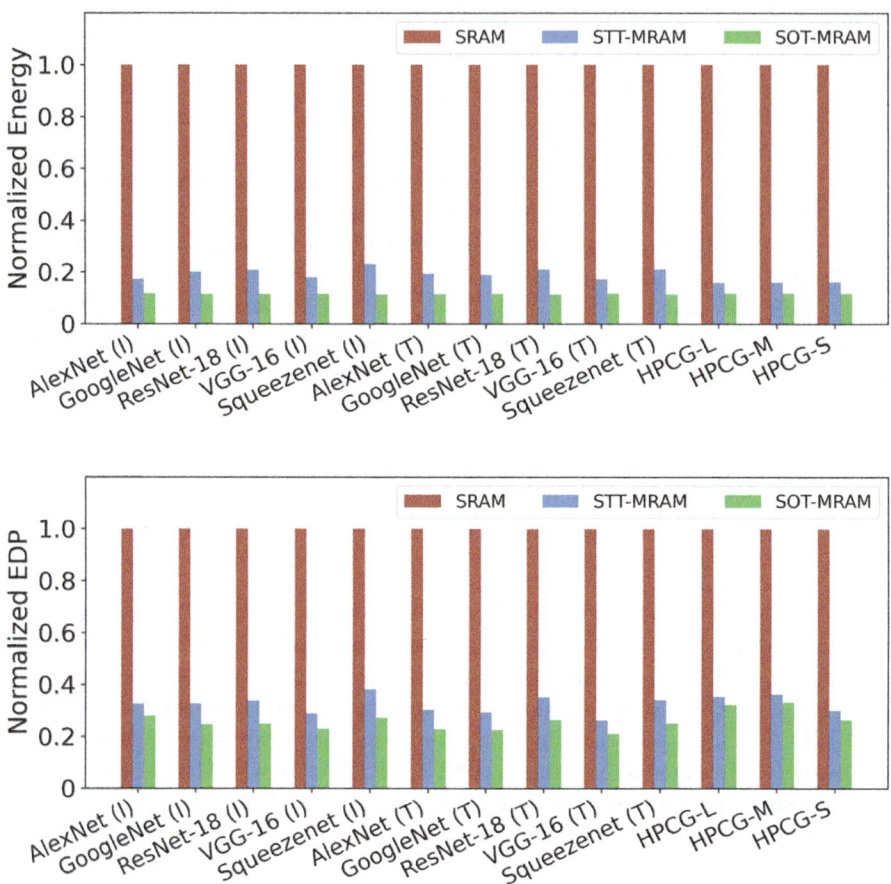

Fig. 5 Iso-capacity (3MB) energy (top chart) and energy-delay product (bottom chart) for NVM-based caches (lower is better) normalized with respect to SRAM-based caches for inference (I) and training (T) stages. DRAM energy and latency are also included in EDP results

MRAM and SOT-MRAM at larger cache capacities. Figure 7 shows that replacing SRAM with STT-MRAM and SOT-MRAM equivalents that fit into the same area significantly reduces the total number of DRAM transactions by 14.6% and 19.8%, respectively, for 1080 Ti GPU.

Figure 8 shows normalized dynamic energy and leakage energy breakdown results for 1080 Ti GPU based on actual platform memory statistics and our MRAM cache models at the same area. We use our iso-area cache parameters in which STT-MRAM (7MB) and SOT-MRAM (10MB) have larger cache capacities for the same area budget with SRAM. We use these cache parameters and profiling results to calculate results for various DNNs for both inference and training workloads and HPCG workloads with various input sizes.

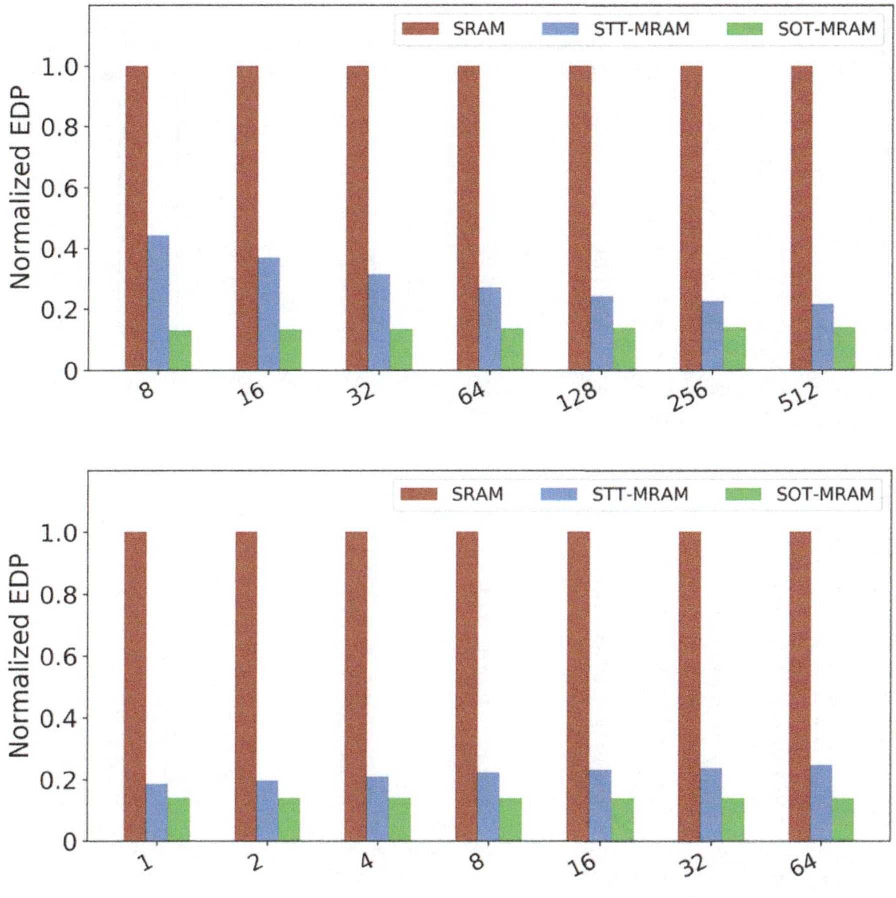

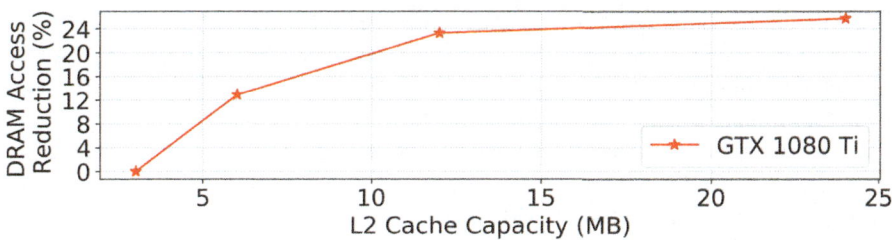

Fig. 6 Impact of batch size on energy-delay product (lower is better) normalized with respect to SRAM by using NVMs with iso-capacity (3MB) for AlexNet for training (top chart) and inference (bottom chart)

Fig. 7 Simulation results for the reduction in the total number of DRAM accesses in percentage

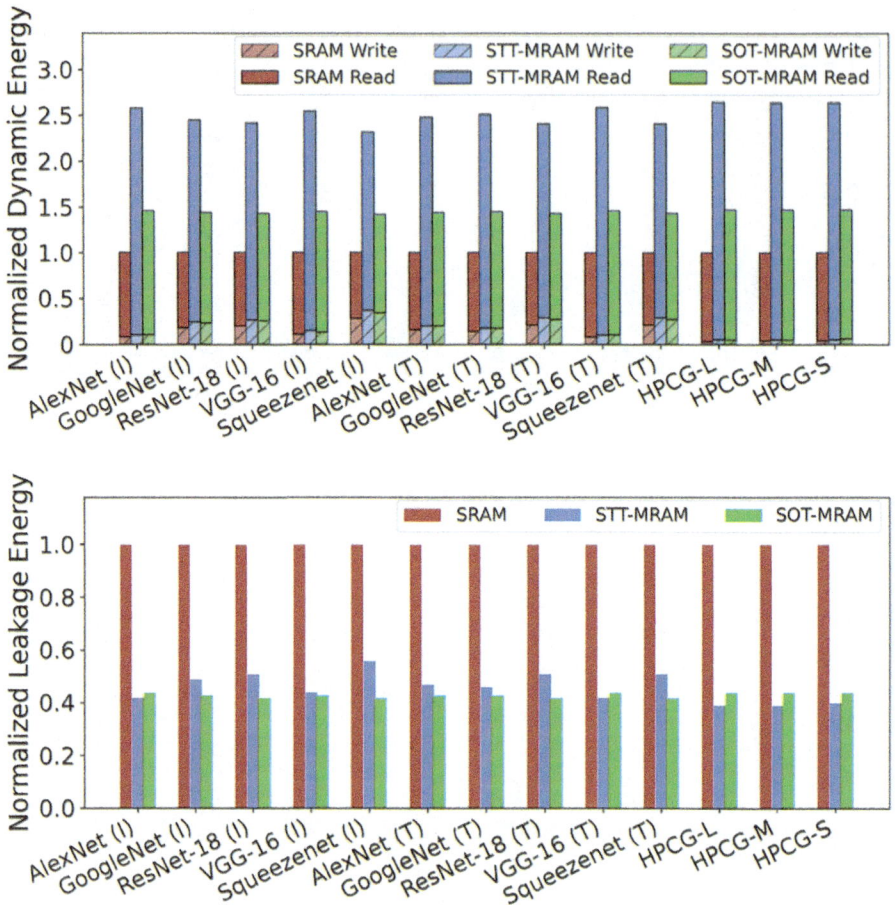

Fig. 8 Dynamic energy (top chart) and leakage energy (bottom chart) (lower is better) normalized with respect to SRAM by using STT-MRAM (7MB) and SOT-MRAM (10MB) with iso-area for inference (I) and training (T) stages

In Fig. 8, we observe that STT-MRAM has 2.5× dynamic energy, whereas SOT-MRAM has 1.5× dynamic energy on average when compared to SRAM baseline. On the other hand, Fig. 8 also shows that STT-MRAM and SOT-MRAM provide 2.2× and 2.3× lower leakage energy on average when compared to SRAM, respectively. Based on this result, STT-MRAM and SOT-MRAM achieve 2× *and* 2.2× *lower energy* when compared to SRAM.

Furthermore, Fig. 9 shows that STT-MRAM and SOT-MRAM provide 1.2× *EDP reduction and* 2.3× *and* 3.3× *larger cache capacity* on average across all workloads when compared to SRAM and off-chip DRAM accesses are not included in the calculations, respectively. When DRAM accesses are included in determining

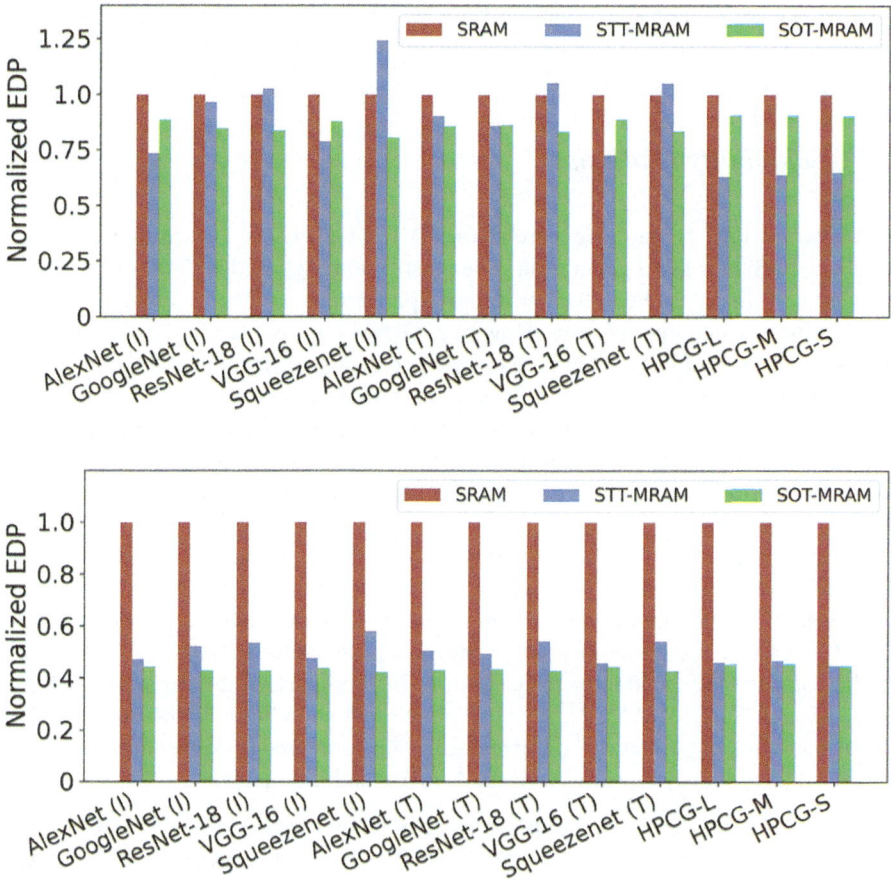

Fig. 9 Iso-area energy-delay product results for STT-MRAM (7MB) and SOT-MRAM (10MB) (lower is better) normalized with respect to SRAM-based caches for inference (I) and training (T) stages without (top chart) and with (bottom chart) DRAM energy and latency

EDP, as shown in Fig. 9, STT-MRAM and SOT-MRAM provide 2× *and* 2.3× *EDP reduction* on average across all workloads when compared to SRAM, respectively.

We show that although the cache latency and energy results for STT-MRAM and SOT-MRAM do not outperform SRAM results at larger cache capacities as shown in Table 2, they do outperform SRAM when costly off-chip DRAM accesses are also considered in EDP calculations. To this end, Chen et al. [13] showed that the normalized energy cost of a global buffer access relative to a MAC operation is 6×, whereas a DRAM access is 200× for a machine learning hardware accelerator. By the same token, the higher cell density of NVM can be exploited to shift the memory traffic from DRAM to L2 cache to further improve power and performance of the overall system. This approach can dramatically reduce the total number of costly

DRAM accesses and reduce data movement, which is a daunting impediment for achieving energy-efficient machine learning hardware [13, 71–74].

4.3 Scalability Analysis

As shown in Fig. 1, the current trend for NVIDIA GPUs is toward increasing L2 size with each new GPU generation. The most recent high-end NVIDIA GPUs have even up to 6MB L2 cache to further improve performance of the system by reducing costly off-chip memory accesses. However, SRAM has a scalability problem due to its high leakage and large bitcell area, which poses a significant challenge to further continue the current GPU trend. To this end, non-volatile memory technologies come to the rescue of future GPU architectures since their PPA scale better as cache capacity increases. Therefore, there is a need for a scalability analysis to project and quantify performance and energy gains that can be achieved by using more scalable memory solutions.

To this end, we perform a scalability analysis by first comparing SRAM, STT-MRAM, and SOT-MRAM for various cache capacities in terms of area, latency, energy results following the *DeepNVM++* framework methodology as described in Fig. 2. Therefore, each memory technology is optimized for EDAP objective at each cache capacity independently to perform a fair comparison among SRAM, STT-MRAM, and SOT-MRAM. Next, we evaluate and show how NVM-based caches behave in terms of performance and energy when compared to conventional SRAM-based caches for deep learning workloads in a scalability analysis.

Area Figure 10a demonstrates the impact of higher cell density of MRAMs on the area of caches compared to SRAM. The area difference between SRAM and the MRAM variants grows significantly as the cache capacity increases. The main reason of this difference comes from the bitcell area difference between SRAM and MRAMs as shown in the last row of Table 1. Particularly for deeply scaled technology nodes wherein interconnects account for a significant portion of parasitics, bigger bitcells translate to longer wires, bigger buffers, and peripheral logic. Therefore, STT-MRAM and SOT-MRAM caches become more area-efficient when compared to SRAM caches as cache capacity increases.

Latency Figure 10b shows that for capacities smaller than 3MB SRAM offers lower read latency, whereas both MRAM variants have lower read latency than SRAM beyond 4MB. In terms of write latency, STT-MRAM has always the highest among all memory technologies due to its inherent device characteristic. In contrast, the write latency of SOT-MRAM becomes increasingly smaller than that of SRAM. Moreover, the write latency of SRAM almost matches that of STT-MRAM at 32MB.

Energy In terms of read access energy, Fig. 10c shows that 7MB is a breakeven point where SOT-MRAM becomes more efficient than SRAM, whereas STT-MRAM clearly has the highest read energy among all memories. Regarding write

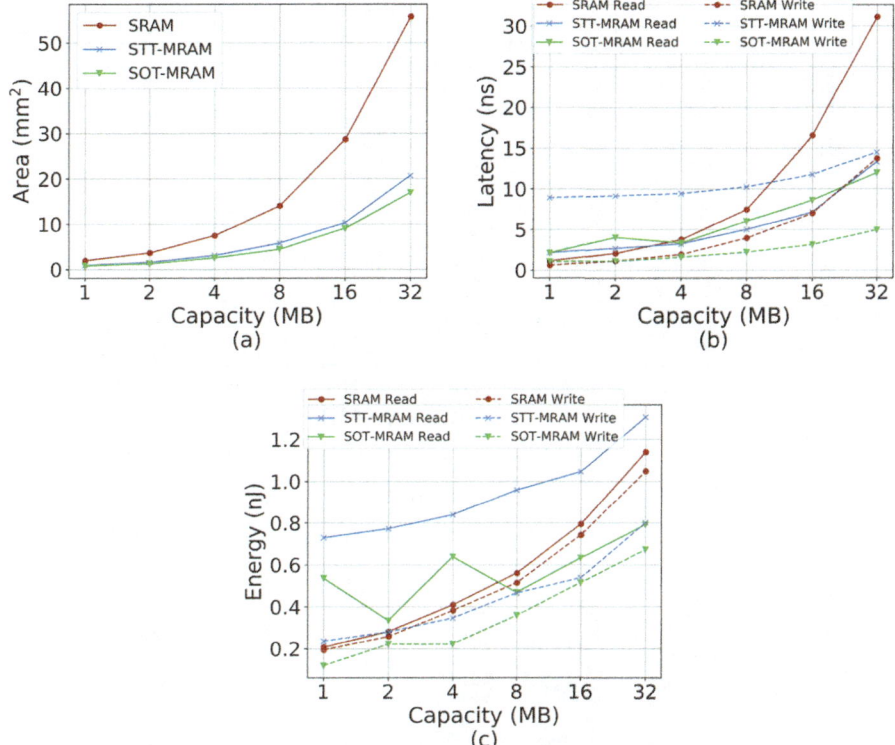

Fig. 10 Cache capacity scaling results for SRAM, STT-MRAM, and SOT-MRAM for (**a**) area, (**b**) latency, and (**c**) energy metrics

access energy, SOT-MRAM is the most efficient option, whereas SRAM consumes the most energy for a write operation beyond 3MB.

Based on these PPA results, we perform a detailed scalability analysis for SRAM, STT-MRAM, and SOT-MRAM. In Figs. 11, 12, 13, we show the normalized energy, latency, and EDP results with respect to SRAM for STT-MRAM and SOT-MRAM for various cache capacities, respectively. As it can be seen, STT-MRAM and SOT-MRAM provide lower energy and latency results as cache capacity increases.

In terms of energy, STT-MRAM and SOT-MRAM provide lower energy as cache capacity increases. Specifically, STT-MRAM and SOT-MRAM caches achieve up to $31.2\times$ *and* $36.4\times$ *energy reduction* as cache capacity increases, respectively. In terms of latency, STT-MRAM and SOT-MRAM have higher latency results for cache capacities up to 4MB, whereas both MRAM variants have lower latency results when compared to SRAM beyond that point. In more detail, SRAM provides up to $3.2\times$ *and* $2\times$ *latency reduction* for small cache capacities when compared to STT-MRAM and SOT-MRAM, respectively. However, STT-MRAM and SOT-MRAM achieve up to $2.1\times$ *and* $2.6\times$ *latency reduction* as cache capacity increases,

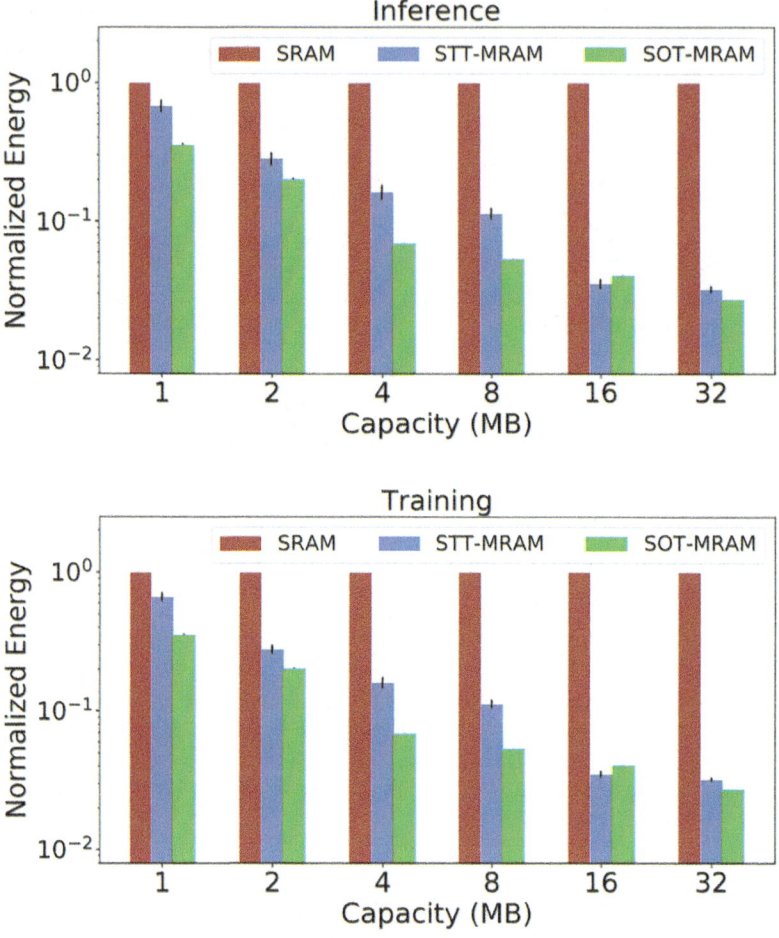

Fig. 11 Mean energy results across all workloads (lower is better) normalized with respect to SRAM for various cache capacities for inference (top chart) and training (bottom chart) stages. Error bars show standard deviation across workloads

respectively. In terms of EDP, we show that STT-MRAM and SOT-MRAM provide up to 65× *and* 95× *EDP reduction* when compared to SRAM, respectively. Therefore, we conclude that for latency-critical applications, SRAM-based caches become a more suitable option when compared to MRAM variants for small cache capacities, whereas MRAMs provide more energy-efficient solutions. Although SRAM provides lower EDP results for smaller cache capacities, STT-MRAM and SOT-MRAM outperform SRAM by orders of magnitude for larger cache capacities due to their better PPA scalability when compared to SRAM. These results show that a significant portion of the overall system energy or latency is saved and can be used for additional on-chip resources or capabilities that are not available now.

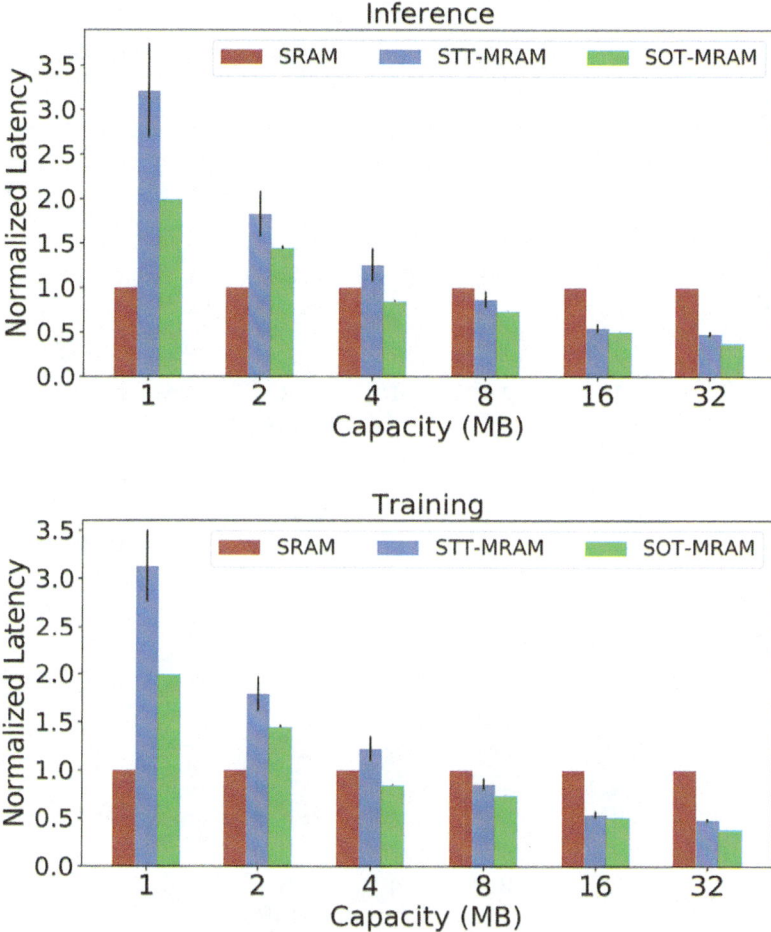

Fig. 12 Mean latency results across all workloads (lower is better) normalized with respect to SRAM for various cache capacities for inference (top chart) and training (bottom chart) stages. Error bars show standard deviation across workloads

5 Discussion

In this section, we discuss the implications of the results shown in this chapter. We also share the potential future directions to guide our community to better explore the use of non-volatile memories for deep learning workloads in different design spaces.

Scalability Is a Major Problem for SRAM As we show in Fig. 10 and Sect. 4.3, one of the key challenges for the current GPU architectures is the scalability problem of SRAM due to its significantly high leakage energy and large area when compared

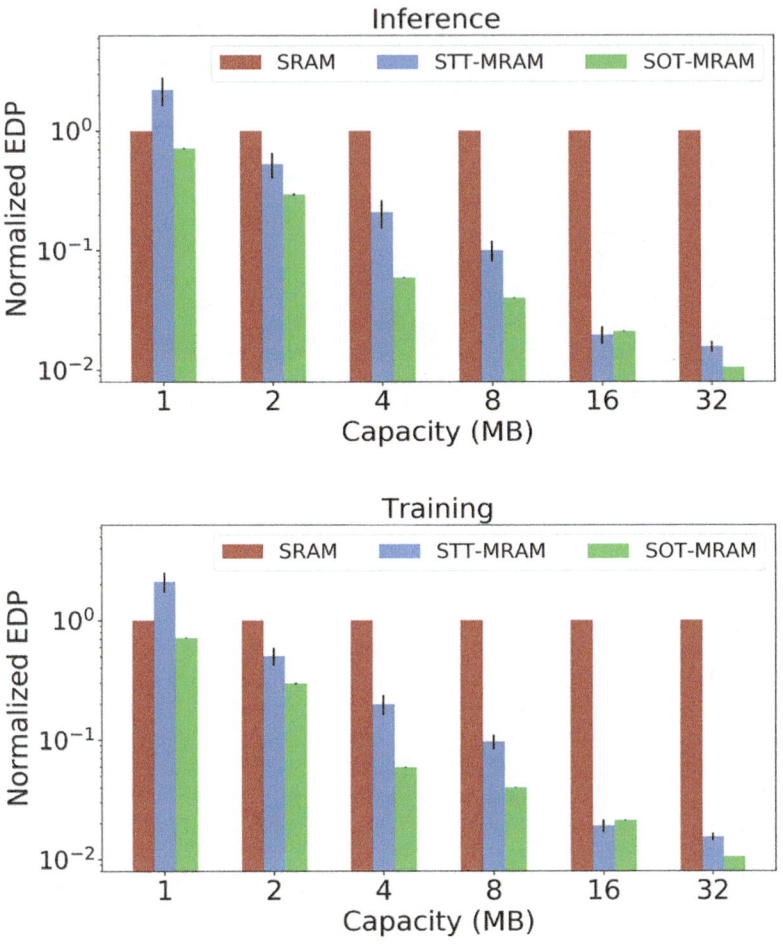

Fig. 13 Mean energy-delay product results across all workloads (lower is better) normalized with respect to SRAM for various cache capacities for inference (top chart) and training (bottom chart) stages. Error bars show standard deviation across workloads

to STT-MRAM and SOT-MRAM. We observe that there is a current trend in GPU architectures toward increasing L2 cache capacity, and we show that SRAM has significant scalability problems in terms of area, latency, and energy. We show that STT-MRAM and SOT-MRAM have promising solutions for larger cache capacities that can maintain the current trend shown in Fig. 1 with increasing performance and energy benefits.

Implications of Dense NVM Caches on Logic Usage Figure 10a shows the area results for SRAM, STT-MRAM, and SOT-MRAM for various cache capacities. We note that STT-MRAM and SOT-MRAM provide increasingly smaller area than SRAM as cache capacity increases. For the same cache capacity, STT-MRAM and SOT-MRAM provide 58% and 65% area reduction on average, respectively. Therefore, the remaining whitespace can be utilized by cramming more processing elements, register files, or L2 cache on the die. This analysis is left for future work.

As CMOS scaling issues limit the affordable improvement of computing systems, our results from device-level simulations to actual GPU profiling show that MRAMs are extremely promising candidates. Particularly, as STT-MRAM and SOT-MRAM fabrication processes become more mature, system-level benefits of STT-MRAM and SOT-MRAM can be maximized, enabling faster and more energy-efficient computation.

Mobile Design Space Exploration for NVM In this chapter, we explore the GPU architecture design space to unveil the potential of non-volatile memories for deep learning workloads. Having said that, we note that inference at the edge devices also becomes a common practice for many service providers such as Google [75], Amazon [76], and Facebook [77] to improve user experience by reducing latency and preserving the private user data on device [78]. To this end, Wu et al. [77] shows that a majority of mobile inference for Facebook workloads run on mobile CPUs. Mobile platforms have various resource constraints such as energy, memory, and computing capabilities. Thus, last-level caches of mobile CPUs or hardware accelerators can also be replaced by STT-MRAM and SOT-MRAM to improve performance and energy by reducing leakage energy and costly off-chip memory accesses due to their non-volatility and higher cell density [79–82]. Therefore, the design space exploration of STT-MRAM and SOT-MRAM for mobile CPUs and hardware accelerators for inference workloads merits further research.

6 Conclusion

In this chapter, we present the first cross-layer analysis framework to characterize, model, and analyze various NVM technologies in GPU architectures for deep learning workloads. Our novel framework can be used to further explore the feasibility of emerging NVM technologies for DL applications for different design choices such as technology nodes, bitcell models, DL workloads, cache configurations, optimization targets, and target platforms.

Our results show that in the iso-capacity case, STT-MRAM and SOT-MRAM provide up to 3.8× *and* 4.7× *EDP reduction and* 2.4× *and* 2.8× *area reduction* when compared to SRAM, respectively. In the iso-area case, STT-MRAM and SOT-MRAM achieve up to 2.2× *and* 2.4× *EDP reduction* and accommodate 2.3× *and* 3.3× *cache capacity* when compared to SRAM, respectively. Finally, we perform a scalability analysis and show that STT-MRAM and SOT-MRAM outperform their SRAM counterpart by orders of magnitude in terms of energy-delay product for large cache capacities. The newly created energy or latency slack can be used for additional on-chip resources or capabilities that are currently not possible.

Acknowledgments This research was supported in part by NSF CCF Grant No. 1815899 and NSF CSR Grant No. 1815780.

References

1. Wulf, W.A., McKee, S.A.: Hitting the memory wall: Implications of the obvious. SIGARCH Comput. Archit. News **23**(1), 20–24 (1995). https://doi.org/10.1145/216585.216588
2. Dennard, R.H., Gaensslen, F.H., Yu, H., Rideout, V.L., Bassous, E., LeBlanc, A.R.: Design of ion-implanted mosfet's with very small physical dimensions. IEEE J. Solid State Circ. **9**(5), 256–268 (1974). https://doi.org/10.1109/JSSC.1974.1050511
3. Murali, S., Mutapcic, A., Atienza, D., Gupta, R., Boyd, S., Benini, L., De Micheli, G.: Temperature control of high-performance multi-core platforms using convex optimization. In: Proceedings of the Conference on Design, Automation and Test in Europe, pp. 110–115 (2008). https://doi.org/10.1109/DATE.2008.4484671
4. Coskun, A.K., Rosing, T.S., Whisnant, K.: Temperature aware task scheduling in MPSoCs. In: 2007 Design, Automation Test in Europe Conference Exhibition, pp. 1–6 (2007). https://doi.org/10.1109/DATE.2007.364540
5. Coskun, A.K., Rosing, T.S., Whisnant, K.A., Gross, K.C.: Static and dynamic temperature-aware scheduling for multiprocessor SoCs. IEEE Trans. Very Large Scale Integr. Syst. **16**(9), 1127–1140 (2008). https://doi.org/10.1109/TVLSI.2008.2000726
6. Tan, M., Le, Q.: EfficientNet: Rethinking model scaling for convolutional neural networks, pp. 6105–6114. PMLR, Long Beach, California, USA (2019). http://proceedings.mlr.press/v97/tan19a.html
7. Touvron, H., Vedaldi, A., Douze, M., Jégou, H.: Fixing the train-test resolution discrepancy. In: Advances in Neural Information Processing Systems (NeurIPS) (2019)
8. Tan, M., Pang, R., Le, Q.V.: Efficientdet: Scalable and efficient object detection. In: 2020 IEEE/CVF Conference on Computer Vision and Pattern Recognition (CVPR) pp. 10778–10787 (2020)
9. Mirhoseini, A., Goldie, A., Yazgan, M., Jiang, J., Songhori, E., Wang, S., Lee, Y.J., Johnson, E., Pathak, O., Bae, S., Nazi, A., Pak, J., Tong, A., Srinivasa, K., Hang, W., Tuncer, E., Babu, A., Le, Q.V., Laudon, J., Ho, R., Carpenter, R., Dean, J.: Chip placement with deep reinforcement learning (2020)
10. Han, S., Mao, H., Dally, W.J.: Deep compression: Compressing deep neural networks with pruning, trained quantization and Huffman coding (2016)
11. Ding, R., Liu, Z., Blanton, R.D.S., Marculescu, D.: Lightening the load with highly accurate storage- and energy-efficient lightnns. ACM Trans. Reconfigurable Technol. Syst. **11**(3) (2018). https://doi.org/10.1145/3270689
12. Chin, T.W., Ding, R., Zhang, C., Marculescu, D.: Towards efficient model compression via learned global ranking. In: Proceedings of the IEEE/CVF Conference on Computer Vision and Pattern Recognition (CVPR) (2020)
13. Chen, Y., Emer, J., Sze, V.: Eyeriss: A spatial architecture for energy-efficient dataflow for convolutional neural networks. In: Proceedings of the 43rd International Symposium on Computer Architecture, pp. 367–379. IEEE Press, Piscataway, NJ, USA (2016). https://doi.org/10.1109/ISCA.2016.40
14. Han, S., Liu, X., Mao, H., Pu, J., Pedram, A., Horowitz, M.A., Dally, W.J.: EIE: Efficient inference engine on compressed deep neural network. In: International Conference on Computer Architecture (ISCA) (2016)
15. Chen, Y.H., Emer, J., Sze, V.: Using dataflow to optimize energy efficiency of deep neural network accelerators. IEEE Micro **37**(3), 12–21 (2017). https://doi.org/10.1109/MM.2017.54
16. Shao, Y., Clemons, J., Venkatesan, R., Zimmer, B., Fojtik, M.R., Jiang, N., Keller, B., Klinefelter, A., Pinckney, N., Raina, P., Tell, S., Zhang, Y., Dally, W., Emer, J., Gray, C.T., Khailany, B., Keckler, S.: Simba: Scaling deep-learning inference with multi-chip-module-

based architecture. In: Proceedings of the 52nd Annual IEEE/ACM International Symposium on Microarchitecture (2019)

17. Inci, A., Marculescu, D.: Solving the non-volatile memory conundrum for deep learning workloads. In: Architectures and Systems for Big Data Workshop in Conjunction with ISCA (2018)

18. Inci, A.F., Isgenc, M.M., Marculescu, D.: Deepnvm: A framework for modeling and analysis of non-volatile memory technologies for deep learning applications. In: Proceedings of the 23rd Conference on Design, Automation and Test in Europe, DATE '20, p. 1295–1298 (2020)

19. Inci, A., Isgenc, M.M., Marculescu, D.: Deepnvm++: Cross-layer modeling and optimization framework of non-volatile memories for deep learning. IEEE Trans. Comput. Aided Des. Integr. Circ. Syst., 1–1 (2021). https://doi.org/10.1109/TCAD.2021.3127148

20. Inci, A., Bolotin, E., Fu, Y., Dalal, G., Mannor, S., Nellans, D., Marculescu, D.: The architectural implications of distributed reinforcement learning on CPU-GPU systems. Preprint (2020). arXiv:2012.04210

21. Inci, A., Isgenc, M.M., Marculescu, D.: Cross-layer design space exploration of NVM-based caches for deep learning. NVMW (2021)

22. Inci, A., Virupaksha, S.G., Jain, A., Thallam, V.V., Ding, R., Marculescu, D.: QAPPA: Quantization-aware power, performance, and area modeling of DNN accelerators. Preprint (2022). arXiv:2205.08648

23. Inci, A., Virupaksha, S.G., Jain, A., Thallam, V.V., Ding, R., Marculescu, D.: QADAM: Quantization-aware DNN accelerator modeling for pareto-optimality. Preprint (2022). arXiv:2205.13045

24. Inci, A., Virupaksha, S.G., Jain, A., Chin, T.W., Thallam, V.V., Ding, R., Marculescu, D.: Quidam: A framework for quantization-aware DNN accelerator and model co-exploration. Preprint (2022). arXiv:2206.15463

25. Chang, M., Rosenfeld, P., Lu, S., Jacob, B.: Technology comparison for large last-level caches (l3cs): Low-leakage SRAM, low write-energy STT-RAM, and refresh-optimized edram. In: 2013 IEEE 19th International Symposium on High Performance Computer Architecture (HPCA), pp. 143–154 (2013). https://doi.org/10.1109/HPCA.2013.6522314

26. Homayoun, H., Veidenbaum, A.: Reducing leakage power in peripheral circuits of l2 caches. In: 2007 25th International Conference on Computer Design, pp. 230–237 (2007). https://doi.org/10.1109/ICCD.2007.4601907

27. Xu, W., Sun, H., Wang, X., Chen, Y., Zhang, T.: Design of last-level on-chip cache using spin-torque transfer RAM (STT RAM). IEEE Trans. Very Large Scale Integr. (VLSI) Syst. **19**(3), 483–493 (2011). https://doi.org/10.1109/TVLSI.2009.2035509

28. Dong, X., Wu, X., Sun, G., Xie, Y., Li, H., Chen, Y.: Circuit and microarchitecture evaluation of 3D stacking magnetic RAM (MRAM) as a universal memory replacement. In: 2008 45th ACM/IEEE Design Automation Conference, pp. 554–559 (2008)

29. List of NVIDIA GPUs: https://en.wikipedia.org/wiki/List-of-Nvidia-graphics-processing-units

30. Kim, J., Chen, A., Behin-Aein, B., Kumar, S., Wang, J., Kim, C.H.: A technology-agnostic MTJ spice model with user-defined dimensions for STT-MRAM scalability studies. In: 2015 IEEE Custom Integrated Circuits Conference (CICC), pp. 1–4 (2015). https://doi.org/10.1109/CICC.2015.7338407

31. Kazemi, M., Rowlands, G.E., Ipek, E., Buhrman, R.A., Friedman, E.G.: Compact model for spin–orbit magnetic tunnel junctions. IEEE Trans. Electron Dev. **63**(2), 848–855 (2016). https://doi.org/10.1109/TED.2015.2510543

32. Dong, X., Xu, C., Xie, Y., Jouppi, N.P.: Nvsim: A circuit-level performance, energy, and area model for emerging nonvolatile memory. IEEE Trans. Comput. Aided Des. Integr. Circ. Syst. **31**(7), 994–1007 (2012). https://doi.org/10.1109/TCAD.2012.2185930

33. Jia, Y., Shelhamer, E., Donahue, J., Karayev, S., Long, J., Girshick, R., Guadarrama, S., Darrell, T.: Caffe: Convolutional architecture for fast feature embedding. In: Proceedings of the 22Nd ACM International Conference on Multimedia, MM '14, pp. 675–678. ACM, New York, NY, USA (2014). https://doi.org/10.1145/2647868.2654889

34. Deng, J., Dong, W., Socher, R., Li, L., Kai Li, Li Fei-Fei: Imagenet: A large-scale hierarchical image database. In: 2009 IEEE Conference on Computer Vision and Pattern Recognition, pp. 248–255 (2009)
35. Bakhoda, A., Yuan, G.L., Fung, W.W.L., Wong, H., Aamodt, T.M.: Analyzing cuda workloads using a detailed GPU simulator. In: 2009 IEEE International Symposium on Performance Analysis of Systems and Software, pp. 163–174 (2009). https://doi.org/10.1109/ISPASS.2009. 4919648
36. Isgenc, M.M.: Enabling design of low-volume high-performance ICs. Ph.D. thesis, Carnegie Mellon University (2019)
37. Isgenc, M.M., Martins, M.G.A., Zackriya, V.M., Pagliarini, S.N., Pileggi, L.: Logic IP for low-cost IC design in advanced CMOS nodes. IEEE Trans. Very Large Scale Integr. (VLSI) Syst. **28**(2), 585–595 (2020). https://doi.org/10.1109/TVLSI.2019.2942825
38. Pagliarini, S.N., Bhuin, S., Isgenc, M.M., Biswas, A.K., Pileggi, L.: A probabilistic synapse with strained MTJs for spiking neural networks. IEEE Trans. Neural Networks Learn. Syst. **31**(4), 1113–1123 (2020). https://doi.org/10.1109/TNNLS.2019.2917819
39. Scheuerlein, R.E.: Magneto-resistive IC memory limitations and architecture implications. In: Seventh Biennial IEEE International Nonvolatile Memory Technology Conference. Proceedings (Cat. No.98EX141), pp. 47–50 (1998). https://doi.org/10.1109/NVMT.1998.723217
40. Zhao, W., Belhaire, E., Mistral, Q., Chappert, C., Javerliac, V., Dieny, B., Nicolle, E.: Macro-model of spin-transfer torque based magnetic tunnel junction device for hybrid magnetic-CMOS design. In: 2006 IEEE International Behavioral Modeling and Simulation Workshop, pp. 40–43 (2006). https://doi.org/10.1109/BMAS.2006.283467
41. Kan, J.J., Park, C., Ching, C., Ahn, J., Xie, Y., Pakala, M., Kang, S.H.: A study on practically unlimited endurance of STT-MRAM. IEEE Trans. Electron Dev. **64**(9), 3639–3646 (2017). https://doi.org/10.1109/TED.2017.2731959
42. Hosomi, M., Yamagishi, H., Yamamoto, T., Bessho, K., Higo, Y., Yamane, K., Yamada, H., Shoji, M., Hachino, H., Fukumoto, C., Nagao, H., Kano, H.: A novel nonvolatile memory with spin torque transfer magnetization switching: spin-RAM. In: IEEE International Electron Devices Meeting, 2005. IEDM Technical Digest., pp. 459–462 (2005)
43. Chi, P., Li, S., Yuanqing Cheng, Yu Lu, Kang, S.H., Xie, Y.: Architecture design with STT-RAM: Opportunities and challenges. In: 2016 21st Asia and South Pacific Design Automation Conference (ASP-DAC), pp. 109–114 (2016)
44. Rasquinha, M., Choudhary, D., Chatterjee, S., Mukhopadhyay, S., Yalamanchili, S.: An energy efficient cache design using spin torque transfer (STT) RAM. In: 2010 ACM/IEEE International Symposium on Low-Power Electronics and Design (ISLPED), pp. 389–394 (2010). https://doi.org/10.1145/1840845.1840931
45. Prenat, G., Jabeur, K., Vanhauwaert, P., Pendina, G.D., Oboril, F., Bishnoi, R., Ebrahimi, M., Lamard, N., Boulle, O., Garello, K., Langer, J., Ocker, B., Cyrille, M., Gambardella, P., Tahoori, M., Gaudin, G.: Ultra-fast and high-reliability SOT-MRAM: From cache replacement to normally-off computing. IEEE Trans. Multi-Scale Comput. Syst. **2**(1), 49–60 (2016). https://doi.org/10.1109/TMSCS.2015.2509963
46. Bishnoi, R., Ebrahimi, M., Oboril, F., Tahoori, M.B.: Architectural aspects in design and analysis of SOT-based memories. In: 2014 19th Asia and South Pacific Design Automation Conference (ASP-DAC), pp. 700–707 (2014)
47. Oboril, F., Bishnoi, R., Ebrahimi, M., Tahoori, M.B.: Evaluation of hybrid memory technologies using SOT-MRAM for on-chip cache hierarchy. IEEE Trans. Comput. Aided Des. Integr. Circ. Syst. **34**(3), 367–380 (2015). https://doi.org/10.1109/TCAD.2015.2391254
48. Li, G., Chen, X., Sun, G., Hoffmann, H., Liu, Y., Wang, Y., Yang, H.: A STT-RAM-based low-power hybrid register file for GPGPUs. In: 2015 52nd ACM/EDAC/IEEE Design Automation Conference (DAC), pp. 1–6 (2015)
49. Wu, X., Li, J., Zhang, L., Speight, E., Rajamony, R., Xie, Y.: Hybrid cache architecture with disparate memory technologies. In: ISCA '09, p. 34–45. Association for Computing Machinery, New York, NY, USA (2009). https://doi.org/10.1145/1555754.1555761

50. Imani, M., Patil, S., Rosing, T.: Low power data-aware STT-RAM based hybrid cache architecture. In: 2016 17th International Symposium on Quality Electronic Design (ISQED), pp. 88–94 (2016). https://doi.org/10.1109/ISQED.2016.7479181
51. Beigi, M.V., Memik, G.: Tapas: Temperature-aware adaptive placement for 3D stacked hybrid caches. In: Proceedings of the Second International Symposium on Memory Systems, MEMSYS '16, p. 415–426. Association for Computing Machinery, New York, NY, USA (2016). https://doi.org/10.1145/2989081.2989085
52. Smullen, C.W., Mohan, V., Nigam, A., Gurumurthi, S., Stan, M.R.: Relaxing non-volatility for fast and energy-efficient STT-RAM caches. In: 2011 IEEE 17th International Symposium on High Performance Computer Architecture, pp. 50–61 (2011)
53. Kuan, K., Adegbija, T.: Energy-efficient runtime adaptable l1 STT-RAM cache design. IEEE Trans. Comput. Aided Des. Integr. Circ. Syst. **39**(6), 1328–1339 (2020). https://doi.org/10.1109/TCAD.2019.2912920
54. Jog, A., Mishra, A.K., Xu, C., Xie, Y., Narayanan, V., Iyer, R., Das, C.R.: Cache revive: Architecting volatile STT-RAM caches for enhanced performance in CMPs. In: DAC Design Automation Conference 2012, pp. 243–252 (2012). https://doi.org/10.1145/2228360.2228406
55. Sun, Z., Bi, X., Li, H., Wong, W., Ong, Z., Zhu, X., Wu, W.: Multi retention level STT-RAM cache designs with a dynamic refresh scheme. In: 2011 44th Annual IEEE/ACM International Symposium on Microarchitecture (MICRO), pp. 329–338 (2011)
56. Wang, J., Dong, X., Xie, Y.: Oap: An obstruction-aware cache management policy for STT-RAM last-level caches. In: 2013 Design, Automation Test in Europe Conference Exhibition (DATE), pp. 847–852 (2013)
57. Sun, G., Dong, X., Xie, Y., Li, J., Chen, Y.: A novel architecture of the 3D stacked MRAM l2 cache for CMPs. In: 2009 IEEE 15th International Symposium on High Performance Computer Architecture, pp. 239–249 (2009). https://doi.org/10.1109/HPCA.2009.4798259
58. Imani, M., Rahimi, A., Kim, Y., Rosing, T.: A low-power hybrid magnetic cache architecture exploiting narrow-width values. In: 2016 5th Non-Volatile Memory Systems and Applications Symposium (NVMSA), pp. 1–6 (2016). https://doi.org/10.1109/NVMSA.2016.7547174
59. Angizi, S., He, Z., Reis, D., Hu, X., Tsai, W., Lin, S.J., Fan, D.: Accelerating deep neural networks in processing-in-memory platforms: Analog or digital approach? In: 2019 IEEE Computer Society Annual Symposium on VLSI (ISVLSI), pp. 197–202 (2019)
60. Reis, D., Gao, D., Angizi, S., Yin, X., Fan, D., Niemier, M., Zhuo, C., Hu, X.S.: Modeling and benchmarking computing-in-memory for design space exploration. In: Proceedings of the 2020 on Great Lakes Symposium on VLSI (2020)
61. Angizi, S., Khoshavi, N., Marshall, A., Dowben, P., Fan, D.: Meram: Non-volatile cache memory based on magneto-electric fets (2020)
62. Seo, Y., Roy, K.: High-density SOT-MRAM based on shared bitline structure. IEEE Trans. Very Large Scale Integr. (VLSI) Syst. **26**(8), 1600–1603 (2018). https://doi.org/10.1109/TVLSI.2018.2822841
63. Krizhevsky, A., Sutskever, I., Hinton, G.E.: Imagenet classification with deep convolutional neural networks. In: Proceedings of the 25th International Conference on Neural Information Processing Systems - Volume 1, NIPS'12, pp. 1097–1105. Curran Associates Inc., Red Hook, NY, USA (2012)
64. Szegedy, C., Wei Liu, Yangqing Jia, Sermanet, P., Reed, S., Anguelov, D., Erhan, D., Vanhoucke, V., Rabinovich, A.: Going deeper with convolutions. In: 2015 IEEE Conference on Computer Vision and Pattern Recognition (CVPR), pp. 1–9 (2015)
65. Simonyan, K., Zisserman, A.: Very deep convolutional networks for large-scale image recognition. CoRR **abs/1409.1556** (2014). http://arxiv.org/abs/1409.1556
66. He, K., Zhang, X., Ren, S., Sun, J.: Deep residual learning for image recognition. In: 2016 IEEE Conference on Computer Vision and Pattern Recognition (CVPR), pp. 770–778 (2016)
67. Iandola, F.N., Han, S., Moskewicz, M.W., Ashraf, K., Dally, W.J., Keutzer, K.: Squeezenet: Alexnet-level accuracy with 50x fewer parameters and <0.5mb model size (2016)
68. Dongarra, J.J., Heroux, M., Luszczek, P.: HPCG benchmark: a new metric for ranking high performance computing systems (2015)

69. NVIDIA CUDA Profiler: https://docs.nvidia.com/cuda/profiler-users-guide/nvprof-overview
70. Redmon, J.: Darknet: Open source neural networks in C. http://pjreddie.com/darknet/ (2013–2016)
71. Chen, Y., Krishna, T., Emer, J.S., Sze, V.: Eyeriss: An energy-efficient reconfigurable accelerator for deep convolutional neural networks. IEEE J. Solid State Circ. **52**(1), 127–138 (2017). https://doi.org/10.1109/JSSC.2016.2616357
72. Gao, M., Pu, J., Yang, X., Horowitz, M., Kozyrakis, C.: Tetris: Scalable and efficient neural network acceleration with 3D memory. SIGARCH Comput. Archit. News **45**(1), 751–764 (2017). https://doi.org/10.1145/3093337.3037702
73. Boroumand, A., Ghose, S., Kim, Y., Ausavarungnirun, R., Shiu, E., Thakur, R., Kim, D., Kuusela, A., Knies, A., Ranganathan, P., Mutlu, O.: Google workloads for consumer devices: Mitigating data movement bottlenecks. In: Proceedings of the Twenty-Third International Conference on Architectural Support for Programming Languages and Operating Systems, ASPLOS '18, p. 316–331. Association for Computing Machinery, New York, NY, USA (2018). https://doi.org/10.1145/3173162.3173177
74. Donato, M., Reagen, B., Pentecost, L., Gupta, U., Brooks, D., Wei, G.: On-chip deep neural network storage with multi-level eNVM. In: 2018 55th ACM/ESDA/IEEE Design Automation Conference (DAC), pp. 1–6 (2018). https://doi.org/10.1109/DAC.2018.8465818
75. Kannan, A., Kurach, K., Ravi, S., Kaufmann, T., Tomkins, A., Miklos, B., Corrado, G., Lukács, L., Ganea, M., Young, P., Ramavajjala, V.: Smart reply: Automated response suggestion for email. CoRR **abs/1606.04870** (2016). http://arxiv.org/abs/1606.04870
76. Tucker, G., Wu, M., Sun, M., Panchapagesan, S., Fu, G., Vitaladevuni, S.: Model compression applied to small-footprint keyword spotting. In: Interspeech 2016, pp. 1878–1882 (2016). https://doi.org/10.21437/Interspeech.2016-1393
77. Wu, C., Brooks, D., Chen, K., Chen, D., Choudhury, S., Dukhan, M., Hazelwood, K., Isaac, E., Jia, Y., Jia, B., Leyvand, T., Lu, H., Lu, Y., Qiao, L., Reagen, B., Spisak, J., Sun, F., Tulloch, A., Vajda, P., Wang, X., Wang, Y., Wasti, B., Wu, Y., Xian, R., Yoo, S., Zhang, P.: Machine learning at facebook: Understanding inference at the edge. In: 2019 IEEE International Symposium on High Performance Computer Architecture (HPCA), pp. 331–344 (2019). https://doi.org/10.1109/HPCA.2019.00048
78. Ghodsi, Z., Veldanda, A., Reagen, B., Garg, S.: Cryptonas: Private inference on a relu budget (2020)
79. Korgaonkar, K., Bhati, I., Liu, H., Gaur, J., Manipatruni, S., Subramoney, S., Karnik, T., Swanson, S., Young, I., Wang, H.: Density tradeoffs of non-volatile memory as a replacement for SRAM based last level cache. In: 2018 ACM/IEEE 45th Annual International Symposium on Computer Architecture (ISCA), pp. 315–327 (2018). https://doi.org/10.1109/ISCA.2018.00035
80. Hankin, A., Shapira, T., Sangaiah, K., Lui, M., Hempstead, M.: Evaluation of non-volatile memory based last level cache given modern use case behavior. In: 2019 IEEE International Symposium on Workload Characterization (IISWC), pp. 143–154 (2019). https://doi.org/10.1109/IISWC47752.2019.9042051
81. Pentecost, L., Donato, M., Reagen, B., Gupta, U., Ma, S., Wei, G.Y., Brooks, D.: MaxNVM: Maximizing DNN storage density and inference efficiency with sparse encoding and error mitigation. In: Proceedings of the 52nd Annual IEEE/ACM International Symposium on Microarchitecture, MICRO '52, p. 769–781. Association for Computing Machinery, New York, NY, USA (2019). https://doi.org/10.1145/3352460.3358258
82. Li, H., Bhargav, M., Whatmough, P.N., Philip Wong, H..: On-chip memory technology design space explorations for mobile deep neural network accelerators. In: 2019 56th ACM/IEEE Design Automation Conference (DAC), pp. 1–6 (2019)

SoC-GANs: Energy-Efficient Memory Management for System-on-Chip Generative Adversarial Networks

Rehan Ahmed, Muhammad Zuhaib Akbar, Muhammad Abdullah Hanif, and Muhammad Shafique

1 Introduction

Deep neural networks (DNNs) process the information artificially based on mathematical models in order to mimic the human-level perception. They are widely used in emerging fields such as robotics, language processing, and computer vision, to name a few. The usage of DNNs is a two-step process: they are first trained, where the training stage tunes the network parameters, and then put in the inference stage, where information from the test data is inferred based on the trained network parameters. Conventionally, *supervised learning* is used to train DNNs [5], but this technique requires a significant amount of labeled data for training. Alternatively, *semi-supervised* and *unsupervised learning* have gained a lot of traction as these techniques can infer information from un-tagged data [1, 6, 10, 11].

Generative adversarial networks (GANs) are the most interesting idea to generate synthetic but realistic examples from the original dataset using unsupervised learning [3]. GANs consist of two neural network models: *generator* and *discriminator* as shown in Fig. 1. The generator model competes against the discriminator model (an adversary) that determines whether a sample generated by the generator belongs to the data distribution of the training samples or not [11]. During the training phase, both the networks are trained as a *two-player game* with the objective to outperform each other. The objective of the generator network is to generate samples from the

R. Ahmed (✉) · M. Z. Akbar
School of Electrical Engineering and Computer Science (SEECS), National University of Sciences and Technology (NUST), Islamabad, Pakistan
e-mail: rehan.ahmed@seecs.edu.pk; makbar.msee16seecs@seecs.edu.pk

M. A. Hanif · M. Shafique
A1-173, Division of Engineering, New York University Abu Dhabi, Saadiyat Island, United Arab Emirates
e-mail: mh6117@nyu.edu; muhammad.shafique@nyu.edu

© The Author(s), under exclusive license to Springer Nature Switzerland AG 2024
S. Pasricha, M. Shafique (eds.), *Embedded Machine Learning for Cyber-Physical, IoT, and Edge Computing*, https://doi.org/10.1007/978-3-031-19568-6_9

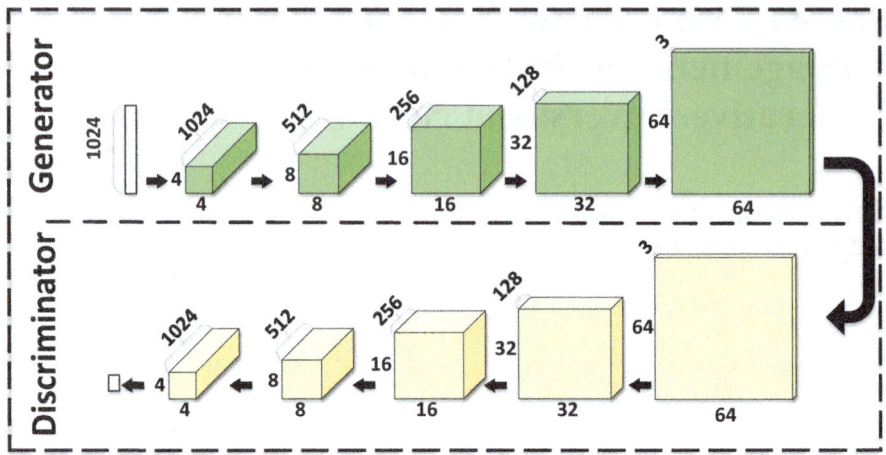

Fig. 1 Deep-convolution-based generative adversarial network (DCGAN) architecture showing the generator and discriminator models

latent space that cannot be detected by the discriminator, while the objective of the discriminator network is to accurately classify examples as either generated (fake) or from the data (real). Both the two models are trained together until the generator model starts to generate the plausible examples, therefore making GANs extremely useful in data generation applications such as text to image synthesis [8], image classifications [7], mobile robots [4], and video prediction [2], to name a few.

GANs are quite computationally expensive in contrast to DNNs. One of the key differences in deep-convolution-based generative adversarial network (DCGAN) is that the max-pooling layer in conventional CNNs is replaced with the *strided convolution*, which is used in the forward-computation phases of the discriminator. The strided convolution operation skips output pixel computation based on a stride size (zero-skipping) that corresponds to a down-sampling process. Similarly, another key operation in DCGANs is *transposed convolution* (de-convolution), which is used in the forward-computation phases of the generator. In transposed convolution, the convolution operation inserts zeros in the input feature map (zero-inserting) that corresponds to up-sampling process. Another type of convolution used in the training phase of GANs is *four-dimensional convolution* that itself is a convolution operation in nature, but there is no accumulation involved after convolving error of one layer with the output of previous layer that results in a four-dimensional output matrix. During weights update in discriminator layers, this operation is used in a similar fashion as that of strided convolution i.e., *skipping zeros operation*, whereas during weights update in generator layers, this operation inserts zeros similar to the transposed convolution. In most cases, these computations are accelerated at the software level by a server equipped with several CPUs and GPUs. But there is a need for more energy-efficient hardware accelerators for GANs as the applications are moved to a mobile form factor.

2   Background: DCGAN Hardware Acceleration and Its Design Challenges

Song et al. [9] proposed a hardware accelerator design that deals with the above mentioned complex computations, namely: strided convolution, transposed convolution, and four-dimensional convolutions. The proposed hardware design consists of two main microarchitectures, namely: **zero-free and output stationary (ZFOST)** and **zero-free and weight stationary (ZFWST)**:

1. **Zero-free and output stationary (ZFOST) Microarchitecture:** The ZFOST microarchitecture, shown in Fig. 2, aims at accelerating *strided convolution* and *transposed convolution*. It is used in forward data pass and backward error pass. The ZFOST microarchitecture is composed of a 4 × 4 processing element (PE) array and an input register array. The PE array is used for processing the output, and the input register array is for feeding the input neurons (i.e., the pixels of an input feature map) to the PEs. The registers shown in green in the register array directly correspond to the respective PEs in the PE array, while the additional

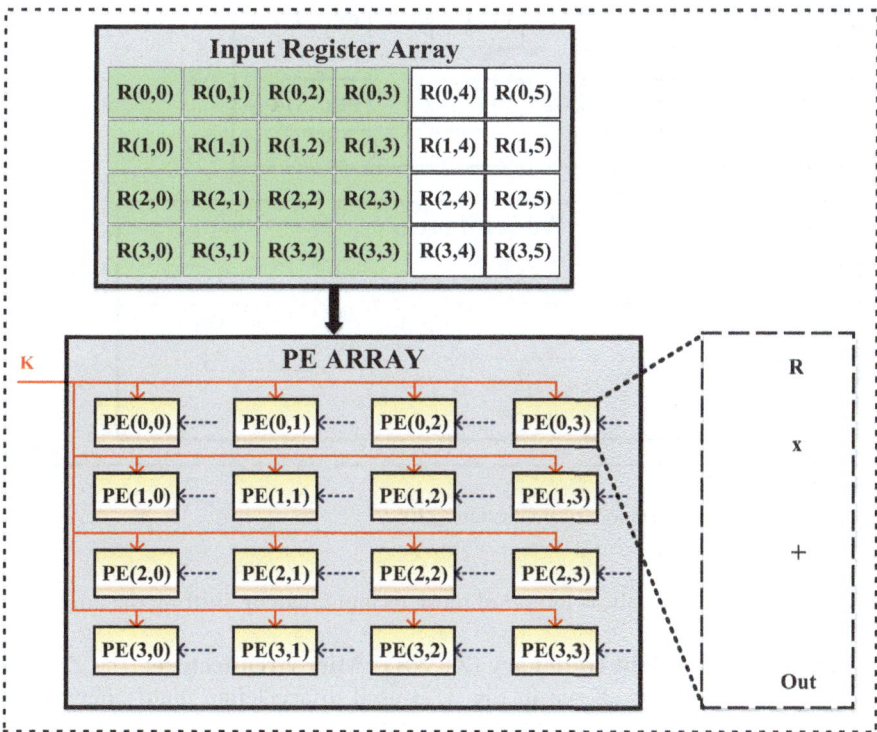

Fig. 2   Zero-free output stationary architecture (ZFOST)

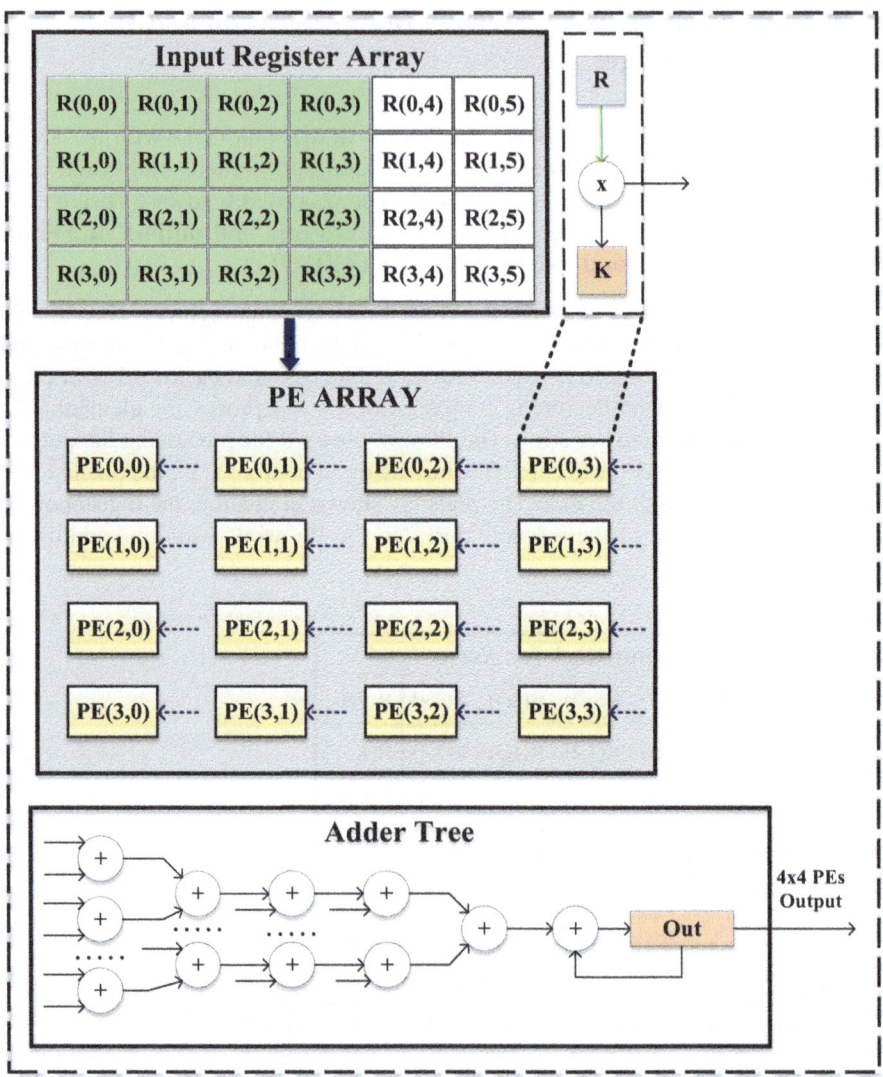

Fig. 3 Zero-free weight stationary architecture (ZFWST)

registers are used to allow temporal reuse of input data by shifting the content of the registers.

2. **Zero-free and weight stationary (ZFWST) Microarchitecture:** The ZFWST microarchitecture, as shown in Fig. 3, is used to accelerate multi-dimensional convolution that is used for backpropagation to update the weights during the training phase. ZFWST microarchitecture consists of a 4×4 PE array, an input register array, and an adder tree. The weights are spatially shared by the PEs

and are fed one at a time. The size of the PE array is kept at 4×4 to match the size of the minimum output feature map in DCGANs. For computations in this microarchitecture, the output neurons (i.e., the pixels of an output feature map) are unrolled, and the spatially neighboring neurons are mapped on the PE array where one output neuron is mapped to one PE and is kept there throughout its computation. The kernel weights are relayed one at a time and are spatially shared by the PEs in the PE array.

In order to avoid bubbles in the computation flow of ZFOST and ZFWST pipelines, an efficient dataflow has also been proposed in [9]. In this flow, the kernel/filter weights are loaded into ZFOST and ZFWST input register arrays in a **type-oriented format** instead of being fetched in a sequential order. The type-oriented format refers to the alignment of data based on its row and column indexes being even or odd in its data class. That is, pixels or weights belonging to even rows and even columns are considered a part of *even–even* type, and similarly others are placed in *even–odd*, *odd–even*, and *odd–odd* categories.

Next, let us take a look at the dataflow for performing computations in the ZFOST and ZFWST microarchitectures.

ZFOST Strided Convolution Dataflow

Figure 4 illustrates the dataflow of strided convolution performed by ZFOST microarchitecture. As depicted, first *even–even* weights (i.e., weights with even row and even column indexes) are processed followed by *even–odd* kernel weights, then *odd–even* kernel weights, and at the end *odd–odd* kernel weights are processed. Since each PE is mapped to an output $O_{(ox,oy)}$, its required input for kernel weight $K_{(kx,ky)}$ can be computed as $I_{(kx+2ox,ky+2oy)}$. Initially, all the inputs marked with "Red" are loaded into the input register array. Then in the next clock cycle, data is shifted in input register array for temporal reuse. Let us take an example, when the kernel weight $K_{(0,0)}$ is provided to all PEs in the first clock cycle for processing,

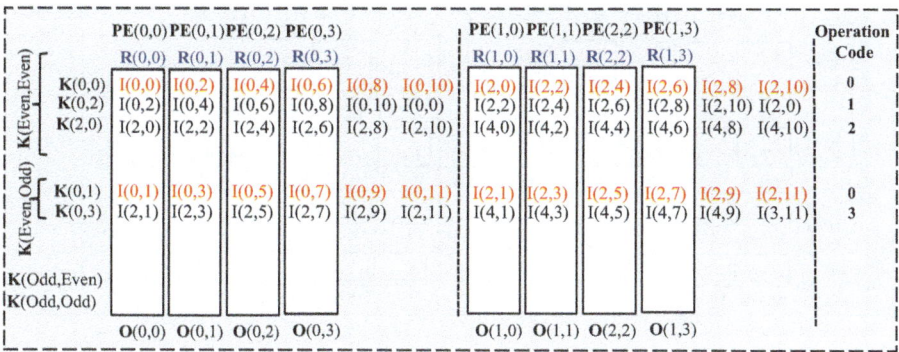

Fig. 4 *Strided convolution dataflow controller using ZFOST.* Two out of four rows of processing element array are shown in this figure. Pixels with the type even–even are processed first followed by the even–odd, odd–even, and finally pixel with odd–odd type are processed. All weight contributes in the computation of each output pixel

then all pixels marked with red are required to be loaded into the input register array, as shown in Fig. 5b. The data is then shifted within the input register array for the next two clock transitions to perform the processing on kernel weights $K_{(0,2)}$ and $K_{(0,4)}$ as illustrated in the dataflow of Fig. 4. For the fourth clock transition when the next *even–even* kernel weight $K_{(2,0)}$ is broadcasted to all PEs, six new data pixels are required in the last row of the input register array, as shown in Fig. 5c.

ZFOST Transposed Convolution Dataflow

Dataflow for the transposed convolution is illustrated in Fig. 6. In transposed convolution, the indexes of non-zero input neurons are spaced due to zero insertion. Therefore, for a particular type of output neurons, only effective pixels belong to the same type. It means when kernel weight $K_{(Even,Even)}$ is being processed, computing only $O_{(Even,Even)}$ is effective. Similarly, kernel weights $K_{(Even,Odd)}$, $K_{(Odd,Even)}$

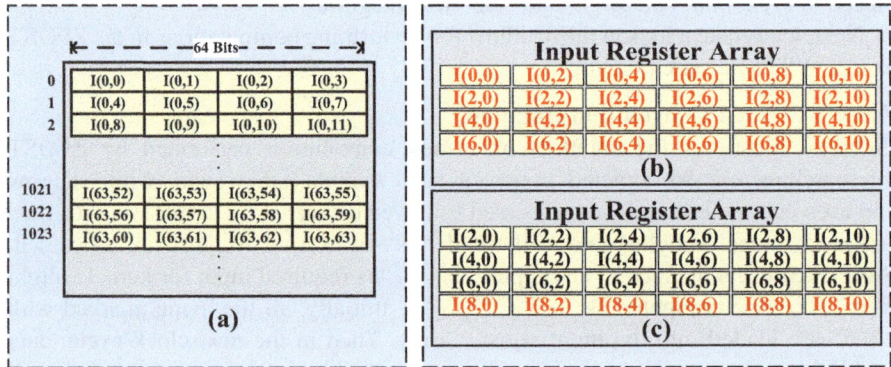

Fig. 5 (**a**) Linear arrangement of data in on-chip/off-chip memory. Data to be loaded in local register of ZFOST (**b**) during processing of K(0,0) and (**c**) during processing of K(2,0)

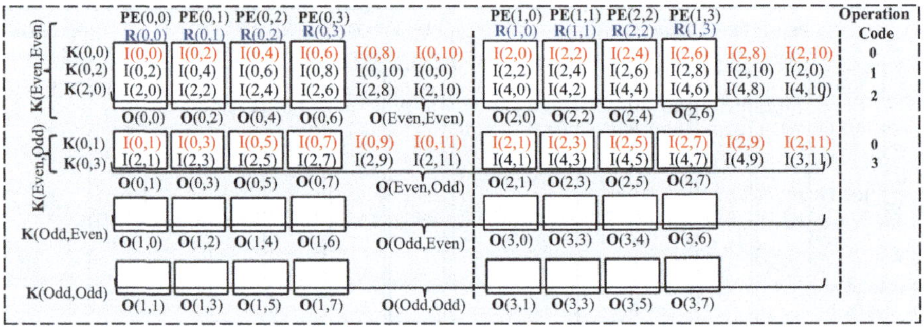

Fig. 6 *Transposed convolution dataflow controller using ZFOST.* Two out of four rows of processing element array are shown in this figure. In this convolution, input image pixels with a certain type contribute only to the output pixels of same type. Even–even type pixels are processed first generating the output pixels of type even–even, followed by even–odd, odd–even, and final odd–odd pixels are processed with kernel weights generating output pixels of odd–odd type

and $K_{(Odd,Odd)}$ contribute only in the computation of $O_{(Even,Odd)}$, $O_{(Odd,Even)}$ and $O_{(Odd,Odd)}$, respectively.

ZFWST Four-Dimensional Convolution Dataflow

The *ZFWST* microarchitecture, shown in Fig. 3, performs four-dimensional convolution. Zeros are inserted in the discriminator kernel during its backward phase as shown in Fig. 7. Similarly, zeros are inserted in the input data of backward phase in generator, as shown in Fig. 8. Therefore, ZFWST is not only responsible to skip zeros in input data, but also in kernel weights. In ZFWST, each PE has stationary kernel weight, and all PEs contribute to one output neuron using an adder tree as illustrated in Fig. 3. The ZFWST microarchitecture uses a similar input register array as that of ZFOST microarchitecture and therefore has similar dataflow. Figures 7 and 8 represent the dataflow of four-dimensional convolution during backward phase of discriminator and generator, respectively.

It is to be noted that the need of loading multiple data points (up to 24 data points) into input register array, and that too in a type-oriented format, from an on-chip memory brings a unique set of challenges. We list the key memory layout and management challenges as follows:

- **Linear on-chip memory does not fulfill the dataflow requirements:** The data in the conventional on-chip memory is stored linearly, which limits the design to feed multiple data points to *ZFOST* and *ZFWST* microarchitectures in a single clock cycle. Therefore, for a 16-*bit* fixed-point system, each location of the on-chip memory will store 4 data point in a cascaded form. In order to load required 24 data points into the input register array during the first clock cycle, a total of 12 read cycles are required to extract the relevant data from the on-chip memory. Moreover, since only *even–even* data is required during this computation, half of the read data will be wasted. Therefore, to implement the dataflow of Figs. 4, 6, 7,

	PE(0,0)	PE(0,1)	PE(0,2)	PE(0,3)			PE(1,0)	PE(1,1)	PE(2,2)	PE(1,3)			Operation Code
	R(0,0)	R(0,1)	R(0,2)	R(0,3)			R(1,0)	R(1,1)	R(2,2)	R(1,3)			
O(0,0)	I(0,0)	I(0,2)	I(0,4)	I(0,6)	I(0,8)	I(0,10)	I(2,0)	I(2,2)	I(2,4)	I(2,6)	I(2,8)	I(2,10)	0
O(0,2)	I(0,2)	I(0,4)	I(0,6)	I(0,8)	I(0,10)	I(0,0)	I(2,2)	I(2,4)	I(2,6)	I(2,8)	I(2,10)	I(2,0)	1
O(2,0)	I(2,0)	I(2,2)	I(2,4)	I(2,6)	I(2,8)	I(2,10)	I(4,0)	I(4,2)	I(4,4)	I(4,6)	I(4,8)	I(4,10)	2
O(0,1)	I(0,1)	I(0,3)	I(0,5)	I(0,7)	I(0,9)	I(0,11)	I(2,1)	I(2,3)	I(2,5)	I(2,7)	I(2,9)	I(2,11)	0
O(0,3)	I(2,1)	I(2,3)	I(2,5)	I(2,7)	I(2,9)	I(2,11)	I(4,1)	I(4,3)	I(4,5)	I(4,7)	I(4,9)	I(3,11)	3
K(0,0)	K(0,1)	K(0,2)	K(0,3)				K(1,0)	K(1,1)	K(2,2)	K(1,3)			

(Row group label: O(Even,Even) spans O(0,0), O(0,2), O(2,0); O(Even,Odd) spans O(0,1), O(0,3); O(Odd,Even); O(Odd,Odd))

Fig. 7 *Four-dimensional convolution dataflow controller for discriminator using ZFWST.* Two out of four rows of processing element array are shown in this figure. Pixels with the type even–even are processed first followed by the even–odd, odd–even, and finally pixel with odd–odd type are processed. Output has been computed with one pixel at a time after processing by all kernel weights in a weight stationary architecture

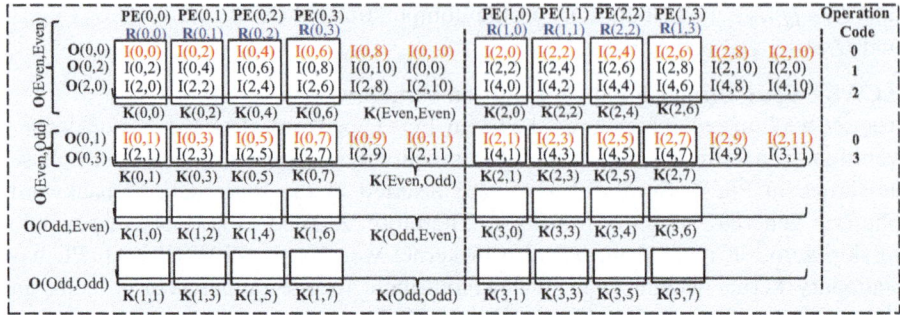

Fig. 8 *Four-dimensional convolution dataflow controller for generator using ZFWST*. Two out of four rows of processing element array are shown in this figure. In this convolution, input image pixels with a certain type contribute only to the output pixels of same type. Being the weight stationary architecture, each pixel of output computed one at a time. Even–even type pixels are processed first generating the output pixels of type even–even, followed by even–odd, odd–even, and final odd–odd pixels are processed with kernel weights generating output pixels of odd–odd type

and 8, *a customized memory architecture is needed, which can provide multiple data points of input feature map from the on-chip memory in one clock cycle without any data wastage as per the requirements of strided, transposed, and four-dimensional convolution.*

- **Multiple type-oriented data points required in a single clock cycle:** *ZFOST* and *ZFWST* microarchitectures require the input data to be loaded in *type-oriented form*, which becomes challenging with conventional linear memory. This requires a customized memory controller that intermediately re-packages the fetched data in type-oriented form before loading it into input register arrays.

3   Memory-Efficient Hardware Architecture for Generative Adversarial Networks (GANs)

Our work addresses the above mentioned memory-related challenges broadly by proposing *a 2-D distributed on-chip memory array* and *Data re-packaging units*, as shown in Fig. 9. The 2-D distributed on-chip memory supports the simultaneous data loading as required by the non-standard GAN convolutions: strided/transposed/four-dimensional convolution. And the data re-packaging units re-arrange the data in type-oriented format before storing it in on-chip memory. It is to be noted that we implemented the zero-free output stationary (ZFOST) and zero-free weight stationary (ZFWST) microarchitectures along with the custom convolution dataflow controllers, *S-CONV*, *T-CONV*, and *W-CONV*, as proposed in [9] and discussed in Sect. 2. Altogether, our proposed distributed on-chip memory, its corresponding distributed memory controller, data re-packaging units, and custom dataflow

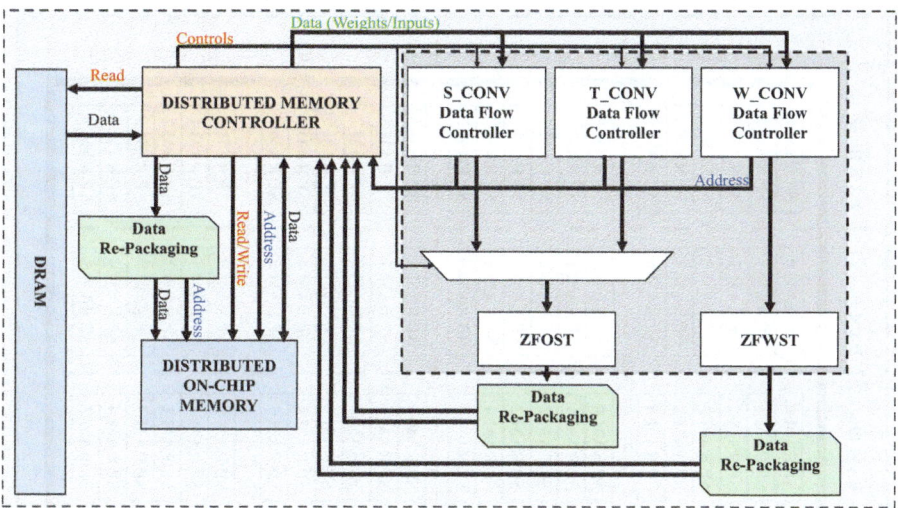

Fig. 9 Propose Architecture of Generative Adversarial Network

controllers make up the complete hardware architecture for GAN acceleration. We describe our proposed hardware blocks in the following sections.

3.1 2-D Distributed On-Chip Memory Array

The proposed 2-D distributed on-chip memory array structure is shown in Fig. 10. The overall structure has two major parts, namely *RAM-Block* and *RAM-Channel*. Each RAM-Block consists of small-sized SRAM blocks that store the same type of input data, in a type-oriented form. For example, input pixels $I_{(even,even)}$ are stored in *RAM-Block* 0. Similarly, input pixels $I_{(even,odd)}$, $I_{(odd,even)}$, $I_{(odd,odd)}$ are stored in *RAM-Blocks* 1, 2, and 3, respectively. Each *RAM-Block* is further divided into four *RAM-Channels*, where each channel stores data required by each row of the input register array in the *ZFOST* microarchitecture. Further, each *RAM-Channel* is divided into six single-port SRAM (SPRAMs) blocks, which is based on the number of columns in the input register array. Therefore, each SPRAM in a *RAM-Block* feeds data to one register of the input register array that enables simultaneous registers loading. For example, during the strided convolution dataflow as shown in Fig. 4, a total of 24 data points are required to be loaded into the local register when $K_{(0,0)}$ is being processed. 6 out of 24 data points, for row-0, row-1, row-2, and row-3 of the local register, are stored in RAM-Block-0 at address-0 of channel-0, channel-1, channel-2, and channel-3, respectively. The SPRAM size is dependent on the maximum size of the input and output feature maps, which can be computed using Eq. (1).

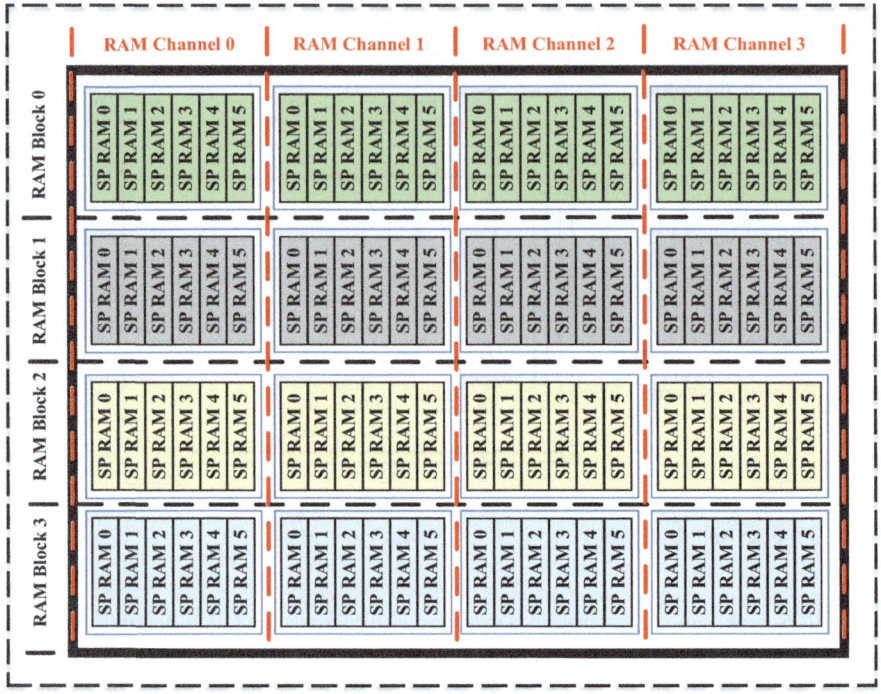

Fig. 10 Distributed on-chip memory design

$$Size_{SPRAM} = \frac{Dim_{image} \times Dim_{image} \times N}{N_{Rams} \times N_{Channels} \times N_{Blocks}}. \tag{1}$$

3.2 Data Re-Packaging Unit

The data re-packaging units re-arrange the data in a type-oriented format before storing it in on-chip memory. In doing so, it translates the input feature address, coming either from the external DRAM or ZFOST/ZFWST microarchitectures, into distributed on-chip memory address. It is to be noted that locating an individual pixel inside the distributed memory requires to compute six elements of its address: *pixel row*, *pixel column*, *RAM-Block* index, *RAM-Channel* index, *RAM-Index* (index of a SPRAM inside a block of the grid), and address of the SPRAM, as depicted in Fig. 11. We discuss the generation of these address components below while referring to notations used in Table 1.

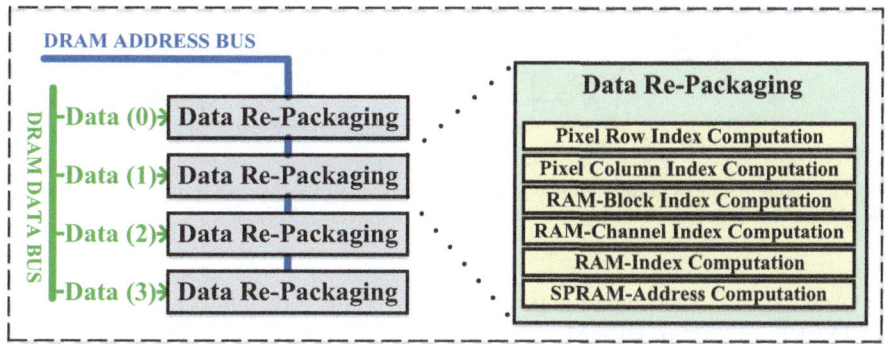

Fig. 11 Data Re-packaging Unit

Table 1 Notations used in the explanation of Re-packaging phases

Symbols	Description
$Size_{SPRAM}$	Size of a SPRAM in distributed memory architecture
N_{RAMs}	The number of SPRAMs in one channel of distributed memory architecture
$N_{Channels}$	The number of channels in one block of distributed memory architecture
N_{Blocks}	The number of blocks in distributed memory architecture
N	The number of bits of one input pixel
P_{row}	Row index of input pixel
P_{col}	Column index of input pixel
Add_{DRAM}	Address of DRAM access
NP_{DRAM}	The number of data point in one DRAM location
dp_x	Data point index out of a total number of data points from one DRAM location
Dim_{image}	Feature map dimension (rows/columns where the numbers of rows and columns are the same)
$Blobk_{Index}$	Block index of IO-Buffer
ISODD(x)	Results 1 if x is an odd number
ISEVEN(x)	Results 1 if x is an even number
$Channel_{Index}$	Channel index of IO-Buffer
$Temp_{RAMIndex}$	Local temporary variable used in computation of RAM index of IO-Buffer
X (mod Y)	Remainder when X is divided by Y
X [B1:B2]	Bit wise selection of X with B1 as MSB and B2 as LSB
X [B1]	Bit wise selection of bit B1 of X
RAM_{Index}	RAM-Index
$RAM_{Address}$	SPRAM Address

3.2.1 Pixel Row Index Computation Block

In this first stage pixel row index computation block, the row index of the input pixel is computed in the following steps:

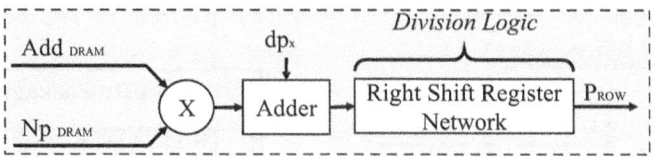

Fig. 12 Input pixel row computation using the information of the data point index, address of the DRAM, and information of the number of data point in one DRAM location

- Address of DRAM location is multiplied with the total number of pixels stored in each location in DRAM.
- The current index of the data point out of the total number of input data points stored in a location of DRAM is added to the result.
- Finally, the final row address result is divided by the dimension of the input feature map to compute the final row address.

Figure 12 shows the block-level diagram for pixel row index computation block.

3.2.2 Pixel Column Index Computation Block

The pixel column index is computed using the following components:

- Pixel row index
- Address of the DRAM from where the data has been fetched
- The total number of input pixels in one location of DRAM
- The dimension of input feature map and the index of input data pixel in the DRAM location

Figure 13 shows the block-level diagram for pixel column index computation using the above information.

3.2.3 RAM-Block Index Computation

The 2-D Distributed Memory Array Block Index Computation, *RAM-Block*, is calculated based on the pixel type (i.e., *even–even*, *even–odd*, *odd–even*, and *odd–odd*). This is achieved by checking the LSBs of the row index and column index. The block-level circuit diagram of RAM-Block index computation is shown in Fig. 14. It generates a *2-bit RAM-Block* index that is 0, 1, 2, and 3, if pixel is *even–even*, *even–odd*, *odd–even*, and *odd–odd*, respectively.

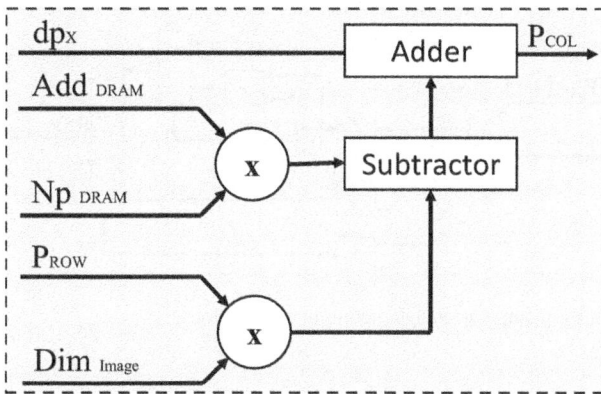

Fig. 13 Input pixel column computation using the information of the data point index, address of the DRAM, information of the number of data point in one DRAM location, computed row in the previous step, and the dimension of the input feature map

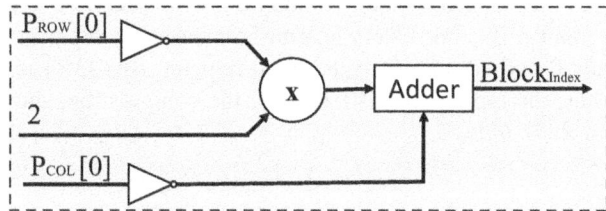

Fig. 14 Block-Index Computation using row index and column index computed in the previous steps

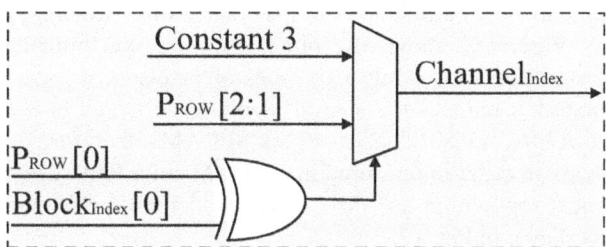

Fig. 15 Channel Index Computation using row index and block index value

3.2.4 RAM-Channel Index Computation

Figure 15 shows the circuit-level diagram for the *RAM-Channel* index computation block. The channel index for each pixel is calculated using the least significant three bits of row index along with the least significant bit of *RAM-Block* index.

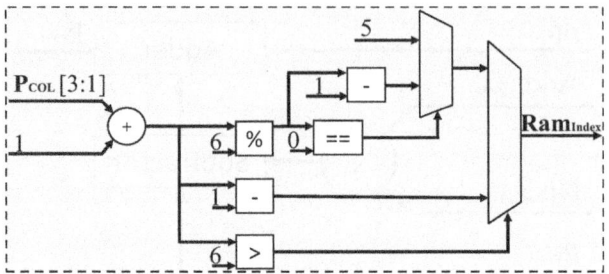

Fig. 16 RAM Index Computation using column index value

3.2.5 RAM Index Computation

Each *RAM-Channel* contains six single-ported SRAMs (SPSRAMs) that are represented by *RAM-Index*, and the index runs from 0 to 5. Figure 16 illustrates the circuit-level diagram for RAM index computation using bits 3 : 1 of the column indexes of the input pixel. For example, consider the input pixel $I(0, 6)$ that will be loaded in row 0 and column 3 of the local register. Bits [3:1] in $(0110)_2$ give $(011)_2 = 3$. Thus, the input pixel $I(0, 6)$ after the computation, shown in Fig. 16, will be stored in *RAM-Index* 2.

3.2.6 SPRAM Address Computation

The size of the input feature map and pixel's column index is used in order to compute the address of a particular SPRAM selected by $Block_{Add}$, $Channel_{Add}$ and RAM_{Index}. Figure 17 shows the block-level diagram implemented for the SPRAM address computation. Only two types of pixels exist for a single row of input feature map that reduces the size of a row to be stored in a *RAM-Channel* to half. As each *RAM-Channel* consists of six SPRAMs, therefore each full row is folded at length six in order to be stored in six RAMs of a *RAM-Channel*. The base address of the next row within a *RAM-Channel* is determined by the first input of the last adder as shown in Fig. 17.

Let us take an example: if the input feature map is of dimension 32 × 32, then one row of pixels will be stored in 3 locations of a SPRAM (from address 0 to address 2). Therefore, when a particular SPRAM is selected again to store a data value, its address must start from the address 3. If the input feature map is greater than 12 pixels, the same SPRAM will contain multiple data values of the sample row. Therefore, the second input of the last adder in Fig. 17 is the address computed on the basis of the column index of the input pixels.

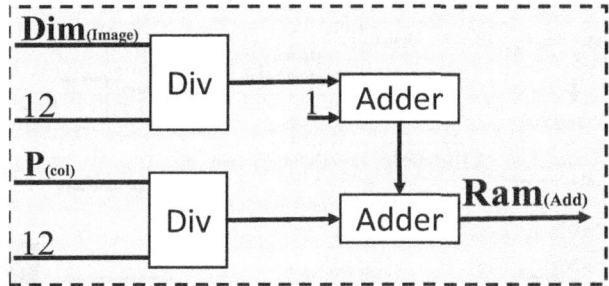

Fig. 17 RAM address computation using the information of input image dimension and computed column index

4 Results and Discussion

4.1 Experimental Setup

We implemented the microarchitecture of the blocks in the proposed GAN hardware architecture, shown in Fig. 9, using Verilog-HDL. The design is synthesized using Xilinx Vivado Design Tool 2017.4 targeting Xilinx Kintex-7 "xc7k410tfbg676-1" FPGA with a clock speed of 200 MHz. The size of an SPRAM inside distributed memory array is set to 32 KB, which is sufficient to hold an image dimension up to 1024 × 1024 pixels where each pixel is 16-bit-wide. In order to verify the functionality of the proposed architecture, a trained DCGAN model [7] is used. We extracted the input feature maps and kernel maps using Matlab tool. These maps are then used to evaluate the design for strided, transposed, and four-dimensional convolution.

Figure 18 depicts the complete tool flow that is used to evaluate the proposed architecture. The weights and input features are pre-loaded in a DRAM linearly that serves as stimuli. The design is evaluated using different image sizes but with a fixed kernel size of 4 × 4. In order to emulate a real-world ASIC-based implementation, the memory configuration used in Vivado Design Tool is also provided to CACTI-p tool by HP to compute the read/write access energy consumption of memory.

We compare the performance of our architecture with 2-D distributed on-chip memory array with the state-of-the-art design [9] containing a conventional on-chip buffer that stores data in a linear format. Figure 19 shows the overall design of a conventional GAN hardware architecture that uses the conventional linear on-chip memory. Each memory location in our architecture can store 4 data points, since the memory data bus is 64-bit-wide and each pixel is 16-bit-wide. So to store an input image of size up to 1024 × 1024 pixels, the total overall memory size becomes 512 KB.

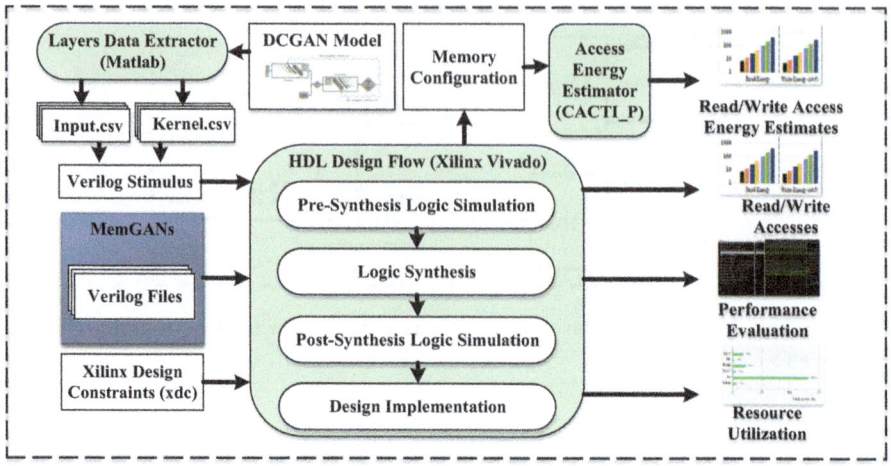

Fig. 18   Tool flow for evaluation of the proposed architecture

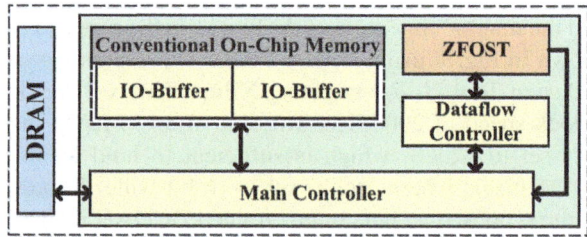

Fig. 19   Conventional Architecture of Generative Adversarial Network

4.2   *Processing Time Evaluation*

Processing time consists of the time it takes data to load from off-chip memory and the time spent in the processing block. Figures 20 and 21 show the processing times of the various convolution blocks, *S-CONV*, *T-CONV*, and *W-CONV*, in both conventional and proposed hardware designs, respectively.

As can be seen, our proposed architecture outperforms the conventional design in processing time of an input feature map. Our proposed architecture supports the ZFOST to achieve 3.65x faster processing time than the baseline, over different image sizes.

Figure 22 shows the breakdown of the overall processing time into two components: *loading time* and *processing time*, in a strided convolution on a input feature map of different sizes. It can be seen that as the dimensions of the input feature map increase, the data loading time dominates in the overall processing of the convolution operation in both proposed and conventional architectures.

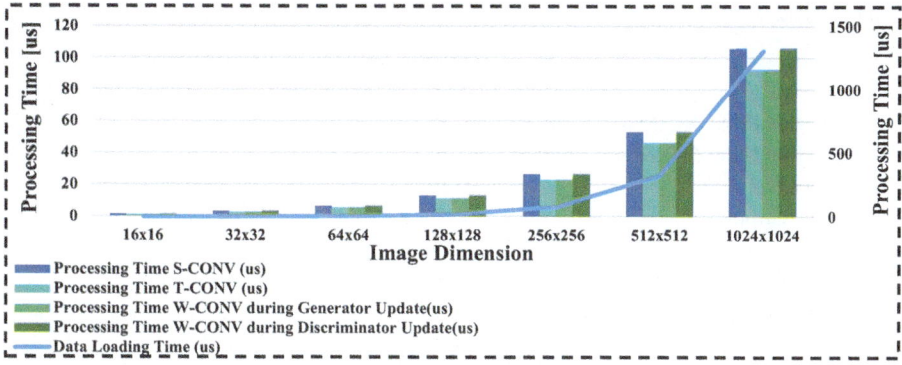

Fig. 20 Performance evaluation of the conventional design of [9]

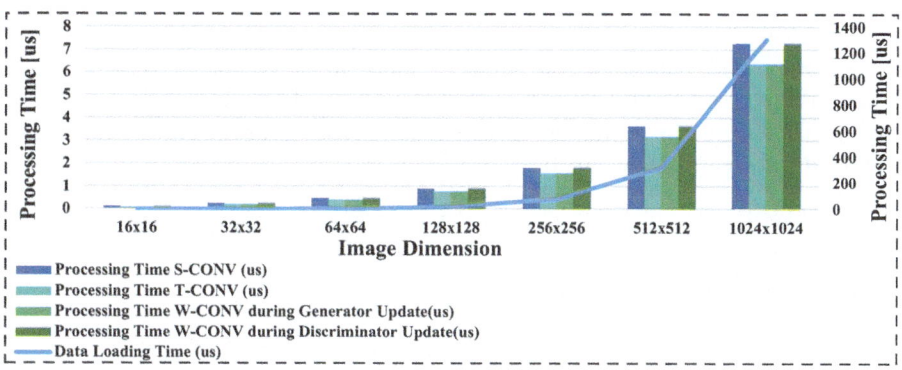

Fig. 21 Performance evaluation of the proposed architecture

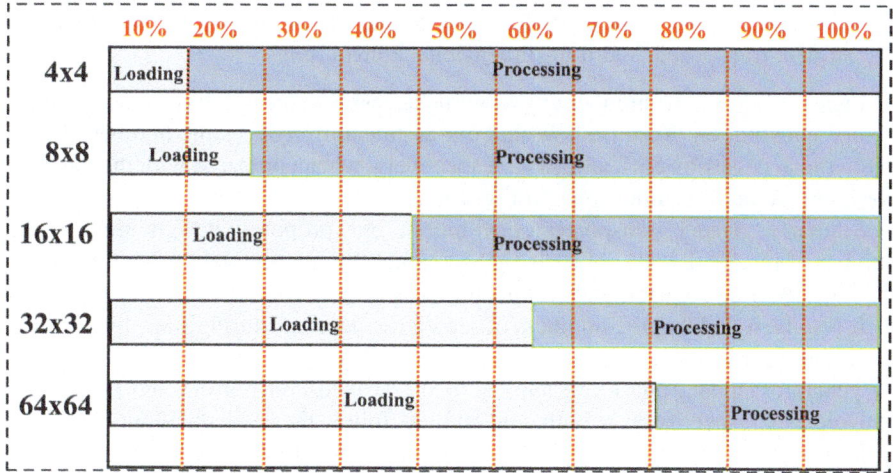

Fig. 22 Contribution of Loading and Processing Times

Table 2 Power breakdown comparison of the proposed design with the conventional design

Module	Power Consumption Breakdown [W]	
	Proposed architecture	Conventional design
Re-packaging Unit (DRAM to IO-Buffer)	0.23	–
Re-packaging Unit (ZFOST to IO-Buffer)	1.3	–
Main controller	0.19	1.48
Dataflow controller	–	1.83
S-CONV controller)	0.58	–
T-CONV controller)	0.61	–
W-CONV controller)	0.73	–
ZFOST	0.10	0.10
ZFWST	0.12	0.12
IOBUFFER (BRAM Blocks in FPGA)	1.74	1.79
Total Power	**5.6**	**5.32**

Our proposed architecture consumes 5% more power compared to the conventional baseline design of Fig. 19. This is because of the additional re-packaging units used in the design. The overall power consumption as well as the power breakdown of both the designs is shown in Table 2.

4.3 Memory Accesses Evaluation

In this experiment, we compare the following:

- The number of read/write accesses to the on-chip memory during strided and transpose convolutions
- The energy consumption of the associated read/write accesses
- The overall time to process the input feature maps of different sizes during strided, transposed, and four-dimensional convolution

Figure 23 shows the number of read/write accesses to the on-chip memory during strided convolution. Results show that our architecture reduces the number of read and write accesses by 85% and 75%, respectively, when compared with the baseline conventional architecture shown in Fig. 19.

Similarly, during transposed convolution, our proposed design reduces the number of read and write accesses by 85% and 80%, respectively, when compared with the baseline conventional architecture as shown in Fig. 24.

It is to be noted that the numbers of read/write accesses during four-dimensional convolution in its discriminator update phase and four-dimensional convolution in its generator update phase are similar to the strided convolution and transposed convolution, respectively, as both convolutions follow the same dataflow.

Figure 25 shows the energy consumption of the read/write accesses during strided convolution. Results show that our architecture reduces the energy consump-

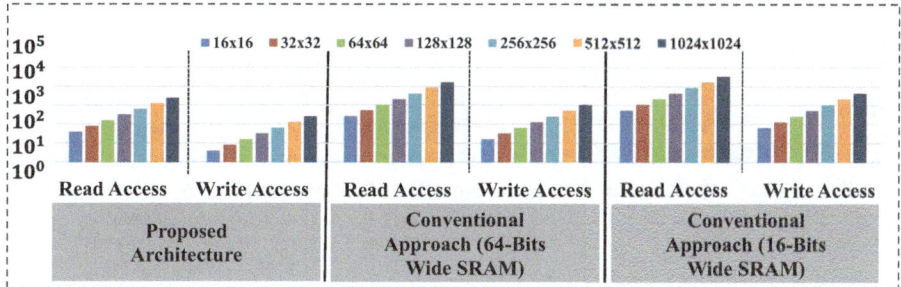

Fig. 23 Comparison between the number of memory accesses during strided convolution considering different memory architectures for various input image sizes

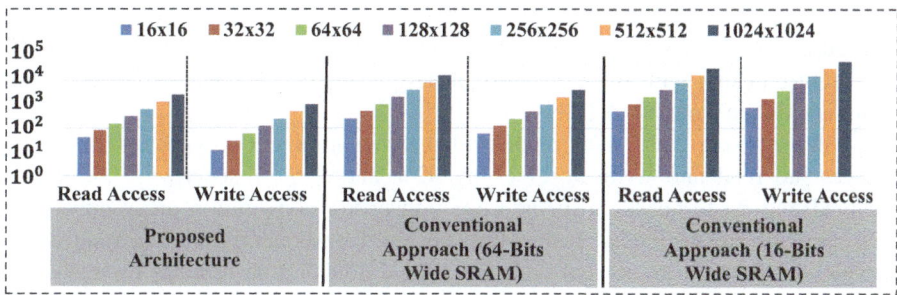

Fig. 24 Comparison between the number of memory accesses during transposed convolution considering different memory architectures for various input image sizes

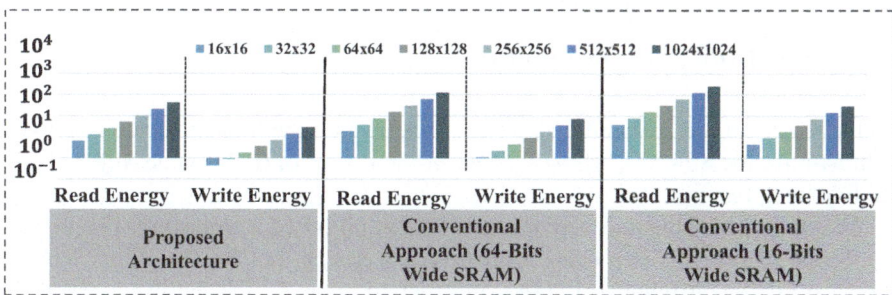

Fig. 25 Comparison between the memory accesses energy during strided convolution considering different memory architectures for various input image sizes

tion of the read and write accesses by 65% and 58%, respectively, when compared with the conventional architecture.

Similarly, our architecture reduces the energy consumption of the read and write accesses in transposed convolution by 65% and 67%, respectively, when compared with the conventional architecture.

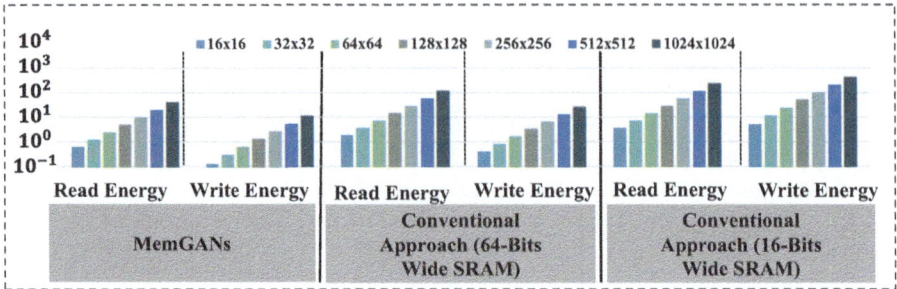

Fig. 26 Comparison between the memory accesses energy during transposed convolution considering different memory architectures for various input image sizes

Figures 23, 24, 25, and 26 also show the comparison of our proposed GAN architecture design with the conventional design of Fig. 19 when the SPRAM bit width is reduced from 64 to 16 bits.

As shown in Fig. 23, in the case of strided convolution, our proposed architecture reduces the read and write accesses by 92% and 93%, respectively. Similarly, the read and write accesses in the case transposed convolution are reduced by 95% and 94%, respectively, as shown in Fig. 24. Consequently, the read and write energy consumption associated with the strided convolution reduces by 82% and 89%, respectively, as shown in Fig. 25, and by 86% and 91%, respectively, during transposed convolution, as shown in Fig. 26.

4.4   Area Utilization Evaluation

As the results of the conventional design with 16-bit-wide SPRAM are worse than with 64-bit-wide SPRAM, we only consider the results of 64-bit-wide SPRAM for the following results.

We implemented the microarchitectures of our proposed architecture (Fig. 9) and the design of [9] (Fig. 19), using Xilinx Vivado design tool for the Xilinx Kintex 7 device (xc7k325tffg900-2 FPGA), in order to compare the hardware resource utilization. As shown in Table 3, our proposed architecture utilizes more resources in comparison to the design of Fig. 19. This is understandably a trade-off between improved performance and energy efficiency.

Table 3 shows the resource utilization of our proposed architecture and conventional design. Our proposed architecture utilizes 1.16x, 1.14x, and 1.14x more look-up tables, flip flops, and DSP blocks on the FPGA when compared with the baseline design. Moreover, the number of BRAMs used by our proposed 2-D on-chip distributed memory array is 6x more. However, this number (i.e., the average number of BRAMs per ZFOST architecture) can be reduced when multiple ZFOST architectures are implemented together.

Table 3 Comparison of Resource Utilization

	Available	Utilized	
		Our proposed architecture	Conventional memory Design
LUT	203,800	77,414	69,317
FF	407,600	15,952	13,469
RAM	445	48	8
DSP	840	98	73

5 Conclusion

In this chapter, we discussed the need for having hardware-based solution for accelerating non-standard convolution operations involved in generative adversarial networks (GANs) using novel 2-D distributed on-chip memory architecture and smart data re-packaging units. The 2-D memory architecture helps in simultaneous loading of multiple pixels into the input register array of the *ZFOST/ZFWST* microarchitectures. Similarly, the re-packaging units provide data organization support by arranging the data in the required *type-oriented format*. Compared with the state-of-the-art design of [9], our proposed hardware architecture, shown in Fig. 9, has the following unique aspects:

- It reduces the number of memory read and write accesses in strided convolution by 85% and 75%, respectively, and by 85% and 80%, respectively, in transposed convolution.
- It reduces the energy consumption during the read and write accesses by 65% and 58%, respectively, during strided convolution, and by 65% and 67%, respectively, during transposed convolution.

Overall, our proposed distributed on-chip memory architecture and data re-organization units achieve 3.65x faster processing time as compared with the state of the art [9]. This shows that by designing complementary memory architectures for the state-of-the-art GAN accelerators, we can further improve their performance and energy efficiency.

References

1. Denton, E.L., Gross, S., Fergus, R.: Semi-supervised learning with context-conditional generative adversarial networks. CoRR, abs/1611.06430 (2016)
2. Ghosh, A., Bhattacharya, B., Chowdhury, S.B.R.: SAD-GAN: synthetic autonomous driving using generative adversarial networks. CoRR, abs/1611.08788 (2016)
3. Goodfellow, I., Pouget-Abadie, J., Mirza, M., Xu, B., Warde-Farley, D., Ozair, S., Courville, A., Bengio, Y.: Generative adversarial nets. In: Advances in Neural Information Processing Systems, pp. 2672–2680 (2014)

4. Lawson, W., Bekele, E., Sullivan, K.: Finding anomalies with generative adversarial networks for a PatrolBot. In Proceedings of the IEEE Conference on Computer Vision and Pattern Recognition Workshops (2017)
5. LeCun, Y., Bengio, Y., Hinton, G.: Deep learning. Nature **521**(7553), 436 (2015)
6. Noroozi, M., Favaro, P.: Unsupervised learning of visual representations by solving jigsaw puzzles. In: European Conference on Computer Vision, pp. 69–84. Springer, New York (2016)
7. Radford, A., Metz, L., Chintala, S.: Unsupervised representation learning with deep convolutional generative adversarial networks. CoRR, abs/1511.06434 (2015)
8. Reed, S., Akata, Z., Yan, X., Logeswaran, L., Schiele, B., Lee, H.: Generative adversarial text to image synthesis. arXiv:1605.05396 (2016)
9. Song, M., Zhang, J., Chen, H., Li, T.: Towards efficient microarchitectural design for accelerating unsupervised GAN-based deep learning. In: Proceedings of the 2018 IEEE International Symposium on High Performance Computer Architecture (HPCA), pp. 66–77. IEEE, New York (2018)
10. Song, M., Zhong, K., Zhang, J., Hu, Y., Liu, D., Zhang, W., Wang, J., Li, T.: In-situ AI: Towards autonomous and incremental deep learning for IoT systems. In Proceedings of the 2018 IEEE International Symposium on High Performance Computer Architecture (HPCA), pp. 92–103. IEEE, New York (2018)
11. Zhu, X., Ghahramani, Z., Lafferty, J.D.: Semi-supervised learning using Gaussian fields and harmonic functions. In: Proceedings of the 20th International Conference on Machine, pp. 912–919 (2003)

Using Approximate DRAM for Enabling Energy-Efficient, High-Performance Deep Neural Network Inference

Lois Orosa, Skanda Koppula, Konstantinos Kanellopoulos, A. Giray Yağlıkçı, and Onur Mutlu

1 Introduction

Deep neural networks (DNNs) [1] are an effective solution to challenges in computer vision [2], speech recognition [3], or medicine [4]. DNNs and their various flavors (e.g., convolutional neural networks [2], transformers [5]) are commonly evaluated in settings with edge devices that demand low energy and real-time responses [6]. Unfortunately, DNNs have high computational and memory demands that make these energy and performance requirements difficult to fulfill. As such, neural networks have been the subject of many recent accelerators and DNN-

L. Orosa
ETH Zürich and Galicia Supercomputing Center (CESGA), Avenida de Vigo s/n, Santiago de Compostela, Spain

S. Koppula
ETH Zürich and Google DeepMind, DeepMind Technologies, London, England

K. Kanellopoulos · A. G. Yağlıkçı · O. Mutlu (✉)
ETH Zürich, Zürich, Switzerland
e-mail: onur.mutlu@inf.ethz.ch

© The Author(s), under exclusive license to Springer Nature Switzerland AG 2024
S. Pasricha, M. Shafique (eds.), *Embedded Machine Learning for Cyber-Physical, IoT, and Edge Computing*, https://doi.org/10.1007/978-3-031-19568-6_10

focused architectures. Recent works (e.g., [7–9]) focus on building specialized architectures for efficient computation scheduling and dataflow to execute DNNs.

Improvements to accelerator efficiency [10], DNN-optimized GPU kernels [11], and libraries designed to efficiently leverage instruction set extensions [12] have improved the computational efficiency of DNN evaluation. However, improving the memory efficiency of DNN evaluation is an on-going challenge [8]. The memory intensity of DNN inference is increasing, and the sizes of state-of-art DNNs have grown dramatically in recent years. The winning model of the 2017 ILSVRC image recognition challenge, ResNeXt [13], contains 837M FP32 parameters (3.3 GB). This is 13.5x the parameter count of AlexNet, the winning model in 2012 [2]. As the machine learning community trends towards larger, more expressive neural networks, we expect off-chip memory problems to bottleneck DNN evaluation.

The focus of recent approximate memory research is to alleviate two main issues (energy and latency) of off-chip DRAM for neural network workloads. First, DRAM has high energy consumption. Prior work on DNN accelerators reports that between 30 and 80% of system energy is consumed by DRAM [7]. Second, DRAM has high latency. A load or store that misses the last level cache (LLC) can take 100x longer time to service compared to an L1 cache hit [14]. Prior work in accelerator design has targeted DRAM latency as a challenge for sparse and irregular DNN inference [15].

To overcome both DRAM energy and latency issues, recent works use three main approaches. First, some works reduce numeric bitwidth, reuse model weights, and use other algorithmic strategies to reduce the memory requirements of the DNN workload [16]. Second, other works propose new DRAM designs that offer lower energy and latency than commodity DRAM [17]. Third, some works propose processing-in-memory approaches that can reduce data movement and access data with lower latency and energy [18]. In this chapter, we discuss an approach that is orthogonal to these existing works: customization of the major operational parameters (e.g., voltage, latency) of *existing* DRAM chips to better suit the intrinsic characteristics of a DNN. The approach is based on two key insights:

(1) DNNs demonstrate remarkable robustness to errors introduced in input, weight, and output data types. This error tolerance allows accurate DNN evaluation on unreliable hardware if the DNN error tolerance is accurately characterized and bit error rates are appropriately controlled.
(2) DRAM manufacturers trade performance for reliability. Prior works show that reducing DRAM supply voltage and timing parameters improves the DRAM energy consumption and latency, respectively, at the cost of reduced reliability, i.e., increased bit error rate.

To exploit these two insights, EDEN[1] was developed: the first framework that improves energy efficiency and performance for DNN inference by using approximate DRAM, which operates with reduced DRAM parameters (e.g., voltage

---

[1] Energy-Efficient **D**eep Neural Network Inference Using Approximate DRAM.

and latency) [19]. EDEN strictly meets a user-specified target DNN accuracy by providing a general framework that (1) uses a new retraining mechanism to improve the accuracy of a DNN when executed on approximate DRAM and (2) maps the DNN to the approximate DRAM using information obtained from rigorous characterizations of the DNN error tolerance and DRAM error properties.

EDEN is based on three key steps. First, EDEN *improves the error tolerance* of the target DNN by retraining the DNN using the error characteristics of the approximate DRAM module. Second, EDEN *profiles* the improved DNN to identify the error tolerance levels of all DNN data (e.g., different layer weights of the DNN). Third, EDEN *maps* different DNN data to different DRAM partitions that best fit each datum's characteristics and accordingly selects the voltage and latency parameters to operate each DRAM partition. By applying these three steps, EDEN can map an arbitrary DNN workload to an arbitrary approximate DRAM module to evaluate a DNN with low energy, high performance, *and* high accuracy.

To show example benefits of the approach, EDEN was run with DNN inference workloads using approximate DRAM with (1) reduced DRAM supply voltage (V_{DD}) to decrease DRAM energy consumption, and (2) reduced DRAM latency to reduce the execution time of latency-bound DNNs. EDEN adjusts the DRAM supply voltage and DRAM latency through interaction with the memory controller firmware. For a target accuracy within 1% of the original DNN, results show that EDEN enables (1) an average DRAM energy reduction of 32% across CPU, GPU, and DNN accelerator (e.g., Tensor Processing Unit [20]) architectures, and (2) cycle reductions of up to 17% when evaluating latency-bound neural networks.

The benefits of EDEN stem from its capacity to run on most hardware platforms in use today for neural network inference, including CPUs, GPUs, FPGAs, and DNN accelerators. Because EDEN is a general approach, its principles can be applied (1) on any platform that uses DRAM, and (2) across memory technologies that can trade-off different parameters (e.g., voltage, latency) at the expense of reliability. Although the evaluation examines supply voltage and access latency reductions, the EDEN framework can be used also to improve performance and energy in other ways: for example, EDEN could increase the effective memory bandwidth by increasing the data bus frequency at the expense of reliability.

In this chapter, we discuss our work EDEN, which makes the following five key contributions:

- We propose the first general framework that increases the energy efficiency and performance of DNN inference by using approximate DRAM that operates with reduced voltage and latency parameters at the expense of reliability. EDEN provides a systematic way to scale main memory parameters (e.g., supply voltage and latencies) while achieving a user-specified DNN accuracy target.
- We discuss how EDEN introduced a methodology to retain DNN accuracy in the presence of approximate DRAM. Evaluation shows that EDEN increases the bit error tolerance of a DNN by 5–10x (depending on the network) through a customized retraining procedure called *curricular retraining*.

- We provide a systematic, empirical characterization of the resiliency of state-of-art DNN workloads to the errors introduced by approximate DRAM. We examine the error resiliency across different numeric precisions, pruning levels, and data types (e.g., DNN layer weights). We find that (1) lower precision levels and DNN data closer to the first and last layers exhibit lower error resiliency, and (2) magnitude-based pruning does not have a significant impact on error resiliency.
- We propose four error models to represent the common error patterns that an approximate DRAM device exhibits. To do so, EDEN characterizes the bit flip distributions that are caused by reduced voltage and latency parameters on eight real DDR4 DRAM modules.
- We evaluate EDEN on multi-core CPUs, GPUs, and DNN accelerators. For a target accuracy within 1% of the original DNN, results show that EDEN enables (1) an average DRAM energy reduction of 21%, 37%, 31%, and 32% in CPU, GPU, and two different DNN accelerator architectures, respectively, across a variety of state-of-the-art networks, and (2) an average (maximum) speedup of 8% (17%) and 2.7% (5.5%) in CPU and GPU architectures, respectively, when evaluating latency-bound neural networks. For a target accuracy the same as the original, EDEN enables 16% average energy savings and 4% average speedup in CPU architectures.

2   Background

2.1   Deep Neural Networks

Artificial neural networks are a type of machine learning model inspired by the structure and activation patterns of neurons in the animal nervous systems [21]. Neural networks work by alternating application of linear and non-linear operations to the data inputs (images, audio signals, etc.) [22]. A convolution operation is one such linear operation, and one of the most common non-linear operation is the ReLU activation function [21]. Neural networks have been used since the 1960s, but it was in the past decade that neural networks—and in particular, deep neural networks (DNNs) [1]—have shown to strongly outperform competing machine learning methods across nearly every large-scale data learning task. The power of neural networks is generally attributed to their lack of hand-crafted input-specific operations, and as such, these models have been shown to be effective across multiple different data modalities.

DNNs are composed of a variety of different layers, including convolutional layers, fully connected layers, attention layers, and pooling layers [1]. Figure 1 shows the three main data types of a DNN layer, and how three DNN layers are connected with each other. Each of these layers is defined by a weight matrix learned via a one-time training process that is executed before the DNN is ready for inference. The three DNN data types that require loads and stores from main

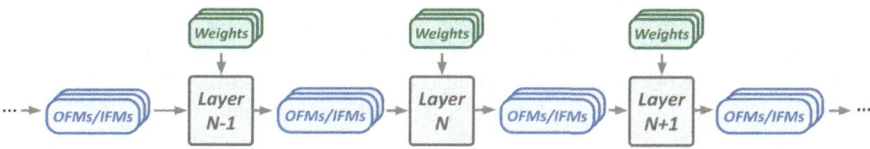

Fig. 1 Example of three DNN layers. Each layer is composed of its weights, input feature maps (IFMs), and output feature maps (OFMs). Adapted from [23]

memory include each layer's input feature maps (IFMs), output feature maps (OFMs), and the weights. Each layer processes its IFMs using the layer's weights, and produces OFMs. The OFMs of a layer are fed to the next layer as the next layer's IFMs. In this work, we explore the introduction of bit errors into the three data types of each layer [19].

The lifecycle of a machine learning model for consumer applications usually involves (1) training the model on large amounts of training data (2) retraining (i.e., fine-tuning) the model to adapt to the target task, (3) optionally, applying modifications to reduce the compute costs of running the model (e.g., through weight quantization or pruning, described below) and (4) freezing the trained model, and using it to yield predictions (i.e., inference, or forward pass) on compute-limited consumer devices. It is common that the model weights are trained once and frozen, while the forward pass may run thousands of times during the production lifetime of the model.

Modern DNNs contain hundreds of layers, providing the DNN with a large number of trainable weights. State-of-art DNNs, as of writing, contain up to half a trillion FP-32 parameters [24], and sometimes require training and inference systems that distribute the storage and compute across many machines [25]. The existence of such a large number of weights is commonly referred to as *overparameterization*, and is, in part, the source of a DNN's accuracy [26]. Overparameterization allows the model to have sufficient learning capacity so that the network can approximate complex input-output functions, and adequately capture high-level semantics (e.g., the characteristics of a cat in an input image). Importantly, overparameterization allows the network to obtain some level of error resilience, generalize across different inputs, and be robust to insignificant changes to the input (e.g., background pixels in an image).

Overparameterization of modern DNNs have advantages and disadvantages. While overparameterization provides DNNs with capacity to learn [1], it also presents significant computational problems, especially on devices that are resource constrained. State-of-the-art DNNs can barely fit in a single commodity server, let alone a consumer mobile device. This problem has spurred research on methods to reduce the computational and storage footprint of these heavy, high-accuracy models, yielding various techniques to help alleviate the challenges. These methods include quantization [27] and pruning [28].

Quantization Quantizing floating point weights and OFMs into low-precision fixed-point numbers can greatly improve performance and energy consumption of DNNs [27]. Prior works demonstrate that it is possible to quantize DNNs to limited numeric precision (e.g., eight-bit integers) without significantly affecting DNN accuracy [27]. In our evaluations, we quantize all DNN models to four different numeric precisions : int4 (4-bit), int8 (8-bit), int16 (16-bit), and FP32 (32-bit).

Pruning Pruning [28] reduces the memory footprint of a DNN by sparsifying the weights and feature maps. This is done by zeroing the lowest magnitude weights and retraining. We study the effects of pruning in EDEN's evaluations.

Training Training is the process of estimating the set of weights that enables the model to best perform a specific task [22]. Neural network training is guided by a training objective specific to that task, also commonly called the loss function. The goal of training is to minimize this loss function. A loss function can intuitively be thought of as an estimate of the error as compared to ground truth real values in a dataset (for tasks in which we have such ground truth labels). For example, in classification tasks, the loss function is commonly the cross-entropy between model predicted categorical distribution $q(x)$ and the data points' true label, given by $p(x)$ $(-\sum_{x \in X} p(x) \log q(x))$, where X is the data points. Training is usually performed with an iterative gradient descent algorithm [29] using a particular training dataset. The training dataset is divided into batches, and each training step corresponds to a single forward and backward pass through the DNN to compute the gradients for that particular batch of data. These gradients dictate how the weights should be modified in order to minimize the loss. A single training "epoch" completes when the entire dataset is passed over once [1]. As mentioned previously, this procedure is usually done once, before the model is frozen and deployed on consumer inference devices.

Additionally, common training-time techniques such as *adding input noise* [30] and *input feature map dropout* [31] try to force the network to *not* rely on any single OFM element and enable robustness in the presence of statistical variance in the IFMs. Additionally, these techniques help combat overfitting that overparameterization might induce on smaller datasets. Inspiration for some of EDEN's mechanisms come from these techniques: robustness to dropout and noise addition may not just be an intentionally applied constraint applied during training—it could also perhaps aid in approximate or unreliable compute environments. EDEN leverages the same core ideas to adapt DNNs and their training procedure to achieve partial error robustness against bit errors caused by approximate DRAM, by taking advantage of weight overparameterization in modern DNNs.

2.2 DRAM Organization and Operation

DRAM Organization A DRAM device is organized hierarchically. Figure 2a shows a *DRAM cell* that consists of a *capacitor* and an *access transistor*. A capacitor encodes a bit value with its charge level. The DRAM cell capacitor is connected to a *bitline* via an access transistor that is controlled by a *wordline*. Figure 2b shows how the DRAM cells are organized in the form of a 2D array (i.e., a *subarray*). Cells in a column of a subarray share a single bitline. Turning on an access transistor causes charge sharing between the capacitor and the bitline, which shifts the bitline voltage up or down based on the charge level of the cell's capacitor. Each bitline is connected to a *sense amplifier* (SA) circuit that detects this shift and amplifies it to a full 0 or 1. The cells that share the same wordline in a subarray are referred to as a DRAM *row*. A *row decoder* drives a *wordline* to enable all cells in a DRAM row. Therefore, charge sharing and sense amplification operate at row granularity. The array of sense amplifiers in a subarray is referred to as *row buffer*. Each subarray typically consists of 512–1024 rows each of which is typically as large as 2–8 KB.

Figure 2c shows the organization of subarrays, banks, and chips that form a *DRAM device*. Each bank partially decodes a given row address and selects the corresponding subarray's *row buffer*. On a read operation, the I/O logic sends the requested portion of the target row from the corresponding subarray's row buffer to the memory controller. A *DRAM chip* contains multiple banks that can operate in parallel. A DRAM device is composed of multiple DRAM chips that share the same command/address bus and are simultaneously accessed to provide high bandwidth and capacity. In a typical system, each memory controller interfaces with a single DRAM bus.

DRAM Operation Accessing data stored in each row follows the sequence of memory controller commands illustrated in Fig. 3. First, the activation command (ACT) activates the row by pulling up the wordline and enabling sense ampli-

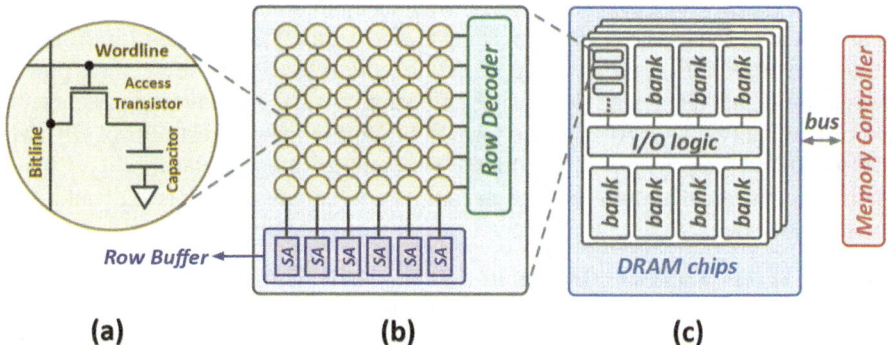

Fig. 2 DRAM organization. Adapted from [23]. (**a**) DRAM Cell. (**b**) DRAM Subarray. (**c**) DRAM device

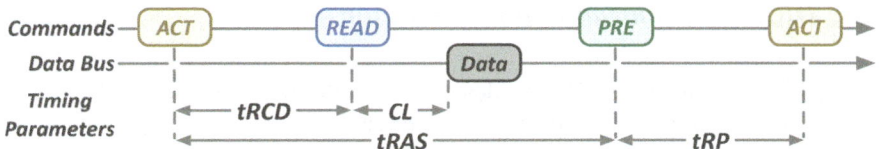

Fig. 3 DRAM read timing. We explore reductions of t_{RCD}, t_{RAS}, and t_{RP} as part of EDEN's evaluation. CL is a characteristic of the device, and not adjustable in the memory controller [32]. Adapted from [23]

fication. After a manufacturer-specified t_{RCD} nanoseconds, the data is reliably sensed and amplified in the *row buffer*. Second, the read command (READ) reads the data from the row buffer to the IO circuitry. After a manufacturer-specified CL nanoseconds, the data is available on the memory bus. Third, the precharge command (PRE) prepares the DRAM bank for activation of another row. A precharge command can be issued a manufacturer-specified t_{RAS} nanoseconds after an activation command, and an activation command can be issued t_{RP} nanoseconds after a precharge command. t_{RCD}, t_{RAS}, t_{RP}, and CL are examples of DRAM timing parameters and their nominal values provided in DRAM DDR4 datasheets are 12.5 ns, 32 ns, 12.5 ns, and 12.5 ns, respectively [32].

2.3   Reducing DRAM Parameters

We build on a large body of work on characterizing DRAM behavior in sub-reliable operation regimes of *supply voltage* and *latency* parameters [33–35].

DRAM Voltage Reduction Voltage reduction is critical to reducing DRAM power consumption since power is proportional to the square of supply voltage (i.e., $V_{DD}^2 \times f$). Prior research [35] shows that reducing voltage increases the propagation delay of signals, which can cause errors when using unmodified timing parameters. One work avoids these errors by increasing the t_{RCD} and t_{RP} latencies [35] to ensure *reliable* operation. In contrast, our goal in this work is to aggressively reduce power consumption and latency by decreasing both supply voltage and timing parameters, which inevitably causes errors in the form of bit flips in the weakest cells of DRAM, making DRAM approximate. Resulting error patterns often exhibit locality. Chang et al. [35] observe that these bit flips accumulate in certain regions (e.g., banks and rows) of DRAM.

DRAM Access Latency Reduction Latency reduction is critical to increase system performance, as heavily emphasized by a recent study on workload-DRAM interactions [36]. Previous works characterize real DRAM devices to find the minimum reliable *row activation (t_{RCD})* and *precharge (t_{RP})* latency values [33, 37]. According to these studies, the minimum DRAM latency values are significantly smaller than the values that datasheets report, due to conservative guardbands introduced by DRAM manufacturers. Further reducing these latency values cause bit flips in weak or unstable DRAM cells.

DRAM Refresh Rate Reduction Other than voltage and latency, previous research also shows that reducing the refresh rate of DRAM chips both can increase performance and reduce energy consumption at the cost of introducing errors [38] that are tolerable by many workloads that can tolerate bit errors.

3 EDEN Framework

To efficiently solve the energy and latency issues of off-chip DRAM for neural network workloads, we propose EDEN. EDEN is the first general framework that improves energy efficiency and performance for neural network inference by using approximate DRAM. EDEN is based on two main insights: (1) neural networks are tolerant to errors, and (2) DRAM timing parameters and voltage can be reduced at the cost of introducing more bit errors.

We first provide an overview of EDEN in Sect. 3.1, and explain EDEN's three steps in Sects. 3.2, 3.3, and 3.4. Finally, Sect. 3.5 explains the changes required by the target DNN inference system to support a DNN generated by EDEN.

3.1 EDEN: A High-Level Overview

EDEN enables the effective execution of DNN workloads using approximate DRAM through three key steps: (1) boosting DNN error tolerance, (2) DNN error tolerance characterization, and (3) DNN-DRAM mapping. These steps are repeated iteratively until EDEN finds the most aggressive DNN and DRAM configuration that meets the target accuracy requirements. EDEN transforms a DNN that is trained on reliable hardware into a device-tuned DNN that is able to run on a system that uses approximate DRAM at a target accuracy level. EDEN allows tight control of the trade-off between accuracy and performance by enabling the user/system to specify the maximal tolerable accuracy degradation. Figure 4 provides an overview of the three steps of EDEN, which we describe next.

1. Boosting DNN Error Tolerance EDEN introduces *curricular retraining*, a new retraining mechanism that boosts a DNN's error tolerance for a target approximate DRAM module. Our curricular retraining mechanism uses the error characteristics of the target approximate DRAM to inject errors into the DNN training procedure and boost the DNN accuracy. The key novelty of curricular retraining is to inject errors at a progressive rate during the training process with the goal of increasing DNN error tolerance while avoiding accuracy collapse with error correction. EDEN boosts the intrinsic bit error tolerance of the baseline DNN by 5–10x. We describe our boosting mechanism in Sect. 3.2.

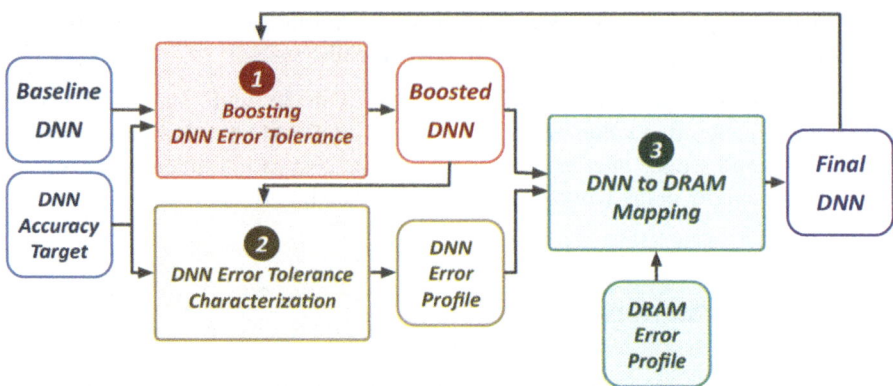

Fig. 4 Overview of the EDEN framework. Adapted from [23]

2. DNN Error Tolerance Characterization EDEN characterizes the error resilience of each *boosted DNN data type* (i.e., IFMs, OFMs, and DNN weights) to identify the limits of bit error tolerance. EDEN measures the effect of bit errors on overall accuracy using the DNN validation dataset. We describe error tolerance characterization in Sect. 3.3.

3. DNN to DRAM Mapping EDEN maps the error tolerance of each DNN data type to a corresponding approximate DRAM partition (e.g., chip, bank, or subarray) in a way that meets the specified accuracy requirements, while maximizing performance. We describe DNN to DRAM mapping in Sect. 3.4.

Together, the three steps of EDEN enable a baseline DNN to become a specialized DNN that is error-tolerant and device-tuned to a target approximate DRAM. EDEN enables energy efficient, high-performance DNN inference on the target approximate DRAM with a user-defined accuracy.

3.2 Boosting DNN Error Tolerance

According to our evaluations, the error tolerance of common DNNs is not sufficient to enable significant DRAM voltage and timing parameter reductions. To overcome this issue, we propose *curricular retraining*, a new retraining mechanism that improves the error tolerance of a DNN when running with approximate DRAM that injects errors into memory locations accessed by the DNN.

The key idea of curricular retraining is based on the observation that introducing high error rates immediately at the beginning of retraining process occasionally causes training divergence and a phenomenon called *accuracy collapse*. To mitigate this problem, *curricular retraining* slowly increases the error rate of the approximate DRAM from 0 to a target value in a step-wise fashion. In our experiments, we observe a good training convergence rate when we increase the error rate every two epochs (i.e., two passes of the entire training dataset). EDEN uses approximate DRAM in the forward pass, and it uses reliable DRAM for the backward pass.

We demonstrate in Sect. 6.4 that our curricular retraining mechanism is effective at improving the accuracy of DNN inference executed on systems with approximate DRAM.

Our experiments show that curricular retraining does not help to improve DNN accuracy on *reliable* DRAM. This implies that introducing bit error is not a regularization technique,[2] but rather, a way of obtaining congruence between the DNN training algorithms and the errors injected by approximate DRAM.

Correcting Implausible Values While executing curricular retraining, a single bit error in the exponent bits of a floating point value can cause *accuracy collapse* in the trained DNN. For example, a bit error in the exponent of a weight creates an enormously large value (e.g., $>10^8$) that propagates through the DNN layers, dominating weights that are significantly smaller (e.g., <10).

To avoid this issue, we propose a mechanism to avoid accuracy collapse caused by bit errors introduced by approximate DRAM. The key idea of our mechanism is to correct the values that are *implausible*. When a value is loaded from memory, our mechanism probabilistically detects that a data type likely contains an error by comparing its value against predefined thresholds. The thresholds of the curricular retraining data types are computed during training of the baseline DNN on DRAM with nominal parameters. Those thresholds usually have rather small values (e.g., most weights in SqueezeNet1.1 are within the range [-5,5]).

Upon detection of an error (i.e., the fact that a value is out of the threshold range) during curricular retraining, EDEN (1) corrects the erroneous value by zeroing the value, and (2) uses the corrected value for curricular retraining.

Our mechanism for correcting implausible values can be implemented in two ways. First, a software implementation that modifies the DNN framework to include extra instructions that correct implausible values resulting from each DNN memory access. Second, a hardware implementation that adds a simple hardware logic to the memory controller that corrects implausible values resulting from each approximate DRAM memory request. Section 5 describes our low-cost hardware implementation.

In our experiments, we find that our mechanism for correcting implausible values increases the tolerable bit error rate from 10^{-7} to 10^{-3} to achieve $<1\%$ accuracy degradation in the eight FP32 DNNs we analyze. We evaluate an alternative mechanism for error correction that saturates an out-of-threshold value (by resetting to the closest threshold value) instead of zeroing it. We observe that saturating obtains lower DNN accuracy than zeroing at the same approximate DRAM bit error rate across all DNN models (e.g., 8% on CIFAR-10 and 7% on ImageNet). We also correct implausible values during the execution of DNN inference to improve the inference accuracy (Sect. 3.5).

---

[2] Regularization is a technique that makes slight modifications to the training algorithm such that the DNN model generalizes better.

3.3   DNN Error Tolerance Characterization

EDEN aims to guarantee that the accuracy of a DNN meets the minimum value required by the user. To this end, EDEN characterizes the boosted DNN (obtained from our boosting mechanism in Sect. 3.2) to find the maximum tolerable bit error rate (BER) by progressively decreasing the approximate DRAM parameters, i.e., voltage and latency. EDEN performs either a *coarse-grained* or a *fine-grained* DNN error tolerance characterization.

Coarse-Grained Characterization EDEN's coarse-grained characterization determines the highest BER that can be applied uniformly to the entire DNN, while meeting the accuracy requirements of the user. This characterization is useful for mapping the DNN to commodity systems (see Sect. 3.4) that apply reduced DRAM parameters to an entire DRAM module (without fine-grained control).

To find the highest BER that satisfies the accuracy goal, our coarse-grained characterization method performs a logarithmic-scale binary search on the error rates. We can use binary search because we found that DNN error tolerance curves are monotonically decreasing. To adjust the BER while doing this characterization, EDEN can either (1) tune the parameters of approximate DRAM or (2) use DRAM error models for injecting bit errors into memory locations (see Sect. 4). EDEN optimizes the error resiliency of a DNN by repeating cycles of DNN error tolerance boosting (Sect. 3.2), coarse-grained DNN characterization, and DNN to DRAM mapping (Sect. 3.4) until the highest tolerable BER stops improving. We evaluate our coarse-grained characterization mechanism in Sect. 6.5.

Fine-Grained Characterization EDEN can exploit variation in the error tolerances of different DNN data types by clustering the data according to its error tolerance level, and assigning each cluster to a different DRAM partition whose error rate matches the error tolerance level of the cluster (see Sect. 3.4). For example, we find that the first and the last convolutional layers have tolerable BERs 2-3x smaller than the average middle layer in a DNN.

To conduct a fine-grained DNN characterization, EDEN searches for the highest tolerable BER of each weight and IFM that still yields an acceptable DNN accuracy. This search space is exponential with respect to the DNN's layer count. To tackle the search space challenge, EDEN employs a DNN data sweep procedure that performs iterations over a list of DNN data types. The mechanism tries to increase the tolerable error rate of a data type by a small amount, and tests if the DNN still meets the accuracy requirements. When a DNN data type cannot tolerate more increase in error rate, it is removed from the sweep list. We evaluate our fine-grained characterization mechanism in Sect. 6.6.

Effect of Pruning EDEN does not include pruning (Sect. 2.1) as part of its boosting routine due to two observations. First, we find that DNN sparsification does not improve the error tolerance. Our experiments show that when we create 10%, 50%, 75%, and 90% sparsity through energy-aware pruning [39], error tolerance of FP32 and int8 DNNs, DNN error tolerance does not improve significantly. Second, the zero values in the network, which increase with pruning, are sensitive to memory error perturbations.

3.4 DNN to DRAM Mapping

After characterizing the error tolerance of each DNN data type, EDEN maps each data type to the appropriate DRAM partition (with the appropriate voltage and latency parameters) that satisfies the data type's error tolerance. Our mechanism aims to map a data type that is very tolerant (intolerant) to errors into a DRAM partition with the highest (lowest) BER, matching the error tolerance of the DNN and the BER of the DRAM partition as much as possible.

DRAM Bit Error Rate Characterization To obtain the BER characteristics of a DRAM device (both in aggregate and for each partition), we perform reduced voltage and reduced latency tests for a number of data patterns. For each voltage level, we iteratively test two consecutive rows at a time. We populate these rows with inverted data patterns for the worst-case evaluation. Then, we read each bit with reduced timing parameters (e.g., tRCD). This characterization requires fine-grained control of the DRAM timing parameters and supply voltage level. EDEN's characterization mechanism is very similar to experimental DRAM characterization mechanisms proposed and evaluated in prior works for DRAM voltage [35] and DRAM latency [33].

Coarse-Grained DNN to DRAM Module Mapping All DNN data types stored within the same DRAM module are exposed to the *same* DRAM voltage level and timing parameters. These parameters are tuned to produce a bit error rate that is tolerable by *all* DNN data types that are mapped to the module.

Under coarse-grained mapping, the application does *not* need to be modified. Algorithms used in DNN inference are oblivious to the DRAM mapping used by the memory controller. The memory controller maps all inference-related requests to the appropriate approximate DRAM module. Data that cannot tolerate bit errors at any reduced voltage and latency levels is stored in a separate DRAM module whose voltage and latency parameters follow the manufacturer specifications.

Coarse-grained mapping can be easily supported by existing systems that allow the modification of V_{dd} and/or $t_{RCD/RP}$ parameters in the BIOS across the entire DRAM module. Section 5 describes the simple hardware changes required to support coarse-grained mapping. We evaluate our coarse-grained mapping mechanism in Sect. 6.5.

Fine-Grained DNN to DRAM Module Mapping DNN data types stored in different DRAM partitions can be exposed to *different* DRAM voltage levels and/or timing parameters. DRAM can be partitioned at chip, rank, bank, or subarray level granularities. Algorithm 1 describes our algorithm for fine-grained mapping of DNN data to DRAM partitions. Our algorithm uses rigorous DRAM characterization and

DNN characterization to iteratively assign DNN data to DRAM partitions in three basic steps. First, our mechanism looks for DRAM partitions that have BERs lower than the tolerable BER of a given DNN data type. Second, we select the DRAM partition with the largest parameter reduction that meets the BER requirements. Third, if the partition has enough space available, our mechanism assigns the DNN data type to the DRAM partition. We evaluate our fine-grained mapping mechanism in Sect. 6.6.

A system that supports fine-grained mapping requires changes in the memory controller (for voltage and latency adjustment) and in DRAM (for only voltage adjustment). We describe the hardware changes required to support fine-grained mapping in Sect. 5.

3.5 DNN Inference with Approximate DRAM

EDEN generates a boosted DNN for running inference in a target system that uses approximate DRAM. EDEN does not require any modifications in DNN inference hardware, framework, or algorithm, except for *correcting implausible values*. Similar to what happens in our curricular retraining (Sect. 3.2), a single bit error in the exponent bits of a floating point value can cause *accuracy collapse* during DNN inference. We use the same mechanism for correcting implausible values in our curricular retraining mechanism (i.e., we zero the values that are outside of a predefined threshold range) to avoid accuracy collapse caused by bit errors introduced by approximate DRAM during DNN inference.

Algorithm 1 Fine-grained DNN to DRAM mapping

```
1   function DNN_to_DRAM_Mapping(DNN_characterization,
        DRAM_characterization):
2       sorted_data = sort_DNN_data(DNN_characterization)
3       for (target_BER, DNN_data) in sorted_data:
4           # Find the DRAM partition that has the least
                voltage/latency at target_BER, and can fit the
                DNN_data
5           for DRAM_partition in DRAM_characterization
6               partition_params =
                    get_voltage_latency(DRAM_partition, target_BER)
7               if DNN_data.size < DRAM_partition.size :
8                   if partition_params < best_parameters:
9                       best_parameters = partition_params
10                      chosen_partition = DRAM_partition
11                      DRAM_partition.size -= DNN_data.size
12          final_mapping[chosen_partition].append(DNN_data)
13      return final_mapping
```

---

4 Enabling EDEN with Error Models

EDEN requires extensive characterization of the target approximate DRAM device for boosting DNN error tolerance (Sect. 3.2), characterization of DNN error tolerance (Sect. 3.3), and mapping of the DNN to the approximate DRAM device (Sect. 3.4). However, applying EDEN in a target system where DNN inference can be performed is not always feasible or practical. For example, a low-cost DNN inference accelerator [7] might perform very slowly when executing our curricular retraining mechanism, because it is *not* optimized for training. Similarly, the target hardware might not be available, or might have very limited availability (e.g., in the pre-production phase of a new approximate hardware design).

To solve this problem and enable EDEN even when target DRAM devices are not available for characterization, we propose to execute the EDEN framework in a system that is different from the target approximate system. We call this idea *EDEN offloading*. The main challenge of offloading EDEN to a different system is how to faithfully emulate the errors injected by the *target* approximate DRAM into the DNN. To address this challenge, we analyze many works that study DRAM error patterns [33–35, 40, 41], and we propose to use four different error models that are representative of most of the error patterns that are observed in real approximate DRAM modules.

EDEN's DRAM Error Models EDEN uses four probabilistic error models that closely fit the error patterns observed in a real approximate DRAM module. Our models contain information about the location of weak cells in the DRAM module, which is used to decide the spatial distribution of bit errors during DNN error tolerance boosting. We create four different types of error models from the data we obtain based on our characterization of existing DRAM devices using SoftMC [42] and a variety of DDR3 and DDR4 DRAM modules. Our error models are consistent with the error patterns observed by prior works [33–35, 40]. In addition, our error models are parameterizable and can be tuned to model individual DRAM chips, ranks, banks, and subarrays from different vendors.

- **Error Model 0:** the bit errors follow a uniform random distribution across a DRAM bank. Several prior works observe that reducing activation latency (t_{RCD}) and precharge latency (t_{RP}) can cause randomly distributed bit flips due to manufacturing process variation at the level of DRAM cells [33, 41]. We model these errors with two key parameters: (1) P is the percentage of *weak cells* (i.e., cells that fail with reduced DRAM parameters), and (2) F_A is the probability of an error in any weak cell. Such uniform random distributions are already observed in prior work [43].
- **Error Model 1:** the bit errors follow a vertical distribution across the bitlines of a DRAM bank. Prior works [33, 35, 41] observe that some bitlines experience more bit flips than others under reduced DRAM parameters due to: (1) manufacturing process variation across sense amplifiers [33, 35], and (2) design-induced latency variation that arises from the varying distance between different bitlines and the

row decoder [41]. We model this error distribution with two key parameters: (1) P_B is the percentage of weak cells in bitline B, and (2) F_B is the probability of an error in the weak cells of bitline B.

- **Error Model 2:** the bit errors follow a horizontal distribution across the wordlines of a DRAM bank. Prior works [33, 35, 41] observe that some DRAM rows experience more bit flips than others under reduced DRAM parameters due to (1) manufacturing process variation across DRAM rows [33, 35], and (2) design-induced latency variation that arises from the varying distance between different DRAM rows and the row buffer [41]. We model this error distribution with two key parameters: (1) P_W is the percentage of weak cells in wordline W, and (2) F_W is the probability of an error in the weak cells of wordline W.
- **Error Model 3:** the bit errors follow a uniform random distribution that depends on the content of the cells (i.e., this is a data-dependent error model). Figure 5 illustrates how the bit error rates depend on the data pattern stored in DRAM, for reduced voltage (top) and reduced t_{RCD} (bottom). We observe that 0-to-1 flips are more probable with t_{RCD} scaling, and 1-to-0 flips are more probable with voltage scaling. Prior works provide rigorous analyses of data patterns in DRAM with reduced voltage [35] and timing parameters [33] that show results similar to ours. This error model has three key parameters: (1) P is the percentage of weak cells, (2) F_{V1} is the probability of an error in the weak cells that contain a 1 value, and (3) F_{V0} is the probability of an error in the weak cells that contain a 0 value.

Model Selection EDEN applies a maximum likelihood estimation (MLE) procedure to determine (1) the parameters (P, F_A, P_B, F_B, P_W, F_W, F_{V1} and F_{V0}) of each error model, and (2) the error model that is most likely to produce the errors observed in the real approximate DRAM chip. In case two models have very similar probability of producing the observed errors, our selection mechanism chooses Error Model 0 if possible, or one of the error models randomly otherwise. Our selection mechanism favors Error Model 0 because we find that it the is error model that performs better. We observe that generating and injecting errors by software with Error Model 0 in both DNN retraining and inference is 1.3x faster than injecting errors with other error models in our experimental setup. We observe that Error Model 0 provides (1) a reasonable approximation of Error Model 1, if $max(F_B) - min(F_B) < 0.05$ and $P_B \approx P$, and (2) a reasonable approximation of Error Model 2, if $max(F_W) - min(F_W) < 0.05$ and $P_W \approx P$.

Handling Error Variations Error rates and error patterns depend on two types of factors. First, factors intrinsic to the DRAM device. The most common intrinsic factors are caused by manufacturer [35], chip, and bank variability [37, 40]. Intrinsic factors are established at DRAM fabrication time. Second, factors extrinsic to the DRAM device that depend on environmental or operating conditions. The most common extrinsic factors are aging [44], data values [45], and temperature [46]. Extrinsic factors can introduce significant variability in the error patterns.

EDEN can capture intrinsic factors in the error model with a unique DRAM characterization pass. However, capturing extrinsic factors in the error model is

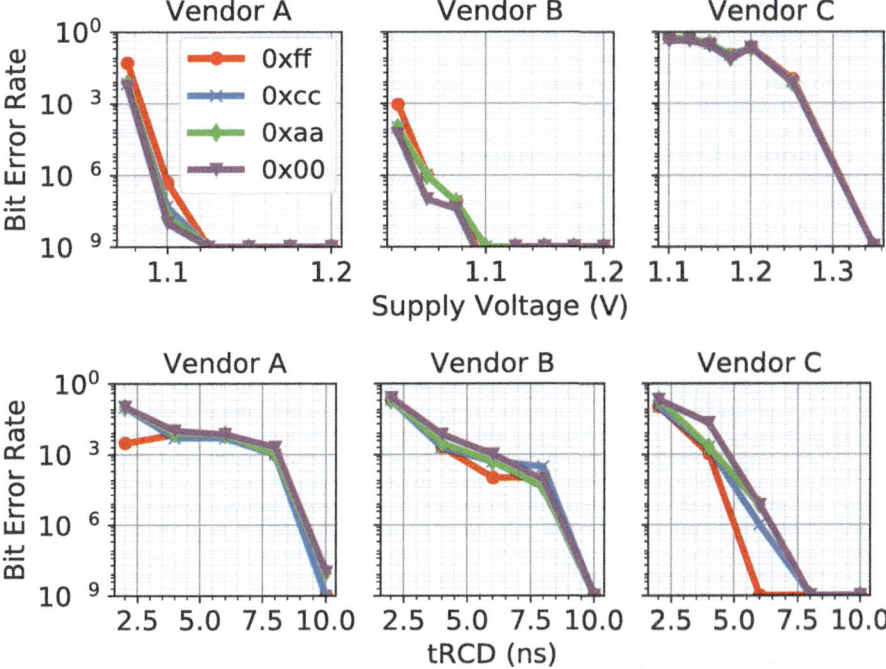

Fig. 5 Bit error rates depend on the data pattern stored in DRAM, with reduced supply voltage [35] and reduced t_{RCD} [33, 35, 37, 41], motivating Error Model 3. Data is based on DDR3 DRAM modules from three major vendors. Adapted from [23]

more challenging. Our DNN models capture three factors extrinsic to the DRAM device.

First, EDEN can capture data-dependent errors by generating different error models for different DNN models (i.e., different IFM and weight values in memory). For each DNN model, EDEN stores the actual weight and IFM values in the target approximate DRAM before characterization to capture data dependencies.

Second, EDEN can capture temperature variations by generating different error models for the same approximate DRAM operating at different temperatures. Errors increase with higher temperatures [46], so the model must match the temperature of DNN inference execution.

Third, EDEN can capture DRAM aging by periodically regenerating new error models. In our experiments with real DRAM modules, we find that the errors are temporally consistent and stable for days of continuous execution (with ±5° C deviations from the profiling temperature), without requiring re-characterization. Prior work [37] reports similar results.

We find in our evaluation that our error models are sufficiently expressive to generate a boosted DNN that executes on real approximate DRAM with minimal accuracy loss (Sect. 6.4). Our four error models are also sufficiently expressive to encompass the bit error models proposed in prior work [47].

5 Memory Controller Support

To obtain the most out of EDEN, we modify the memory controller to (1) correct implausible values during both curricular retraining and DNN inference, (2) support coarse-grained memory mapping, and (3) support fine-grained memory mapping.

Hardware Support for Correcting Implausible Values We correct implausible values that cause accuracy collapse during both curricular retraining (Sect. 3.2) and DNN inference (Sect. 3.5). Our mechanism (1) compares a loaded value to an upper-bound and a lower-bound threshold, and (2) sets the value to zero (i.e., supplies the load with a zero result) in case the value is out of bounds. Because these operations are done for *every* memory access that loads a DNN value, it can cause significant performance degradation if performed in software. To mitigate this issue, we incorporate simple hardware logic in the memory controller that we call *bounding logic*. Our bounding logic (1) compares the exponent part of the loaded floating point value to DNN-specific upper-bound and lower-bound thresholds, and (2) zeros the input value if the value is out of bounds. In our implementation, the latency of this logic is only 1 cycle and its hardware cost is negligible.

Enabling Coarse-Grained Mapping Coarse-grained mapping applies the same voltage and timing parameters to the entire DRAM for executing a particular DNN workload. However, different DNN workloads might require applying different sets of DRAM parameters to maximize energy savings and performance. In many existing commodity systems, the memory controller sets the DRAM voltage and the timing parameters at start-up, and it is not possible to change them at runtime. To overcome this limitation, the memory controller requires minimal hardware support for changing the DRAM parameters of each DRAM module at runtime.

Enabling Fine-Grained Mapping Fine-grained mapping applies different voltage and/or timing parameters to different DRAM partitions.

To apply different voltages to different memory partitions, EDEN (1) adopts the approach used by Voltron [35] to implement a robust design for voltage scaling at the bank granularity based on modest changes to the power delivery network, and (2) tracks which memory partition is operating at what voltage. To implement this mechanism in commodity DDR4/LPDDR4 chips with 16/32 banks, EDEN requires at most 32B of metadata to represent all 8-bit voltage step values.

To apply different timing parameters to different memory partitions, EDEN requires memory controller support for (1) configuring the target memory partition to operate at specific timing parameters, and (2) tracking which memory partition is operating at what latency. For the timing parameter we tested in our evaluation

(t_{RCD}), 4-bits are enough to encode all possible values of the parameter with enough resolution.

It is sufficient for EDEN to split DRAM into at most 2^{10} partitions, because most commonly used DNN architectures have at most 1024 different types of error-resilient IFMs and weights. EDEN requires 1 KB of metadata to support 2^{10} partitions. To support mappings at subarray level granularity (i.e., the finest supported granularity), EDEN needs a larger amount of metadata. For example, for an 8 GB DDR4 DRAM module with 2048 subarrays, EDEN needs to store 2 KB of metadata.

6 DNN Accuracy Evaluation

In this section, we evaluate EDEN's ability to improve DNN accuracy in approximate DRAM. We explain our methodology (Sect. 6.1), evaluate the accuracy of our error models (Sect. 6.2), evaluate the error tolerance of the DNN baselines (Sect. 6.3), and analyze the accuracy of our curricular retraining mechanism (Sect. 6.4).

6.1 Methodology

We use an FPGA-based infrastructure running SoftMC [42] to reduce DRAM voltage and timing parameters. SoftMC allows executing memory controller commands on individual banks, and modifying t_{RCD} and other DRAM timing parameters. We perform all our experiments at room temperature. Using this infrastructure, we can obtain characteristics of real approximate DRAM devices. However, our infrastructure also has some performance limitations caused by delays introduced with SoftMC's FPGA buffering, host-FPGA data transmission, and instruction batching on the FPGA.

To overcome these performance limitations, we emulate real approximate DRAM modules by using the error models described in Sect. 4. To ensure that our evaluation is accurate, we validate our error models against real approximate DRAM devices (Sect. 6.2).

We incorporate EDEN's error models into DNN inference libraries by following the methodology described in Fig. 6. We create a framework on top of PyTorch [48] that allows us to modify the loading of weights and IFMs. Our PyTorch implementation (1) injects errors into the original IFM and weight values using our DRAM error models, and (2) applies our mechanism to correct implausible values caused by bit errors in IFMs and weights (Sect. 3.2). Our DRAM error models are implemented as custom GPU kernels for efficient and simple integration into PyTorch. This simulation allows us to obtain DNN accuracy estimates 80–90x faster than with the SoftMC infrastructure.

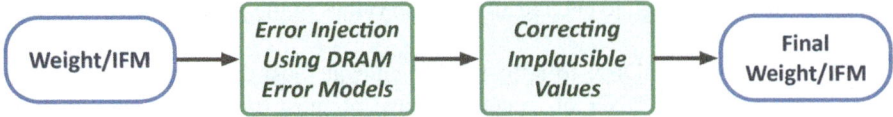

Fig. 6 Methodology to incorporate DRAM error models in the DNN evaluation framework. Adapted from [23]

Table 1 DNN models used in our evaluations. The listed total model size and summed IFM+weight sizes are for the FP32 variant of each model

Model	Dataset	Model size	IFM+Weight size
ResNet101	CIFAR10	163.0 MB	100.0 MB
MobileNetV2	CIFAR10	22.7 MB	68.5 MB
VGG-16	ILSVRC2012	528.0 MB	218.0 MB
DenseNet201	ILSVRC2012	76.0 MB	439.0 MB
SqueezeNet1.1	ILSVRC2012	4.8 MB	53.8 MB
AlexNet	CIFAR10	233.0 MB	208.0 MB
YOLO	MSCOCO	237.0 MB	360.0 MB
YOLO-Tiny	MSCOCO	33.8 MB	51.3 MB
LeNet*	CIFAR10	1.65 MB	2.30 MB

* we use this small model in some evaluations where the experimental setup does not support large models

DNN Baselines We describe the DNN baselines that we use in the evaluation of the three EDEN steps (Sects. 3.2, 3.3, and 3.4). Table 1 lists the eight modern and commonly used DNN models we evaluate. We target both small (e.g., CIFAR-10) and large-scale (e.g., ILSVRC2012) image classification datasets. ResNet101 [49], VGG-16 [50], and DenseNet201 [51] models are top-five winners of past ImageNet ILSVRC competitions. We use Google MobileNetV2 [6] to test smaller, mobile-optimized networks that are widely used on mobile platforms, and SqueezeNet [16] to test embedded, real-time applications. Table 1 also shows the summed sizes of all IFMs and weights of each network for processing one input, which is a good indicator of the memory intensity of each DNN model.

Table 2 shows the accuracy we obtain in our experiments for our baseline networks across four different numeric precisions (int4, int8, int16 and FP32), using *reliable* commodity DRAM. We quantize using the popular symmetric linear DNN quantization scheme [52]. This quantization scheme applies weight-dependent affine scaling to linearly map weights into the range $[-2^{b-1}, 2^{b-1} - 1]$, where b is the target model weight bit precision. YOLO and YOLO-Tiny's framework only support int8 and FP32 numeric precisions.

Two of the models, DenseNet201 and SqueezeNet1.1, suffer from accuracy collapse at 4-bit precision. We did not use hyper-parameter tuning in our baselines or subsequent experiments. All results use the default DNN architectures and learning rates.

Table 2 Baseline accuracies of the networks used in our evaluation with reliable DRAM memory (no bit errors) using different numeric precisions

Model	int4	int8	int16	FP32
ResNet101 [49]	89.11%	93.14%	93.11%	94.20%
MobileNetV2 [6]	51.00%	70.44%	70.46%	78.35%
VGG-16 [50]	59.05%	70.48%	70.53%	71.59%
DenseNet201 [51]	0.31%	74.60%	74.82%	76.90%
SqueezeNet1.1 [16]	8.07%	57.07%	57.39%	58.18%
AlexNet [2]	83.13%	86.04%	87.21%	89.13%
YOLO* [53]	–	44.60%	–	55.30%
YOLO-Tiny* [53]	–	14.10%	–	23.70%
LeNet [21]	–	61.30%	–	67.40%

* these models use mean average precision (mAP) instead of the accuracy metric

6.2   Accuracy Validation of the Error Models

EDEN uses errors obtained from real DRAM devices to build and select accurate error models. We profile the DRAM (1) before running DNN inference, and (2) when the environmental factors that can affect the error patterns change (e.g., when temperature changes). We find that an error model can be accurate for many days if the environmental conditions do not change significantly, as also observed in prior work [37, 41, 54].

We derive our probabilistic error models (Sect. 4) from data obtained from eight real DRAM modules. We use the same FPGA infrastructure as the one described in Sect. 6.1. We find that complete profiling of a 16-bank, 4 GB DDR4 DRAM module takes under 4 minutes in our evaluation setup.

We validate our error models by comparing the DNN accuracy obtained after injecting bit errors using our DRAM error models to the accuracy obtained with each real approximate DRAM module. Figure 7 shows an example of the DNN accuracy obtained using DRAM modules from three major vendors with reduced voltage and t_{RCD}, and the DNN accuracy obtained using our Error Model 0. We use Error Model 0 because it is the model that fits better the errors observed in the three tested DRAM modules. Our main observation is that the DNN accuracy obtained with our model is very similar to that obtained with real approximate DRAM devices. We conclude that our error models mimic very well the errors observed in real approximate DRAM devices.

6.3   Error Tolerance of Baseline DNNs

To better understand the baseline error tolerance of each DNN (before boosting the error tolerance), we examine the error tolerance of the baseline DNNs. This also shows us how differences in quantization, best-fit error model, and BER can potentially affect the final DNN accuracy.

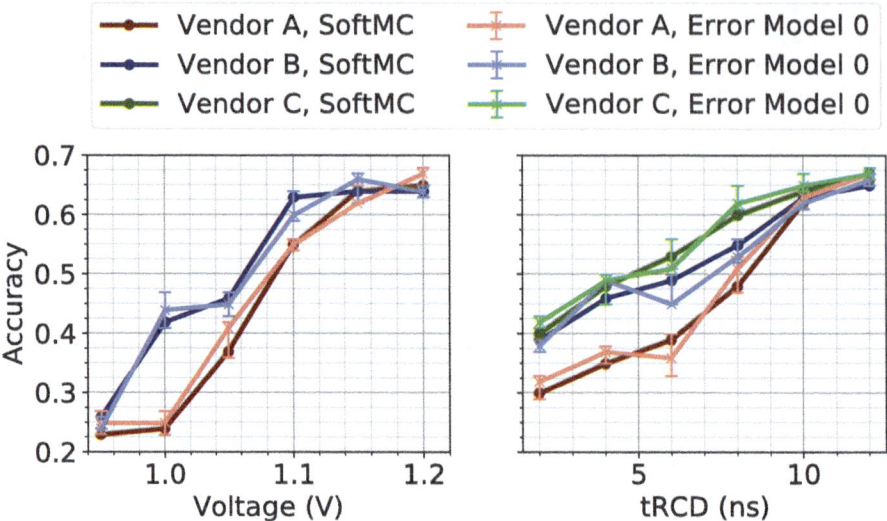

Fig. 7 LeNet/CIFAR-10 accuracies obtained using real approximate DRAM devices (via SoftMC) and using our Error Model 0. Error bars show the 95% confidence interval of Error Model 0. Adapted from [23]

Figure 8 shows the accuracy of ResNet101 at different precision levels and BERs using all four error models. We see that all DNNs exhibit an accuracy drop at high BER ($>10^{-2}$), but different error models cause the drop-off for all DNNs to be higher or lower. This is rooted in how each error model disperses bit errors into the DNN IFMs and weights. A good example of this is Error Model 1, which exhibits the most early and extreme drop-offs, especially for FP32 DNNs. We find that the cause of this is that, in our experimental setup, IFMs and weights are aligned in DRAM, so the MSBs of different DNN data types are mapped to the same bitline B. If the percentage of weak cells in bitline B (P_B) is high, the DNN suffers many MSB failures. However, Error Model 0 distributes these weak cell failures uniformly and randomly across the bank, causing far fewer MSB failures. In general, the way in which each error model captures the distribution of weak cells across data layout in memory greatly affects its impact on the error curve.

Quantization Precision also affects the error model and the error tolerance curve. For example, in Error Model 2, we observe that the int-4 DNN has the weakest error tolerance curve. We find that this is because Error Model 2 clusters weak cells along a row: a large number of neighboring 4-bit values end up corrupted when Error Model 2 indicates a weak wordline. This is in contrast to larger precisions, which might have numbers distributed more evenly across rows, or error models that do not capture error locality (e.g., Error Model 0). In general, we find that clusters of erroneous values cause significant problems with accuracy (the errors compound faster as they interact with each other in the DNN). Such locality of errors is more

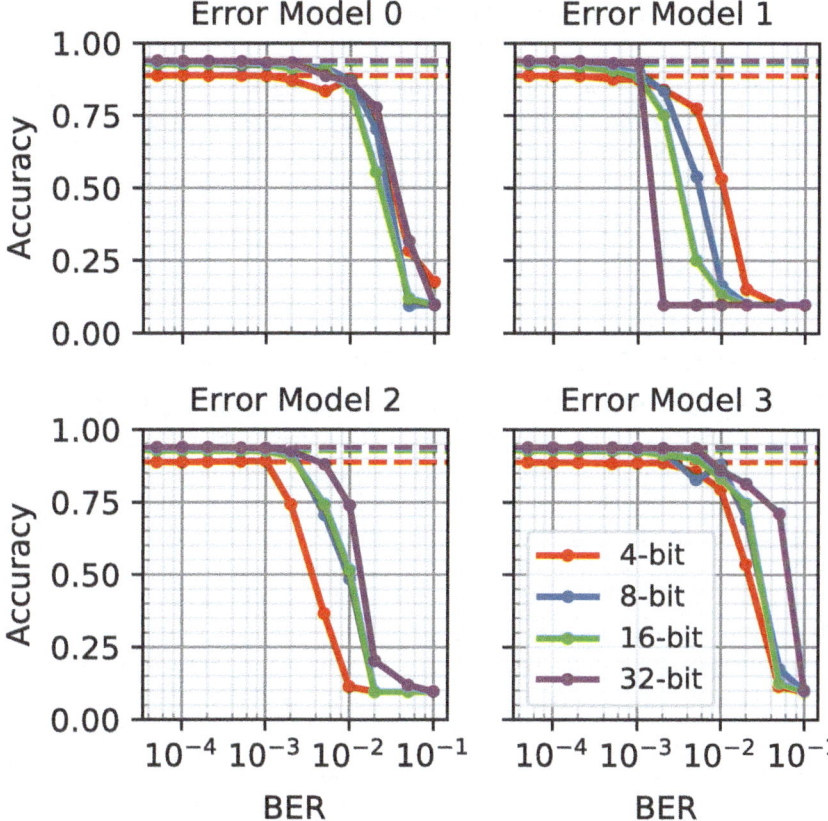

Fig. 8 ResNet101 accuracy across different BERs (*x*-axis) and quantization levels when we use four error models to inject bit errors. We fit the parameters of the error models to the errors observed by reducing $tRCD$ in a real DRAM device from Vendor A. Adapted from [23]

common in low-bitwidth precisions and with spatial correlation-based error models (Error Models 1 and 2).

DNN Size We observe that larger DNNs (e.g., VGG16) are more error resilient. Larger models exhibit an accuracy drop-off at higher BER ($> 10^{-2}$) as compared to smaller models (e.g., SqueezeNet1.1, $< 10^{-3}$). These results are not plotted.

Accuracy Collapse We can observe the accuracy collapse phenomenon caused by implausible values (see Sect. 3.2) when we increase the bit error rate over 10^{-6} in large networks. These implausible values propagate, and in the end, they cause accuracy collapse in the DNN.

6.4 Curricular Retraining Evaluation

We run DNN inference on real DRAM devices using the boosted DNN model generated by our curricular retraining mechanism. To our knowledge, this is the first demonstration of DNN inference on real approximate memory. We also evaluate our curricular retraining mechanism using our error models (see Sect. 4).

Experimental Setup We evaluate curricular retraining using real DRAM devices by running LeNet [21] on the CIFAR-10 [55] validation dataset. We use SoftMC [42] to scale V_{DD} and t_{RCD} on an FPGA-based infrastructure connected to a DDR3 DRAM module. We also evaluate curricular retraining using our error models by running ResNet [49] on the CIFAR-10 validation dataset.

Results with Real DRAM Figure 9 shows the accuracy of (1) baseline LeNet without applying any retraining mechanism (Baseline), and (2) LeNeT boosted with our curricular retraining mechanism (Boosted), as a function of DRAM supply voltage and t_{RCD}. We make two observations. First, EDEN's boosted LeNet allows a voltage reduction of ∼0.25 V and a t_{RCD} reduction of 4.5 ns, while maintaining accuracy values equivalent to those provided by nominal voltage (1.35 V) and nominal t_{RCD} (12.5 ns). Second, the accuracy of baseline LeNet decreases very quickly when reducing voltage and t_{RCD} below the nominal values. We conclude that our curricular retraining mechanism can effectively boost the accuracy of LeNeT on approximate DRAM with reduced voltage and t_{RCD}.

Results with Error Models Figure 10 (left) shows an experiment that retrains ResNet101 with two different models: (1) a good-fit error model (that closely matches the tested device) and (2) a poor-fit error model. We make two observations. First, retraining using a poor-fit error model (red), yields little improvement over the baseline (no retraining, green). Second, retraining with a good-fit error model (blue)

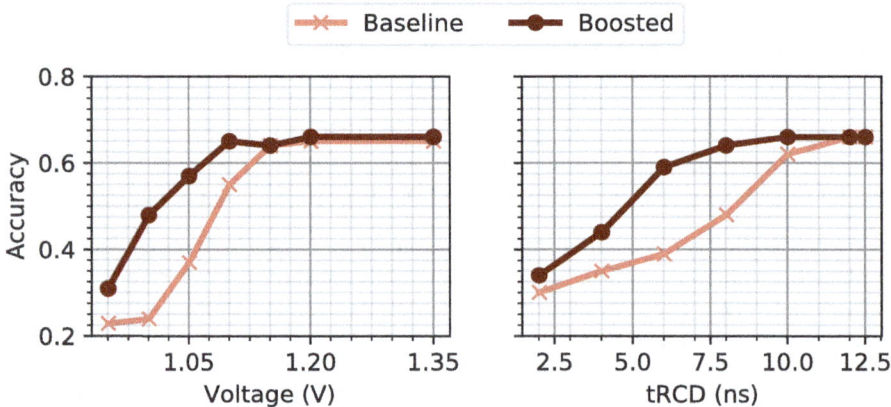

Fig. 9 LeNet accuracy using baseline and boosted DNNs. Adapted from [23]

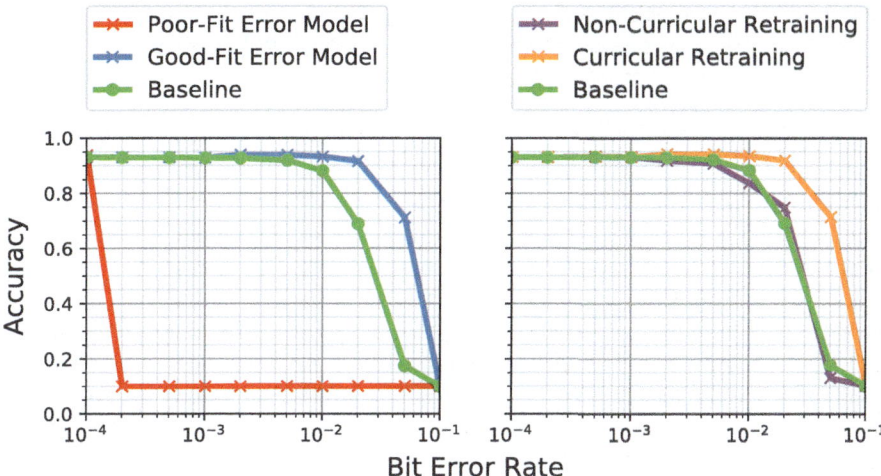

Fig. 10 Accuracy of boosted ResNet101 DNNs in presence of memory errors. Left: accuracy of poor-fit and good-fit error models. Right: accuracy of non-curricular and curricular retraining using a good-fit error model. Adapted from [23]

improves BER at the 89% accuracy point by >10x (shifting the BER curve right). We conclude that using a good-fit error model in the retraining mechanism is critical to avoid accuracy collapse.

Figure 10 (right) shows the effectiveness of our curricular retraining mechanism using a good-fit error model. We make two observations. First, the accuracy of the DNN with regular retraining (purple) collapses, compared to the baseline DNN (no retraining, green). Second, the DNN trained with our curricular retraining (orange) exhibits a boosted error tolerance. We conclude that our curricular retraining mechanism is effective at boosting the DNN accuracy in systems that use approximate DRAM.

Running this retraining process for 10–15 epochs is sufficient to boost tolerable BERs by 5–10x to achieve the same DNN accuracy as the baseline DNN executed in DRAM with nominal parameters. For our ResNet101 on CIFAR-10 with an NVIDIA Tesla P100, this one-time boosting completes within 10 min.

6.5 Coarse-Grained DNN Characterization and Mapping

In this section, we show the results of EDEN's coarse-grained DNN characterization (see Sect. 3.3) and how the target DNN model maps to an approximate DRAM with optimized parameters for a target accuracy degradation of <1%.

Characterization Table 3 shows the DNN's maximum tolerable BER for eight DNN models with FP32 and int8 numeric precisions.

Table 3 Maximum tolerable BER for each DNN using EDEN's coarse-grained characterization, and DRAM parameter reduction to achieve the maximum tolerable BER. Nominal parameters are $V_{DD} = 1.35V$ and $t_{RCD} = 12.5\,ns$

Model	FP32			int8		
	BER	ΔV_{DD}	Δt_{RCD}	BER	ΔV_{DD}	Δt_{RCD}
ResNet101	4.0%	$-0.30\,V$	$-5.5\,ns$	4.0%	$-0.30\,V$	$-5.5\,ns$
MobileNetV2	1.0%	$-0.25\,V$	$-1.0\,ns$	0.5%	$-0.10\,V$	$-1.0\,ns$
VGG-16	5.0%	$-0.35\,V$	$-6.0\,ns$	5.0%	$-0.35\,V$	$-6.0\,ns$
DenseNet201	1.5%	$-0.25\,V$	$-2.0\,ns$	1.5%	$-0.25\,V$	$-2.0\,ns$
SqueezeNet1.1	0.5%	$-0.10\,V$	$-1.0\,ns$	0.5%	$-0.10\,V$	$-1.0\,ns$
AlexNet	3.0%	$-0.30\,V$	$-4.5\,ns$	3.0%	$-0.30\,V$	$-4.5\,ns$
YOLO	5.0%	$-0.35\,V$	$-6.0\,ns$	4.0%	$-0.30\,V$	$-5.5\,ns$
YOLO-Tiny	3.5%	$-0.30\,V$	$-5.0\,ns$	3.0%	$-0.30\,V$	$-4.5\,ns$

We observe that the maximum tolerable BER demonstrates significant variation depending on the DNN model. For example, YOLO tolerates 5% BER and SqueezeNet tolerates only 0.5%. We conclude that (1) the maximum tolerable BER highly depends on the DNN model, and (2) DNN characterization is required to optimize approximate DRAM parameters for each DNN model.

Mapping EDEN maps each DNN model to an approximate DRAM module that operates with the maximum reduction in voltage (ΔV_{DD}) and t_{RCD} (Δt_{RCD}) that leads to a BER below the maximum DNN tolerable BER for that DNN model. Table 3 shows the maximum reduction in DRAM voltage (ΔV_{DD}) and t_{RCD} (Δt_{RCD}) that causes a DRAM BER below the maximum tolerable BER, for a target DRAM module from vendor A. The nominal DRAM parameters for this DRAM module are $V_{DD} = 1.35$ V and $t_{RCD} = 12.5\,ns$. We make two observations. First, the tolerable BER of a network is directly related to the maximum tolerable V_{DD} and t_{RCD} reductions. Second, the reductions in V_{DD} and t_{RCD} are very significant compared to the nominal values. For example, EDEN can reduce voltage by 26% and t_{RCD} by 48% in YOLO while maintaining the DNN accuracy to be within 1% of the original accuracy.

6.6   Fine-Grained DNN Characterization and Mapping

Characterization We characterize the ResNet101 DNN model with our fine-grained DNN characterization procedure (see Sect. 3.3). For each IFM and weight, we iteratively increase the bit error rate until we reach the maximum tolerable BER of the data type for a particular target accuracy degradation. We perform a full network retraining in each iteration. To reduce the runtime of our procedure, we sample 10% of the validation set during each inference run to obtain the accuracy estimate. We also bootstrap the BERs to the BER found in coarse-grained DNN

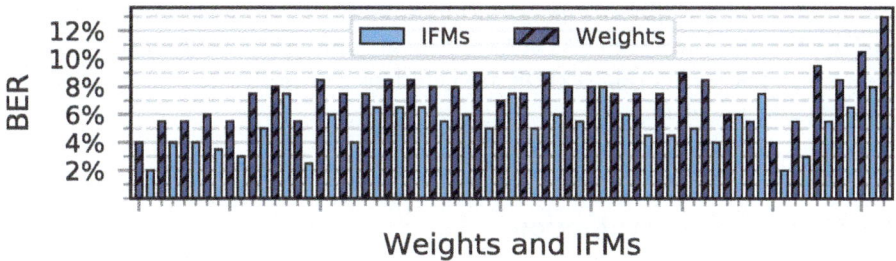

Fig. 11 Fine-grained characterization of the tolerable BERs of ResNet101 IFMs and weights. Deeper layers are on the right. Adapted from [23]

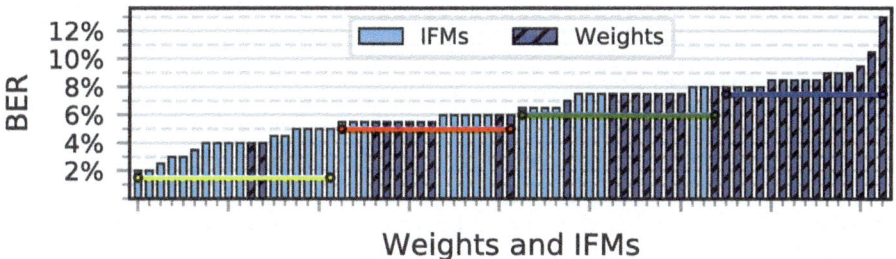

Fig. 12 Mapping of ResNet101 IFMs and weights into four partitions with different V_{DD} values (colored horizontal lines). Adapted from [23]

characterization and use a linear scale in 0.5 increments around that value. For ResNet101, this one-time characterization completes in one hour using an Intel Xeon CPU E3-1225.

Figure 11 shows the maximum tolerable BER for each IFM and weight in ResNet101 obtained with our fine-grained DNN characterization method (Sect. 3.3), assuming a maximum accuracy loss of <1%. Each bar in the figure represents the BER tolerance of an IFM or weight, and they are ordered by their depth in the DNN, going deeper from left to right. We make three observations. First, fine-grained characterization enables individual IFMs and weights to tolerate up to 3x BER (13% for the last weight) of the maximum tolerable BER of the coarse-grained approach (4% for ResNet101 in Table 3). Second, weights usually tolerate more errors than IFMs. Third, the maximum tolerable BER is smaller in the first layers than in the middle layers of the DNN. We conclude that fine-grained DNN characterization enables a significant increase in the maximum tolerable BER compared to coarse-grained characterization.

Mapping We map each individual IFM or weight into different DRAM partitions based on (1) the BER tolerance of each IFM and weight, and (2) the BER of each DRAM partition, using our algorithm in Sect. 3.4. Figure 12 shows an example that maps the ResNet101 IFMs and weights from Fig. 11 into 4 different DRAM partitions with different voltage parameters that introduce different BERs (four horizontal colored bars), following the algorithm in Sect. 3.4.

We conclude that the wide range of tolerable BERs across all ResNet101 data types enables the use of both (1) DRAM partitions with significant voltage reduction (e.g., horizontal red line), and (2) DRAM partitions with moderate voltage reduction (e.g., horizontal blue line).

7 System-Level Evaluation

We evaluate EDEN in three different DNN inference architectures: CPUs, GPUs, and inference accelerators.

7.1 CPU Inference

Experimental Setup We evaluate EDEN on top of a multi-core OoO CPU using the simulated core configuration listed in Table 4. We use ZSim [56] and Ramulator [57] to simulate the core and the DRAM subsystem, respectively. We use DRAMPower [58] to estimate energy consumption for DDR4 devices. We use a 2-channel, 32-bank 8GB DDR4-2133 DRAM device.

We use twelve different inference benchmarks: eight from the Intel OpenVINO toolkit [12] and four from the AlexeyAB-fork of the DarkNet framework. For each DNN, we study the FP32 and the int 8-quantized variant. We use 8-bit quantization in our baselines, because it is commonly used for production CPU workloads. We evaluate EDEN's coarse-grained DNN characterization procedure and target a <1% accuracy degradation. Table 3 lists the reduced V_{DD} and t_{RCD} values.

DRAM Energy Figure 13 shows the DRAM energy savings of EDEN, compared to a system with DRAM operating at nominal voltage and nominal latency. We make two observations. First, EDEN achieves significant DRAM energy savings across different DNN models. The average DRAM energy savings is 21% across all workloads, and 29% each for YOLO and VGG. Second, the DRAM energy savings

Table 4 Simulated system configuration

Cores	2 Cores @ 4.0 GHz, 32nm, 4-wide OoO,
	Buffers: 18-entry fetch, 128-entry decode,
	128-entry reorder buffer,
L1 Caches	32 KB, 8-way, 2-cycle, Split Data/Instr.
L2 Caches	512 KB per core, 8-way, 4-cycle, Shared Data/Instr.,
	Stream Prefetcher
L3 Caches	8 MB per core, 16-way, 6-cycle, Shared Data/Instr.,
	Stream Prefetcher
Main memory	8GB DDR4-2133 DRAM, 2 channels, 16 banks/channel

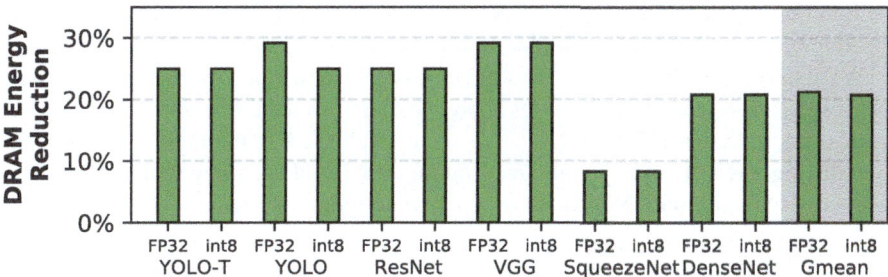

Fig. 13 DRAM energy savings of EDEN. We use FP32 and quantized int8 networks. Adapted from [23]

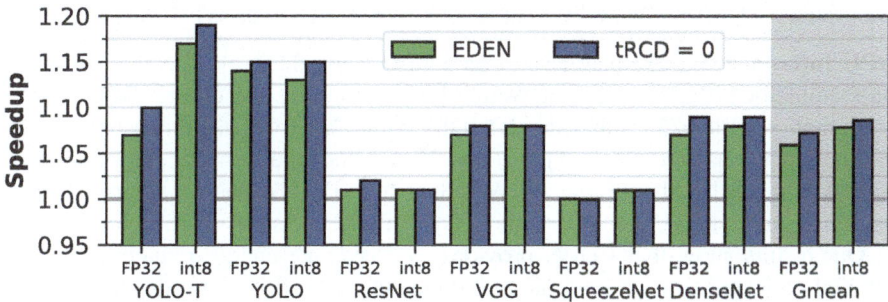

Fig. 14 Speedup of EDEN over baseline and versus a system with ideal activation latency. We use FP32 and an quantized int8 networks. Adapted from [23]

for FP32 and int8 are roughly the same, because the voltage reduction is very similar for both precisions (see Table 3).

We also perform evaluations for a target accuracy that is the same as the original. Our results show that EDEN enables an average DRAM energy reduction of 16% (up to 18%).

We conclude that EDEN is effective at saving DNN inference energy by reducing voltage while maintaining the DNN accuracy within 1% of the original.

Performance Figure 14 shows the speedup of EDEN when we reduce t_{RCD}, and the speedup of a system with a DRAM module that has ideal $t_{RCD} = 0$, compared to a system that uses DRAM with nominal timing parameters. We make three observations. First, YOLO DNNs exhibit high speedup with EDEN, reaching up to 17% speedup. The results of YOLO are better than the average because YOLO is more sensitive to DRAM latency. This is because some steps in YOLO (e.g., Non-Maximum Suppression, confidence, and IoU thresholding) perform arbitrary indexing into matrices that lead to random memory accesses, which cannot easily be predicted by the prefetchers. Second, the average speedup of EDEN (8%) is very close to the average speedup of the ideal system with $t_{RCD} = 0$ (10%). Third, we find that SqueezeNet1.1 and ResNet101 exhibit very little maximum theoretical speedup because they are not bottlenecked by memory latency.

We also perform evaluations for a target accuracy that is the same as the original. Our results show that EDEN enables an average performance gain of 4% (up to 7%).

We conclude that EDEN is effective at improving DNN inference performance by reducing DRAM latency while maintaining the DNN accuracy within 1% of the original, especially on DNNs that are sensitive to memory latency.

7.2 Accelerators

We evaluate EDEN on three different accelerators: GPU, Eyeriss [7], and TPU [20].

GPU Inference We evaluate EDEN on a GPU using the cycle-accurate GPGPU-Sim simulator [59]. We use GPUWattch [60] to evaluate the overall GPU energy consumption. Table 5 details the NVIDIA Titan X GPU model we use in our evaluation. We use the reduced t_{RCD} and V_{DD} values that provide <1% accuracy degradation (as listed in Table 3). We adapt four DarkNet-based binaries to run inference on the FP32/int8 YOLO and YOLO-Tiny DNNs.

Our results show that EDEN provides 37% average energy reduction (41.7% for YOLO-Tiny, and 32.6% for YOLO) compared to a GPU that uses DRAM with nominal parameters.

Our results also show that EDEN provides 2.7% average speedup (5.5% for the YOLO-Tiny, and 0% for YOLO) compared to a GPU that uses DRAM with nominal parameters. DRAM with ideal $tRCD$ ($t_{RCD} = 0$) provides 6% speedup for YOLO-Tiny and 2% speedup for YOLO. These results indicate that (1) the YOLO DNN family is not DRAM latency bound in our evaluation configuration, and (2) EDEN can achieve close to the ideal speedup of zero activation latency when the DNN is latency bound.

Neural Network Inference Accelerators We evaluate EDEN on Eyeriss [7] and Google's Tensor Processing Unit (TPU) [20] using the cycle-accurate SCALE-Sim simulator [61]. We use DRAMPower [58] to obtain DRAM energy consumption from memory traces produced by SCALE-Sim. We use the built-in int8 AlexNet and YOLO-Tiny models and their accelerator-specific dataflows. We use DRAM parameters that yield a maximum accuracy loss of 1% (Table 3). Table 6 details the

Table 5 Simulated NVIDIA Titan X GPU configuration

Shader core	28 SMs, 1417 MHz, 32 SIMT Width,
	64 Warps per SM, 4 GTO Schedulers per Core
Private L1 cache	24 KB per SMM, Cache Block Size 128B
Shared memory	96 KB, 32 Banks. Shared L2 Cache: 3 MB
Main memory	GDDR5, 2500 MHz, 6 channels, 24 chips

Table 6 Simulated Eyeriss and TPU configurations

	Eyeriss	TPU
Array	12×14 PEs	256×256 PEs
SRAM buffers	324 KB	24 MB
Main memory	4GB DDR4-2400	4GB DDR4-2400
	4GB LPDDR3-1600	4GB LPDDR3-1600

configuration of the Eyeriss and TPU inference accelerators. Eyeriss has an array of 12×14 processing elements (PEs) with a 324 KB SRAM buffer for all data types (i.e., IFMs, weights and OFMs), and the TPU has an array of 256×256 PEs with a 24 MB SRAM buffer for all data types. We evaluate both accelerators with DDR4 and LPDDR3 DRAM configurations, using AlexNet and YOLO-Tiny workloads.

Our results show that reducing the voltage level in DDR4 DRAM leads to significant DRAM energy reductions on both Eyeriss and TPU accelerators. EDEN provides (1) 31% average DRAM energy savings on Eyeriss (31% for YOLO-Tiny, and 32% for AlexNet), and (2) 32% average DRAM energy savings on TPU (31% for YOLO-Tiny, and 34% for AlexNet).

Our results with a reduced voltage level in LPDDR3 are similar to those with DDR4. EDEN provides an average DRAM energy reduction of 21% for both Eyeriss and TPU accelerators running YOLO-Tiny and AlexNet. By using the accelerator/network/cache/DRAM energy breakdown provided by the Eyeriss evaluations on AlexNet [7], we estimate that EDEN can provide 26.8% system-level energy reduction on fully connected layers and 7% system-level energy reduction on convolutional layers.

Our results with reduced t_{RCD} in LPDDR3 and DDR4 show that Eyeriss and TPU exhibit no speedup from reducing t_{RCD}. We observe that prefetchers are very effective in these architectures because the memory access patterns in the evaluated DNNs are very predictable.

8   Related Work

To our knowledge, EDEN is the first paper to propose a general framework that reduces energy consumption and increases performance of DNN inference by using approximate DRAM with reduced voltage and latency. EDEN introduces a new methodology to improve DNN's tolerance to approximate DRAM errors which is based on DNN error tolerance characterization and a new curricular retraining mechanism. We demonstrate the effectiveness of EDEN by using error patterns that occur in real approximate DRAM devices.

In this section, we discuss closely related work on (1) approximate computing hardware for DNN workloads, and (2) modifying DRAM parameters.

Approximate Computing Hardware for DNN Workloads Many prior works propose to use approximate computing hardware for executing machine learning workloads [62–74]. All these works propose techniques for improving DNN tolerance for different types of approximate hardware mechanisms and error injection rates. Compared to these works, EDEN is unique in (1) being the first work to use approximate DRAM with reduced voltage and latency, (2) being the first demonstration of DNN inference using error characterization of real approximate DRAM devices, (3) using a novel curricular retraining mechanism that is able to customize the DNN for tolerating *high* error rates injected by the target approximate DRAM, and (4) mapping each DNN data type to a DRAM partition based on the error tolerance of the DNN data type and the bit error rate of the DRAM partition. We classify related works on approximate hardware for DNN workloads into six categories.

First, works that reduce DRAM refresh to save DNN energy [62–64]. RANA [63] and St-DRC [62] propose to reduce DRAM refresh rate in the embedded DRAM (eDRAM) memory of DNN accelerators. Nguyen et al. [64] propose to apply similar refresh optimization techniques to off-chip DRAM in DNN accelerators. These mechanisms use customized retraining mechanisms to improve the accuracy of the DNN in the presence of a moderate amount of errors.

Second, works that study the error tolerance of neural networks to uniform random faults in SRAM memory [65, 66]. For example, Li et al. [65] analyze the effect of various numeric representations on error tolerance. MoRS [66] is an approximate undervolting fault model using real faults extracted from experimental undervolting studies on SRAMs to build the model. The authors inject the faults generated by MoRS into the on-chip memory of the DNN accelerator to evaluate the resilience of the system under the test.

Third, works that study approximate arithmetic logic in DNN workloads [67, 68]. ThUnderVolt [67] proposes to underscale the voltage of arithmetic elements. Salami et al. [68] present fault-mitigation techniques for neural networks that minimize errors in faulty registers and logic blocks with pruning and retraining.

Fourth, works that study approximate emerging memory technologies for neural network acceleration. Panda et al. [69] and Kim [70] propose neuromorphic accelerators that use spintronics and memristors to run a proof-of-concept fuzzy neural network.

Fifth, works that study the effects of approximate storage devices on DNN workloads [71, 72]. Qin et al. [71] study the error tolerance of neural networks that are stored in approximate non-volatile memory (NVM) media. The authors study the effects of turning the ECC off in parts of the NVM media that store the neural network data. Wen et al. [72] propose to mitigate the effects of unreliable disk reads with a specialized ECC variant that aims to mitigate error patterns present in weights of shallow neural networks.

Sixth, works that study the intrinsic error resilience of DNNs by injecting randomly distributed errors in DNN data [73, 74]. These works assume that the errors can come from any component of the system (i.e., they do not target a specific approximate hardware component). Marques et al. [73] study the accuracy of DNNs under different error injection rates and propose various error mitigation techniques. This work uses a simple probabilistic method to artificially inject errors into the DNN model. ApproxANN [74] uses an algorithm that optimizes the DNN accuracy by taking into account the error tolerance and the criticality of each component of the network.

Approximate Computing Hardware for Other Workloads There are applications other than neural network inference that can tolerate errors, thus they can benefit from approximate memory [75, 76]. Bharti et al. [75] propose an heterogeneous SRAM structure for low-power multimedia applications in smartphones. Nguyen et al. [76] propose an approximate DRAM architecture for improving the performance of neural network training via refreshing only the most critical bits [76].

Modifying DRAM Parameters Many prior works study the effects of modifying DRAM parameters on reliability, performance, and energy consumption. We already discuss some prior works that reduce DRAM voltage, access latency, and refresh rate in Sect. 2.3. EDEN leverages the characterization techniques introduced in Voltron [35] and Flexible-Latency DRAM [33] to perform the DRAM characterization required to map a DNN to approximate DRAM with reduced voltage and reduced latency (Sect. 3.4). We classify other related works that modify DRAM parameters into three categories.

First, works that aim to characterize and reduce energy consumption at reduced supply voltage levels [35, 77]. David et al. [77] propose memory dynamic voltage and frequency scaling (DVFS) to reduce DRAM power. Voltron [35] studies voltage reduction in real DRAM devices in detail and proposes solutions to reduce voltage reliably based on observed error characteristics and system performance requirements.

Second, works that investigate DRAM characteristics under reduced access latency [33, 37]. Adaptive-Latency DRAM [37] characterizes the guardbands present in timing parameters defined by DRAM manufacturers, and exploits the extra timing margins to reliably reduce DRAM latency across different chips and temperatures. Flexible-Latency DRAM [33] analyzes the spatial distribution of reduced-latency-induced cell failures, and uses this information to reliably access different regions of DRAM with different timing parameters.

Third, works that aim to reduce DRAM latency by modifying the microarchitecture of DRAM or the memory controller (e.g., [17]). These works reduce latency without introducing bit errors.

Approximate Solid-State Memories There are several works that propose to reuse faulty solid-state memories [78, 79]. Chenlin et al. [78] propose to increase NAND Flash memory lifetime via reusing faulty memory blocks that contain uncorrectable errors to store approximate data. Jevdjic et al. [79] propose a novel and efficient methodology to compute bit-level reliability requirements for encoded videos by tracking visual and metadata dependencies within encoded bitstreams.

9   Discussion and Challenges

In this section we discuss the potential impact of EDEN on future works, and we describe the main challenges for EDEN to be commonly accepted by industry

9.1   *Discussion*

EDEN has inspired several works on approximate computing [80–82], error tolerance in neural networks [83–85], improving energy efficiency on neural network inference [66, 86], improving energy efficiency on neural network training [76], or make machine learning algorithms more secure [87]. We believe that EDEN will continue to have impact and inspire future work for five main reasons.

First, large data centers and computing networks are growing very fast due to their increasing demand. Although the many efforts on making computing and memory more energy-efficient, the overall energy consumption is increasing significantly due to the increasing number of available computing resources. Thus, reducing the energy and power consumption of memory devices when running error-tolerant applications would have a large impact on global energy consumption. Other computing networks like cryptocurrency mining can also benefit from the EDEN framework, as they can potentially be somewhat tolerant to errors. It is estimated that the energy consumption of bitcoin alone is in the order of 100 TWh per year, which is larger than the energy consumption of some countries.

Second, data movement between the CPU and main memory is a major obstacle against improving performance, scalability, and energy efficiency in modern systems [88]. To solve this issue, many papers propose different Processing-in-Memory (PIM) mechanisms that place computing units close to memory [88–90]. The power budget in memory systems is very limited, thus, reducing the energy required for accessing memory would enable to increase the in-memory computing capabilities.

Third, the number of internet of things (IoT) and edge devices has been increasing significantly in last few years, and they will continue to grow in the upcoming years. These devices have a very limited power budget, and they are usually powered by batteries, or they harvest energy from the environment, thus reducing energy consumption is critically important. One effective way to reduce the energy budget of these devices is to reduce the DRAM voltage or timing parameters in cases where the workload can tolerate bit errors. Many of the applications in IoT and edge devices are inherently tolerant to noise. For example, IoT devices acquire data with sensors, which are subject to inherent variations and noises (e.g., measurement noise, small differences in a few pixels, etc.), and this data might be processed by DNNs.

Fourth, many emerging and new applications that can tolerate bit errors can benefit from some ideas proposed by EDEN. Some examples of common and very used applications are (1) image processing algorithms and applications [91],

(2) signal processing algorithms [92], (3) genome analysis [93], (4) financial analysis [94], or (5) big data analysis [95].

Fifth, modern DRAM devices are becoming fundamentally less reliable and insecure due to the RowHammer vulnerability [46, 96], where repeatedly accessing (i.e., hammering) a DRAM row can cause bit flips in physically nearby rows. RowHammer is caused by the reduction of DRAM cell size and cell-to-cell spacing. Some works propose RowHammer attacks to gain unrestricted access to the system (e.g., [97]), or collapse the accuracy of DNNs (e.g., [98]). The EDEN framework can be useful to retrain the DNN in the presence of RowHammer-induced bit flips with the goal of avoiding accuracy collapse.

We conclude that EDEN is tackling a problem that is of paramount importance in a global context. EDEN already demonstrates the advantages in the context of DNN inference workloads, but it can inspire similar frameworks for other emerging ML workloads, and new frameworks for other approximate applications.

9.2 Challenges

EDEN has great potential to contribute and inspire future work in low-power computing using approximate DRAM. However, there are two main challenges that need to be solve for this approach to be commonly accepted by industry. First, the characterization of errors in memory devices. Each device are unique, and the bit errors caused by reducing the voltage levels or timing parameters depend on process variation, which is unique to each device. To ensure that an approximate application works reliably in the presence of errors, we must ensure a maximum bit error rate, which can be obtained only by characterizing each memory device individually. There are some works that contribute to simplifying the error characterization errors [46], but there is still work to do for making it a practical solution. Second, once the errors are characterized for a particular device, the error patters might change for two main reason. First, the aging of the memory device [44] might cause variations in the bit error rate and error patterns. Because there is no thorough long-term DRAM aging study that determines how bit errors change with time, it is not possible to assess the impact of aging. Second, variable retention time [99], which makes that the retention time of some cells can change at runtime, might need to be considered for some critical applications. For example, DeepHammer [98] shows that flipping only a few bits can collapse the accuracy on some DNN models.

EDEN can also be applied to other memory technologies other than DRAM. To this end, it is important to understand the different failure mechanisms and error characteristics of each particular memory technology. There are several works that analyze the errors of NAND Flash memories (e.g., [78]), or SRAMs (e.g., [100]), but there are no experimental error studies on real devices from many new memory technologies that allow us to assess the viability of applying a framework similar to EDEN on those technologies.

We conclude that EDEN have some important challenges to address before being commonly adopted by industry, but we believe that the potential benefits of EDEN are large enough to be considered as an efficient way to reduce energy consumption on applications that can tolerate bit errors.

10 Conclusion

This chapter introduces EDEN, the first general framework that enables energy-efficient and high-performance DNN inference via approximate DRAM, while strictly meeting a target DNN accuracy. EDEN uses an iterative mechanism that profiles the DNN and the target approximate DRAM with reduced voltage and timing parameters. EDEN improves DNN accuracy with a novel curricular retraining mechanism that tolerates high bit error rates. We evaluate EDEN in both simulation and on real hardware. Our evaluation shows that EDEN enables (1) an average DRAM energy reduction of 21%, 37%, 31%, and 32% in CPU, GPU, Eyeriss, and TPU architectures, respectively, across a variety of state-of-the-art DNNs, and (2) average (maximum) performance gains of 8% (17%) in CPUs and 2.7% (5.5%) in GPUs, for latency-bound DNNs. We expect that the core principles of EDEN generalize well across different memory devices, memory parameters, and memory technologies. We hope that EDEN (1) enables further research and development on the use of approximate memory for machine learning workloads and (2) inspires new research and development on the use of approximate memory for other workloads and algorithms that tolerate bit errors.

References

1. LeCun, Y., Bengio, Y., Hinton, G.: Deep learning. Nature **521**(7553), 436–444 (2015)
2. Krizhevsky, A., Sutskever, I., Hinton, G.E.: ImageNet classification with deep convolutional neural networks. In: NIPS (2012)
3. Hinton, G., Deng, L., Yu, D., Dahl, G.E., Mohamed, A.-r., Jaitly, N., Senior, A., Vanhoucke, V., Nguyen, P., Sainath, T.N.: Deep neural networks for acoustic modeling in speech recognition: the shared views of four research groups. IEEE Signal Process. Mag. **29**(6), 82–97 (2012)
4. Ramachandran, N., Hong, S.C., Sime, M.J., Wilson, G.A.: Diabetic retinopathy screening using deep neural network. Clin. Exp. Ophthalmol. **46**(4), 412–416 (2018)
5. Vaswani, A., Shazeer, N., Parmar, N., Uszkoreit, J., Jones, L., Gomez, A.N., Kaiser, Ł., Polosukhin, I: Attention is all you need. In: Advances in neural information processing systems (2017)
6. Sandler, M., Howard, A., Zhu, M., Zhmoginov, A., Chen, L.-C.: MobileNetV2: the next generation of on-device computer vision networks. In: CVPR (2018)
7. Chen, Y.-H., Krishna, T., Emer, J.S., Sze, V.: Eyeriss: an energy-efficient reconfigurable accelerator for deep convolutional neural networks. In: JSSC (2017)
8. Han, S., Liu, X., Mao, H., Pu, J., Pedram, A., Horowitz, M.A., Dally, W.J.: EIE: efficient inference engine on compressed deep neural network. In: ISCA (2016)

9. Chen, T., Du, Z., Sun, N., Wang, J., Wu, C., Chen, Y., Temam, O.: DianNao: a small-footprint high-throughput accelerator for ubiquitous machine-learning. In: ASPLOS (2014)
10. Chen, Y.-H., Yang, T.-J., Emer, J., Sze, V.: Eyeriss v2: a flexible accelerator for emerging deep neural networks on mobile devices. In: JETCAS (2019)
11. Chetlur, S., Woolley, C., Vandermersch, P., Cohen, J., Tran, J., Catanzaro, B., Shelhamer, E.: cuDNN: efficient primitives for deep learning, arXiv (2014)
12. Kozlov, A., Osokin, D.: Development of real-time ADAS object detector for deployment on CPU. In: IntelliSys (2019)
13. Xie, S., Girshick, R., Dollár, P., Tu, Z., He, K.: Aggregated residual transformations for deep neural networks. In: CVPR (2017)
14. Levinthal, D.: Performance analysis guide for Intel Core i7 processor and Intel Xeon 5500 processors. https://software.intel.com/sites/products/collateral/hpc/vtune/performance_ analysis_guide.pdf (2009)
15. Ueyoshi, K., Ando, K., Hirose, K., Takamaeda-Yamazaki, S., Kadomoto, J., Miyata, T., Hamada, M., Kuroda, T., Motomura, M.: QUEST: A 7.49 TOPS multi-purpose log-quantized DNN inference engine stacked on 96 MB 3D SRAM using inductive-coupling technology in 40nm CMOS. In: ISSCC (2018)
16. Iandola, F.N., Han, S., Moskewicz, M.W., Ashraf, K., Dally, W.J., Keutzer, K.: SqueezeNet: AlexNet-level accuracy with 50x fewer parameters and <0.5 mb model size, arXiv (2016)
17. Wang, Y., Tavakkol, A., Orosa, L., Ghose, S., Ghiasi, N.M., Patel, M., Kim, J.S., Hassan, H., Sadrosadati, M., Mutlu, O.: Reducing DRAM latency via charge-level-aware look-ahead partial restoration. In: MICRO (2018)
18. Seshadri, V., Lee, D., Mullins, T., Hassan, H., Boroumand, A., Kim, J., Kozuch, M.A., Mutlu, O., Gibbons, P.B., Mowry, T.C.: Ambit: in-memory accelerator for bulk bitwise operations using commodity DRAM technology. In: MICRO (2017)
19. Koppula, S., Orosa, L., Yağlıkçı, A.G., Azizi, R., Shahroodi, T., Kanellopoulos, K., Mutlu, O.: EDEN: Enabling energy-efficient, high-performance deep neural network inference using approximate DRAM. In: Proceedings of the 52nd Annual IEEE/ACM International Symposium on Microarchitecture (MICRO) (2019)
20. Jouppi, N.P., Young, C., Patil, N., Patterson, D., Agrawal, G., Bajwa, R., Bates, S., Bhatia, S., Boden, N., Borchers, A., et al.: In-datacenter performance analysis of a tensor processing unit. In: ISCA (2017)
21. LeCun, Y., Bottou, L., Bengio, Y., Haffner, P.: Gradient-based learning applied to document recognition. Proc. IEEE **86**(11), 2278–2324 (1998)
22. Goodfellow, I., Bengio, Y., Courville, A.: Deep Learning. MIT Press, New York (2016)
23. Koppula, S., Orosa, L., Yağlıkçı, A.G., Azizi, R., Shahroodi, T., Kanellopoulos, K., Mutlu, O.: EDEN: Enabling energy-efficient, high-performance deep neural network inference using approximate DRAM. arXiv (2019)
24. Fedus, W., Zoph, B., Shazeer, N.: Switch transformers: scaling to trillion parameter models with simple and efficient sparsity, arXiv preprint arXiv:2101.03961 (2021)
25. Lepikhin, D., Lee, H., Xu, Y., Chen, D., Firat, O., Huang, Y., Krikun, M., Shazeer, N., Chen, Z.: GShard: scaling giant models with conditional computation and automatic sharding, arXiv preprint arXiv:2006.16668 (2020)
26. Du, S.S., Lee, J.D.: On the power of over-parametrization in neural networks with quadratic activation, arXiv (2018)
27. Hubara, I., Courbariaux, M., Soudry, D., El-Yaniv, R., Bengio, Y.: Quantized neural networks: training neural networks with low precision weights and activations. In: IMLR (2017)
28. Cun, Y.L., Denker, J.S., Solla, S.A.: Optimal brain damage. In: NIPS (1990)
29. Robbins, H., Monro, S.: A stochastic approximation method. In: The Annals of Mathematical Statistics (1951)
30. Lashgari, E., Liang, D., Maoz, U.: Data augmentation for deep-learning-based electroencephalography. J. Neurosci. Methods **346**, 108885 (2020)

31. Srivastava, N., Hinton, G., Krizhevsky, A., Sutskever, I., Salakhutdinov, R.: Dropout: a simple way to prevent neural networks from overfitting. J. Mach. Learn. Res. **15**(1), 1929–1958 (2014)
32. JEDEC Standard: DDR4 SDRAM specification (JESD79-4) (2012)
33. Chang, K.K., Kashyap, A., Hassan, H., Ghose, S., Hsieh, K., Lee, D., Li, T., Pekhimenko, G., Khan, S., Mutlu, O.: Understanding latency variation in modern DRAM chips: experimental characterization, analysis, and optimization. In: SIGMETRICS (2016)
34. Kim, J.S., Patel, M., Hassan, H., Orosa, L., Mutlu, O.: D-RaNGe: using commodity DRAM devices to generate true random numbers with low latency and high throughput. In: HPCA (2019)
35. Chang, K.K., Yağlıkçı, A.G., Ghose, S., Agrawal, A., Chatterjee, N., Kashyap, A., Lee, D., O'Connor, M., Hassan, H., Mutlu, O.: Understanding reduced-voltage operation in modern DRAM devices: experimental characterization, analysis, and mechanisms. In: Proceedings of the ACM on Measurement and Analysis of Computing Systems (SIGMETRICS) (2017)
36. Ghose, S., Li, T., Hajinazar, N., Senol Cali, D., Mutlu, O.: Demystifying complex workload-DRAM interactions: an experimental study. In: Proceedings of the ACM on Measurement and Analysis of Computing Systems (SIGMETRICS) (2019)
37. Lee, D., Kim, Y., Pekhimenko, G., Khan, S., Seshadri, V., Chang, K., Mutlu, O.: Adaptive-latency DRAM: optimizing DRAM timing for the common-case. In: HPCA (2015)
38. Liu, J., Jaiyen, B., Veras, R., Mutlu, O.: RAIDR: retention-aware intelligent DRAM refresh. In: ISCA (2012)
39. Yang, T.-J., Chen, Y.-H., Sze, V.: Designing energy-efficient convolutional neural networks using energy-aware pruning. In: CVPR (2017)
40. Kim, J.S., Patel, M., Hassan, H., Mutlu, O.: The DRAM latency PUF: quickly evaluating physical unclonable functions by exploiting the latency-reliability tradeoff in modern commodity DRAM devices. In: HPCA (2018)
41. Lee, D., Khan, S., Subramanian, L., Ghose, S., Ausavarungnirun, R., Pekhimenko, G., Seshadri, V., Mutlu, O.: Design-induced latency variation in modern DRAM chips: characterization, analysis, and latency reduction mechanisms. In: Proceedings of the ACM on Measurement and Analysis of Computing Systems (SIGMETRICS) (2017)
42. Hassan, H., Vijaykumar, N., Khan, S., Ghose, S., Chang, K., Pekhimenko, G., Lee, D., Ergin, O., Mutlu, O.: SoftMC: a flexible and practical open-source infrastructure for enabling experimental DRAM studies. In: HPCA (2017)
43. Hamamoto, T., Sugiura, S., Sawada, S.: On the retention time distribution of dynamic random access memory (DRAM). IEEE Trans. Electron Devices **45**(6), 1300–1309 (1998)
44. Fieback, M.: DRAM reliability: aging analysis and reliability prediction model (2017)
45. Khan, S., Lee, D., Mutlu, O.: PARBOR: an efficient system-level technique to detect data-dependent failures in DRAM. In: DSN (2016)
46. Orosa, L., Yaglikci, A.G., Luo, H., Olgun, A., Park, J., Hassan, H., Patel, M., Kim, J.S., Mutlu, O.: A deeper look into RowHammer's sensitivities: experimental analysis of real DRAM chips and implications on future attacks and defenses. In: MICRO (2021)
47. Patel, M., Kim, J.S., Hassan, H., Mutlu, O.: Understanding and modeling on-die error correction in modern DRAM: an experimental study using real devices. In: DSN (2019)
48. Paszke, A., Gross, S., Chintala, S., Chanan, G., Yang, E., DeVito, Z., Lin, Z., Desmaison, A., Antiga, L., Lerer, A.: Automatic differentiation in PyTorch. In: NIPS-W (2017)
49. He, K., Zhang, X., Ren, S., Sun, J.: Deep residual learning for image recognition. In: CVPR (2016)
50. Simonyan, K., Zisserman, A.: Very deep convolutional networks for large-scale image recognition. arXiv (2014)
51. Iandola, F., Moskewicz, M., Karayev, S., Girshick, R., Darrell, T., Keutzer, K.: DenseNet: implementing efficient convNet descriptor pyramids. arXiv (2014)
52. Lin, D., Talathi, S., Annapureddy, S.: Fixed point quantization of deep convolutional networks. In: ICML (2016)
53. Redmon, J., Farhadi, A.: YOLO9000: better, faster, stronger. arXiv (2017)

54. Orosa, L., Wang, Y., Sadrosadati, M., Kim, J.S., Patel, M., Puddu, I., Luo, H., Razavi, K., Gómez-Luna, J., Hassan, H., et al.: CODIC: a low-cost substrate for enabling custom in DRAM functionalities and optimizations. In: ISCA (2021)
55. The CIFAR-10 Dataset. https://www.cs.toronto.edu/~kriz/cifar.html
56. Sanchez, D., Kozyrakis, C.: ZSim: fast and accurate microarchitectural simulation of thousand-core systems. In: ISCA (2013)
57. Kim, Y., Yang, W., Mutlu, O.: Ramulator: a fast and extensible DRAM simulator. IEEE Comput. Archit. Lett. **15**(1), 45–49 (2015)
58. Chandrasekar, K., Weis, C., Li, Y., Akesson, B., Wehn, N., Goossens, K., DRAMPower: open-source DRAM power and energy estimation tool (2012)
59. Bakhoda, A., Yuan, G.L., Fung, W.W., Wong, H., Aamodt, T.M.: Analyzing CUDA workloads using a detailed GPU simulator. In: ISPASS (2009)
60. Leng, J., Hetherington, T., ElTantawy, A., Gilani, S., Kim, N.S., Aamodt, T.M., Reddi, V.J.: GPUWattch: enabling energy optimizations in GPGPUs. In: ISCA (2013)
61. Samajdar, A., Zhu, Y., Whatmough, P.N., Mattina, M., Krishna, T.: SCALE-Sim: systolic CNN accelerator. arXiv (2018)
62. Nguyen, D.-T., Ho, N.-M., Chang, I.-J.: St-DRC: Stretchable DRAM refresh controller with no parity-overhead error correction scheme for energy-efficient DNNs. In: DAC (2019)
63. Tu, F., Wu, W., Yin, S., Liu, L., Wei, S.: RANA: towards efficient neural acceleration with refresh-optimized embedded DRAM. In: ISCA (2018)
64. Nguyen, D.T., Kim, H., Lee, H.-J., Chang, I.-J.: An approximate memory architecture for a reduction of refresh power consumption in deep learning applications. In: ISCAS (2018)
65. Li, G., Hari, S.K.S., Sullivan, M., Tsai, T., Pattabiraman, K., Emer, J., Keckler, S.W.: Understanding error propagation in deep learning neural network (DNN) accelerators and applications. In: SC (2017)
66. Yüksel, İ.E., Salami, B., Ergin, O., Unsal, O.S., Kestelman, A.C.: MoRS: an approximate fault modelling framework for reduced-voltage SRAMs. IEEE Trans. Comput. Aided Des. Integr. Circuits Syst. **41**(6), 1663–1673 (2021)
67. Zhang, J., Rangineni, K., Ghodsi, Z., Garg, S.: Thundervolt: enabling aggressive voltage underscaling and timing error resilience for energy efficient deep learning accelerators. In: DAC (2018)
68. Salami, B., Unsal, O., Cristal, A.: On the resilience of RTL NN accelerators: fault characterization and mitigation. arXiv (2018)
69. Panda, P., Sengupta, A., Sarwar, S.S., Srinivasan, G., Venkataramani, S., Raghunathan, A., Roy, K.: Cross-layer approximations for neuromorphic computing: from devices to circuits and systems. In: DAC (2016)
70. Kim, Y.: Energy efficient and error resilient neuromorphic computing in VLSI, Ph.D. dissertation. MIT, New York (2013)
71. Qin, M., Sun, C., Vucinic, D.: Robustness of neural networks against storage media errors. arXiv (2017)
72. Shi, W., Wen, Y., Liu, Z., Zhao, X., Boumber, D., Vilalta, R., Xu, L.: Fault resilient physical neural networks on a single chip. In: CASES (2014)
73. Marques, J., Andrade, J., Falcao, G.: Unreliable memory operation on a convolutional neural network processor. In: SiPS (2017)
74. Zhang, Q., Wang, T., Tian, Y., Yuan, F., and Xu, Q.: ApproxANN: an approximate computing framework for artificial neural network. In: DATE (2015)
75. Bharti, P.K., Surana, N., Mekie, J.: Power and area efficient approximate heterogeneous 8T SRAM for multimedia applications. In: VLSID (2019)
76. Nguyen, D.-T., Min, C.-H., Ho, N.-M., Chang, I.-J.: DRAMA: an approximate DRAM architecture for high-performance and energy-efficient deep training system. In: ICCAD (2020)
77. David, H., Fallin, C., Gorbatov, E., Hanebutte, U.R., Mutlu, O.: Memory power management via dynamic voltage/frequency scaling. In: ICAC (2011)

78. Ma, C., Zhou, Z., Han, L., Shen, Z., Wang, Y., Chen, R., Shao, Z.: Rebirth-FTL: lifetime optimization via approximate storage for NAND flash memory. In: IEEE Transactions on Computer-Aided Design of Integrated Circuits and Systems (2021)
79. Jevdjic, D., Strauss, K., Ceze, L., Malvar, H.S.: Approximate storage of compressed and encrypted videos. In: ASPLOS (2017)
80. Salami, B., Onural, E.B., Yuksel, I.E., Koc, F., Ergin, O., Kestelman, A.C., Unsal, O., Sarbazi-Azad, H., Mutlu, O.: An experimental study of reduced-voltage operation in modern FPGAs for neural network acceleration. In: DSN (2020)
81. Felzmann, I., Fabrício Filho, J., Wanner, L.: Risk-5: controlled approximations for RISC-V. IEEE Trans. Comput. Aided Des. Integr. Circuits Syst. 39(11), 4052–4063 (2020)
82. Larimi, S.S.N., Salami, B., Unsal, O.S., Kestelman, A.C., Sarbazi-Azad, H., Mutlu, O.: Understanding power consumption and reliability of high-bandwidth memory with voltage underscaling. In: DATE (2021)
83. Buschjäger, S., Chen, J.-J., Chen, K.-H., M. Günzel, Hakert, C., Morik, K., Novkin, R., Pfahler, L., Yayla, M.: Margin-maximization in binarized neural networks for optimizing bit error tolerance. In: DATE (2021)
84. Putra, R.V.W., Hanif, M.A., Shafique, M.: Respawn: energy-efficient fault-tolerance for spiking neural networks considering unreliable memories. In: ICCAD (2021)
85. Ponzina, F., Peón-Quirós, M., Burg, A., Atienza, D.: E 2 CNNs: ensembles of convolutional neural networks to improve robustness against memory errors in edge-computing devices. IEEE Trans. Comput. 70(8), 1199–1212 (2021)
86. Jafri, S.M., Hassan, H., Hemani, A., Mutlu, O.: Refresh triggered computation: improving the energy efficiency of convolutional neural network accelerators. ACM Trans. Archit. Code Optim. (TACO) 18(1), 1–29 (2020)
87. Xu, Q., Arafin, M.T., Qu, G.: MIDAS: model inversion defenses using an approximate memory system. In: AsianHOST (2020)
88. Boroumand, A., Ghose, S., Kim, Y., Ausavarungnirun, R., Shiu, E., Thakur, R., Kim, D., Kuusela, A., Knies, A., Ranganathan, P., Mutlu, O.: Google workloads for consumer devices: mitigating data movement bottlenecks. In: ASPLOS (2018)
89. Ahn, J., Hong, S., Yoo, S., Mutlu, O., Choi, K.: A scalable processing-in-memory accelerator for parallel graph processing. In: ISCA (2015)
90. Nai, L., Hadidi, R., Sim, J., Kim, H., Kumar, P., Kim, H.: GraphPIM: enabling instruction-level PIM offloading in graph computing frameworks. In: HPCA (2017)
91. Parker, J.R.: Algorithms for Image Processing and Computer Vision. Wiley, New York (2010)
92. Van Drongelen, W., Signal Processing for Neuroscientists. Academic Press, New York (2018)
93. Alser, M., Shahroodi, T., Gómez-Luna, J., Alkan, C., Mutlu, O.: SneakySnake: a fast and accurate universal genome pre-alignment filter for CPUs, GPUs and FPGAs. Bioinformatics 36(22–23), 5282–5290 (2020)
94. Ozbayoglu, A.M., Gudelek, M.U., Sezer, O.B.: Deep learning for financial applications: a survey. Appl. Soft Comput. 93, 106384 (2020)
95. Chen, M., Mao, S., Liu, Y.: Big data: a survey. Mobile Networks and Applications 19(2), 171–209 (2014)
96. Kim, J.S., Patel, M., Yağlıkçı, A.G., Hassan, H., Azizi, R., Orosa, L., Mutlu, O.: Revisiting RowHammer: an experimental analysis of modern DRAM devices and mitigation techniques. In: ISCA (2020)
97. de Ridder, F., Frigo, P., Vannacci, E., Bos, H., Giuffrida, C., Razavi, K.: SMASH: synchronized many-sided rowhammer attacks from JavaScript. In: USENIX Security (2021)
98. Yao, F., Rakin, A.S., Fan, D.: DeepHammer: depleting the intelligence of deep neural networks through targeted chain of bit flips. In: USENIX Security 20 (2020)
99. Qureshi, M.K., Kim, D.-H., Khan, S., Nair, P.J., Mutlu, O.: AVATAR: a Variable-Retention-Time (VRT) aware refresh for DRAM systems. In: DSN (2015)
100. Neggaz, M.A., Alouani, I., Lorenzo, P.R., Niar, S.: A reliability study on CNNs for critical embedded systems. In: ICCD (2018)

Part III
Emerging Substrates for Embedded Machine Learning

On-Chip DNN Training for Direct Feedback Alignment in FeFET

Fan Chen

1 Introduction

Deep neural networks (DNNs) are at the heart of latest revolutions in various artificial intelligence (AI) applications, such as computer vision [1, 2], natural language processing [3, 4], autonomous systems [5], and precision health [6]. A DNN is first trained with labeled data to perform a desired task (such as image classification or object detection) through a *training* process. To obtain an acceptable accuracy, training typically needs to run hundreds or thousands of iterations. Then the developed model can be deployed for *inference* tasks. The execution of DNN, especially its training process, requires intensive computing and huge memory storage. For instance, AlexNet [1]—a medium-sized DNN—involves 62 million parameters, 2.2 billion operations, and $>130\,\text{MB}$ memory storage, to perform training on a single RGB image with only 224×224 pixels. Moreover, a recent analysis [7] shows that the amount of computation used in DNN training has constantly increased by $300,000\times$ from AlexNet (2012) to AlphaGo Zero (2018), yielding a 3.4-month doubling period.

The ever-increasing computing requirements of DNN models motivated the latest wide adoption of domain-specific accelerators [8–18] that provide two to three orders of magnitude performance improvement compared to general-purpose CPUs and GPUs through intensive data reuse and specifically designed memory hierarchy. However, the majority of these accelerators is designed only for DNN inference and lacks basic support for the DNN training. The reasons are twofold. First, the *de facto* training method, error backpropagation (BP) [19], involves complex compute phases and sophisticated data dependency. BP requires all weights and

F. Chen (✉)
Intelligent Systems Engineering Department, Indiana University Bloomington, Bloomington, IN, USA
e-mail: fc7@iu.edu

intermediate data to be stored in memory so that they can be sequentially consumed in the subsequent error backpropagation paths. In this way, weight updates are non-local and depend on upstream layers, which make training parallelization extremely challenging and greatly limit the continuous improvement of system computing performance. Second, the algorithmic complexity incurs significant overheads on hardware in terms of computing units, memory, and control circuits. Although some of the latest accelerators attempted to address the training requirements, with efforts from both industry (e.g., Google TPU [9] V3) and academia proposals [11, 12], their power consumption can reach 200~250 Watt.

The recent proposed accelerators using resistive random access memory (ReRAM)-based in-memory computing circuitry [13] demonstrated a great potential in low-power DNN acceleration because they can typically provide 1000× to 10,000× energy reduction in executing massive matrix–vector multiplication (MVM)—the dominant operator in DNNs. However, the power and area efficiency of such mix-signal accelerators [14–18] suffer from the significant overhead of analog-to-digital (A/D) conversion. Essentially, the CMOS analog-to-digital converters (ADCs) and digital-to-analog converters (DACs) account for >60% of the energy consumption and ~30% of the chip footprint in a typical ReRAM-based DNN accelerator [15]. Furthermore, ReRAM-based accelerators still lack efficient training support due to the inherent algorithmic complexity of the backpropagation algorithm.

In this chapter, we present our recent research [20] on efficient and low-power accelerator architecture for DNN training. We set out to address the aforementioned challenges by combining innovations in training algorithm, circuits, and architecture. We analyze the recently proposed direct feedback alignment (DFA) [21], which replaces the sequential error information used in BP with a random matrix, thus avoiding the need to store weights. More importantly, DFA provides an opportunity for parallel layer updates. Previous studies have shown that DFA can be applied to various deep learning tasks [21–23] with accuracy closed to fine-tuned BP. Based on our analysis, we propose a customized design to support DNN training with DFA. We leverage the unique features of ferroelectric field-effect transistors (FeFETs) and implemented a holistic accelerator including a FeFET-based low-power ADC, a random number generator, and a matrix–vector multiplication engine. The following summarize our contributions:

- We investigate DFA's potential in overcoming the limitations of long-range data dependency in the BP algorithm and identify the two major architectural challenges for deploying DFA on hardware systems: (1) a low-cost on-chip random number generator and (2) an efficient computing pipeline that supports parallel training operations.
- We exploit the unique features of FeFET, such as low-power operations, stochastic polarization switching, and tunable threshold voltage, and propose a holistic training accelerator that includes an MVM engine, a low-power ADC, and a real-time random number generator, all of which are implemented by FeFET.

- We use a diverse set of DNN applications with distinct benchmarks to evaluate the effectiveness of the proposed accelerator architecture. Our experimental results show that the proposed design achieves 1.3× speedup and 2.5x× improvement on power efficiency compared with the state-of-the-art ReRAM-based DNN accelerator.

2 Background

2.1 DNN Training Methods

A deep neural network (DNN) is a learning algorithm that deploys a feed-forward function for *inference* and a backward process for *training*. The backpropagation (BP) algorithm [19] has achieved great success in training supervised DNNs and has been used as the *de facto* method due to theoretical simplicity and proven performance. We show the backpropagation and direct feedback alignment algorithm in Fig. 1. BP and DFA share the same forward inference path but utilize distinct error backward propagation paths as highlighted in blue. The forward inference function at layer l can be formulated as the matrix–vector multiplication between the input vector h_{l-1} and weight matrix W_l, followed by a nonlinear activation $f(\cdot)$. Note that here we omit bias in the computation of each layer for simplicity. The final output is calculated by using a task-specific *cost function*, which essentially quantifies the difference between the actual output generated by the neural network

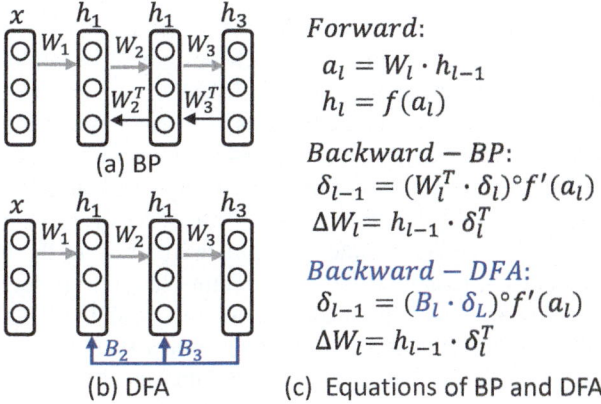

Forward:
$$a_l = W_l \cdot h_{l-1}$$
$$h_l = f(a_l)$$

Backward − BP:
$$\delta_{l-1} = (W_l^T \cdot \delta_l)^\circ f'(a_l)$$
$$\Delta W_l = h_{l-1} \cdot \delta_l^T$$

Backward − DFA:
$$\delta_{l-1} = (B_l \cdot \delta_L)^\circ f'(a_l)$$
$$\Delta W_l = h_{l-1} \cdot \delta_l^T$$

(a) BP

(b) DFA

(c) Equations of BP and DFA

Fig. 1 An comparison between backpropagation (BP) and direct feedback alignment (DFA). Gray arrows indicate the forward path shared by BP and DFA. Black arrows and blue arrows, respectively, indicate the backward paths of BP and DFA. (**a**) Backpropagation. (**b**) Direct feedback alignment. (**c**) The mathematical formulas

and an expected target value. Below we compare the backward error propagation paths in BP and DFA.

BP-Based Error Propagation The error propagation follows the chain rule of derivatives [19], where the error for a specific layer, i.e., δ_l, is first multiplied with the transposed weight matrix of the current layer, i.e., W_l^T, and then multiplied by the derivatives of the current input to obtain the error vector for the previous layer. Because the inputs and weights of all the layers are required in the calculation of the backward path, these values need to be stored in memory, which causes significant memory overhead. Moreover, the sequential nature of the BP backward execution prevents the parallellism of error updates. For a DNN model with L layers, a training on a single input typically takes $3L$ logical cycles if we assume the computation of each layer consumes one cycle.

DFA-Based Error Propagation The gradient signal $W_l^T \cdot \delta_l$ from the l-th layer is replaced by $B_l \cdot \delta_L$ [21], where B_l is a random matrix with appropriate shape, and δ_L is the global error signal from the output layer. In this way, the error calculation of a single layer is independent of other layers, and hence, the error propagation can be processed in parallel, offering a great potential in reducing latency. In addition, there is no need to store the weights of each layer, which also saves memory requirements. Recent works [21–23] have demonstrated the applicability of DFA to various tasks including computer vision, recommendation systems, and natural language processing.

2.2 DNN Acceleration in Resistive Memory

An example of resistive memory (ReRAM)-based analog DNN acceleration is illustrated in Fig. 2a. The key approach of such designs is to implement matrix–vector multiplication (MVM) units using ReRAM crossbars. We denote each memory cell as a circle, whose conductance is pre-programmed to represent the weights of DNNs. The input feature map values are converted to the read voltages and applied onto the horizontal wordlines. According to Ohm's law, the current sensed across each memory cell is the product of the input element and the weight value stored in the cell. Therefore, the accumulated current at the bit lines (BL) aggregates all the current along the BL, representing the dot product between the inputs vector and the stored weight vector in a column. If all the WLs are activated in an $N \times N$ analog crossbar, the multiplication between a $1 \times N$ vector and an $N \times N$ matrix can be processed in $O(1)$ time. It is worth mentioning that the calculation in ReRAM crossbars is analog in nature; hence, digital-to-analog converters (DACs) and analog-to-digital converters (ADCs) are needed at the input and output to ensure communication with other digital components on the chip.

Various DNN accelerators [14–18] leveraging ReRAM-based matrix–vector multiplication engines have been recently proposed. Such designs typically can provide up to two orders of magnitude performance improvement [15, 16] compared to

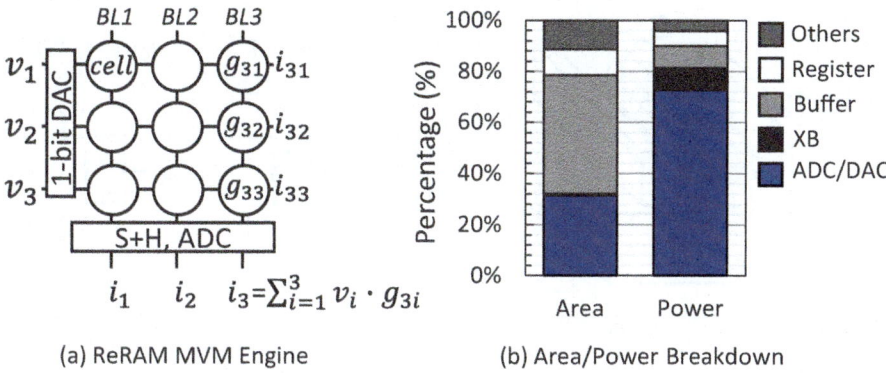

(a) ReRAM MVM Engine   (b) Area/Power Breakdown

Fig. 2 ReRAM-based DNN acceleration. (**a**) An ReRAM MVM engine. (**b**) The area and power breakdown in a ReRAM-based CNN accelerator [15]

conventional CMOS-based ASICs. Despite the supreme performance enhancement, the unacceptable overhead of ADC/DAC makes such design unsuitable for DNN training. As shown in Fig. 2b, ~72% of the total system power is consumed by the CMOS ADCs and DACs. Latest work shows that the power-hungry A/D conversion still accounts for >25% of the power and ~50% of the chip footprint even with the advanced carbon nanotube field-effect transistors (CNFETs)-based low-power ADCs [18] in a monolithic 3D integrated ReRAM accelerator.

2.3 Ferroelectric Field-Effect Transistor

As the memory industry marches along Moore's Law, they are facing the problem of reduced power efficiency and increased unit cost, which makes the realization of high-efficiency devices at advanced process nodes a very big challenge. To address these problems, the ferroelectric field-effect transistor (FeFET) is currently gaining significant momentum because of their scalability, fast speed, and low-power operations [24].

Figure 3a illustrates the structure of a FeFET device, which is implemented by adding an extra ferroelectric (FE) layer, e.g., HfO_2 [24], in the gate stack of a conventional MOSFET. By applying a positive (negative) gate-to-source voltage, i.e., V_G, the polarization of the FE layer can be set (reset) to the positive (negative) direction and retained under the subsequent withdrawal of V_G. In this case, the polarization of a FeFET device can be dynamically tuned, resulting in a controllable threshold voltage (i.e., V_{th}) and drain current (i.e., I_{DS}), as demonstrated in Fig. 3b. Based on these unique features, previous work [25] proposed to utilize the programmable V_{th} of a FeFET device to represent logic "1" and "0." By leveraging the similar design concept with ReRAM crossbar-based designs, a DNN accelerator using FeFET crossbar is constructed and achieves significantly reduced processing

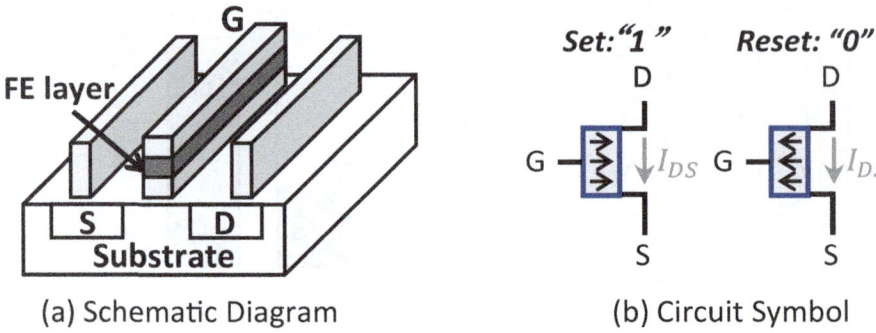

Fig. 3 FeFET basics. (**a**) The schematic diagram of a FeFET device. (**b**) The circuit symbol of a FeFET device

latency and power consumption due to the fact that FeFET can be programmed with a shorter write pulse (i.e., several nanoseconds) with significantly reduced pulse amplitude (i.e., 3 to 5 volts).

A practical FeFET device features $> 10^6$ on/off current ratio [26]. By adjusting the pulse amplitude and width, it is possible to exploit partial polarization states to record two or more bits of information in a single cell, leading to multi-level cell FeFET. In addition, FeFET devices exhibit some unique features: (1) The device characteristics are closely related to the thickness of FE layer (T_{FE}) [27]. Specifically, a sharper and wider switching hysteresis can be achieved by increasing the FE thickness. (2) The I_{DS}-V_G transition curve exhibits an abrupt and stochastic switching in scaled devices [28]. Using these features, we explored the wide range of applications of FeFET, including FeFET MVM engine, FeFET-based RNG, and FeFET-based ADC unit.

3 An FeFET-Based DNN Training Accelerator Architecture for Direct Feedback Alignment

In this section, we first present the overview of the proposed accelerator architecture, followed by detailed discussions on each novel feature, including the FeFET-based on-chip random number generator (RNG) and a low-power ADC leveraging FE layer conductance tuning. At last, we describe how to integrate the above innovations into a $(L + 2)$-stage pipeline capable of processing L-layer DNNs with high throughput. We also report the hardware overhead and compare its cost with previous ReRAM-based DNN accelerators.

3.1 Overall Architecture

The proposed architecture [20] follows the processing-in-memory strategy and can be used as a standalone accelerator and communicate with a general processor via a software–hardware interface. The overall architecture of our design is illustrated in Fig. 4. At a high level, it is composed of a 64 KB eDRAM buffer for input/output storage, a controller that orchestrates the computing flow, an input/output interface that communicates with off-chip DRAM, and a number of in situ processing engines (PE) connected via an on-chip mesh.

Each PE contains a few MVM engines implemented with FeFET crossbars (*XB*). Note that we adopt a 2-transistor (2T) FeFET-based device [27] for enhanced reliability. To facilitate the communication between the analog signal within each PE and other digital components on the chip, we particularly design an FeFET-based ADC unit and attach it to the bit lines of each crossbar. The input end of each PE is equipped with multiple 1-bit DACs for sequential digital-to-analog conversion. The PE also has (1) multiple sample-and-hold units (*S&H*) that convert the output currents into a voltage and send the voltage to ADCs; (2) several shift-and-add units (*S&A*) that aggregate the outputs from XBs; (3) an activation unit (*Act*) that implements the activation function; and (4) simple algorithm and logic units (*sALU*) that provide simple pre- and post-processing functions such as element-wise addition and scalar multiplication. Each PE has 2.5 KB register for input/output storage (*IR/OR*). To support random number generation for DFA-based

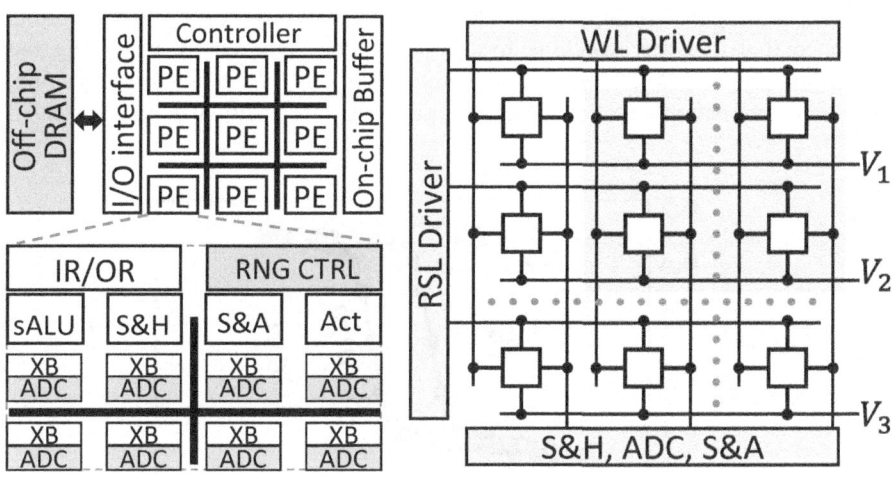

(a) Architecture Overview   (b) FeFET-based MVM engine

Fig. 4 The overall architecture of the proposed FeFET-based DNN training accelerator [20]. (**a**) The architecture overview. (**b**) The FeFET-based matrix–vector multiplication engine

error propagation, a counter-based RNG control circuit is particularly designed and attached to the PEs that reserved for backward calculation.

CNN mapping follows the standard scheme used in mainstream CNN accelerators [15, 16]. As illustrated in Fig. 4b, we store the CNN weights into the conductance states of FeFET devices. The input vectors are converted into voltages and sequentially applied onto the data lines (*DL*). The read lines (RL) aggregate the currents passing through all the cells on the same RL, representing the dot product between the DNN weights and the input vector. We adopt a conservative 4-bit per-cell FeFET device model [25] and set the crossbar size to 128×128.

3.2 FeFET Switching Characterization

We adopt the SPICE model proposed in [26] for the FeFET device simulation. The kinetic coefficient ρ is set to 0.01 (calculated by considering the polarization switching time $\sim$200 ps), and the ferroelectric layer thickness T_{FE} is set to 10 nm and 8.6 nm.

The simulated $I_{DS} - V_{GS}$ curve is shown in Fig. 5. A hysteresis loop indicates that a FeFET can be programmed to a *reset* or a *set* state with a write pulse with appropriate amplitude and duration. In addition, FeFET devices exhibit two unique features. First, the thickness of FE layer (T_{FE}) plays a crucial role in the device switching characteristics [27]. As shown in Fig. 5, a device with a 8.6 nm FE layer switch states at a lower voltage compared with a device with a 10 nm FE layer. We also validate this feature in a wider range of T_{FE}, and experimental results confirmed that a wider switching hysteresis can be achieved by increasing the FE thickness. In the following, we refer this feature as *FE layer tuning*. Second, the I_{DS}-V_G transition curve exhibits an abrupt and stochastic switching in scaled

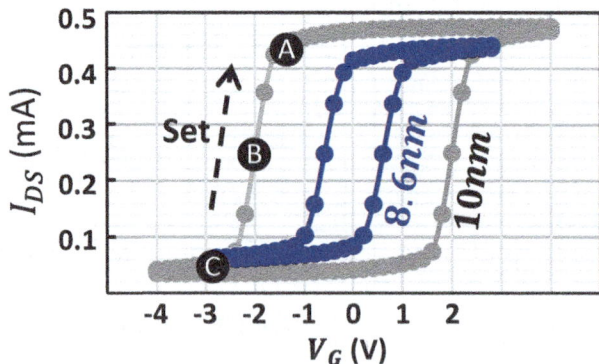

Fig. 5 Simulated $I_{DS} - V_{GS}$ in FeFET devices. Two devices with respective 8.6 nm and 10 nm ferroelectric layer thickness are simulated

devices [28]. More specifically, the slope of the switching curve in a FeFET with a 8.6 nm FE layer is sharper than that in a FeFET with a 10 nm FE layer. By exploiting these features, we explored the applications of FeFET in distinct circuits, including a FeFET-based random number generator (RNG), a FeFET-based ADC unit, and a FeFET MVM engine.

3.3   FeFET-Based Random Number Generator

For DFA-based training accelerators, random number generator (RNG) is an essential module as the layer-wise error map is replaced by a suitably sized matrix of random numbers. To avoid intensive latency and energy overhead caused by off-chip random number access, it is important to identify a convenient on-chip entropy source and design the corresponding circuit to generate a random bit stream with high throughput and stability. Based on our previous experimental results and the recent research on a FeFET model [28], we explored how to use the random characteristics in a scaled FeFET device for random number generation.

The key idea is to utilize the abrupt jump of V_{TH} in scaled FeFET devices. We first program a small FeFET device with $W/L = 80\,\text{nm}/30\,\text{nm}$ to the *set* state; then we apply a gate voltage V_G close to the median value (i.e., point B in Fig. 5). The measured output I_{DS} hence demonstrated random outputs as shown in Fig. 6a. However, the output random bits are biased and tend to generate more "0's". To address the unbalanced output bits, we adopt the output probability tracking scheme

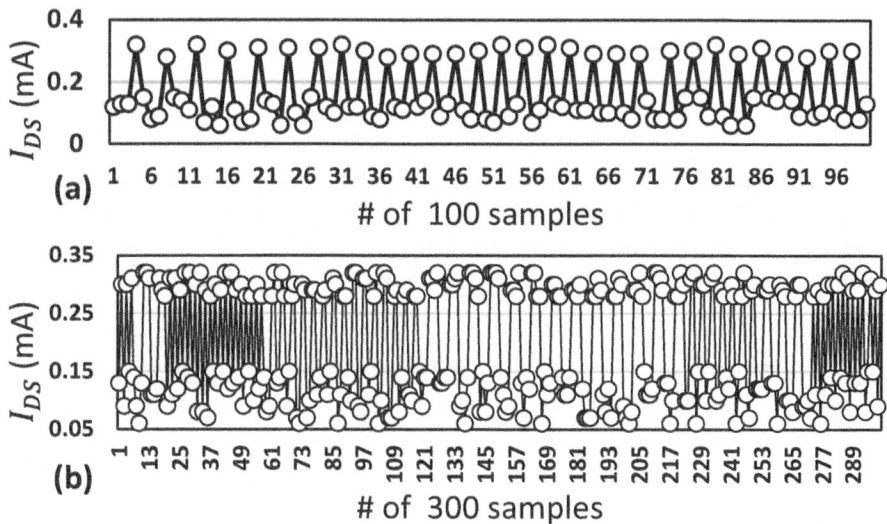

Fig. 6 I_{DS} extracted in a (**a**) baseline and (**b**) optimized FeFET-based random number generator

proposed in [29] and implement an counter-based output track control circuitry as detailed in Fig. 4c. We use a 8-bit counter to calculate the output probability of each consecutive 256-bit segment. The counter output is used as a feedback control signal to adjust and fine-tune the write voltage applied for the following segments. The extracted I_{DS} currents with output tracking are shown in Fig. 6b. As it shows, after an initial locking period for adjustment, the output bits exhibit unbiased randomness with evenly distributed "0's" and "1's," which meets the random requirement of DFA training.

3.4 Low-Power ADC Based on FE Layer Tuning

The output of analog crossbar-based processing engines needs to be converted into digital signals through ADCs. In order to reduce the large cost of CMOS ADC, we proposed FeFET-based ADC unit by utilizing the tunability of FE layer conductance. As we illustrated in previous section, the FeFET switching hysteresis is closely related to the thickness of the ferroelectric layer (T_{FE}) [27]. By engineering the FE layer conductance with appropriate program pulses, the V_{th} can be fine-tuned, resulting in a shift in the I_{DS}-V_G curve. Based on this observation, we can build a FeFET-based low-power ADC to accelerate analog-to-digital conversion. The proposed ADC unit can be integrated with the FeFET dot-product units for efficient and low-power MVM processing.

An example of a 9-cell design is demonstrated in Fig. 7. We select multi-level cell FeFET devices with T_{FE} of 2.5 nm, 2.4 nm, and 2.3 nm. We conservatively adopt a 3-level per-cell FeFET device based on the SPICE model [26] for preliminary study. In the ADC array, the difference of V_{th} is reflected by the stored polarization degree of the FE layer in the FeFET devices. For instance, *TFE2.5S1* and *TFE2.5S3* denote the FeFET device with a 2.5 nm FE layer, respectively, in the *reset* and *set* states, while *TFE2.5S2* represent the FeFET device with the same T_{FE} but in a partial *set* state.

The simulated I_{DS}-V_G of the 9-level ADC is shown in Fig. 8. Devices with the same T_{FE} demonstrated a similar I_{DS}-V_G curve with different V_{th}; therefore, the sensing current can be classified into three groups. The thicker T_{FE} is, the wider and more gradual the transition slop is. To work with the FeFET-based ADC, the read

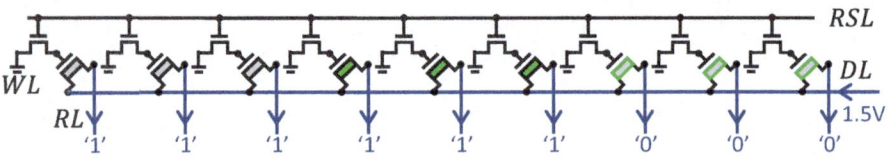

Fig. 7 A 9-bit ADC FeFET array implemented with multi-level cell FeFET devices with T_{FE} of 2.5 nm, 2.4 nm, and 2.3 nm

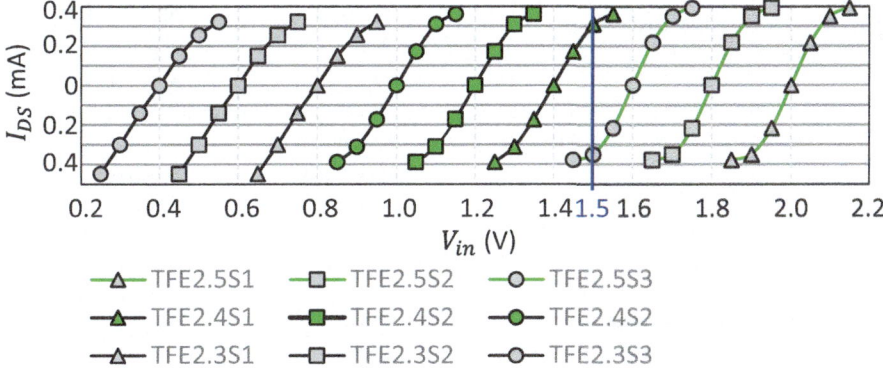

Fig. 8 Input voltage vs. sensing current in a 9-bit ADC FeFET array implemented with multi-level cell FeFET devices with T_{FE} of 2.5 nm, 2.4 nm, and 2.3 nm

select line (*RSL*) is asserted (V_{DD}), and the write line (*WL*) is driven to 0V (*GND*). The input voltage is then applied to data line (*DL*). Due to the different V_{th}, the read current sensed at the read lines (RL) is different. For instance, when the input voltage is within the range of 1.4 V~1.6 V, an higher-than-threshold current will be sensed on FeFET cells with a FE layer less than 2.5 nm, while a lower current is detected on devices with a 2.5 nm FE layer. We encode high/low current with logic "1"/"0" accordingly; therefore, an 9-bit output "111111000" is recorded with an appropriately set current threshold.

In our ongoing work, we are exploring reliable high-precision ADC design based on the preliminary implementation. In general, there is a trade-off between T_{FE}, ADC resolution, and ADC frequency. As T_{FE} increases, the slope of the I_{DS}-V_G curve becomes flatter, and hence, higher ADC resolution can be achieved since more bits can be represented using a single device. However, the ADC frequency decreases with the increase of T_{FE} because a longer transition time is required when I_{DS}-V_G curve becomes flatter. To balance the trade-off, we conduct a design space exploration by studying the conductance behavior in response to pulse schemes and FE layer thickness. For our preliminary work [20], we used 16-level FeFET devices with 4 different FE layer thickness. We summarize the optimal device parameters for simulation and the corresponding ADC parameters in Table 1.

3.5 Pipeline

Previous training accelerators implemented a pipeline for DNN training processing by exploiting the intrinsic parallelism in batch-based training. Typically, a batch of inputs (e.g., 32, 64, 128) are processed using the same weights. The parameters are only updated at the end of each batch. For BP-based training method, a DNN with

Table 1 Parameters of the optimal 16-level FeFET ADC

	Parameters	Specification
Device	W/L	500 nm/300 nm
	T_{FE}	2, 2.2, 2.4, 2.6 nm
	Kinetic coefficient ρ	0.01
ADC	Resolution	6
	Frequency	20 MHz
	Number	8*12*168
	Total power	1.6 W
	Total area	0.1 mm^2
CMOS ADC [15]	Total power	32.3 W
	Total area	19.4 mm^2

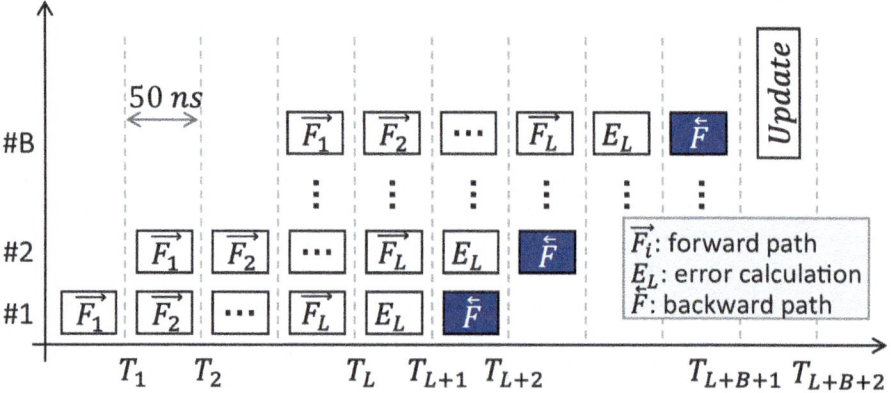

Fig. 9 The training pipeline of a DNN model using DFA

L layers requires $(2L + B + 1)$ [16, 17] cycles for processing a batch of inputs with a size of B. Specifically, within a batch, the forward computation for the first input requires L cycles, while the backward computation requires $L+1$ cycles. Then there will be $(B - 1)$ cycles until the end of batch. Finally, there is one cycle required to apply all weight updates within the batch. Therefore, the total number of cycles to process N inputs is $(N/B)(2L + B + 1)$.

This work leverages DFA and is implemented in a reduced processing pipeline as shown in Fig. 9. Since DFA allows the errors to be propagated directly from the last layer to all the processing layers, the backward computation can be processed in parallel in one cycle using the FeFET-based PE with RNG support. The overall number of cycles to process N inputs is $(N/B)(L + B + 2)$, achieving $\sim 2\times$ reduction in processing time. More importantly, the processing cycle discussed here is a logical cycle, which may require several physical cycles depending on the customized implementation. Though the processing time of each stage varies in ReRAM-based accelerators, the cycle time of the pipeline is essentially bottlenecked by the ADCs. Previous work [15] uses a 8-bit 1.28 giga-samples-per-second (GSps) ADC shared among 128 BLs in the same ReRAM crossbar, resulting in a 100 ns cycle time. This work employs the FeFET-based 6-bit ADC unit with a 50 ns cycle time, providing $2\times$ latency reduction for a single processing stage of the pipeline.

4 Evaluation

4.1 Experimental Setup

Benchmark We evaluate the proposed design architecture using MNIST [30], CIFAR-10, and CIFAR-100 [31]. The networks used in our evaluation are LeNet-5 [32] on MNIST for simple hand-written digits, AlexNet [1], and CaffeNet [33] on CIFAR-10/CIFAR-100 for complex classification tasks. To demonstrate the applicability of DFA and the proposed architecture in a broader range of applications, we also include two recommender systems, Deep Factorization Machines (DeepFM) [34] and Adaptative Factorization Network (AFN) [35], into our evaluation. Both DeepFM and AFN are evaluated on Criteo dataset [36]. All models are trained in TensorFlow. We summarize the typologies of the networks and evaluated dataset in Table 2. We quantized both the activations and weights of all CNNs with 8 bit based on the training accuracy analysis in the following section.

Scheme We compare the proposed design against five counterparts shown in Table 3. We selected Intel Xeon E5-2630 V3, 8-core CPU, an Nvidia GTX 1080 GPU, a Xilinx Virtex7 FPGA [37], one ASIC chip Google TPU [9], and a ReRAM-based DNN training accelerator [16]. For TPU, we employ four chips to ensure the high throughput required for training, but they consume more computing power. To support DFA training, we equip the above counterparts with the state-of-the-art CMOS RNG [38] for random matrix generation. The runtimes for CPU/GPU are measured by TensorFlow, and the energy costs are measured on real hardware. The FPGA numbers are scaled and calculated based on the original paper. We build an in-house simulator to model the performance of TPU.

Table 2 Models under evaluation (C: convolutional layer; F: fully connected layer)

Name	Database	Topology
LeNet-5 [32]	MNIST	2C,3F
AlexNet [1]	CIFAR-10	5C,3F
	CIFAR-100	
CaffeNet [33]	CIFAR-10	5C,3F
	CIFAR-100	
DeepFM [34]	Criteo [36]	Fully connected embedding layer
AFN [35]		

Table 3 The scheme comparison (normalized to 32nm)

Name	Description	*Power (W)*
CPU	Intel Xeon E5-2630 V3	85
GPU	Nvidia Tesla P100	250
FPGA [37]	Xilinx Virtex7 VX485T	40
TPU [9]	4-chip ASIC	384
PipeLayer [16]	ReRAM PIM	168

Table 4 Trade-off between resolution and accuracy

	LeNet	AlexNet	CaffeeNet	DeepFM	AFN
Float	1	1	1	1	1
16 bit	0.99	0.98	0.985	0.99	0.99
8 bit	0.987	0.97	0.977	0.97	0.90
6 bit	0.95	0.93	0.935	0.92	0.88
4 bit	0.93	0.88	0.90	0.85	0.75
2 bit	0.78	0.69	0.71	0.66	0.60

For PipeLayer [16] and the proposed architecture, the DAC, S&H, S&A, maxpool, activation logic designs are all adopted from an existing ReRAM DNN accelerator [15] implemented in 32 nm process technology. We use CACTI 6.5 [39] at 32 nm to model energy and area for all buffers and on-chip interconnects. Other digital circuits (e.g., peripherals for TRNG) all are modeled and estimated using Cadence Virtuoso with 32 nm PTM CMOS model [40]. We adopt the SPICE model in [26] for FeFET simulation. The kinetic coefficient ρ is set to 0.01 (calculated by considering the polarization switching time $\sim$200 ps), and the ferroelectric layer thickness T_{FE} is set to 10.5 nm, 8.6 nm, and 6.6 nm. We build a simulator based on NVSim [41] to evaluate the inference throughput, power, and energy consumption of the ReRAM-based accelerator and the proposed FeFET-based accelerator.

4.2 Experimental Results

Training Accuracy To deploy DFA training onto the fix-point FeFET crossbars, we conducted a set of experiments to explore the trade-off between numerical precision and model training accuracy. We quantize both the weights and activations to fix-point values on the five evaluation models. We normalized the model accuracy to float resolution in original implementation [23] and show the results in Table 4. We see that DFA training with 16-bit precision exhibits negligible accuracy reduction among all the evaluated CNNs and recommendation systems. A slight ($<$0.09) accuracy degradation is observed in 8-bit training. The accuracy of all models drops sharply when trained with less than 6-bit precision. In order to ensure training accuracy while reducing computing and memory requirements, we implement all the candidate networks in the following discussion with 8-bit precision.

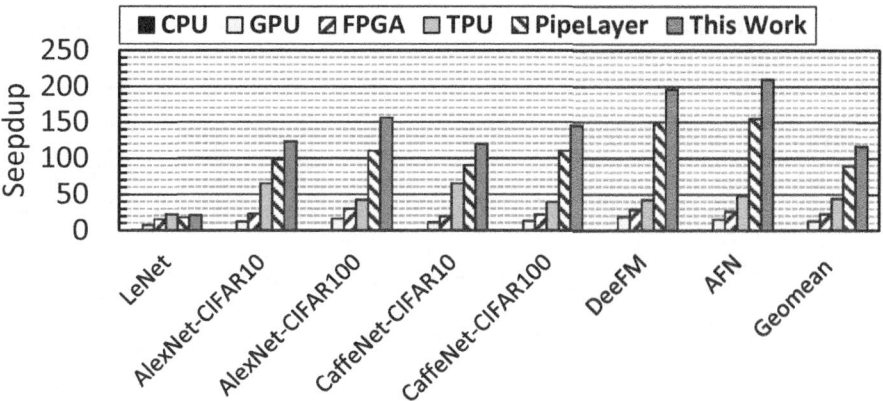

Fig. 10 The comparison on performance of different designs. Results are normalized to CPU

Performance Figure 10 compares the training performance throughput among different designs. In general, customized DNN accelerators achieve significant performance speedup compared with general-purpose CPU and GPU. The main reason is that accelerators implement dedicated hardware and data management suitable for the processing patterns in DNNs, especially for large matrix–vector multiplications, which provides a huge potential for reducing the running time. As hardware resource increases, the accelerators can achieve higher performance. For instance, the 4-chip TPU achieves better performance compared with both GPU and FPGA accelerators in all the benchmarks. The proposed design achieves the best performance in all the benchmarks. Compared with GPU, FPGA, and TPU, our design achieves, respectively, $8.9\times$, $5.1\times$, and $2.7\times$ speedup. Compared against the ReRAM-based accelerator, the performance is improved by $1.3\times$, which is mainly due to shortened pipeline cycle time and reduced ADC overhead.

Power Efficiency We show the normalized power efficiency among different hardware platforms in Fig. 11. Although the performance of TPU is better than the FPGA design, its power efficiency is $5\times$ lower than that of FPGA. As shown in Table 3, the power consumption of the 4-chip TPU is $\sim 9\times$ higher than that of the FPGA design, which excludes TPU from low-power DNN applications. The ReRAM-based accelerator demonstrated a similar power efficiency as FPGA design. The proposed architecture achieves $2.2\times$, $11.5\times$, $2.5\times$ improvement on power efficiency compared with FPGA, TPU, and ReRAM-based accelerators, respectively.

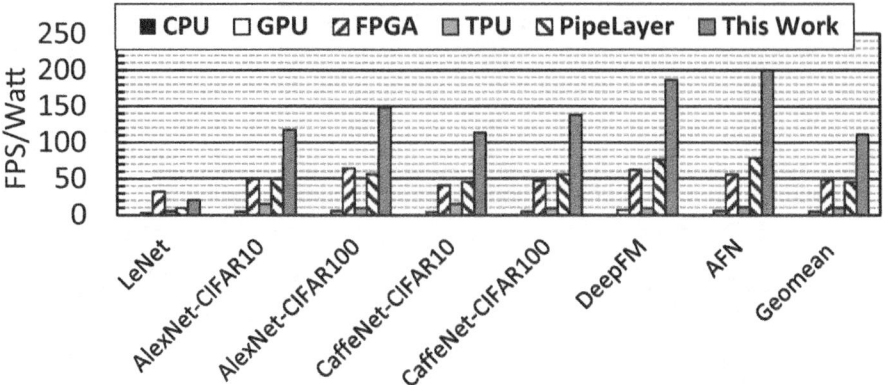

Fig. 11 The comparison on power efficiency of different designs. Results are normalized to CPU

5   Conclusion

In this chapter, we review our recent research efforts on efficient DNN training accelerators. We investigate the potential of the emerging direct alignment feedback training algorithm in overcoming the limitations of current backpropagation-based training and identify the major challenges for deploying direct alignment feedback training algorithm on hardware computing systems. We then present a customized FeFET-based accelerator architecture consisting of a FeFET-based random number generator, a low-power FeFET-based analog–digital converter, and an efficient $(L + 2)$-stage pipeline for training of an L-layer deep neural network. Our experimental results show that the proposed design is suitable for the training of a broader range of deep neural networks and achieves $1.3\times$ speedup and $2.5\times$ improvement on power efficiency compared with the state-of-the-art deep neural network training accelerator.

References

1. Krizhevsky, A., Sutskever, I., Hinton, G.E.: ImageNet classification with deep convolutional neural networks. In: Advances in Neural Information Processing Systems, vol. 25, Curran Associates, Inc., New York (2012)
2. Simonyan, K., Zisserman, A.: Very deep convolutional networks for large-scale image recognition. In: Proceedings of the 3rd International Conference on Learning Representations, ICLR 2015, San Diego, CA, USA, May 7–9, 2015, Conference Track Proceedings (2015)
3. Vaswani, A., Shazeer, N., Parmar, N., Uszkoreit, J., Jones, L., Gomez, A.N., Kaiser, Ł., Polosukhin, I.: Attention is All You Need, vol. 30 (2017)
4. Devlin, J., Chang, M., Lee, K., Toutanova, K.: BERT: pre-training of deep bidirectional transformers for language understanding. In: Proceedings of the 2019 Conference of the North American Chapter of the Association for Computational Linguistics: Human Language Technologies, NAACL-HLT 2019, Minneapolis, MN, USA, June 2–7, 2019, vol. 1 (Long and Short Papers), pp. 4171–4186. Association for Computational Linguistics, New York (2019)

5. Bojarski, M., Yeres, P., Choromanska, A., Choromanski, K., Firner, B., Jackel, L.D., Muller, U.: Explaining how a deep neural network trained with end-to-end learning steers a car. CoRR, vol. abs/1704.07911 (2017)
6. Boža, V., Brejová, B., Vinař, T.: DeepNano: deep recurrent neural networks for base calling in MinION nanopore reads. PloS One **12**(6), e0178751 (2017)
7. AI and Compute (2018). https://openai.com/blog/ai-and-compute/
8. Chen, Y., Chen, T., Xu, Z., Sun, N., Temam, O.: DianNao Family: Energy-Efficient Hardware Accelerators for Machine Learning, vol. 59, pp. 105–112, ACM, New York (2016)
9. Jouppi, N.P., Young, C., Patil, N., Patterson, D.A., Agrawal, G., Bajwa, R., Bates, S., Bhatia, S., Boden, N., Borchers, A., Boyle, R., Cantin, P., Chao, C., Clark, C., Coriell, J., Daley, M., Dau, M., Dean, J., Gelb, B., Ghaemmaghami, T.V., Gottipati, R., Gulland, W., Hagmann, R., Ho, C.R., Hogberg, D., Hu, J., Hundt, R., Hurt, D., Ibarz, J., Jaffey, A., Jaworski, A., Kaplan, A., Khaitan, H., Killebrew, D., Koch, A., Kumar, N., Lacy, S., Laudon, J., Law, J., Le, D., Leary, C., Liu, Z., Lucke, K., Lundin, A., MacKean, G., Maggiore, A., Mahony, M., Miller, K., Nagarajan, R., Narayanaswami, R., Ni, R., Nix, K., Norrie, T., Omernick, M., Penukonda, N., Phelps, A., Ross, J., Ross, M., Salek, A., Samadiani, E., Severn, C., Sizikov, G., Snelham, M., Souter, J., Steinberg, D., Swing, A., Tan, M., Thorson, G., Tian, B., Toma, H., Tuttle, E., Vasudevan, V., Walter, R., Wang, W., Wilcox, E., Yoon, D.H.: In-datacenter performance analysis of a tensor processing unit. In: Proceedings of the 44th Annual International Symposium on Computer Architecture, ISCA 2017, Toronto, ON, Canada, June 24–28, 2017, pp. 1–12. ACM, New York (2017)
10. Chen, Y., Krishna, T., Emer, J.S., Sze, V.: 14.5 Eyeriss: An energy-efficient reconfigurable accelerator for deep convolutional neural networks. In: 2016 IEEE International Solid-State Circuits Conference, ISSCC 2016, San Francisco, CA, USA, January 31–February 4, 2016, pp. 262–263. IEEE, New York (2016)
11. Venkataramani, S., Ranjan, A., Banerjee, S., Das, D., Avancha, S., Jagannathan, A., Durg, A., Nagaraj, D., Kaul, B., Dubey, P., Raghunathan, A.: ScaleDeep: A scalable compute architecture for learning and evaluating deep networks. In: Proceedings of the 44th Annual International Symposium on Computer Architecture, ISCA 2017, Toronto, ON, Canada, June 24–28, 2017, pp. 13–26. ACM, New York (2017)
12. Jain, A., Phanishayee, A., Mars, J., Tang, L., Pekhimenko, G.: Gist: Efficient data encoding for deep neural network training. In: Proceedings of the 45th ACM/IEEE Annual International Symposium on Computer Architecture, ISCA 2018, Los Angeles, CA, USA, June 1–6, 2018, pp. 776–789. IEEE Computer Society, New York (2018)
13. Hu, M., Strachan, J.P., Li, Z., Grafals, E.M., Davila, N., Graves, C., Lam, S., Ge, N., Yang, J.J., Williams, R.S.: Dot-product engine for neuromorphic computing: programming 1T1M crossbar to accelerate matrix-vector multiplication. In: Proceedings of the 53rd Annual Design Automation Conference, DAC 2016, Austin, TX, USA, June 5–9, 2016, pp. 19:1–19:6. ACM, New York (2016)
14. Fujiki, D., Mahlke, S.A., Das, R.: In-memory data parallel processor. In: Proceedings of the Twenty-Third International Conference on Architectural Support for Programming Languages and Operating Systems, ASPLOS 2018, Williamsburg, VA, USA, March 24–28, 2018, pp. 1–14. ACM, New York (2018)
15. Shafiee, A., Nag, A., Muralimanohar, N., Balasubramonian, R., Strachan, J.P., Hu, M., Williams, R.S., Srikumar, V.: ISAAC: A convolutional neural network accelerator with in-situ analog arithmetic in crossbars. In: Proceedings of the 43rd ACM/IEEE Annual International Symposium on Computer Architecture, ISCA 2016, Seoul, South Korea, June 18–22, 2016, pp. 14–26. IEEE Computer Society, New York (2016)
16. Song, L., Qian, X., Li, H., Chen, Y.: PipeLayer: A pipelined ReRAM-based accelerator for deep learning. In: Proceedings of the 2017 IEEE International Symposium on High Performance Computer Architecture, HPCA 2017, Austin, TX, USA, February 4–8, 2017, pp. 541–552. IEEE Computer Society, New York (2017)

17. Chen, F., Song, L., Chen, Y.: ReGAN: A pipelined ReRAM-based accelerator for generative adversarial networks. In: 23rd Asia and South Pacific Design Automation Conference, ASP-DAC 2018, Jeju, Korea (South), January 22–25, 2018, pp. 178–183. IEEE, New York (2018)
18. Chen, F., Song, L., Li, H., Chen, Y.: Marvel: A vertical resistive accelerator for low-power deep learning inference in monolithic 3D. In: Design, Automation & Test in Europe Conference and Exhibition, DATE 2021, Grenoble, France, February 1–5, 2021, pp. 1240–1245. IEEE, New York (2021)
19. Rumelhart, D.E., Hinton, G.E., Williams, R.J.: Learning representations by back-propagating errors. Nature **323**(6088), 533–536 (1986)
20. Chen, F.: PUFFIN: an efficient DNN training accelerator for direct feedback alignment in FeFET. In: IEEE/ACM International Symposium on Low Power Electronics and Design, ISLPED 2021, Boston, MA, USA, July 26–28, 2021, pp. 1–6. IEEE, New York (2021)
21. Nøkland, A.: Direct feedback alignment provides learning in deep neural networks. In: Advances in Neural Information Processing System, pp. 1037–1045 (2016).
22. Lillicrap, T.P., Cownden, D., Tweed, D.B., Akerman, C.J.: Random synaptic feedback weights support error backpropagation for deep learning. Nat. Commun. **7**(1), 1–10 (2016)
23. Launay, J., Poli, I., Boniface, F., Krzakala, F.: Direct feedback alignment scales to modern deep learning tasks and architectures. In: Advances in Neural Information Processing Systems, vol. 33 (2020), pp. 9346–9360
24. Müller, J., Böscke, T., Müller, S., Yurchuk, E., Polakowski, P., Paul, J., Martin, D., Schenk, T., Khullar, K., Kersch, A., et al.: Ferroelectric hafnium oxide: A CMOS-compatible and highly scalable approach to future ferroelectric memories. In: 2013 IEEE International Electron Devices Meeting, pp. 10–8. IEEE, New York (2013)
25. Jerry, M., Chen, P.-Y., Zhang, J., Sharma, P., Ni, K., Yu, S., Datta, S.: Ferroelectric FET analog synapse for acceleration of deep neural network training. In: Proceeding of the 2017 IEEE International Electron Devices Meeting (IEDM), pp. 6–2. IEEE, New York (2017)
26. Aziz, A., Ghosh, S., Datta, S., Gupta, S.K.: Physics-based circuit-compatible SPICE model for ferroelectric transistors. IEEE Electron Device Lett. **37**(6), 805–808 (2016)
27. George, S., Ma, K., Aziz, A., Li, X., Khan, A., Salahuddin, S., Chang, M.-F., Datta, S., Sampson, J., Gupta, S., et al.: Nonvolatile memory design based on ferroelectric FETs. In: Proceedings of the 53rd Annual Design Automation Conference, pp. 1–6 (2016)
28. Deng, S., Yin, G., Chakraborty, W., Dutta, S., Datta, S., Li, X., Ni, K.: A comprehensive model for ferroelectric FET capturing the key behaviors: Scalability, variation, stochasticity, and accumulation. In: 2020 IEEE Symposium on VLSI Technology. IEEE, New York, pp. 1–2 (2020)
29. Choi, W.H., Lv, Y., Kim, J., Deshpande, A., Kang, G., Wang, J.-P., Kim, C.H.: A magnetic tunnel junction based true random number generator with conditional perturb and real-time output probability tracking. In: 2014 IEEE International Electron Devices Meeting. IEEE, New York, pp. 12–5 (2014)
30. LeCun, Y., et al.: The MNIST Database of Handwritten Images (2012)
31. Krizhevsky, A., Hinton, G., et al.: Learning Multiple Layers of Features from Tiny Images (2009)
32. LeCun, Y., Bottou, L., Bengio, Y., Haffner, P.: Gradient-based learning applied to document recognition. Proc. IEEE **86**(11), 2278–2324 (1998)
33. Jia, Y., Shelhamer, E., Donahue, J., Karayev, S., Long, J., Girshick, R., Guadarrama, S., Darrell, T.: Caffe: Convolutional architecture for fast feature embedding. In: Proceedings of the 22nd ACM International Conference on Multimedia, pp. 675–678 (2014)
34. Guo, H., Tang, R., Ye, Y., Li, Z., He, X.: DeepFM: A factorization-machine based neural network for CTR prediction. In: Proceedings of the Twenty-Sixth International Joint Conference on Artificial Intelligence, IJCAI 2017, Melbourne, Australia, August 19–25, 2017, pp. 1725–1731, ijcai.org (2017)
35. Cheng, W., Shen, Y., Huang, L.: Adaptive factorization network: Learning adaptive-order feature interactions. In: Proceedings of the AAAI Conference on Artificial Intelligence, pp. 3609–3616. AAAI Press, New York (2020)

36. Criteo Dataset. http://labs.criteo.com/downloads/2014-kaggle-displayadvertising-challenge-dataset/
37. Zhang, C., Li, P., Sun, G., Guan, Y., Xiao, B., Cong, J.: Optimizing FPGA-based accelerator design for deep convolutional neural networks. In; Proceedings of the 2015 ACM/SIGDA International Symposium on Field-Programmable Gate Arrays, Monterey, CA, USA, February 22–24, 2015, pp. 161–170. ACM, New York (2015)
38. Kim, E., Lee, M., Kim, J.: 8.2 8Mb/s 28Mb/mJ robust true-random-number generator in 65nm CMOS based on differential ring oscillator with feedback resistors. In: Proceedings of the 2017 IEEE International Solid-State Circuits Conference, ISSCC 2017, San Francisco, CA, USA, February 5–9, 2017, pp. 144–145. IEEE, New York (2017)
39. Wilton, S.J., Jouppi, N.P.: CACTI: An enhanced cache access and cycle time model. IEEE J. Solid State Circuits $31(5)$, 677–688 (1996)
40. Predictive Technology Model. http://ptm.asu.edu/ (2015)
41. Dong, X., Xu, C., Xie, Y., Jouppi, N.P.: NVSim: A circuit-level performance, energy, and area model for emerging nonvolatile memory. IEEE Trans. Comput. Aided Des. Integr. Circuits Syst. $31(7)$, 994–1007 (2012)

Platform-Based Design of Embedded Neuromorphic Systems

M. L. Varshika and Anup Das

1 Introduction

Neuromorphic computing is emerging as an attractive candidate for power-constrained systems such as embedded devices and edge nodes. Neuromorphic computing operates on the design principles of the central nervous system in primates. It has the potential to drive the development of a more distributed, scalable, and efficient computing paradigm. Historically, the term neuromorphic computing was coined in the late '80s to describe a type of analog computing hardware that mimics the architecture of a mammalian brain [69]. Initially, the primary goal of neuromorphic computing was to emulate the physical properties of neurons and synapses exploiting the physics of analog complementary metal–oxide–semiconductor (CMOS) electronics. This is to understand and reproduce the efficiency of neural computing systems. Today, neuromorphic computing addresses a broader range of computing systems designed using digital, mixed-signal (analog/digital) CMOS electronics, and novel emerging non-volatile memory (NVM) technology elements. Yet, in all neuromorphic systems, the aim is to build architectures that can execute machine learning applications designed using spiking neural networks (SNNs). SNNs represent the third and more bio-inspired generation of neural networks [64]. SNNs enable powerful computations due to their spatio-temporal information encoding capabilities [78]. SNNs can implement different machine learning approaches such as supervised learning [100], unsupervised learning [24], reinforcement learning [59], and lifelong learning [86].

In an SNN, neurons are connected via synapses. A neuron can be implemented as an integrate-and-fire (IF) logic [15], which is illustrated in Fig. 1 (left). Here, an

M. L. Varshika (✉) · A. Das
Electrical and Computer Engineering, Drexel University, Philadelphia, PA, USA
e-mail: lm3486@drexel.edu; anup.das@drexel.edu

© The Author(s), under exclusive license to Springer Nature Switzerland AG 2024
S. Pasricha, M. Shafique (eds.), *Embedded Machine Learning for Cyber-Physical, IoT, and Edge Computing*, https://doi.org/10.1007/978-3-031-19568-6_12

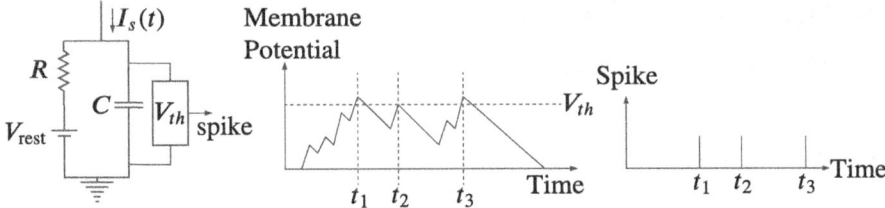

Fig. 1 A leaky-integrate-and-fire (LIF) neuron with current input $U(t)$ (left). The membrane potential over time of the neuron (middle). The spike output of the neuron representing its firing times (right). Each firing time represents the time instance when the membrane potential crosses the firing threshold

input current spike $U(t)$ from a pre-synaptic neuron raises the membrane voltage of a post-synaptic neuron. When this voltage crosses a threshold V_{th}, the IF logic emits a spike, which propagates to its post-synaptic neurons. Figure 1 (middle) illustrates the membrane voltage due to input spike trains. Moments of threshold crossing, i.e., the firing times, are illustrated in Fig. 1 (right).

Over the past decades, there has been a significant amount of progress made on neuromorphic computing, on both the software (e.g., application and algorithm [3, 14, 22, 27, 31, 36, 46, 82, 88, 111]) and hardware (e.g., architecture and technology [17, 39, 43, 67, 87]) fronts. These architectures and algorithms differ from conventional computing paradigms for their memory and communication structures and computational properties. While traditional von Neumann architectures have one or more central processing units physically separated from the main memory, neuromorphic architectures exploit co-localization of memory and compute, near and in-memory computation [49]. Alongside to the tremendous progress in devising novel neuromorphic computing architectures, there have been many recent works that address how to map and compile (trained) SNN models for efficient execution in neuromorphic hardware [2, 4–8, 10–12, 23, 26, 28, 34, 40, 41, 45, 48, 52, 61–63, 81, 90, 91, 95, 102, 105, 110].

To cope with the growing complexity of neuromorphic systems, challenges in integrating emerging NVM technologies, and faster time-to-market pressure, efficient design methodologies are needed. Here, we discuss one such methodology, that of platform-based design.

2 Platform-Based Design Methodology

Platform-based design has emerged as an important design style for the electronics industry [35, 55, 74, 84, 85, 89]. Platform-based design separates parts of the system design process such that they can be independently optimized for different metrics such as performance, power, cost, and reliability. Platform-based design methodology can also be adopted for neuromorphic system design [10], where

Fig. 2 Illustration of the
platform-based design
methodology. Here, the
hardware design space
exploration (DSE) is
performed independent of the
application and mapping
(application allocation on
hardware) DSEs. A design
point is obtained by pruning
the design spaces of these
explorations

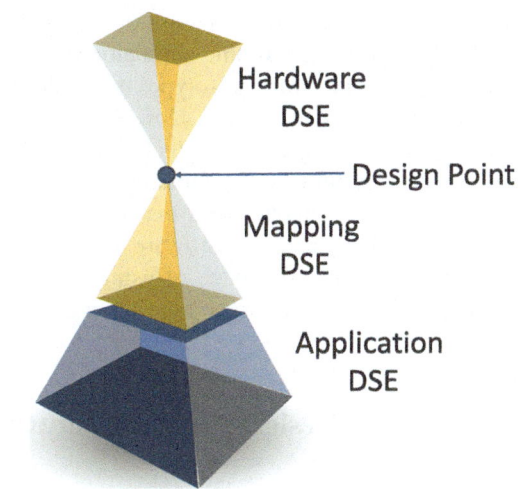

the software can be optimized independently from the underlying neuromorphic
hardware platform. Figure 2 shows this design methodology for a general electronic
system design. Here, the hardware design space is explored to generate a platform
that satisfies the target design cost. This could include a combination of recurring
and non-recurring design costs. Alongside the hardware development, the software
design space is also explored. Here, the software optimization includes allocation
of tasks of a given application to the hardware computing units for a specific design
objective.

As in a conventional computing system, the abstractions for a neuromorphic
system include: (1) the *application software*, (2) the *system software*, and (3) the
hardware [33, 51, 77]. In the context of neuromorphic computing, the application
software includes applications designed using different SNN topologies such as
multi-layer perceptron (MLP) [27], convolutional neural network (CNN) [71] and
recurrent neural network (RNN) [31], and bio-inspired learning algorithms such
as spike timing-dependent plasticity (STDP) [19], long-term plasticity (LTP) [25],
and FORCE [72]. The system software includes the equivalent of a compiler
and a runtime manager to execute SNN applications on the hardware. Finally,
the hardware abstraction includes the platform, which consists of a neuromorphic
hardware.

We focus on the system software abstraction and its design space exploration.
A key optimization objective is the performance of machine learning workloads
on the hardware. Here, we distinguish between application-level and system-
level performance metrics. Examples of application-level metrics include accuracy,
peak signal-to-noise ratio (PSNR), and structural similarity index measure (SSIM).
System-level metrics include throughput, latency, energy, and reliability.

A key component of the software design space exploration is to estimate/evaluate
the system-level performance. The way to obtain the most realistic performance
estimates is measuring it on the real hardware. However, this is often not available

until late in the design process. An alternative is simulating the workload on a cycle-accurate neuromorphic simulator such as NeuroXplorer [10]. However, this is rather slow. Hardware prototypes on a field-programmable gate array (FPGA) are also a viable alternative [23]. However, the high synthesis time makes the design space exploration infeasible. Finally, transaction-level simulators can also be used to estimate system-level performance. Here, high-level abstractions are often used to represent the behavior of a machine learning workload. Such abstractions can then be simulated using a hardware architecture and mapping of the application to the hardware. The advantage of using such a simulator is that it is faster than a cycle-accurate simulator or synthesizing a prototype on FPGA.

3   Software Design Space Exploration

Cycle-accurate neuromorphic simulators are those that accurately model the behavior of a neuromorphic system on a cycle-by-cycle basis. We focus on one such simulator—NeuroXplorer [10]. To understand the basics of this simulator and how it is used for the software design space exploration within the platform-based design methodology, we focus on the generic tile-based architecture of a neuromorphic system [83], where tiles are interconnected via a shared interconnect. A tile may include: (1) a neuromorphic core, which implements neuron and synapse circuitries, (2) peripheral logic to encode and decode spikes into address event representation (AER), and (3) a network interface to send and receive AER packets from the interconnect. Switches are placed on the interconnect to route AER packets to their destination tiles.

NVM devices present an attractive option for implementing synaptic storage due to their demonstrated potential for low-power multi-level operations and high integration densities [13, 50, 73, 99, 104]. Recently, several NVMs are being explored for neuromorphic computing: Oxide-based Resistive Random Access Memory (ReRAM) [107], phase change memory (PCM) [108], ferroelectric RAM [1], and Spin-Transfer Torque Magnetic or Spin-Orbit-Torque RAM (STT- and SoT-MRAM) [47]. Figure 3 shows a neuromorphic hardware with tiles (C) and switches (S). Without loss of generality, we show each tile as a crossbar, where NVM cells are organized in a two-dimensional grid formed using horizontal and vertical wires.

The figure also illustrates a small example of implementing an SNN on a crossbar. Synaptic weights w_1 and w_2 are programmed as conductance of NVM cells P1 and P2, respectively. The output spike voltages, v_1 from N1 and v_2 from N2, inject currents into the crossbar, which are obtained by multiplying a presynaptic neuron's output spike voltage with the NVM cell's conductance (Ohm's law). Current summations along columns are performed in parallel (Kirchhoff's current law), and they implement the sum $\sum_j w_i v_i$ (i.e., neuron excitations).

Figure 4 shows the detailed architecture of the NeuroXplorer. It includes an architecture simulator, which can be configured to simulate a specific neuromorphic architecture such as TrueNorth [38], Loihi [37], DYNAPs [70], and μBrain [98].

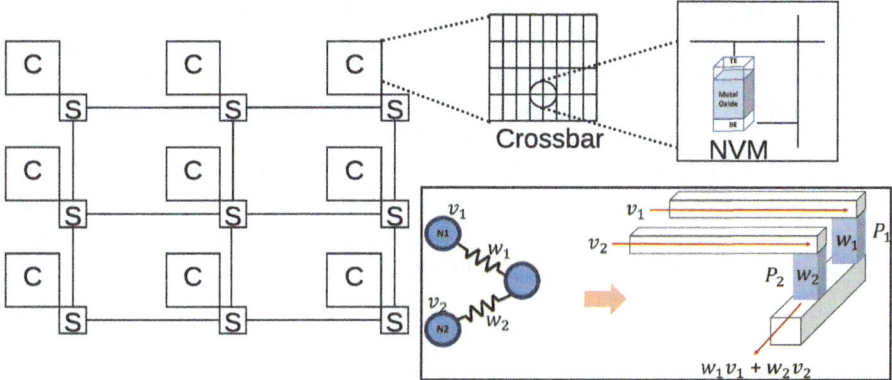

Fig. 3 A representative tile-based neuromorphic hardware [93] (left). Each tile is a neuromorphic core, which can map a few neurons and synapses. In its simplest form, a neuromorphic core can be implemented as a crossbar as shown in the figure. An NVM device may be connected at the intersection of a bitline and a wordline to store synaptic weight. The bottom-right corner illustrates the mapping of a 2×1 neural network on a crossbar

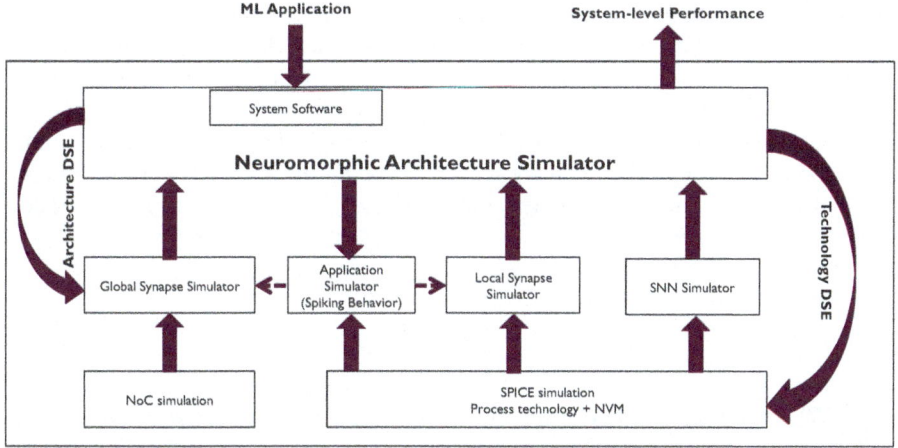

Fig. 4 A detailed architecture of the NeuroXplorer simulator [10]. It consists of an architecture simulator at the frontend and is integrated with hardware-level simulators at the backend. NeuroXplorer can perform both architectural DSE and technology DSE

Internally, the architecture simulator uses models for: (1) local synapses, i.e., the tiles, and (2) global synapses, i.e., the shared interconnect.

The system software component is what facilitates the design space exploration. It consists of two steps—clustering and mapping. An SNN application typically consists of many neurons and synapses, well beyond the capacity of a single core. Therefore, the application must first be partitioned into clusters, where each cluster consists of a subset of neurons and synapses of the application. Recently, several graph-based SNN partitioning approaches are proposed. Most of these are based on the Kernighan–Lin Graph Partitioning algorithm [54], minimizing the inter-cluster spike communication. Once clusters are generated from an application, they are mapped to different cores of the hardware. Mapping of a cluster to a core involves allocating pre-synaptic neurons to the input of the core, post-synaptic neurons to the output of the core, and programming the synaptic weights connecting a pre- to a post-synaptic neuron as the conductance of an NVM cell in the core.

In the following, we discuss how the system software component is incorporated inside a design space exploration (DSE) framework to perform system-level performance, energy, and reliability optimizations.

3.1 Performance-Oriented DSE

The problem of mapping neuron clusters to the neuromorphic architecture shares some similarity with the task mapping problem in multiprocessors, which has been investigated extensively in the past [20, 29, 30, 32, 53, 68, 75, 76, 89]. However, the difference is that while high latency of the hardware can impact the application's execution time (i.e., its real-time properties) in the case of multiprocessor systems, high latency for neuromorphic architectures can create distortion (change) of its inter-spike interval (ISI) and spike disorder, leading to additional impact on the application's accuracy.

In fact, when clusters are mapped to cores, inter-cluster spike communications are mapped on to the shared interconnect of the hardware. Considering a mesh-based two-dimensional interconnect architecture, the average latency experienced by spikes on the interconnect is

$$L = \sum_{i=1}^{N_s} [(h_i - 1) * l_w + h_i * l_s]/N_s, \tag{1}$$

where N_s is the total number of spikes on the shared interconnect, h_i is the number of hops a spike traverses between the source and destination, l_w is the interconnect segment delay, and l_s is the delay of the hop.

To analyze the impact of latency on application performance, Fig. 5 shows the accuracy numbers achieved via an ad hoc mapping of the clusters to the cores. Compared to an accuracy of 86% obtained through software-based simulations, the accuracy on the 6×6 neuromorphic hardware (36 crossbars with 25 neurons per crossbar) is only 66.7%—a loss of $\approx 20\%$. This loss is due to the latency on the hardware, which delays some spikes more than others.

Figure 6 summarizes how the latency, ISI distortion, spike disorder increase as we increase the size of the neuromorphic architecture. We observe that as we increase the number of crossbars in the hardware, latency, ISI, and spike disorder increase. This is because with an increase in the number of crossbars, spike traffic on the shared interconnect increases, which increases the congestion and delays some

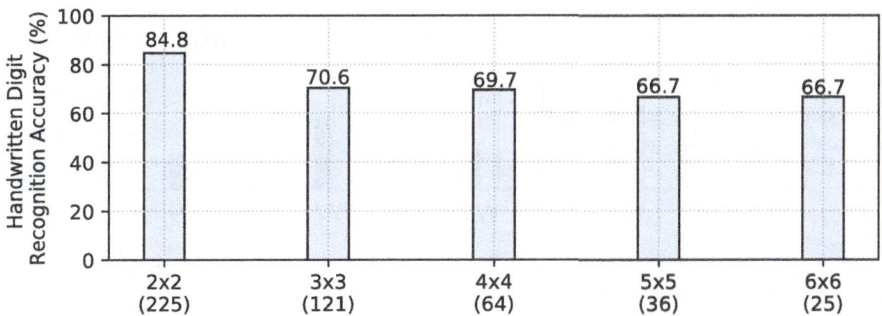

Fig. 5 Handwritten-digit recognition accuracy on different neuromorphic hardware configurations. This accuracy is lower than the 86% accuracy obtained via software simulation

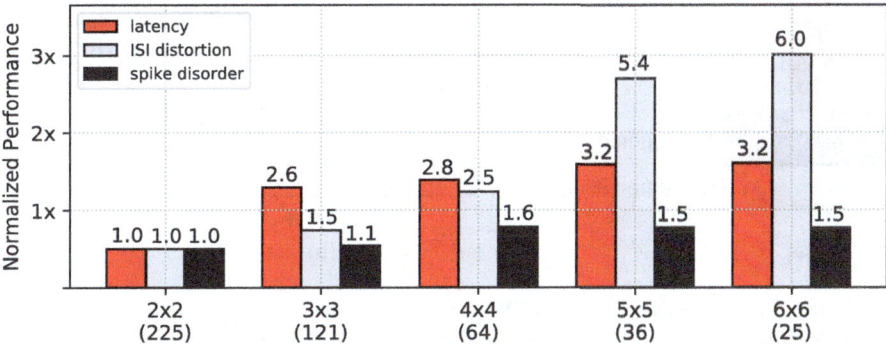

Fig. 6 Latency, inter-spike interval (ISI) distortion, and spike disorder for five neuromorphic hardware configurations, normalized with respect to the baseline configuration of 2×2

spikes more than others. When we use a hardware with 36 small crossbars arranged in a 6×6 mesh, we observe a significant increase of latency (average 3.2x), ISI distortion (average 6x), and spike disorder (average 1.5x) compared to the baseline configuration of using 4 large crossbars.

In our recent work [6], we propose SpiNeMap to place cluster to cores minimizing the average spike latency on the interconnect. SpiNeMap operates in two steps. In Step 1 (SpiNeCluster), we use a heuristic-based clustering algorithm to partition SNNs into local and global synapses, with local synapses mapped within crossbars, and global synapses to the shared interconnect. SpiNeCluster minimizes spikes on the shared interconnect, reducing spike congestion and ISI distortion. In Step 2 (SpiNePlacer), we use an instance of the particle swarm optimization (PSO) to place clusters on physical crossbars within the tiles in the hardware, optimizing energy consumption and spike latency on the shared interconnect.

Figure 7 shows the spike latency of SpiNeMap normalized to a baseline mechanism for ten workloads. These workloads are defined in [6]. The baseline mechanism randomly places clusters to cores without considering spike latency on the shared interconnect.

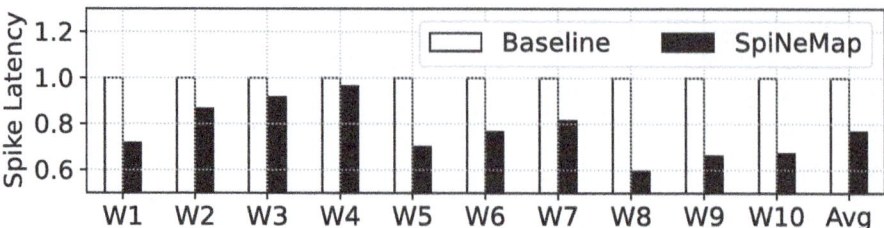

Fig. 7 Spike latency of SpiNeMap, normalized to baseline. High spike latency leads to larger inter-spike interval distortion, which directly impacts the latency

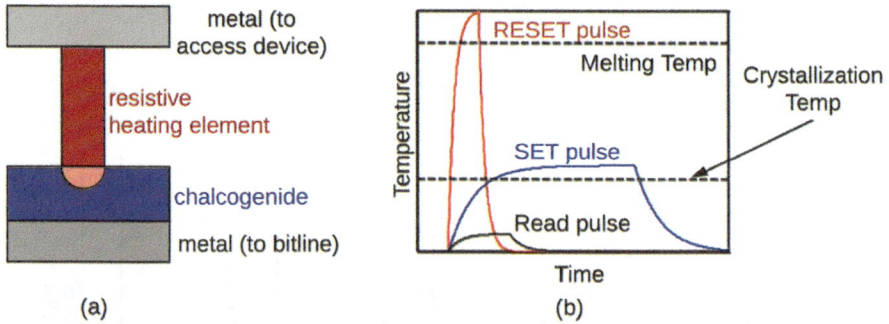

Fig. 8 (**a**) A phase change memory (PCM) cell designed using chalcogenide material that can be either in amorphous or in crystalline state, and (**b**) current needed for SET (amorphous to crystalline transition), RESET (crystalline to amorphous transition), and read (identify the state) operations

3.2   Energy-Oriented DSE

There are several sources of energy consumption in an NVM-based neuromorphic hardware. In [102], we have formulated the detailed energy consumption in a neuromorphic hardware considering phase-change memory (PCM), an emerging memory technology that has shown significant potential as synapse in a neuromorphic hardware [18]. To introduce this technology, Fig. 8a illustrates a chalcogenide semiconductor alloy that is used to build a PCM cell. The amorphous phase (logic "0") has higher resistance than its crystalline phase (logic "1"). When used in these two stable states, a PCM cell can implement a binary synapse. However, with precise control of the crystallization process, a PCM cell can be programmed in a partially crystallized state. This way, the PCM cell can implement a multi-bit synapse.

A phase change in a PCM cell is induced via Joule's heating by injecting current into the resistor-chalcogenide junction and heating the chalcogenide alloy. Figure 8b shows the different current values needed to program and read in a PCM device. Therefore, depending on whether a PCM device in a crossbar is programmed to SET state, RESET state, or one of the intermediate states, different amount of energy will be required to propagate a spike through the device.

We formulate the spike energy through a PCM synaptic device in a crossbar as

$$e_{synapse} = I_{prog}^2 \cdot t_{spk} \cdot \left(R_{ON} + \frac{1}{w} \right), \tag{2}$$

where w is the conductance of the PCM cell, I_{prog} is the programming current, t_{spk} is the spike duration (typically in ms), and R_{ON} is the ON resistance of the access transistor connecting the PCM cell to the bitline and wordline in a crossbar.

Inside a crossbar, the programming current (I_{prog}) can vary considerably due to bitline and wordline parasitics. With technology scaling, the value of parasitic resistances along the bitline and wordline of a current path increases [42]. The unit wordline (bitline) parasitic resistance ranges from approximately 2.5Ω (1Ω) at 65nm node to 10Ω (3.8Ω) at 16nm node. The values of these unit parasitic resistances are expected to scale further reaching $\approx 25\Omega$ at 5nm node [42]. This increase in the value of unit parasitic resistance increases the voltage drop, which increases the current variation in a crossbar. Figure 9 shows the current variation in a 128×128 PCM crossbar. We observe that the programming current is higher

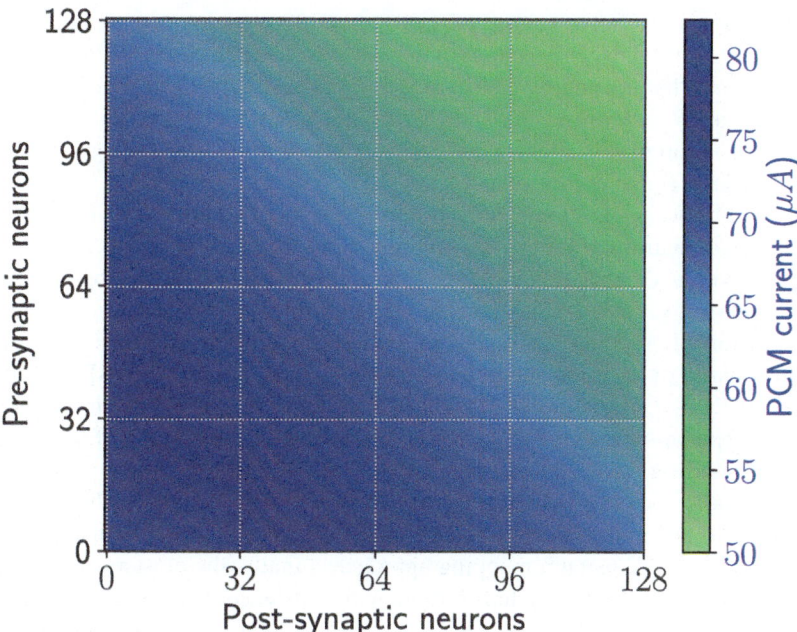

Fig. 9 Current map in a 128×128 crossbar. There are fewer parasitics on the current path at the bottom-left corner of the crossbar. So the IR drop is lower and, therefore, the current is higher. There are more parasitics on the current path at the top-right corner. So, the IR drop is higher and, therefore, the current is lower. Overall, current varies within a crossbar

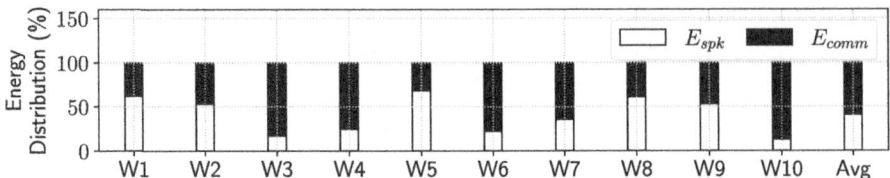

Fig. 10 Total energy distributed into spike propagation energy, which is the energy consumed in propagating spikes via the synaptic cells of a crossbar, and communication energy, which is the energy consumed in propagating spikes via the shared interconnect

for spikes propagating through a synaptic cell located at the bottom-left corner than through a synaptic cell located at the top-right corner.

Spike propagation energy, together with the energy needed to communicate spikes over the interconnect, can then be used to compute the total energy. Therefore, that energy consumption depends on: (1) how an SNN model is partitioned into clusters (determines the number of neurons and synapses in each cluster), (2) how the clusters are mapped to the cores (determines the hop distances), and 3) how the neurons and synapses of a cluster are placed inside each core (determines the propagation current, I_{prog}).

Using the total energy formulation, Fig. 10 shows the distribution of the two energy components—core energy (E_{spk}) and communication energy (E_{comm}) for ten different workloads. Description to these workloads can be found in [102].

We observe that the energy distribution is workload-dependent. For some workloads such as W6, W7, and W10, the communication energy dominates the total energy consumption. For other workloads such as W5, the core energy dominates the total energy consumption.

Figure 11 shows our framework to perform energy-aware DSE. The left subfigure shows a neuromorphic system comprising of the application layer, the system software layer, and the hardware layer. The application layer at the top consists of the user space to run machine learning applications. In this illustration, we show the execution of AlexNet for ImageNet classification. The hardware layer at the bottom consists of the neuromorphic hardware such as TrueNorth [38], Loihi [37], and DYNAPs [70]. At the middle is the system software layer, which interacts with both the application and hardware layers. The system software performs energy optimization using the iterative approach shown to the right.

The workflow of the system software involves clustering a machine learning application to generate clustered SNN graph. Next, the clusters are mapped to the tiles of the hardware using a mapping approach. Finally, the clusters are placed to crossbars using the placement step. Although the clustering step could potentially be incorporated inside the iterative loop, we placed it outside to limit the complexity of the design space exploration. In fact, clustering of applications is an NP-hard problem as shown in SpiNeMap [6]. Our clustering approach uses the graph partitioning algorithm of SpiNeMap, minimizing: (1) inter-cluster communication (similar to SpiNeMap) and (2) maximizing cluster utilization (similar to Decom-

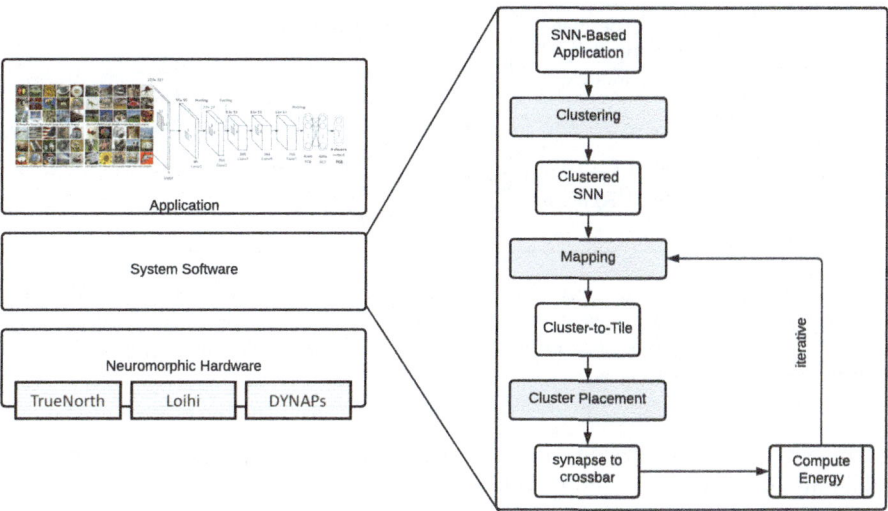

Fig. 11 Our energy-aware system software for neuromorphic hardware. A typical hardware is represented as three-layer architecture (left), with the application at the top, the system software at the middle, and the hardware at the bottom. The hardware layer may have DYNAPs [70], Loihi [37], or TrueNorth [38] hardware. The system software is elaborated to the right. An SNN-based application is first partitioned to for clusters that can be mapped to cores. Next, using an iterative approach, the mapping is refined over time to achieve a desired objective

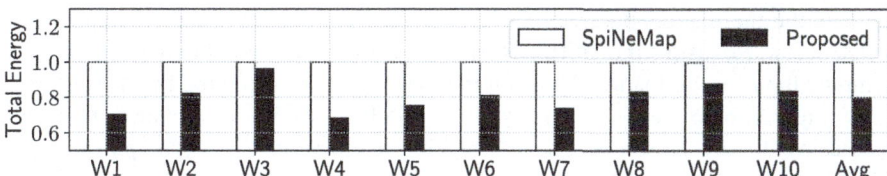

Fig. 12 Total energy normalized to SpiNeMap [6]. The proposed platform-based design is able to reduce the total energy consumption significantly using the iterative mapping approach

posedSNN [9]). The iterative approach is based on a Hill Climbing heuristic, which is described in [102].

Figure 12 reports the total energy consumption for each workload for the proposed energy-aware DSE normalized to SpiNeMap. The workloads are defined in [102]. We observe that the proposed DSE reduces energy consumption by 20%.

3.3 Reliability-Oriented DSE

Using NVMs such as PCM as synaptic devices in a crossbar leads to several reliability issues due to high temperature and current requirements for these devices.

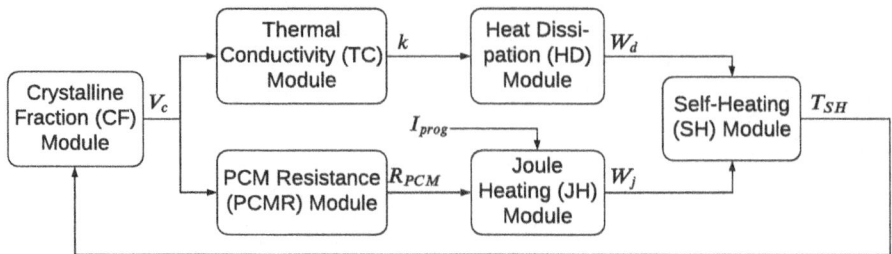

Fig. 13 Iterative approach to calculating the self-heating temperature of a PCM cell during amorphization. The PCM cell is initially programmed in its crystalline state. Then a current is injected, which triggers a raise in the temperature. Through Joule's heating, the PCM cell transitions to its amorphous state

Here, we show how design space explorations can be performed to improve reliability of NVMs in a neuromorphic hardware.

In our recent work [103], we use the phenomenological endurance model [97] to compute the endurance of a PCM cell as a function of its self-heating temperature obtained during amorphization of its crystalline state. Figure 13 shows the iterative approach to compute this self-heating temperature (T_{SH}) [109].

At start of the amorphization process, the temperature of a PCM cell is equal to the ambient temperature T_{amb}. Subsequently, the PCM temperature is computed iteratively as follows. For a given crystalline fraction V_C of the GST material within the cell, the thermal conductivity k is computed using the TC Module, and PCM resistance R_{PCM} using the PCMR Module. The thermal conductivity is used to compute the heat dissipation W_d using the HD Module, while the PCM resistance is used to compute the Joule heating in the GST W_j for the programming current I_{prog} using the JH Module. The self-heating temperature T_{SH} is computed inside the SH Module using the Joule heating and the heat dissipation. Finally, the self-heating temperature is used to compute the crystallization fraction V_c using the CF Module. The iterative process terminates when the GST is amorphized, i.e., $V_c = 0$. We now describe these steps:

- **Crystallization Fraction (CF) Module:** CF represents the fraction of solid in a GST during the application of a reset current. V_c is computed using the Johnson–Mehl–Avrami (JMA) equation as

$$V_c = \exp\left[-\alpha \times \frac{(T_{SH} - T_{amb})}{T_m} \times t\right],\tag{3}$$

  where t is the time, $T_m = 810K$ is the melting temperature of the GST material [66, 109], T_{amb} is the ambient temperature computed using [101], and $\alpha = 2.25$ is a fitting constant [66, 109].
- **Thermal Conductivity (TC) Module:** TC of the GST is computed as [60]

$$k = (k_a - k_c) \times V_c + k_a,\tag{4}$$

where $k_a = 0.002 \text{WK}^{-1}\text{cm}^{-1}$ for amorphous GST, and $k_c = 0.005 \text{WK}^{-1}\text{cm}^{-1}$ for crystalline GST [66, 109].

- **PCM Resistance (PCMR) Module:** The effective resistance of the PCM cell is given by

$$R_{PCM} = R_{set} + (1 - V_c) \times (R_{reset} - R_{set}), \tag{5}$$

where $R_{set} = 10\text{K}\Omega$ in the crystalline state of the GST and $R_{reset} = 200\text{K}\Omega$ in the amorphous state.

- **Heat Dissipation (HD) Module:** Assuming heat is dispersed to the surrounding along the thickness of the PCM cell, HD is computed as [58]

$$W_d = \frac{kV}{l^2}(T_{SH} - T_{amb}), \tag{6}$$

where $l = 120$ nm is the thickness and $V = 4 \times 10^{-14} \text{cm}^3$ is the volume of GST [66, 109].

- **Joule Heating (JH) Module:** The heat generation in a PCM cell due to the programming current I_{prog} is

$$W_j = I_{prog}^2 \times R_{PCM}. \tag{7}$$

- **Self-heating (SH) Module:** The SH temperature of a PCM cell is computed by solving an ordinary differential equation as [109]

$$T_{SH} = \frac{I_{prog}^2 R_{PCM} l^2}{kV} - \left[1 - \exp\left(-\frac{kt}{l^2 C}\right)\right] + T_{amb}, \tag{8}$$

where $C = 1.25 \text{JK}^{-1}\text{cm}^{-3}$ is the heat capacity of the GST [66, 109].

The endurance of a PCM cell is computed as [97]

$$\text{Endurance} \approx \frac{t_f}{t_s}, \tag{9}$$

where t_f and t_s are, respectively, the failure time and the switching time. In this model, to switch memory state of a PCM cell, an ion (electron) must travel a distance d across insulating matrix (the gate oxide) upon application of the programming current I_{prog}, which results in the write voltage V across the cell. Assuming thermally activated motion of an ion with activation energy U_s and local self-heating thermal temperature T_{SH}, the switching speed can be approximated as

$$t_s = \frac{d}{v_s} \approx \frac{2d}{fa} \exp\left(\frac{U_s}{k_B T_{SH}}\right) \exp\left(-\frac{qV}{2k_B T_{SH}}\frac{a}{d}\right), \tag{10}$$

where $d = 10$nm, $a = 0.2$nm, $f = 10^{13}$Hz, and $U_s = 2$eV [97].

The failure time is computed considering that the endurance failure mechanism is due to thermally activated motion of ions (electrons) across the same distance d but with higher activation energy U_F, so that the average time to failure is

$$t_f = \frac{d}{v_f} \approx \frac{2d}{fa} \exp\left(\frac{U_f}{k_B T_{SH}}\right) \exp\left(-\frac{qV}{2k_B T_{SH}}\frac{a}{d}\right), \tag{11}$$

where $U_f = 3ev$ [97].

The endurance, which is the ratio of average failure time and switching time, is given by

$$\text{Endurance} \approx \frac{t_f}{t_s} \approx \exp\left(\frac{\gamma}{T_{SH}}\right), \tag{12}$$

where $\gamma = 1000$ is a fitting parameter [97].

Figure 14 plots the temperature and endurance maps of a 128×128 crossbar at 65nm process node with $T_{amb} = 298K$. The PCM cells at the bottom-left corner have higher self-heating temperature than at the top-right corner. This asymmetry in the self-heating temperature creates a wide distribution of endurance, ranging from 10^6 cycles for PCM cells at the bottom-left corner to 10^{10} cycles at the top-right corner. These endurance values are consistent with the values reported for recent PCM chips from IBM [16].

Figure 15 shows a high-level overview of the proposed design space exploration, consisting of three abstraction layers—the application layer, system software

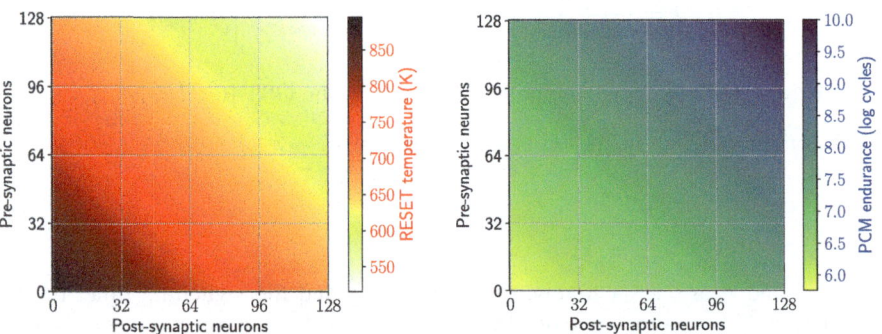

(a) Thermal map for PCM RESET operations in a 128x128 crossbar.

(b) Endurance map of the PCM cells in a 128x128 crossbar.

Fig. 14 Temperature and endurance map of a 128×128 crossbar at 65nm process node with the ambient temperature T_{amb} set to 298K. Temperature varies widely within a crossbar (**a**). Bottom-left corner is at higher temperature than the top-right corner due to difference in the parasitic elements. This thermal variation leads to endurance variation within the crossbar (**b**)

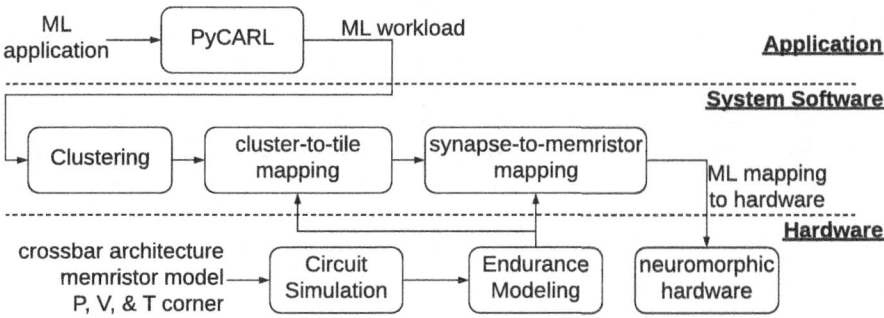

Fig. 15 High-level overview of the reliability (endurance)-aware design space exploration. A machine learning application is first analyzed using PyCARL [2] to exact workload information. This is then used to cluster the workload. Following this, clusters are mapped to tiles, and synapses within a cluster are placed to the NVM cells of crossbar. Device characterization data are used to do these mappings

layer, and hardware layer. A machine learning application is first simulated using PyCARL [2], which uses CARLsim [21] for training and testing of SNNs. PyCARL estimates spike times and synaptic strength on every connection in an SNN. This constitutes the workload of the machine learning application. The proposed framework maps and places neurons and synapses of a workload to crossbars of a neuromorphic hardware, improving the effective lifetime. To this end, a machine learning workload is first analyzed to generate clusters of neurons and synapses, where each cluster can fit on a crossbar. It uses the Kernighan–Lin Graph Partitioning algorithm of SpiNeMap [6] to partition an SNN workload, minimizing the spike latency. Next, it uses an instance of PSO to map the clusters to the cores of a hardware, maximizing the minimum effective lifetime of PCM devices in each core's crossbar. Synapses of a cluster are implemented on PCM using a synapse-to-memristor mapping, ensuring that those with higher activation are mapped to PCM cells with higher endurance, and vice versa.

Figure 16 compares the effective endurance lifetime obtained using the proposed DSE compared to SpiNeMap for 10 workloads. These workloads are described in [103]. We observe that the effective inference lifetime of the proposed framework is higher than SpiNeMap by an average 3.5x.

Limited write endurance is not the only reliability issue in a PCM crossbar. In our recent works [8, 92–94], we show that elevated voltages and currents needed to operate PCM cause aging of CMOS-based transistors in each neuron and synapse circuit in the hardware, drifting the transistor's parameters from their nominal values. Aggressive device scaling increases power density and temperature, which accelerates the aging, challenging the reliable operation of neuromorphic systems.

One important aging mechanism at scaled technology nodes is the bias temperature instability (BTI). This is a failure mechanism in a CMOS device where positive charges are trapped at the oxide–semiconductor boundary underneath the

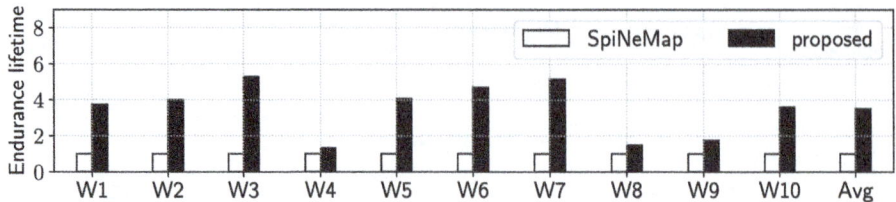

Fig. 16 Effective lifetime for the evaluated applications. The effective lifetime is defined as the number of inference operations that can be successfully performed between two successive reprogramming of the synaptic cells in a crossbar [79, 80, 96]

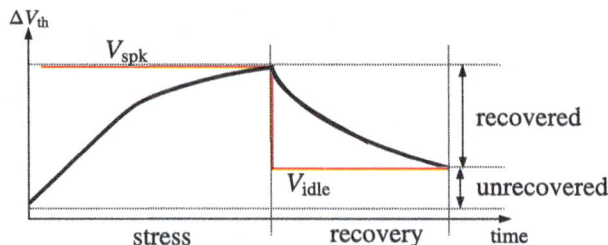

Fig. 17 Demonstration of threshold voltage degradation of a CMOS transistor due to bias temperature instability (BTI) aging. When a stress voltage is applied, the CMOS transistor parameters drift from their nominal values, thereby shifting the threshold voltage. Upon removal of the stress voltage, the threshold voltage recovers partially. The amount of unrecovered threshold voltage depends on the time duration for which the transistor is exposed to the stress voltage

gate [44]. BTI manifests as: (1) decrease in drain current and transconductance and (2) increase in off current and threshold voltage.

Recent works such as [44, 56, 57, 106] suggest that BTI is the collective response of two independent defects—the *as-grown hole traps* (AHTs) and *generated defects* (GDs). AHTs and a small proportion of GDs can be recovered by annealing at high temperatures if the BTI stress voltage is removed (*de-stress*). Figure 17 illustrates the stress and recovery of the threshold voltage of a CMOS transistor on application of a high (V_{spk}) and a low voltage (V_{idle}). We observe that both stress and recovery depend on the time of exposure to the corresponding voltage level. This implies that when a neuron is idle, the BTI aging of the neuron recovers from stress.

Figure 18 shows the shift in threshold voltage of a NMOS transistor in a neuron for continuous usage with a constant firing rate of 50Hz.

BTI aging can also be incorporated in the design space exploration with the objective of improving the lifetime.

Figure 19 reports the mean time to failure (MTTF) of the proposed reliability-oriented design space exploration normalized to SpiNeMap for 10 workloads, which are described in [93]. We observe that the average MTTF is 18% higher using the reliability-oriented exploration.

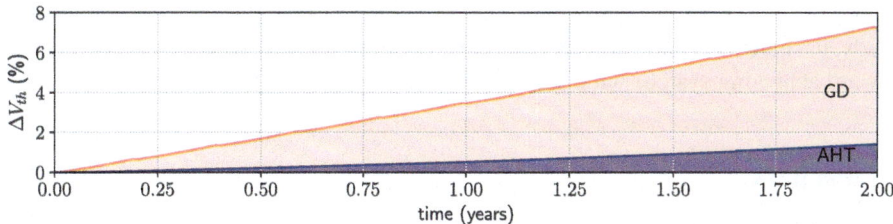

Fig. 18 Simulation of the long-term impact of BTI aging on the threshold voltage. If CMOS transistors are continuously exposed to the stress voltage, the threshold voltage shift can be as high as 10% after 2 years of operation. A significant portion of this drift (indicated by GD) cannot be recovered even after removing the stress voltage

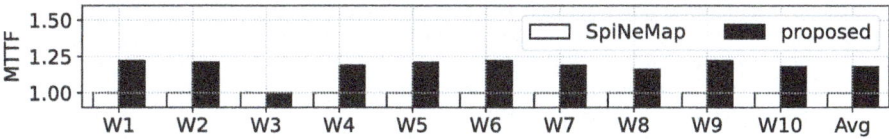

Fig. 19 MTTF normalized to SpiNeMap (higher is better)

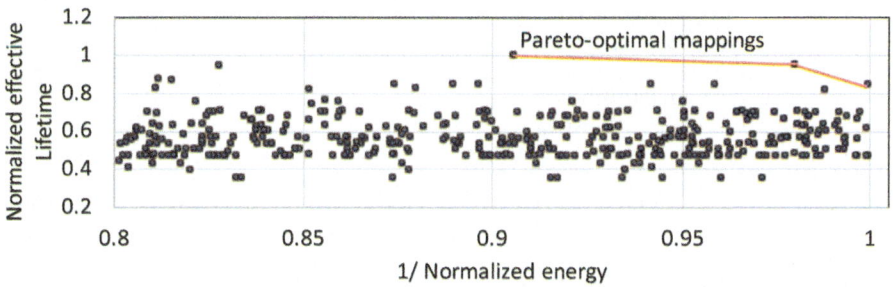

Fig. 20 Mapping explorations for one of the workloads

4 Summary

To summarize, we show that the system software framework of a neuromorphic hardware can be fine-tuned to improve performance, energy, and reliability without requiring any changes to the underlying hardware or its interface. These optimization objectives can also be combined. For instance, Fig. 20 shows the normalized effective endurance lifetime and the normalized energy of the mappings explored using the PSO algorithm of [103] for one of the workloads. The figure shows the mappings that are Pareto optimal with respect to endurance lifetime and energy.

Therefore, software-based optimizations can be performed orthogonal to any hardware- and technology-oriented optimization, e.g., [65]. We conclude that through platform-based design, the system software can proceed independently of the hardware. Later in the design stage when the hardware platform becomes ready, the hardware and software optimization mechanisms can work independently toward achieving system-wide performance, energy, and reliability goals.

References

1. Arimoto, Y., Ishiwara, H.: Current Status of Ferroelectric Random-Access Memory. Mrs Bulletin (2004)
2. Balaji, A., Adiraju, P., Kashyap, H.J., Das, A., Krichmar, J.L., Dutt, N.D., Catthoor, F.: PyCARL: A PyNN interface for hardware-software co-simulation of spiking neural network. In: IJCNN (2020)
3. Balaji, A., Corradi, F., Das, A., Pande, S., Schaafsma, S., Catthoor, F.: Power-accuracy trade-offs for heartbeat classification on neural networks hardware. In: JOLPE (2018)
4. Balaji, A., Das, A.: A framework for the analysis of throughput-constraints of SNNs on neuromorphic hardware. In: ISVLSI (2019)
5. Balaji, A., Das, A.: Compiling spiking neural networks to mitigate neuromorphic hardware constraints. In: IGSC Workshops (2020)
6. Balaji, A., Das, A., Wu, Y., Huynh, K., Dell'anna, F.G., Indiveri, G., Krichmar, J.L., Dutt, N.D., Schaafsma, S., Catthoor, F.: Mapping spiking neural networks to neuromorphic hardware. In: TVLSI (2020)
7. Balaji, A., Marty, T., Das, A., Catthoor, F.: Run-time mapping of spiking neural networks to neuromorphic hardware. In: JSPS (2020)
8. Balaji, A., Song, S., Das, A., Dutt, N., Krichmar, J., Kandasamy, N., Catthoor, F.: A framework to explore workload-specific performance and lifetime trade-offs in neuromorphic computing. In: CAL (2019)
9. Balaji, A., Song, S., Das, A., Krichmar, J., Dutt, N., Shackleford, J., Kandasamy, N., Catthoor, F.: Enabling resource-aware mapping of spiking neural networks via spatial decomposition. In: ESL (2020)
10. Balaji, A., Song, S., Titirsha, T., Das, A., Krichmar, J., Dutt, N., Shackleford, J., Kandasamy, N., Catthoor, F.: NeuroXplorer 1.0: An extensible framework for architectural exploration with spiking neural networks. In: ICONS (2021)
11. Balaji, A., Ullah, S., Das, A., Kumar, A.: Design methodology for embedded approximate artificial neural networks. In: GLSVLSI (2019)
12. Balaji, A., Wu, Y., Das, A., Catthoor, F., Schaafsma, S.: Exploration of segmented bus as scalable global interconnect for neuromorphic computing. In: GLSVLSI (2019)
13. Bez, R., Pirovano, A.: Non-volatile memory technologies: emerging concepts and new materials. Materials Science in Semiconductor Processing (2004)
14. Bohte, S.M., Kok, J.N., La Poutré, J.A.: SpikeProp: Backpropagation for networks of spiking neurons. In: ESANN (2000)
15. Burkitt, A.N.: A review of the integrate-and-fire neuron model: I. Homogeneous synaptic input. Biological Cybernetics (2006)
16. Burr, G.W., Brightsky, M.J., Sebastian, A., Cheng, H.Y., Wu, J.Y., Kim, S., Sosa, N.E., Papandreou, N., Lung, H.L., Pozidis, H., et al.: Recent progress in phase-change memory technology. In: JETCAS (2016)
17. Burr, G.W., Shelby, R.M., Sebastian, A., Kim, S., Kim, S., Sidler, S., Virwani, K., Ishii, M., Narayanan, P., Fumarola, A., Sanches, L.L., Boybat, I., Le Gallo, M., Moon, K., Woo, J., Hwang, H., Leblebici, Y.: Neuromorphic computing using non-volatile memory. Adv. Phys. X (2017)
18. Burr, G.W., Shelby, R.M., Sebastian, A., Kim, S., Kim, S., Sidler, S., Virwani, K., Ishii, M., Narayanan, P., Fumarola, A., et al.: Neuromorphic computing using non-volatile memory. Adv. Phys. X (2017)

19. Caporale, N., Dan, Y.: Spike timing–dependent plasticity: a hebbian learning rule. Annu. Rev. Neurosci. (2008)
20. Ceng, J., Castrillón, J., Sheng, W., Scharwächter, H., Leupers, R., Ascheid, G., Meyr, H., Isshiki, T., Kunieda, H.: MAPS: An integrated framework for MPSoC application parallelization. In: Design Automation Conference (DAC), pp. 754–759 (2008)
21. Chou, T., Kashyap, H., Xing, J., Listopad, S., Rounds, E., Beyeler, M., Dutt, N., Krichmar, J.: CARLsim 4: An open source library for large scale, biologically detailed spiking neural network simulation using heterogeneous clusters. In: IJCNN (2018)
22. Corradi, F., Pande, S., Stuijt, J., Qiao, N., Schaafsma, S., Indiveri, G., Catthoor, F.: Ecg-based heartbeat classification in neuromorphic hardware. In: 2019 International Joint Conference on Neural Networks (IJCNN), pp. 1–8. IEEE (2019)
23. Curzel, S., Agostini, N.B., Song, S., Dagli, I., Limaye, A., Tan, C., Minutoli, M., Castellana, V.G., Amatya, V., Manzano, J., et al.: Automated generation of integrated digital and spiking neuromorphic machine learning accelerators. In: ICCAD (2021)
24. Dan, Y., Poo, M.m.: Spike timing-dependent plasticity of neural circuits. Neuron **44**(1) (2004)
25. Daoudal, G., Debanne, D.: Long-term plasticity of intrinsic excitability: learning rules and mechanisms. Learning & Memory (2003)
26. Das, A.: Real-time scheduling of machine learning operations on heterogeneous neuromorphic SoC. In: MEMOCODE (2022)
27. Das, A., Catthoor, F., Schaafsma, S.: Heartbeat classification in wearables using multi-layer perceptron and time-frequency joint distribution of ECG. In: CHASE (2018)
28. Das, A., Kumar, A.: Dataflow-based mapping of spiking neural networks on neuromorphic hardware. In: GLSVLSI (2018)
29. Das, A., Kumar, A., Veeravalli, B.: Energy-aware communication and remapping of tasks for reliable multimedia multiprocessor systems. In: International Conference on Parallel and Distributed Systems (ICPADS), pp. 564–571. IEEE (2012)
30. Das, A., Kumar, A., Veeravalli, B.: Fault-tolerant network interface for spatial division multiplexing based Network-on-Chip. In: ReCoSoC (2012)
31. Das, A., Pradhapan, P., Groenendaal, W., Adiraju, P., Rajan, R., Catthoor, F., Schaafsma, S., Krichmar, J., Dutt, N., Van Hoof, C.: Unsupervised heart-rate estimation in wearables with Liquid states and a probabilistic readout. Neural Networks (2018)
32. Das, A., Singh, A.K., Kumar, A.: Energy-aware dynamic reconfiguration of communication-centric applications for reliable MPSoCs. In: ReCoSoC (2013)
33. Das, A., Walker, M.J., Hansson, A., Al-Hashimi, B.M., Merrett, G.V.: Hardware-software interaction for run-time power optimization: A case study of embedded linux on multicore smartphones. In: ISLPED (2015)
34. Das, A., Wu, Y., Huynh, K., Dell'Anna, F., Catthoor, F., Schaafsma, S.: Mapping of local and global synapses on spiking neuromorphic hardware. In: DATE (2018)
35. Das, A.K., Kumar, A., Veeravalli, B., Catthoor, F.: Reliable and Energy Efficient Streaming Multiprocessor Systems. Springer (2018)
36. Davies, M.: Benchmarks for progress in neuromorphic computing. Nat. Mach. Intell. (2019)
37. Davies, M., Srinivasa, N., Lin, T.H., et al.: Loihi: A neuromorphic manycore processor with on-chip learning. IEEE Micro (2018)
38. Debole, M.V., Taba, B., Amir, A., et al.: TrueNorth: Accelerating from zero to 64 million neurons in 10 years. Computer (2019)
39. Esser, S.K., Appuswamy, R., Merolla, P., Arthur, J.V., Modha, D.S.: Backpropagation for energy-efficient neuromorphic computing. NeurIPS (2015)
40. Fang, H., Mei, Z., Shrestha, A., Zhao, Z., Li, Y., Qiu, Q.: Encoding, model, and architecture: systematic optimization for spiking neural network in FPGAs. In: ICCAD (2020)
41. Fang, H., Taylor, B., Li, Z., Mei, Z., Li, H.H., Qiu, Q.: Neuromorphic algorithm-hardware codesign for temporal pattern learning. In: DAC (2021)
42. Fouda, M.E., Eltawil, A.M., Kurdahi, F.: Modeling and analysis of passive switching crossbar arrays. In: TCAS I (2017)
43. Furber, S.: Large-scale neuromorphic computing systems. In: JNE (2016)

44. Gao, R., Ji, Z., Manut, A.B., Zhang, J.F., Franco, J., Hatta, S.W.M., Zhang, W.D., Kaczer, B., Linten, D., Groeseneken, G.: NBTI-Generated defects in nanoscaled devices: Fast characterization methodology and modeling. In: TED (2017). https://doi.org/10.1109/TED.2017.2742700

45. Hu, M., Li, H., Chen, Y., Wu, Q., Rose, G.S., Linderman, R.W.: Memristor crossbar-based neuromorphic computing system: A case study. In: TNNLS (2014)

46. Hu, Y., Tang, H., Pan, G.: Spiking deep residual networks. In: TNNLS (2018)

47. Huai, Y., et al.: Spin-transfer torque MRAM (STT-MRAM): Challenges and prospects. AAPPS Bulletin (2008)

48. Huynh, P.K., Varshika, M.L., Paul, A., Isik, M., Balaji, A., Das, A.: Implementing spiking neural networks on neuromorphic architectures: A review. arXiv (2022)

49. Indiveri, G., Liu, S.C.: Memory and information processing in neuromorphic systems. Proc. IEEE **103**(8), 1379–1397 (2015)

50. Jeong, H., Shi, L.: Memristor devices for neural networks. J. Phys. D Appl. Phys. (2018)

51. Jerraya, A.A., Bouchhima, A., Pétrot, F.: Programming models and HW-SW interfaces abstraction for multi-processor SoC. In: DAC (2006)

52. Ji, Y., Zhang, Y., Li, S., Chi, P., Jiang, C., Qu, P., Xie, Y., Chen, W.: NEUTRAMS: Neural network transformation and co-design under neuromorphic hardware constraints. In: MICRO (2016)

53. Jiashu, L., Das, A., Kumar, A.: A design flow for partially reconfigurable heterogeneous multi-processor platforms. In: IEEE International Symposium on Rapid System Prototyping (RSP), pp. 170–176 (2012)

54. Kernighan, B.W., Lin, S.: An efficient heuristic procedure for partitioning graphs. Bell Syst. Tech. J. (1970)

55. Keutzer, K., Newton, A.R., Rabaey, J.M., Sangiovanni-Vincentelli, A.: System-level design: Orthogonalization of concerns and platform-based design. In: TCAD (2000)

56. Kraak, D., Agbo, I., Taouil, M., Hamdioui, S., Weckx, P., Cosemans, S., Catthoor, F.: Degradation analysis of high performance 14nm FinFET SRAM. In: DATE (2018). https://doi.org/10.23919/DATE.2018.8342003

57. Kraak, D., Taouil, M., Agbo, I., Hamdioui, S., Weckx, P., Cosemans, S., Catthoor, F.: Parametric and Functional Degradation Analysis of Complete 14-nm FinFET SRAM. In: TVLSI (2019). https://doi.org/10.1109/TVLSI.2019.2902881

58. Kwong, K.C., Li, L., He, J., Chan, M.: Verilog-A model for phase change memory simulation. In: ICSICT (2008)

59. Lee, K., Kwon, D.S.: Synaptic plasticity model of a spiking neural network for reinforcement learning. Neurocomputing (2008)

60. Liao, Y.B., Lin, J.T., et al.: Temperature-based phase change memory model for pulsing scheme assessment. In: ICICDT (V) (2008)

61. Lin, C.K., Wild, A., Chinya, G.N., Lin, T.H., Davies, M., Wang, H.: Mapping spiking neural networks onto a manycore neuromorphic architecture. In: PLDI (2018)

62. Liu, C., Yan, B., Yang, C., Song, L., Li, Z., Liu, B., Chen, Y., Li, H., Wu, Q., Jiang, H.: A spiking neuromorphic design with resistive crossbar. In: DAC (2015)

63. Liu, X., Wen, W., Qian, X., Li, H., Chen, Y.: Neu-NoC: A high-efficient interconnection network for accelerated neuromorphic systems. In: ASP-DAC (2018)

64. Maass, W.: Networks of spiking neurons: The third generation of neural network models. Neural Networks (1997)

65. Mallik, A., Garbin, D., Fantini, A., Rodopoulos, D., Degraeve, R., Stuijt, J., Das, A., Schaafsma, S., Debacker, P., Donadio, G., et al.: Design-technology co-optimization for OxRRAM-based synaptic processing unit. In: VLSIT (2017)

66. Marcolini, G., Giovanardi, F., Rudan, M., Buscemi, F., Piccinini, E., Brunetti, R., Cappelli, A.: Modeling the dynamic self-heating of PCM. In: ESSDERC (2013)

67. Marković, D., Mizrahi, A., Querlioz, D., Grollier, J.: Physics for neuromorphic computing. Nat. Rev. Phys. (2020)

68. Marwedel, P., Bacivarov, I., Lee, C., Teich, J., Thiele, L., Xu, Q., Kouveli, G., Ha, S., Huang, L.: Mapping of applications to mpsocs. In: International Conference on Hardware/Software Codesign and System Synthesis (CODES+ ISSS), pp. 109–118 (2011)
69. Mead, C.: Neuromorphic electronic systems. Proc. IEEE (1990)
70. Moradi, S., Qiao, N., Stefanini, F., Indiveri, G.: A scalable multicore architecture with heterogeneous memory structures for dynamic neuromorphic asynchronous processors (DYNAPs). In: TBCAS (2017)
71. Moyer, E.J., Das, A.: Machine learning applications to DNA subsequence and restriction site analysis. In: SPMB (2020)
72. Nicola, W., Clopath, C.: Supervised learning in spiking neural networks with FORCE training. Nature Communications (2017)
73. Nishi, Y.: Challenges and opportunities for future non-volatile memory technology. Current Appl. Phys. (2011)
74. Nuzzo, P., Sangiovanni-Vincentelli, A.L., Bresolin, D., Geretti, L., Villa, T.: A platform-based design methodology with contracts and related tools for the design of cyber-physical systems. Proc. IEEE (2015)
75. Odendahl, M., Castrillon, J., Volevach, V., Leupers, R., Ascheid, G.: Split-cost communication model for improved MPSoC application mapping. In: International Symposium on System on Chip (SoC), pp. 1–8 (2013)
76. Onnebrink, G., Walbroel, F., Klimt, J., Leupers, R., Ascheid, G., Murillo, L.G., Schürmans, S., Chen, X., Harn, Y.: DVFS-enabled power-performance trade-off in MPSoC SW application mapping. In: International Conference on Embedded Computer Systems: Architectures, Modeling, and Simulation (SAMOS), pp. 196–202 (2017)
77. Patterson, D.A., Hennessy, J.L.: Computer Organization and Design ARM Edition: The Hardware Software Interface. Morgan Kaufmann (2016)
78. Paugam-Moisy, H., Bohte, S.M.: Computing with spiking neuron networks. Handbook of Natural Computing (2012)
79. Paul, A., Das, A.: Design technology co-optimization for neuromorphic computing. In: IGSC Workshops (2021)
80. Paul, A., Song, S., Titirsha, T., Das, A.: On the mitigation of read disturbances in neuromorphic inference hardware. IEEE Des. Test (2022)
81. Paul, A., Tajin, M.A.S., Das, A., Mongan, W., Dandekar, K.: Energy-efficient respiratory anomaly detection in premature newborn infants. Electronics (2022)
82. Perez-Nieves, N., Goodman, D.: Sparse spiking gradient descent. NeurIPS (2021)
83. Rajendran, B., Sebastian, A., Schmuker, M., Srinivasa, N., Eleftheriou, E.: Low-power neuromorphic hardware for signal processing applications: A review of architectural and system-level design approaches. Signal Proc. Mag. (2019)
84. Sangiovanni-Vincentelli, A., Carloni, L., De Bernardinis, F., Sgroi, M.: Benefits and challenges for platform-based design. In: DAC (2004)
85. Sangiovanni-Vincentelli, A., Martin, G.: Platform-based design and software design methodology for embedded systems. IEEE Des. Test (2001)
86. Schmidgall, S., Hays, J.: Stable lifelong learning: Spiking neurons as a solution to instability in plastic neural networks. Preprint (2021). arXiv:2111.04113
87. Schuman, C.D., Potok, T.E., Patton, R.M., Birdwell, J.D., Dean, M.E., Rose, G.S., Plank, J.S.: A survey of neuromorphic computing and neural networks in hardware. Preprint (2017). arXiv:1705.06963
88. Sengupta, A., Ye, Y., Wang, R., Liu, C., Roy, K.: Going deeper in spiking neural networks: VGG and residual architectures. Front. Neurosci. (2019)
89. Singh, A.K., Das, A., Kumar, A.: Energy optimization by exploiting execution slacks in streaming applications on multiprocessor systems. In: DAC (2013)
90. Song, S., Balaji, A., Das, A., Kandasamy, N., Shackleford, J.: Compiling spiking neural networks to neuromorphic hardware. In: LCTES (2020)

91. Song, S., Chong, H., Balaji, A., Das, A., Shackleford, J., Kandasamy, N.: DFSynthesizer: Dataflow-based synthesis of spiking neural networks to neuromorphic hardware. In: TECS (2021)
92. Song, S., Das, A.: A case for lifetime reliability-aware neuromorphic computing. In: MWSCAS (2020)
93. Song, S., Das, A., Kandasamy, N.: Improving dependability of neuromorphic computing with non-volatile memory. In: EDCC (2020)
94. Song, S., Hanamshet, J., Balaji, A., Das, A., Krichmar, J., Dutt, N., Kandasamy, N., Catthoor, F.: Dynamic reliability management in neuromorphic computing. In: JETC (2021)
95. Song, S., Mirtinti, L.V., Das, A., Kandasamy, N.: A design flow for mapping spiking neural networks to many-core neuromorphic hardware. In: ICCAD (2021)
96. Song, S., Titirsha, T., Das, A.: Improving inference lifetime of neuromorphic systems via intelligent synapse mapping. In: ASAP (2021)
97. Strukov, D.B.: Endurance-write-speed tradeoffs in nonvolatile memories. Appl. Phys. A Mater. Sci. Proc. (4) (2016)
98. Stuijt, J., Sifalakis, M., Yousefzadeh, A., Corradi, F.: μBrain: An event-driven and fully synthesizable architecture for spiking neural networks. Front. Neurosci. (2021)
99. Suzuki, K., Swanson, S.: A survey of trends in non-volatile memory technologies: 2000–2014. In: IMW (2015)
100. Tavanaei, A., Ghodrati, M., Kheradpisheh, S.R., Masquelier, T., Maida, A.: Deep learning in spiking neural networks. Neural Networks (2019)
101. Titirsha, T., Das, A.: Thermal-aware compilation of spiking neural networks to neuromorphic hardware. In: LCPC (2020)
102. Titirsha, T., Song, S., Balaji, A., Das, A.: On the role of system software in energy management of neuromorphic computing. In: CF (2021)
103. Titirsha, T., Song, S., Das, A., Krichmar, J., Dutt, N., Kandasamy, N., Catthoor, F.: Endurance-aware mapping of spiking neural networks to neuromorphic hardware. In: TPDS (2021)
104. Varshika, M.L., Corradi, F., Das, A.: Nonvolatile memories in spiking neural network architectures: Current and emerging trends. Electronics **11**(10), 1610 (2022)
105. Varshika, M.L., et al.: Design of many-core big little μBrains for energy-efficient embedded neuromorphic computing. In: DATE (2022)
106. Weckx, P., Kaczer, B., Kukner, H., Roussel, J., Raghavan, P., Catthoor, F., Groeseneken, G.: Non-Monte-Carlo methodology for high-sigma simulations of circuits under workload-dependent BTI degradation-application to 6T SRAM. In: IRPS (2014). https://doi.org/10.1109/IRPS.2014.6860671
107. Wong, H.S.P., Lee, H.Y., Yu, S., Chen, Y.S., Wu, Y., Chen, P.S., Lee, B., Chen, F.T., Tsai, M.J.: Metal-oxide RRAM. Proc. IEEE (2012)
108. Wong, H.S.P., Raoux, S., Kim, S., Liang, J., Reifenberg, J.P., Rajendran, B., Asheghi, M., Goodson, K.E.: Phase change memory. Proc. IEEE (2010)
109. Xi, L., Zhitang, S., Daolin, C., et al.: An spice model for phase-change memory simulations. J. Semicond. (9) (2011)
110. Yan, B., Liu, C., Liu, X., Chen, Y., Li, H.: Understanding the trade-offs of device, circuit and application in ReRAM-based neuromorphic computing systems. In: IEDM (2017)
111. Yin, B., Corradi, F., Bohté, S.M.: Accurate and efficient time-domain classification with adaptive spiking recurrent neural networks. Nat. Mach. Intell. (2021)

Light Speed Machine Learning Inference on the Edge

Febin P. Sunny, Asif Mirza, Mahdi Nikdast, and Sudeep Pasricha

1 Introduction

Over the last decade, machine learning (ML) applications have become increasingly prevalent, with many emerging applications, such as autonomous transportation, medical prognosis, real-time speech translation, network anomaly detection, and audio/video synthesis. This prevalence is fueled by the emergence of sophisticated and powerful machine learning models over the past decade, such as deep neural networks (DNNs) and convolutional neural networks (CNNs). More sophisticated CNN models usually warrant deeper models with higher connectivity, which in turn increase the compute power and the memory requirement necessary to train and deploy them. Such increasing complexity also necessitates that the underlying hardware platforms consistently deliver better performance while satisfying strict power requirements. This endeavor to achieve high performance per watt has driven hardware architects to design custom accelerators for deep learning, e.g., Google's TPU [1] and Intel's Movidius [2], with much higher performance per watt than CPUs and GPUs. The performance-per-watt requirement still remains a challenge in resource-constrained environments, where computational power, energy expenditure, and available memory are often limited, such as many embedded devices. Binarized neural networks (BNNs) [3, 4] can reduce memory and computational requirements of DNN and CNN models while offering competitive accuracies with full precision models. As such, they are a possible solution to the performance requirement challenge, when executed on custom accelerators.

F. P. Sunny (✉) · A. Mirza · M. Nikdast · S. Pasricha
Department of Electrical and Computer Engineering, Colorado State University, Fort Collins, CO, USA
e-mail: Febin.Sunny@colostate.edu; asifmirz@rams.colostate.edu; Mahdi.Nikdast@colostate.edu; sudeep@colostate.edu

© The Author(s), under exclusive license to Springer Nature Switzerland AG 2024
S. Pasricha, M. Shafique (eds.), *Embedded Machine Learning for Cyber-Physical, IoT, and Edge Computing*, https://doi.org/10.1007/978-3-031-19568-6_13

Exploring more efficient hardware accelerator platforms is another potential solution to reduce performance per watt for neural-network processing. Conventional electronic accelerator platforms face fundamental limits in the post-Moore era where the high costs and diminishing performance improvements with semiconductor-technology scaling prevent significant improvements in future product generations [5]. Moving data in accelerators is a well-known bottleneck in these accelerators, due to the bandwidth and latency limitations of electronic interconnects, which puts limits on achievable performance and energy savings [6]. A solution to the data-movement bottleneck has presented itself in the form of silicon photonics technology, which enables ultra-high bandwidth, low-latency, and energy-efficient communication [7–30]. CMOS-compatible optical interconnects have already replaced metallic ones for light speed data transmission at almost every level of computing and are now actively being considered for chip-scale integration [8]. Recent research work has also shown that it is also possible to use optical components to efficiently perform computation, e.g., matrix-vector multiplication [31–33]. Due to the emergence of both chip-scale optical communication and computation, it is now possible to conceive photonic integrated circuits (PICs) that offer low latency and energy-efficient optical domain data transport and computation.

Despite the benefits of utilizing photonics for computation and communication, there are several challenges that must be addressed before photonic accelerators become truly viable. One of the main obstacles that impacts the robustness and reliability of photonic accelerators is the sensitivity of photonic devices to fabrication process and thermal variations. These variations introduce undesirable crosstalk, optical phase shifts, frequency drifts, tuning overheads, and photodetection current mismatches, which adversely affect the reliable and robust operation of photonic accelerators. In order to correct the impact of variations, thermo-optic (TO) or electro-optic (EO) tuning circuits are often used, which have notable power overheads. Because of the phase-change effects, it has on photonic devices, tuning mechanisms may also be used to control weight/activation imprinting via microring resonators (MRs). But the high latency of operation (in μs range [34]) of TO tuning can limit the achievable throughput and parallelism in photonic accelerators.

In this chapter, we discuss *ROBIN* [35], a novel optical-domain BNN accelerator that addresses the challenges highlighted above by optimizing electro-optic components across the device, circuit, and architecture layers. *ROBIN* combines novel device- and circuit-level techniques to achieve more efficient fabrication-process-variation (FPV) correction in optical devices, which helps with reducing energy and improving accuracy in BNNs that utilize these devices. Additionally, circuit-level tuning enhancements for inference latency reduction and an optimized architecture-level design help improve performance and also energy consumption compared to the state-of-the-art. The novel contributions from [35] include the following:

- The design of a novel optical-domain BNN accelerator architecture that is robust to fabrication-process variations (FPVs) and thermal variations and utilizes efficient wavelength reuse and a modular structure to enable high-throughput and energy-efficient execution across BNN models.

- A novel integration of heterogeneous optical microring resonator (MR) devices; we also conduct design space exploration for these MR designs to determine device characteristics for efficient BNN execution.
- An enhanced tuning circuit to simultaneously support large thermal-induced resonance shifts and high-speed, low-loss device tuning to compensate for FPVs.
- A comprehensive comparison with state-of-the-art BNN and non-BNN accelerator platforms from the optical and electronic domains, to demonstrate the potential of our BNN accelerator platform.

The rest of this chapter is organized as follows: Sect. 2 briefly explores the related works in the field of BNN acceleration. Sect. 3 gives a brief overview of noncoherent optical computation for photonic accelerators similar to ours. Sect. 4 provides an overview of BNNs and the partially binarized approach we have adopted for better accuracy in models. Sect. 5 describes the *ROBIN* architecture and our optimization efforts in tuning circuits, photonic devices, and photonic system level. Details of the experiments conducted, simulation setup, and the obtained results are provided in Sect. 6. Finally, Sect. 7 presents some concluding remarks.

2 Background and Related Work

Silicon-photonic-based DNN accelerator architectures are becoming increasingly prominent with significant interest from both academic and industrial research communities [36]. This growth in interest can be attributed to the previously discussed benefits of photonic acceleration over electronic acceleration. Optical DNN accelerator architectures can be broadly classified into two types: coherent architectures and noncoherent architectures. Coherent architectures use a single wavelength to operate and imprint weight/activation parameters onto the electrical field amplitude of the light wave [37, 38]. These architectures mainly use on-chip optical interferometer devices called Mach-Zehnder interferometers (MZIs). For imprinting the parameters, optical phase-change mechanisms are introduced to MZI devices. These mechanisms use heating or carrier injection to change the refractive index in the MZI structure. Weighting occurs with electrical field amplitude attenuation proportional to the weight value, and phase modulation that is proportional to the sign of the weight. The weighted signals are then accumulated with cascaded optical combiners, through coherent interference. Here the term coherent refers to the physical property of the wave, where it is possible for waves of the same wavelength to interfere constructively or destructively. Noncoherent architectures, such as [32, 33, 39–41], use multiple wavelengths. These architectures are referred to as non-coherent architectures as they use different optical wavelengths, the interaction among which can be noncoherent. A large number of neuron operations can be represented simultaneously in noncoherent architectures by using wavelength-division multiplexing (WDM) or dense WDM (DWDM). In these architectures, parameter values are imprinted on to the signal amplitude directly, and to manipulate individual wavelengths, wavelength-selective devices such as microring resonators (MRs) or microdisks are used. The optical signal power is controlled, for imprinting parameter values, by controlling the optical loss in these devices through tuning mechanisms (Sect. 5.1). The broadcast and weight (B&W) protocol [42] is typically employed for setting and updating the weight and activation values. The *ROBIN* architecture we present in this chapter is a noncoherent architecture, i.e., it uses multiple wavelengths that are routed to photonic computation units in waveguides using WDM in accordance with the B&W protocol. The growing interest in noncoherent architectures can be attributed to the limitations in scalability, phase encoding noise, and phase error accumulation in coherent architectures [36, 43].

For optical DNN acceleration using noncoherent mechanics, [39] introduced a photonic accelerator for CNNs where all the layers of CNN models are implemented using connected photonic convolution units. In these units, MRs are used to tune wavelength amplitudes to desired kernel values. Another such work, in [40], utilizes microdisks instead of MRs due to the lower area and power consumption they offer. But microdisks use "whispering gallery mode" resonance which is inherently lossy due to the tunneling ray attenuation phenomenon [44], which reduces reliability and energy-efficiency with microdisks. There are very few works which focus on implementations of BNN accelerators using silicon photonics. The work in [45] proposed an MR-based accelerator for discretized neural network acceleration, with an encoding scheme to enable positive and negative product considerations. The authors in [46] leveraged microdisks for implementing an accelerator with a design similar to [40]. This work considered an accelerator for fully binarized neural networks, i.e., both weights and activations and considered to be single-bit parameters. Because of this simplification, [46] was able to utilize energy-efficient photonic XOR and population count operations instead of conventional multiply and accumulate operations. The work also made use of photonic nonvolatile memory and claimed operating frequencies of up to 50 GHz. All of these existing works on noncoherent optical-domain DNN/BNN acceleration have several shortcomings. They suffer from susceptibility to fabrication-process variations (FPVs) and thermal crosstalk, which are not addressed in these architectures. Microsecond granularity thermo-optic tuning latencies further can reduce the speed and efficiency of optical computing [34], which is also not considered when analyzing accelerator performance. We address these crucial shortcomings as part of our *ROBIN* optical-domain BNN accelerator architecture in this work.

In this work, we aim to ensure the robustness of the architecture against process and thermal variations by using MR design-space exploration and photonic tuning-circuit optimizations, which will be further explained in Sect. 5. We also utilize the broadband capabilities of the key photonic device in our work, microring resonators (MRs), to perform batch normalization folding, which moves batch normalization operations from the electrical domain to the photonic domain. Section 5.3 further details the modular architectural design aiming at ensuring wavelength reuse, to reduce VCSEL usage and splitter losses and waveguide length reduction. We also explore how the architecture performs in the presence of FPVs and how we may further reduce energy consumption in terms of device tuning in this scenario, in Sect. 6.2.

3 Overview of Noncoherent Optical Computation

Noncoherent optical accelerators leverage the low-latency and energy-efficient optical computation for multiply and accumulate (MAC) operations, which consumes substantial computational power and incurs high latencies in electronic accelerators. These accelerators typically utilize the B&W protocol with multiple wavelengths. Figure 1a (from [47]) gives an overview of a B&W-based optical MAC unit. The figure depicts a recurrent MAC unit which is employed repeatedly to compute different layers of a neural network model. The layer parameters such as weights or activations can be imprinted on to the wavelengths using the MRs that are tuned to modify the optical signal amplitude to represent those values. The MRs are placed in MR banks where multiple parameters can be imprinted onto wavelengths simultaneously. In the MR banks, each MR is tuned to a specific optical wavelength and can be used to alter the amplitude of the wavelength to represent the imprinted parameter. There can be separate wavelengths which carry positive and negative parameters, as discussed in [45]; these parameters are summed using balanced photodetectors (BPDs), as shown in Fig. 1a.

The output from the MAC unit is passed on to a Mach-Zehnder Modulator (MZM) which tunes the output from a designated laser diode (LD) to this output. Multiple MZMs and LDs are used to generate the outputs from multiple MAC units; these are collected and multiplexed using an arrayed waveguide grating (AWG)-based optical multiplexer (MUX). The output from the MUX, now embedded with parameters for the next layer, is passed back into the MAC units, through splitters. Devices such as electro-optic modulators (not depicted) may be used to implement nonlinearities after the MAC operation. Unfortunately, the static nature of the hardware limits the size of the neural network model that can be accelerated using such a configuration. This configuration would also require a large number of splitters, which can cause increased optical losses and thus higher laser power requirement to compensate for the losses, as the size of an accelerator using this B&W configuration increases.

MRs and other on-chip optical resonators such as microdisks are crucial components in such noncoherent MAC configurations, as they impact the reliability and efficiency of the operation performed. Figure 1b depicts an MR bank and its output spectrum along with the free spectral range (FSR). Factors such as fabrication-process variations (FPVs) and thermal variations which impact the MR critical dimensions and hence the effective refractive index (n_{eff}) of the device can cause a drift in the resonant wavelength ($\Delta\lambda_{MR}$) [48]. This drift can introduce errors into optical computation and is thus usually corrected with TO or EO tuning

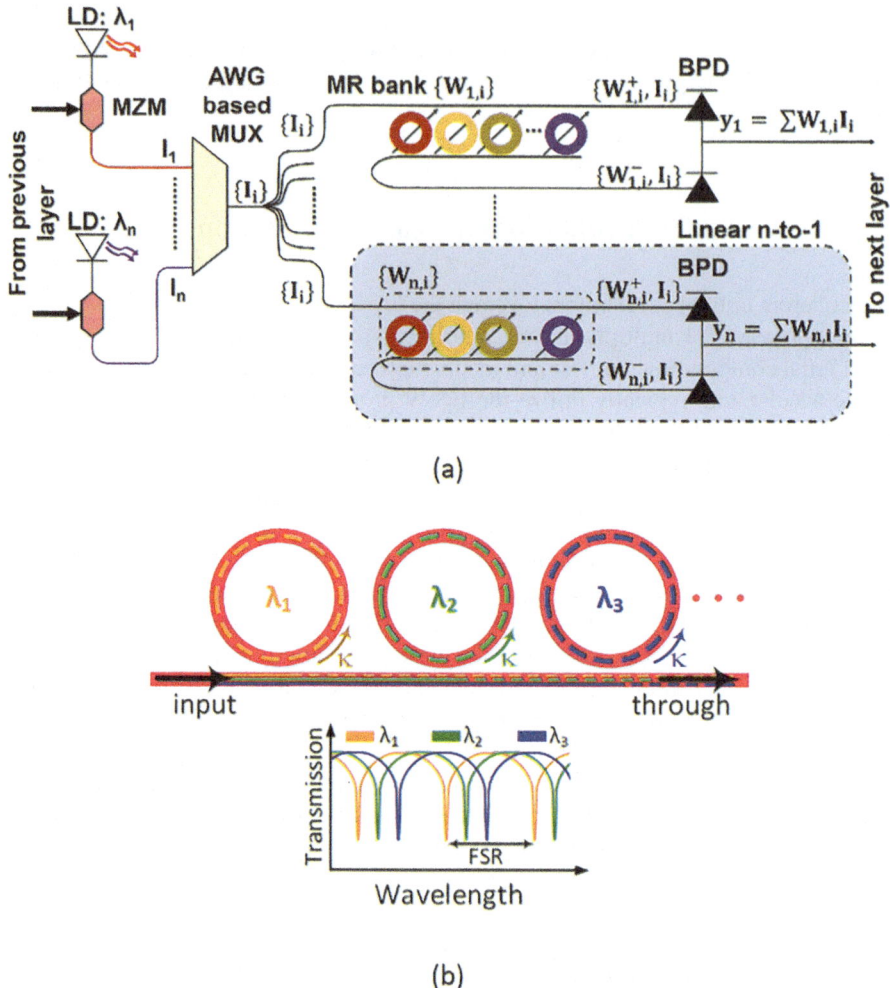

Fig. 1 (**a**) A recurrent noncoherent B&W MAC-based design [47]; (**b**) an MR bank consisting of MRs with individual resonant wavelength (λi) coupled to the MRs at crossover coupling (κ) and the output spectrum, showing free spectral range (FSR)

circuits. While EO offers faster tuning (~ns range) and consumes lesser power (~4 μm/nm), it also has a smaller tuning range [49]. TO tuning, on the other hand, consumes higher power (~27 mW/FSR) and has higher tuning latency (~μs range) [34] but offers a larger tuning range. Because of the larger correction capacity, TO is often preferred over EO despite its higher latency and power consumption. Therefore, as the number of MRs increases—when considering larger CNN or MLP models—the tuning power consumption also increases. This also creates increased wavelength requirements per waveguide and calls for longer waveguides to host the MRs, causing increased laser power consumption to supply the wavelengths and to compensate for the propagation losses in the longer waveguides. Also, more MRs and more wavelengths increase optical crosstalk and also introduce thermal crosstalk due to the larger number of TO tuners employed. To counteract these challenges and ensure better weight resolution, crosstalk mitigation strategies must also be considered.

To design an effective optical-domain BNN accelerator, all of these considerations must be taken into account. This highlights the need for (i) better device optimizations to tolerate variations; (ii) efficient and low-latency tuning mechanisms; (iii) and a scalable architecture design, which is optimized for energy efficiency, area, and throughput. The work in [35] addressed all of these concerns for an efficient BNN accelerator implementation in the photonic domain.

4   Binarized Neural Networks

BNNs [3] are types of DNNs (or CNNs) where both weights and activation parameters only use binary values, and the binary values are utilized during both inference and training using backpropagation. In the light of the discussion of various noises in photonic accelerator architectures, it is to be noted that the binary nature of weights in BNNs makes them resilient to small perturbations which can usually lead to gross classification errors in DNNs. Inspired by the seminal work on efficiently training BNNs [3], recent efforts either explore how BNN accuracy can be improved, apply BNNs to different application domains, or explore how BNNs can be implemented efficiently in hardware to leverage their low computation power and memory requirements in resource constrained environments.

BNNs utilize the sign function to convert real valued weights to $+1$ or -1. But this typically leads to complications in training as the gradient for the sign function always results in a zero. A heuristic called straight through estimator (STE), introduced in [50], can be used to circumvent this issue. STEs approximate the gradient by bypassing the gradient of the layer, by turning it into an identity function. The gradient thus obtained is used for updating real valued weights, using standard optimization strategies such as Adam or stochastic gradient descent (SGD). This process is utilized for activation parameters as well. Also, the use of batch normalization (BN) layers in BNNs has been shown to lead to several benefits [4]. The gain (γ) and bias (β) terms of the BN layer not only help condition the values during training, which speeds up BNN training, but also helps to improve accuracy in BNNs.

Inference accuracy in BNNs can be increased by considering partially binarized BNNs, where selected layers have their parameters at higher precision. The last layer is usually not binarized to avoid severe loss in accuracy. With detailed analysis of the model, critical layers can be identified and can be kept at higher precision, for better accuracy, at the cost of increased resource (computation, memory) utilization. We conduct a BNN accuracy analysis to determine the appropriate

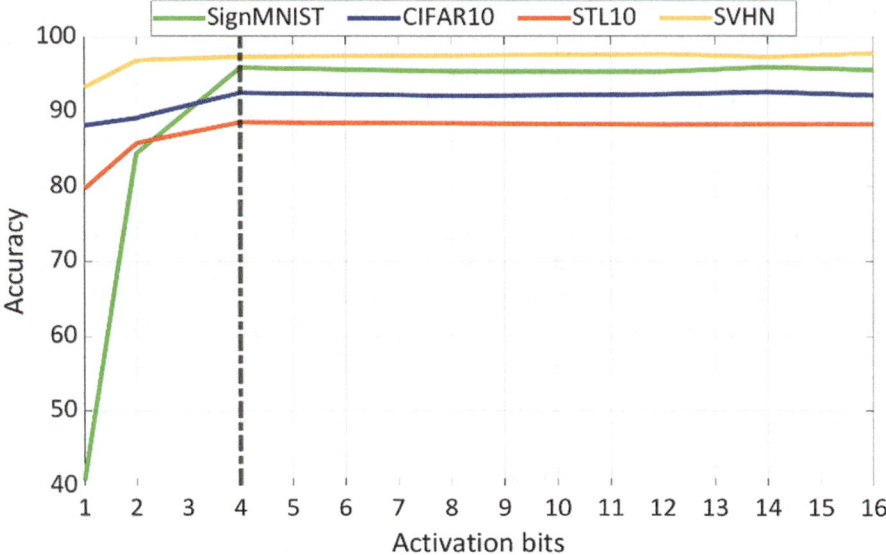

Fig. 2 The accuracy sensitivity study conducted by varying activation parameter precision (number of bits). Weights are kept as binary values in all cases. The study was performed across four different models and their datasets (described later in Sect. 6.1)

activation parameter precision in considered models, which is required to determine the digital-to-analog converter (DAC) resolution in our accelerator architecture. In this analysis, weight parameters were restricted to binary (1-bit) values, but the bit precision level of the activations was altered from 1 bit to 16 bits. During BNN training, we ensured that we only binarize weights during the forward and backward propagations but not during the parameter update step, because keeping good precision weights during the updates is necessary for SGD to work at all (as parameter changes are usually tiny during gradient descent). After training, all weights were in binary format, while the precision of input activations was varied. Figure 2 shows the results of varying activation precision across four different models and their datasets (described later in Sect. 6.1). We observed that the accuracy had notable change initially as activations bits were increased, but this gain in accuracy soon saturated. Based on the results, we consider binary (1-bit) weights with 4-bit activations and thus use 4-bit DACs in our architecture.

5 *ROBIN* Architecture

In this section, we describe the various optimization considerations at device, circuit, and architecture level used for designing the *ROBIN* architecture.

5.1 Tuning Circuit Design

A tuning circuit design is essential for fast and accurate operation of MRs in our BNN accelerator. The presence of fabrication-process variations (FPVs) can impact MR operations by altering their resonant wavelength (λ_{MR}) from the originally designed values. The errors caused by this shift can be significantly reduced by using an appropriate MR tuning circuit. The tuning circuit employed can be either thermo-optic (TO) or electro-optic (EO) tuning circuits. Thermo-optic(TO)-based tuning mechanisms use microheaters to change the temperature in the proximity of a microring resonator (MR), which then alters the effective index (n_{eff}) of the MR. This in turn changes the λ_{MR} of the device. Such a change in resonant wavelength ($\Delta\lambda_{MR}$) can help compensate for fabrication process and thermal variations in MRs. The electro-optic (EO)-based tuning mechanisms in an MR is based on the depletion and injection of carriers on a PN diode. However, only small shifts in an MR's resonant wavelength can be compensated using this mechanism (i.e., EO has a limited correction range). TO tuning is preferred to compensate for large shifts in MR's resonant wavelength. However, one has to compromise on latency ($\sim\mu s$ range) and power consumption, which is higher than for EO tuning. To reduce *ROBIN*'s reliance on TO tuning, which entails high overheads, the possibility of a hybrid tuning mechanism was explored. In this hybrid tuning mechanism, both TO and EO tuning are used to compensate for $\Delta\lambda_{MR}$. Such a tuning method has been proposed earlier [51] and can be easily transferred to an optimized MR (as discussed in Sect. 5.2) for hybrid tuning in our architecture. Such a mechanism would significantly reduce the overhead caused just by TO tuning.

To reduce the power overhead of TO tuning in such a hybrid approach, we adapt a method called thermal eigenmode decomposition (TED), which was first proposed in [52] that involves collectively tuning all the MRs in an MR bank. By doing so we can cancel the effect of crosstalk (i.e., undesired phase shift) in MRs with much lower power consumption. The amount of phase crosstalk induced from one MR on another MR, placed adjacent to each other, can be modeled using the trend in Fig. 3 (pink line). In this figure, as the distance between two devices (MRs) increases, the amount of phase crosstalk between them reduces. Correspondingly, as an example we calculate the tuning power compensation for an MR bank consisting of 10 MRs and different radii placed at a distance (d) from each other. A few important trends to observe from Fig. 3 are (i) as the radius of an MR increases, tuning power compensation for $\Delta\lambda_{MR}$ increases; (ii) without TED (collective tuning of MRs), the tuning power consumption is high, indicating that each MR would require more power to compensate for respective shifts in resonant wavelength ($\Delta\lambda_{MR}$); (iii) by employing TED, we see a significant reduction in tuning power consumption: 51% (radius of 1.5 μm) and 41% (radius of 5 μm) when MRs are placed at a distance of 5 and 7 μm apart from each other, respectively. Though placing MRs further close to each other would yield better compensation in power, one must take into account the placement and routing of tuning circuit for each MR in an MR bank. Additional power reduction can be obtained by performing device level

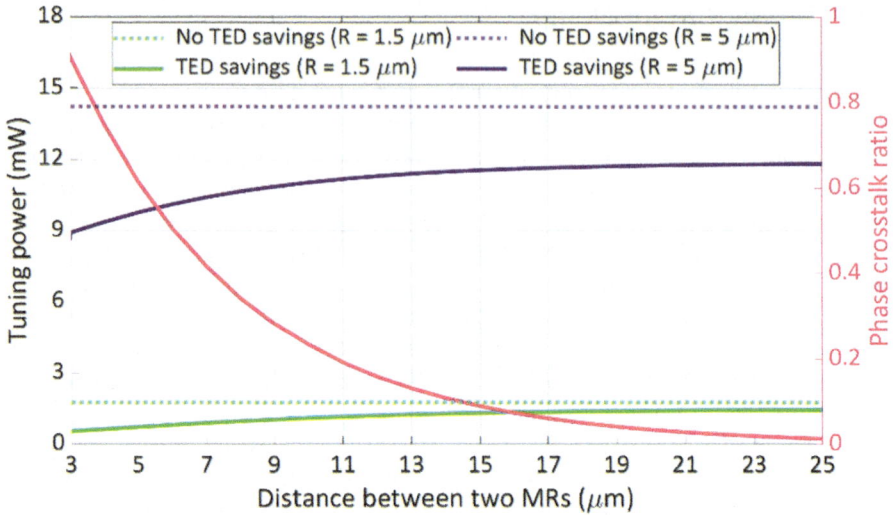

Fig. 3 Tuning power compensation in a block of 10 MRs placed with and without considering thermal eigenmode decomposition (TED) for different MR radius. The orange line represents phase crosstalk ratio variation with distance between MRs

optimizations, as designing MRs tolerant to FPVs would reduce the total power used to compensate for fabrication variations.

5.2 Device-Level Optimization

We explore different MR designs to accommodate different needs in our *ROBIN* architecture such as multi-bit precision for activation values, single-bit precision for weight value representation, and batch normalization.

5.2.1 Fabrication-Process Variation Resilience

FPVs cause undesirable changes in device critical dimensions (e.g., width and thickness), which cause resonant wavelength shifts ($\Delta\lambda_{MR}$). To address $\Delta\lambda_{MR}$, we explore the impact of change in device parameters such as waveguide width, thickness, gap between input and ring waveguide, and radius using our in-house MR device-exploration tool. We map the behavior of different changes in the waveguide width, thickness, and radius in MRs due to FPVs. Figure 4a shows one of our design exploration results where we understand and observe the behavior of resonant resonant-wavelength shift slopes due to change variations in the waveguide width, thickness, and radius represented by orange, green, and blue lines, respectively.

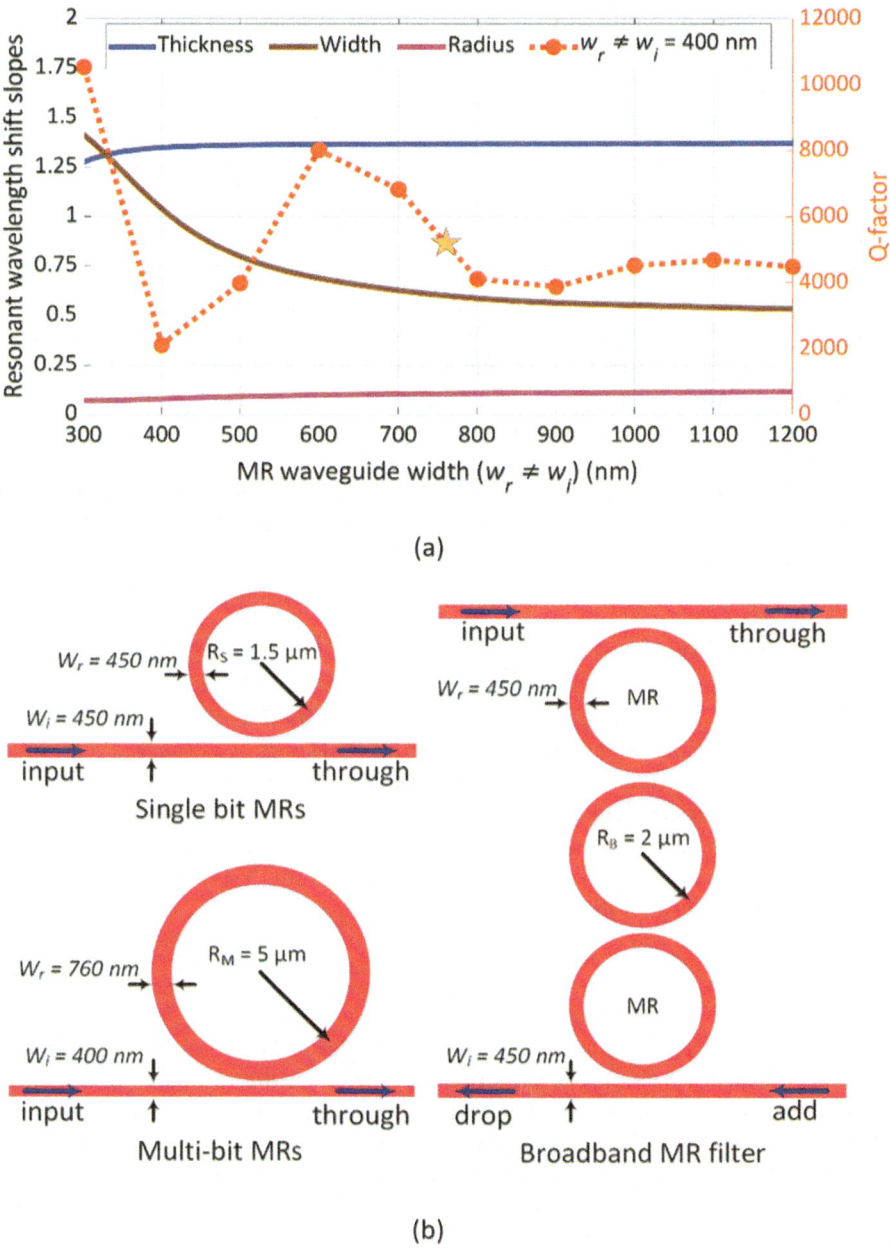

Fig. 4 (**a**) Resonant-wavelength shift slopes with respect to changes in waveguide width, thickness, and radius and corresponding cross-over coupling (κ), when the input waveguide (w_i) is set to 400 nm the marked point represents our selected MR design; (**b**) the different MR designs considered in this work

Resonant wavelength shift slope due to change in waveguide width ($\partial \lambda_{MR} / \partial w$) can be given as:

$$\frac{\partial \lambda_{MR}}{\partial w} = \left| \frac{(\Delta \lambda_{MR} (\lambda, w + \epsilon_W, t, R) - \Delta \lambda_{MR} (\lambda, w - \epsilon_W, t, R))}{2\epsilon_W} \right| \tag{1}$$

In Eq. (1), ϵ_w denotes a small change in waveguide width and $\Delta \lambda_{MR}$ depends on changes in width (w), thickness (t), and radius (R). Similarly, $\partial \lambda_{MR} / \partial(t, R)$ can also be approximated.

Figure 4a clearly shows that the impact of resonant-wavelength shift reduces as we increase the waveguide width, whereas the impact of thickness and radius variations remains constant. From the conducted experiments, $\Delta \lambda_{MR}$ is more sensitive to changes in waveguide width, hence the impact of $\Delta \lambda_{MR}$ reduces as the waveguide width is increased. We employ Lumerical MODE [53], an Eigen mode solver to calculate these shifts in resonant wavelengths. One can easily overcome higher-order mode excitation by employing adiabatic designs [53] and waveguide tapers [54] in MRs with wider waveguides. Such a design translates to lesser tuning-power consumption due to FPVs.

5.2.2 Multi-Bit Precision MRs

As discussed in Sect. 4, increasing the number of bits used to capture activations in a model can boost the model accuracy in BNNs. However, we observed that there is not a significant accuracy boost beyond 4-bit activation values; hence, we explore MR designs which can achieve a resolution of 4 bits. To achieve a resolution of 4 bits, we have to take into consideration how the optical signals from MRs impact each other due to crosstalk. We consider calculations from [55] to define the amount of noise from one MR on the other:

$$\phi (i, j) = \frac{\delta^2}{(\lambda_i - \lambda_j)^2 + \delta^2}, \tag{2}$$

where $\phi(i - j)$ describes the noise content from the jth MR present in the signal from the ith MR, $(\lambda_i - \lambda_j)$ is the difference between the resonant wavelengths (λ_i, λ_j), and $\delta = \lambda/(2 \cdot Q - factor)$. The quality factor or Q-factor is a measure of the sharpness of the resonance relative to the central frequency of a microring resonator (MR) that impacts the optical channel spacing, crosstalk, bandwidth, and other factors in the MR [56]. A sharper resonance (i.e., a higher Q-factor) can result in increased susceptibility to noise, as even a small change in the central frequency of the MR (due to perturbance) can lead to large losses. This limits the achievable resolution of the parameters being represented. Thus, smaller Q factors are preferred. However, too small a Q-factor can also lead to larger device dimensions and higher optical crosstalk, which in turn can lead to larger losses and higher tuning power requirements. Q-factor in an MR is defined as follows:

$$Q - \text{factor} = \frac{\lambda_{\text{MR}}}{\text{FWHM}}, \tag{3}$$

where FWHM is the full width at half maximum of a resonance spectrum which can be defined for an all-pass ring resonator (see Fig. 4) as follows:

$$\text{FWHM} = \frac{(1-ra)\lambda_{\text{MR}}^2}{\pi n_g L \sqrt{ra}}, \tag{4}$$

where r is the self-coupling coefficient and a is the single-amplitude transmission, including both the propagation loss in the ring and the loss in the couplers; this can be written as $a = e^{-\alpha L}$, where α is power attenuation coefficient. L is round trip length or the circumference of the MR. In this chapter, we assume a lossless coupler in our designed MRs, hence $|\kappa|^2 + |r|^2 = 1$, where κ is the cross-over coupling coefficient. For ideal cases with zero attenuation, $a \approx 1$. Based on the above equations, the noise power component can thus be calculated as:

$$P_{\text{noise}} = \sum_j^{(n-1)} \phi(i, j) P_{\text{in}}[i] \tag{5}$$

For power intensity (P_{in}) of 1, the resolution can be computed as:

$$\text{Resolution} = \frac{1}{\max|P_{\text{noise}}|}, \tag{6}$$

To achieve a bit resolution of at least 4 bits, we need MRs with a Q-factor of ≈5000 (from Eq. (6)) while being tolerant to FPVs. Q-factor is highly sensitive to losses and change in dimensions of MR. In order to achieve the specific Q-factor value, we select the following MR dimensions: input waveguide width of 400 nm and ring waveguide width of 760 nm and radius (RM) of 5 μm. This MR design, as shown in Fig. 4a (magenta line), provides improved tolerance to FPV, desirable Q-factor, and smaller area consumption. Such an MR design with Q-factor of 5000 allows enough levels of distinction between bits by slightly changing intensity and helps easily detect optical signal at the output port satisfying the requirement for multi-bit precision of activation values.

5.2.3   Single-Bit MRs

In our architecture, we represent weight values with a single bit, and this requires just two levels of precision with the output signal from an MR. An MR of high Q factor may be used here, as we do not have to have high resolution here. Compact ring designs with high Q-factor have been proposed in [57, 58]. The work in [58] proposes an MR design with radius 1.5 μm to achieve a high Q-factor of 46,000 without the consideration of sidewall roughness while maintaining low bending loss ≈7 cm^{-1}. Similarly, an adiabatic MR structure of radius 3 μm is designed in [57] to avoid higher order mode excitation where a high Q-factor of 27,000 is achieved. These works indicate that such high Q-factor rings can be designed.

For one-bit weight representation in *ROBIN*, we design a ring of radius 1.5 μm, as shown in Fig. 4b, with input waveguide (w_i) and ring waveguide (w_r) width both set to 450 nm, to achieve a Q-factor of 25,000 that corresponds to a bit resolution of 1 from Eq. (6). These designs allow our architecture to save on area and tuning power consumption. We acknowledge that FPVs are an inevitable part of the fabrication process. However, since we just need to differentiate between two levels of operations, we do not explore for designs that are tolerant towards FPVs, for single-bit MRs.

5.2.4 Broadband MRs

Batch normalization (BN) layers can be considered essential in BNNs as they add complexity to the models, via the gain (γ) and bias (β) terms of the layer. These terms are learned during the training process along with the normalization parameters of the batch mean (μ) and standard deviation (σ). During the training phase, these terms are dynamic, but during inference they have static values. This allows for a hardware implementation of a photonic version of batch normalization folding, where we may tune weights as per the following equation:

$$w_{\text{fold}} = \gamma \cdot \frac{W}{\sqrt{\sigma^2 + \epsilon}} = C_{\text{fold}} \cdot W \tag{7}$$

There is a similar equation for bias terms as well, but since BNN models benefit from batch normalization after every layer, these will be normalized out and hence can be ignored. The above constant, C_{fold}, is applied to every weight term and hence is a participant in every matrix multiplication operation, i.e.:

$$\text{Input}_{l+1} = f\left(A_l \cdot (w_{\text{fold}})_l\right) = C_{\text{fold}} \cdot f\left(A_l \cdot W_l\right) \tag{8}$$

In Eq. (8), Input_{l+1} refers to the input to the $(l+1)$th layer, $f()$ is nonlinear activation function, A_l is the activation of lth layer, and W_l is the weights from lth layer. This operation can be applied to partial sums as well, and can be implemented using a broadband photonic device with its gain tuned to reflect C_{fold}.

For implementing the photonic batch normalization, a broadband device is preferred as this allows simultaneous gain tuning of all the wavelengths in the waveguide efficiently, both area and energy wise. Hence, the last type of MRs we consider are broadband MRs that are needed for batch normalization (BN) layers due to their relevance in BNNs. A large passband can be achieved by cascading several MRs and properly selecting the design parameters of MRs [59]. We explore such a higher order MR, or cascaded MR filter, to achieve a wide passband. The work presented in [60] explores a possibility for passband widths ranging from 6.25 Ghz to a maximum of 3 Thz. This work explores different design parameters of a higher-order filter while evaluating different losses such as insertion, propagation, and coupling loss in higher order MRs. A 0.5 nm resonant wavelength shift of MR was reported for a fabrication error of 10 nm showing that such a design is tolerant to FPVs.

A third-order MR-based switching device with radius of 2 μm shown in Fig. 4b fits the requirement for broadband MR. The coupling coefficients at the input (κ_i^2) is 0.53, and coupling at higher order rings is 0.2. The propagation loss of 25 dB/cm has been reported and insertion loss of the two elements in higher-order filter are 4.35 dB and 0.36 dB, respectively [59]. Having such a design, one can achieve a flat-top passband with bandwidth width of at least 3 THz. Employing this broadband MR can help us apply the batch normalization parameter C_{fold} on all the available resonant wavelengths in the bank. Having a large bandwidth such as 2.5 Thz allows us to conveniently tune up to 20 different wavelengths.

5.3 Architecture Design

An overview of the *ROBIN* accelerator architecture is shown in Fig. 5. The optical device and tuning circuit optimizations from the previous subsections are utilized within the optical binary vector dot product (VDP) units. We use banks of heterogeneous MRs (described in Sect. 5.2) to imprint activation parameters, weights, and the BN layer constants onto optical signals. Multiple such VDP units are composed together to form the overall architecture, as shown in the figure, which is then used to accelerate a given BNN model. We utilize a photonic summation unit for summing the partial sum outputs from our VDPs, before passing the partial sums on to the electronic control unit (ECU), as shown in Fig. 5. We also rely on the ECU for fetching parameters from the global memory, decomposing them to lower dimensional vectors, distributing these vectors among the VDP units, and implementing nonlinear activations functions and pooling layers. We describe the working of the *ROBIN* architecture in more detail in the following subsections.

5.3.1 Decomposing Vector Operations

To map convolution (CONV) and fully connected (FC) layers from BNN models to our accelerator, we first need to decompose large vector sizes into smaller ones, so they can be mapped to the VDP array in our architecture. This decomposition approach can be explained as follows.

In CONV layers, a filter performs convolution on a patch (e.g., 2 × 2 elements) of the activation matrix in a channel to generate an element of the output matrix. The operation can be represented as:

$$KA = Y. \tag{9}$$

Assuming a 2 × 2 filter kernel and weight matrices, Eq. (9) can be rewritten as:

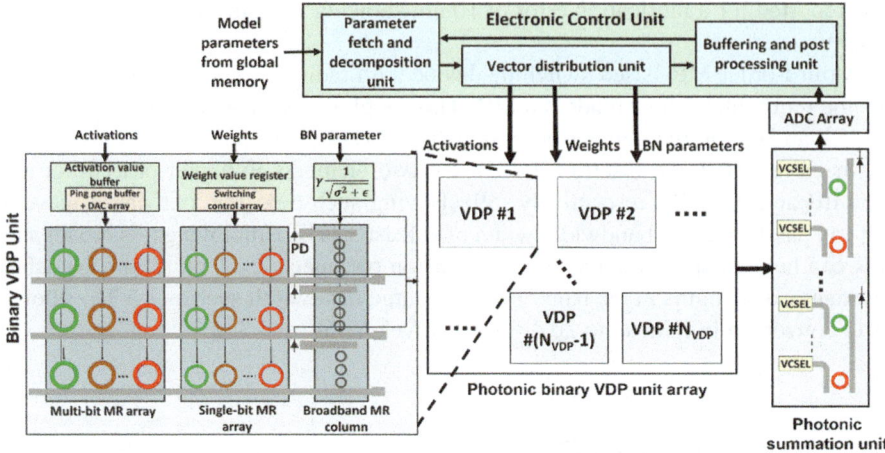

Fig. 5 An overview of the ROBIN architecture, showing the electronic control unit, the photonic vector dot product (VDP) unit array, and the photonic summation unit, along with a detailed view of the VDP unit internal structure

$$\begin{bmatrix} k_1 & k_2 \\ k_3 & k_4 \end{bmatrix} \begin{bmatrix} a_1 & a_2 \\ a_3 & a_4 \end{bmatrix} = k_1 a_1 + k_2 a_2 + k_3 a_3 + k_4 a_4, \tag{10}$$

Rewriting Eq. (10) as a vector dot product, we have:

$$\begin{bmatrix} k_1 & k_2 & k_3 & k_4 \end{bmatrix} \cdot \begin{bmatrix} a_1 \\ a_2 \\ a_3 \\ a_4 \end{bmatrix} = k_1 a_1 + k_2 a_2 + k_3 a_3 + k_4 a_4, \tag{11}$$

Once we can represent the operation as a vector dot product, it is easy to see how it can be decomposed into partial sums. For example:

$$\begin{bmatrix} k_1 & k_2 \end{bmatrix} \cdot \begin{bmatrix} a_1 \\ a_2 \end{bmatrix} = k_1 a_1 + k_2 a_2 = PS_1, \tag{12a}$$

$$\begin{bmatrix} k_3 & k_4 \end{bmatrix} \cdot \begin{bmatrix} a_3 \\ a_4 \end{bmatrix} = k_3 a_3 + k_4 a_4 = PS_2, \tag{12b}$$

$$PS_1 + PS_2 = Y. \tag{12c}$$

In FC layers, typically much larger dimension matrix-vector multiplication operations are performed between input activation vectors and weight matrices. Therefore, we have:

$$A \cdot W = \begin{bmatrix} a_1 \\ a_2 \\ \vdots \\ a_n \end{bmatrix} \cdot \begin{bmatrix} w_1 & w_2 & \cdots & w_n \end{bmatrix}, \tag{13}$$

$$A \cdot W = \begin{bmatrix} a_1 \cdot w_1 + a_1 \cdot w_2 + \cdots + a_1 \cdot w_n \\ a_2 \cdot w_1 + a_2 \cdot w_2 + \cdots + a_2 \cdot w_n \\ \vdots \\ a_n \cdot w_1 + a_n \cdot w_2 + \cdots + a_n \cdot w_n \end{bmatrix}. \tag{14}$$

In Eq. (13), a_1 to a_n represent column vectors of activations (A), and w_1 to w_n represent row vectors of weight matrix (W). The resulting vector is a summation of dot products of vector elements Eq. (14). Similar to the decomposition of CONV operation, these can then be decomposed into lower dimensional dot products.

5.3.2 Vector Dot Product (VDP) Unit Design

As discussed in Sect. 5.3.1, we decompose matrix operations to lower dimensional vector dot product operations. These vector dot product operations are executed optically within our VDP units. The heterogeneous MR designs combined with optical circuit-level optimizations for area and power consumption are utilized to design VDP units (Fig. 5) suited for accelerating both CONV and FC layers without compromising on accelerator throughput. For representing weight values, we use high Q-factor, small radius single-bit MRs described in Sect. 5.2.2. The smaller radius contributes to lower tuning power and helps reduce propagation loss along the VDP waveguide. This is possible due to the binarized nature of weight matrices in BNNs. For activation values, we consider MRs with slightly lower Q-factor, for better resolution, as discussed in Sect. 5.2.1. Optical BN layer implementation requires simultaneously tuning all the wavelengths in the waveguide to the batch normalization constant, and for this, we use third order MR filters, as described in Sect. 5.2.3. The combination of these heterogeneous designs allows the VDP units to be highly energy efficient. We also make use of electronic buffering in the VDP units to reduce the digital to analog converter (DAC) usage. In particular, we make use of ping-pong buffers, which allow us to use a single DAC array to feed the activation devices in all the waveguides in a VDP unit. As weight values are single-bit values, we can use simple switching circuits to essentially turn the MR tuning circuits on or off depending on the value of the weight parameters.

In designing a VDP unit, there are several important parameters that must be carefully considered: number of higher resolution MRs for activation representation (N_A), number of single-bit MRs for weight representation (N_W), and number of broadband MRs (N_B) for batch normalization folding implementation. Thus, the total number of MRs per waveguide NMR $= N_A + N_W + N_B$. The number of required DACs is equal to N_A. By the mathematical property of the dot product operation, N_A must be equal to N_W. The number of waveguides to which we distribute the MRs is denoted as N_{WG}. The maximum size of the vector that can be represented in a VDP unit is given by $N_{WG} * N_A$. We divide this vector across multiple waveguides to reduce power consumption, as this allows us to reuse wavelengths and reduce the overall laser power consumption, as discussed next, in Sect. 5.3.3. Multiple VDP units work concurrently on parameters from the same layer and generate partial sums simultaneously, for efficient parallelization and to increase the throughput of the accelerator. The total VDP unit count used in *ROBIN* is N_{VDP}. Thus, the VDP and architecture design process can be considered as an optimization problem where we try to explore N_{VDP}, N_{WG}, N_A ($= N_W$), and NB values while trying to maximize throughput and minimize area and power consumption. We present results of this architecture exploration analysis in Sect. 6.3.

5.3.3 Optical Wavelength Reuse in VDP Units

Prior works on optical accelerator design typically considers a separate wavelength to represent each individual element of a vector. As the size of the vectors being mapped increase, this approach leads to an increase in the total number of lasers needed in the laser bank, which in turn increases power consumption. Beyond employing the decomposition approach discussed above, we also consider wavelength reuse per VDP unit to minimize laser power. In this approach, within VDP units, the vectors assigned from the electronic control unit (ECU) are further decomposed into smaller sized vectors for which dot products can be performed using MRs in parallel, in each arm of the VDP unit. By decomposing the mapped vectors further, same wavelengths can be reused across arms within a VDP to reduce the number of unique wavelengths required from the laser. Photodetectors (PDs) perform summation of the element-wise products to generate partial sums from decomposed vector dot products. The partial sums from the decomposed operations are then converted back to the optical domain by VCSELs (bottom right of Fig. 5), multiplexed into a single waveguide, and accumulated using another PD, before being sent for buffering. Thus, our approach leads to an increase in the number of PDs and splitters compared to other accelerators but significantly reduces both the number of MRs per waveguide and the overall laser power consumption. The reduction in overall power consumption is also assisted by the fact that PDs do not consume significant power.

In each arm within a VDP unit, we can use a maximum of 15 MRs per bank for a total of 30 MRs per arm. The choice of MRs per arm considers not only the thermal crosstalk and layout spacing issues and the benefits of wavelength reuse (as discussed earlier), but also the fact that optical splitter losses become non-negligible as the number of MRs per arm increase, which in turn increases laser power requirements. Thus, the selection of MRs per arm within a VDP unit must be carefully adjusted to balance parallelism within/across arms and laser power overheads.

5.3.4 *ROBIN* Pipelining and Scheduling

The pipeline and schedule of operations during BNN model execution on the *ROBIN* accelerator are shown in Fig. 6. The electronic control unit (ECU) for the accelerator communicates with the global memory and retrieves the trained weights for the model being accelerated. The weights are stored in SRAM-based buffers. Considering the vector granularity of the VDP units, latency of operation of the photonic core, and the parameter sizes (4-bit activation bits and binary weight parameters), we can calculate the memory bandwidth necessary. From our analyses (presented in Sect. 6.3), we found that our architecture needs a maximum bandwidth of 93.75 GB/s at the ECU to photonic core interface. This is a reasonable bandwidth assumption for an SRAM-based memory with operating frequency $\geq$2.5 GHz and a read width of 250 bits. Previous works, such as [61], have explored similar SRAM systems but for a much higher bandwidth requirement at 250 GB/s. The lower bandwidth requirement for our system can be attributed to the smaller parameter sizes, while the work in [61] considered 16-bit precision for the neural network parameters. Memory interfaces which exceed the necessary bandwidth are already available commercially: e.g., NVIDIA Tesla K20M GPUs have 320-bit memory interfaces at 2.6 GHz which can operate every half clock cycle to provide a bandwidth of 208 GB/s.

These weight matrices are decomposed to lower dimensional vectors and are distributed to the VDPs by the ECU's vector decomposition unit. The decomposition

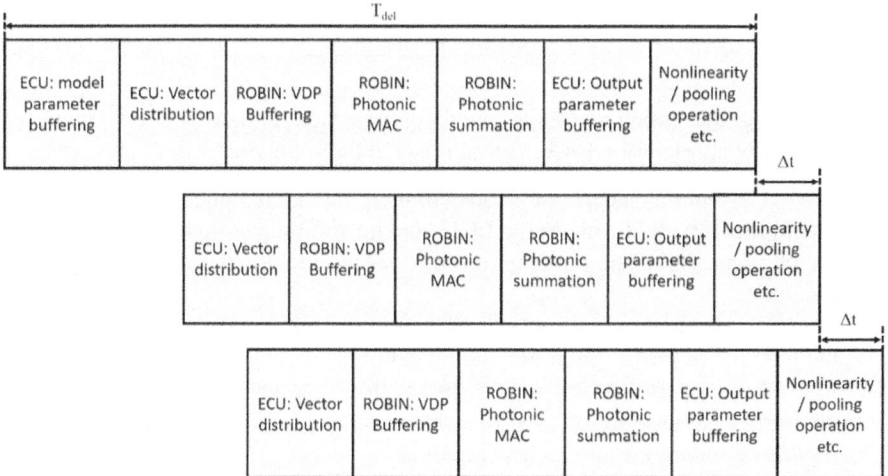

Fig. 6 Pipelined scheduling of operations during BNN execution on the ROBIN accelerator

operation is described by the left-hand side of Eqs. 10, 11, and 12. As described in the equations, the vector decomposition unit converts matrices to vectors (row-wise conversion for weight matrices and column-wise conversion for activation matrices) and then those vectors into sub-vectors. The size of the sub-vectors depends on the granularity of the VDP units. The received vectors are buffered in the VDP units and are fed into the DAC array through a ping-pong buffer so that they can keep the MAC operation running continuously. The partial sums generated are passed on to the photonic summation unit, the output from which is passed on to the ECU. The ECU buffers the sums and calculates inputs that are then passed on to the next layer by subjecting the parameters to nonlinearities (activation functions) and performing other layer specific operations, like pooling.

The model parameter buffering stage is not repeated every pipeline operation, but must be repeated as the parameters buffered in the buffers in ECU are depleted (i.e., distributed to VDP units). As such, the total time required by *ROBIN* to perform inference acceleration for a given model can be given as:

$$
\begin{gathered}
\text{Total time of operation} = \\
T_{\text{del}} + \Delta t \times X + (\text{ECU parameter buffering delay}) \times x,
\end{gathered}
\tag{15}
$$

where

$$
\Delta t = \text{local buffer operation delay} + \text{vector distribution delay}, \tag{16}
$$

$$
X = \frac{\text{Total number of parameters in the model}}{N_{\text{w}} \times N_{\text{VDP}}}, \tag{17}
$$

$$
x = \frac{(\text{Parameters buffered in ECU})}{N_{\text{w}} \times N_{\text{VDP}}}. \tag{18}
$$

Comparing our pipeline to the pipeline presented in the previous work on photonic BNN acceleration [46], we can observe the following differences:

(i) *ROBIN*'s pipeline takes into consideration model parameter retrieval from global memory, buffering in the ECU, and how these parameters are utilized in the photonic core. The pipeline in [46] does not include these operations in its pipeline:

(ii) *ROBIN*'s pipeline considers both ECU and photonic core operation, whereas the pipeline in [46] is photonic system centric.

(iii) *ROBIN* utilizes photonic batch normalization folding which does not require an extra step, whereas in [46] this operation is performed electronically and requires a separate stage in their pipeline.

6 Experiments and Results

6.1 Simulation Setup

Several simulation studies were conducted to evaluate the effectiveness of the *ROBIN* BNN accelerator. The optimized heterogeneous MR designs, the tuning circuit optimizations, and architectural level considerations discussed so far were included in our simulation considerations.

The operation of the *ROBIN* architecture was simulated using a custom Python simulator to estimate its performance in terms of power, frames per second (FPS) performance, and energy consumption. For analyzing the inference accuracy across different activation precision and the impact of FPV noise on the inference accuracy, we used Tensorflow 2.3 along with Qkeras [62]. Figure 7 shows the training accuracy versus epoch graph of the models described in Table 1, to illustrate the accuracy and loss across the epochs.

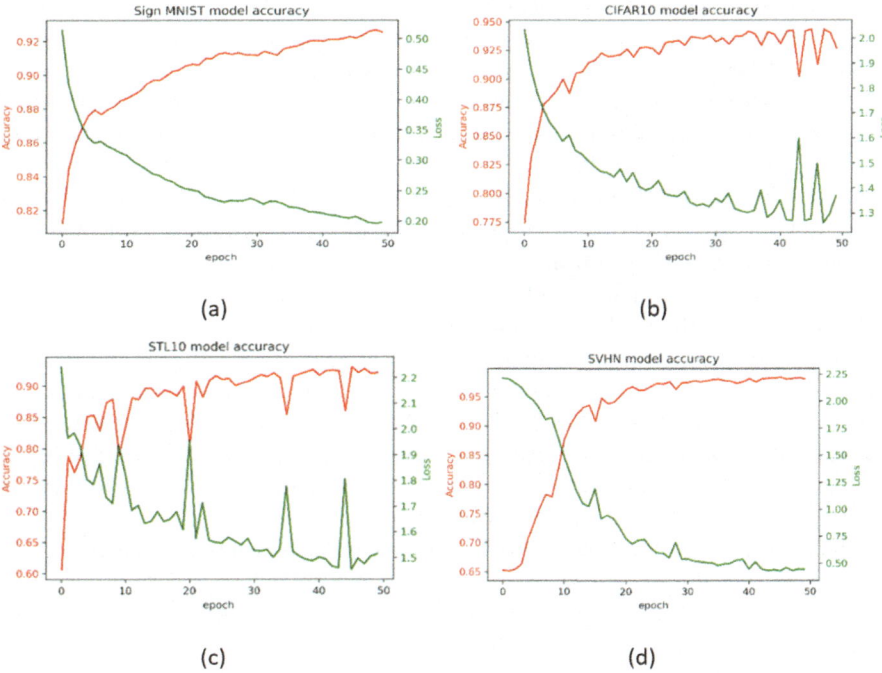

Fig. 7 The training accuracy vs epoch for the BNN models considered for (**a**) Sign MNIST, (**b**) CIFAR10, (**c**) STL10, and (**d**) SVHN datasets. (**a**) shows top-1 accuracy, while (**b–d**) show top-5 accuracy

Table 1 Models and datasets used for evaluations

Model no.	CONV layers	FC Layers	BN layers	Parameters	Datasets
1	2	2	3	60,642	Sign MNIST
2	6	3	6	1,546,570	CIFAR10
3	6	3	7	13,570,186	STL10
4	6	2	6	552,362	SVHN

Table 2 Parameters considered for analysis of photonic accelerators

Devices	Latency	Power
EO tuning [49]	20 ns	4 μW/nm
TO tuning [34]	4 μs	27.5 mW/FSR
VCSEL [70]	10 ns	0.66 mW
TIA [71]	0.15 ns	7.2 mW
Photodetector [72]	5.8 ps	2.8 mW
DAC [69]	0.33 ns	59.7 mW
ADC [68]	24 ns	62 mW

We compare *ROBIN* with DEAP-CNN [39] and HolyLight [40], two recent optical DNN accelerators from prior work, along with LightBulb [46], which is an optical BNN accelerator, as well as numbers reported from several electronic DNN and BNN accelerators. For simulating the operation of optical accelerators, we considered optical signal losses due to various factors: signal propagation loss (1 dB/cm [8]), splitter loss (0.13 dB [63]), combiner loss (0.9 dB [64]), MR through loss (0.02 dB [65]), MR modulation loss (0.72 dB [66]), microdisk loss (1.22 dB [67]), EO tuning loss (6 dB/cm [49]), and TO tuning loss (1 dB/cm [34]). We also considered the ADC design from [68] and the 4-bit DAC from [69] in our analyses. The analysis of the optical accelerators (DEAP-CNN [39], HolyLight [40], and LightBulb [46]) follows the modeling methodology we have adopted for *ROBIN*, where we factor in power consumption and delays associated with photonic devices used in these accelerators. A summary of the power and latency considerations for our analyses is given in Table 2. These power and latency values were used in our simulations and latency of operation of our architecture. In order to give better perspective on the architecture's performance, a comparison for inference time on *ROBIN* and a conventional CPU is presented in Sect. 6.5.

To calculate laser power consumption, we use the following power model:

$$P_{laser} - S_{detector} \geq P_{photo-loss} + 10 \times \log_{10} N_{\lambda}, \tag{19}$$

where P_{laser} is the laser power in dBm, $S_{detector}$ is the PD sensitivity in dBm, and $P_{photo-loss}$ is the total loss encountered by the optical signal, due to all of the factors discussed above.

6.2 Fabrication-Process Variation Analysis

FPV in optical devices is corrected using TED tuning in our architecture, as discussed in Sect. 5. At the system level, this tuning leads to significant power consumption overhead, and any avenue to further reduce tuning power consumption becomes important. We conduct an FPV noise injection analysis, where we inject noise, modeled using FPV data, into the MR devices into our *ROBIN* accelerator, during the inference phase. This experiment was conducted to *(i)* study the impact of FPV induced noise on BNN models mapped to our accelerator; *(ii)* determine how effective TED tuning is in such scenarios; and *(iii)* uncover any opportunities for further power minimization.

To analyze the impact of FPV on the model and how TED tuning compensates for it, we first consider the effect of FPV on the shift in resonant wavelength ($\Delta\lambda_{MR}$) in MRs. Resonant-wavelength shift in an MR can be modeled from [73] as:

$$\Delta\lambda_{MR} = \frac{\partial\lambda_{MR}}{\partial w}\sigma_w + \frac{\partial\lambda_{MR}}{\partial t}\sigma_t + \frac{\partial\lambda_{MR}}{\partial R}\sigma_R, \tag{20}$$

where $\sigma_{w,t,R}$ are the associated standard deviations for waveguide width, thickness, and radius variations and $\frac{\partial\lambda_{MR}}{\partial(w,t,r)}$ is the rate of change in the MR resonant wavelength considering the variations in the waveguide width, thickness, and radius represented in Eq. (20). We generate virtual FPV maps for the accelerator layout with a mean (μ) of 0 and standard deviation ($\sigma_{(w,t,R)}$) of 4.9, 1.5, and 0.75 nm for waveguide width, thickness, and radius, respectively. These standard deviation values are experimentally obtained based on real fabricated MR devices through our collaboration with CEA-Leti. Using these values, we are able to derive $\Delta\lambda_{MR}$ using Eq. (20). So, the current resonant wavelength (λ'_{MR}) of the FPV affected MR becomes:

$$\lambda'_{MR} = \lambda_{MR} + \Delta\lambda_{MR}, \tag{21}$$

Due to a shift in λ_{MR}, the transmission of the wavelength through the MR is impacted. The intensity of the wavelength at the through port is given by the following equation from [56].

$$T = \frac{I_{out}}{I_{in}} = \frac{a^2 - 2ra\cos\phi + r^2}{1 - 2ar\cos\phi + ra^2}, \tag{22}$$

In Eq. (22), $\phi = \beta L$, with L being the roundtrip length and β the propagation constant $\beta = 2\pi/\lambda$ of the circulating mode and r^2 is the self-coupling coefficient of an MR. A detailed analyses for the calculation of r using super mode theory is presented in [74]. The output intensity from the MR is important, as for noncoherent MAC units, the parameter values are encoded onto the signal intensity, and a change in expected output can be seen as perturbation or noise source.

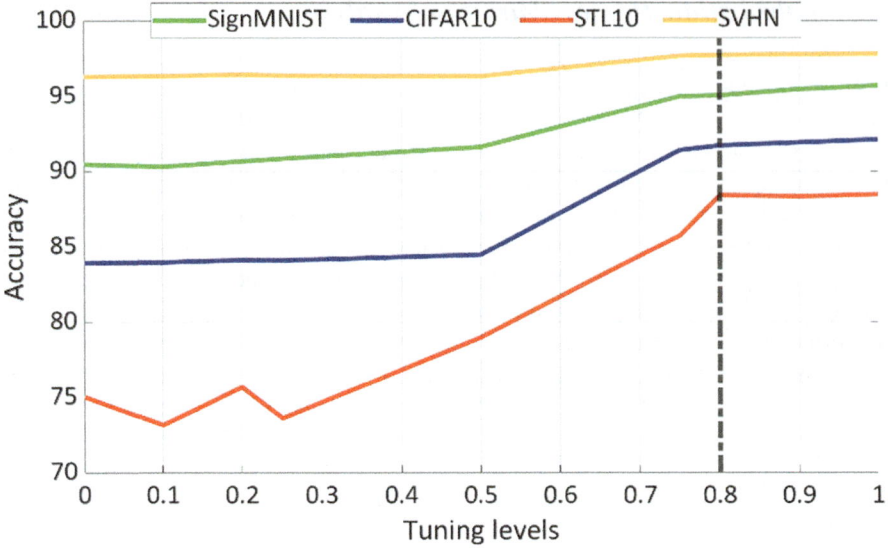

Fig. 8 Inference accuracy versus level of tuning applied. At 80% tuning, the inference accuracy saturates, rendering further tuning unnecessary, and providing an opportunity to save tuning power

The noise injection was modeled using Eqs. (20, 22), where we consider the resonant-wavelength shift ($\Delta\lambda_{MR}$) in MRs due to FPV and its impact on the parameters imprinted on the MRs. From our analysis using the FPV data from our device fabrications with CEA-Leti, and Eq. (18), we are able to obtain the mean and standard deviation values for $\Delta\lambda_{MR}$ in a wafer. The values calculated are $\mu = -0.1461$ nm and $\sigma = 24.417$ nm . Using these values, 50 $\Delta\lambda_{MR}$ maps for the accelerator were generated and then using Eq. (22) the perturbation to the parameters imprinted on to the devices were modeled. Noise injection to the models was performed at inference time using TensorFlow.

Figure 8 shows the results of this experiment, where we explored the impact of FPV-induced noise in the four BNN models and the effect of TED tuning for FPV compensation. We expected that the better the devices were tuned, the better the accuracy that would be exhibited by the accelerator. But it was observed that the model's accuracy can be sustained without perfectly tuning the devices. Figure 8 shows that at 80% FPV correction through tuning, the BNN retains appreciable inference accuracy. Thus, there is not a significant accuracy benefit to tune beyond the 80% level; this allows for a 20% reduction in tuning power requirement. This reduction in tuning power is factored into our architecture level analysis, which is presented next.

6.3 ROBIN Architecture Optimization Analysis

In this section, we show results of our exploration of the parameters discussed in Sect. 5.3.2. As mentioned in Sect. 5.3.2, we try to optimize N_{VDP}, N_{WG}, N_A, and N_B to reduce area and power consumption while trying to obtain the best throughput (frames per second or FPS) possible. N_B was fixed to be 1 per waveguide, allowing us to have up to 20 wavelengths in the same waveguide with a channel spacing of 1 nm, which in turn allows us to tune all the MRs simultaneously to the BN layer parameters. We then explored N_{VDP}, N_{WG}, and N_A, with the goal of optimizing power, area, and FPS. The result of this exploration analysis is shown in Fig. 9 in the form of a scatter plot. From this analysis, we identified two configurations for *ROBIN*, where one is optimized for FPS/Watt, with lowest area and power consumption (energy optimized *ROBIN* or *ROBIN-EO*), and another with the best FPS but with higher area and power consumption (performance optimized *ROBIN* or *ROBIN-PO*). In terms of (N_A, N_{VDP}, N_{WG}), these configurations can be represented as (10, 50, 10) for *ROBIN-EO* and (50, 200, 10) for *ROBIN-PO*. These configurations were compared against other optical and electronic DNN/BNN accelerator platforms, to showcase their efficiency of operation. The results for these comparisons with other accelerators are presented in the following section.

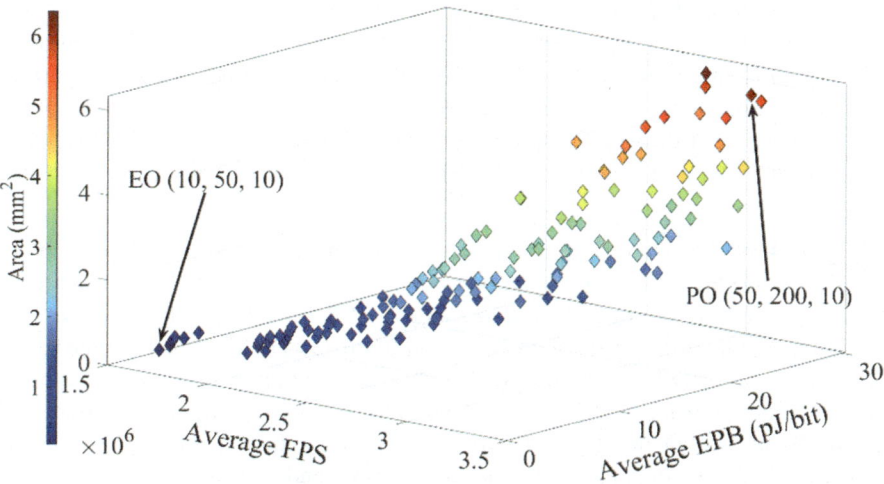

Fig. 9 Scatterplot of average FPS vs. average EPB vs. area of various ROBIN configurations. The configuration with highest FPS/Watt (energy optimized or EO) and the one with best FPS (performance optimized or PO) are specified

6.4 Comparison with State-of-the-Art Optical and Electronic DNN/BNN Accelerators

We compared *ROBIN-EO* and *ROBIN-PO* against various electronic and optical neural network acceleration platforms. For optical DNN accelerator platforms, we selected DEAP-CNN [39] and HolyLight [40]. The electronic accelerator platforms considered are GPU (Nvidia Tesla P100), SIGMA [75], Edge TPU [76], DaDianNao [77], and FPGA implementation of Null Hop [78]. We also compare *ROBIN* against the best-known previous photonic BNN accelerator, LightBulb [46].

When compared to LightBulb, *ROBIN* has the following differences:

(i) *ROBIN* is designed to accelerate partially binarized neural networks, as opposed to fully binarized neural networks as in [46], for obtaining better accuracies.
(ii) *ROBIN* utilizes photonic batch normalization folding for faster, energy-efficient batch normalization layer operation whereas [46] relies on an electronic implementation of the batch normalization operation.
(iii) *ROBIN* has various circuit- and device-level optimizations in place to counteract thermal and process variations, which also ensure high-throughput and energy-efficient operation, whereas [46] does not take into account thermal and process variations and the necessary tuning latency and energy consumption overheads needed to counter them.
(iv) Architecture-level optimizations in *ROBIN* ensure lower power consumption in terms of tuning and laser power; these considerations are not part of the architecture proposed in [46]. We also compare against electronic BNN accelerators FBNA [79] and FINN [80]. We used the GOPS and power consumption parameters from [8, 81] to simulate inference on the electronic platforms.

Figure 10 shows the power comparison across the accelerators from prior work and the two *ROBIN* variants. It can be observed that *ROBIN-PO* has substantially higher power consumption than *ROBIN-EO*, as *ROBIN-PO* is focused on FPS performance rather than energy conservation. *ROBIN-PO* has a much larger vector granularity per VDP unit along with substantially higher VDP unit count to maximize parallelism, when compared to *ROBIN-EO*. The larger unit count and the waveguide count in *ROBIN-PO* drive its power requirements higher. On the other hand, it can be observed that the energy- and area-efficient *ROBIN-EO* has comparable power consumption to that of edge and mobile electronic neural network accelerators.

In Fig. 11, we compare the energy-per-bit values (EPB) across the various BNN accelerators considered in this work. We can observe that both the *ROBIN* variants perform significantly better than the optical accelerators in comparison. This lower EPB is owing to the meticulous device, circuit, and architecture level optimizations we have considered in our architecture, which takes into account various losses and delays at the architecture level and counteracts them. The heterogeneous MRs

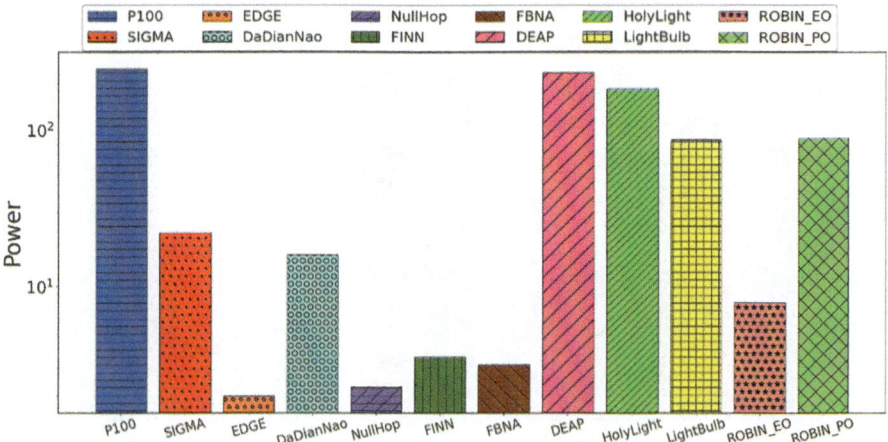

Fig. 10 Power consumption comparison among variants of ROBIN versus other optical accelerators (DEAP-CNN, Holylight, LightBulb) and electronic accelerator platforms (P100, SIGMA, EdgeTPU, DaDianNao, Null Hop, FINN, and FBNA)

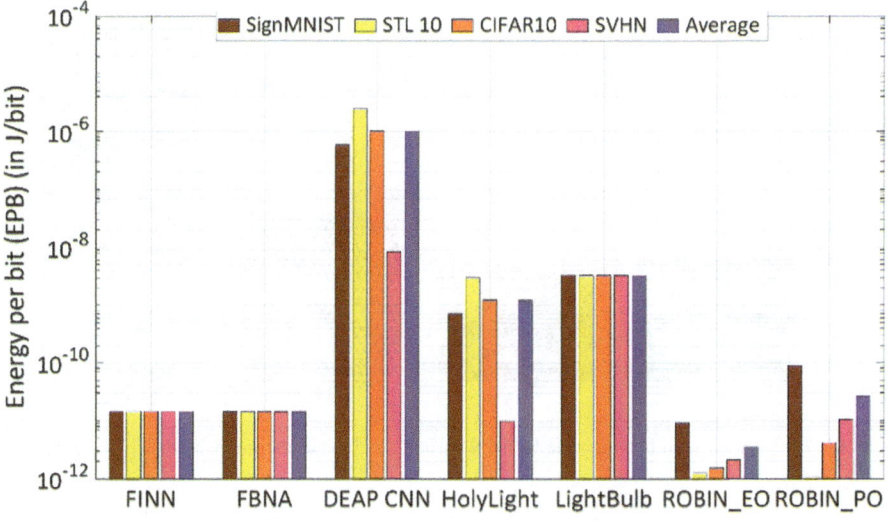

Fig. 11 EPB comparison between electrical BNN accelerators, optical accelerators, and the ROBIN variants

used in *ROBIN* provide energy and area benefits, and the utilization of TED for collectively tuning MRs provides further energy benefits on top of the 20% reduction we obtained from the analysis in Sect. 6.3. TED also allows for closer placement of MRs, which in turn helps reduce propagation delays. This reduction is also impacted by the faster inputs to DAC arrays enabled by local buffering and ping-pong buffers in the VDP units.

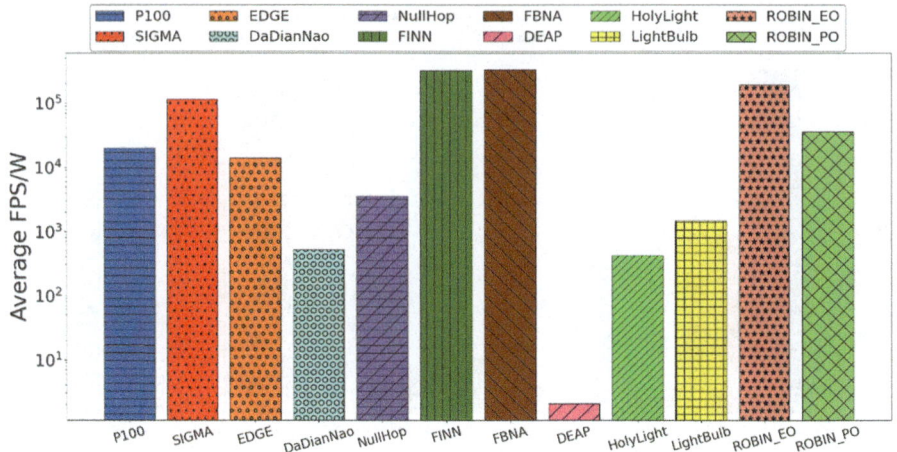

Fig. 12 Average FPS/Watt among different accelerator platforms, visualized

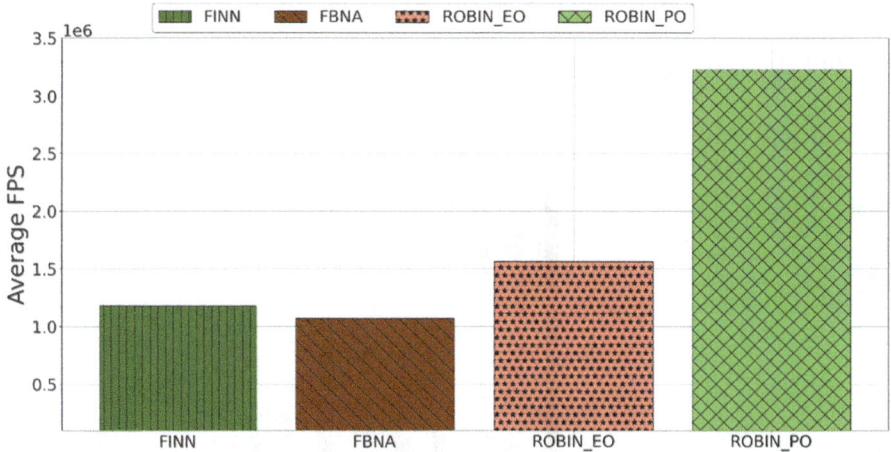

Fig. 13 FPS comparison between the ROBIN variants and the electronic BNN accelerators

Finally, in Fig. 12 we present the average FPS/Watt comparison between the various accelerator platforms. Both the *ROBIN* variants perform well against the accelerator platforms to which they were compared against. *ROBIN-EO* outperforms all other platforms other than FBNA and FINN. This is owing to the extremely low power consumption reported by these BNN accelerators. However, the *ROBIN* variants display superior FPS performance with respect to these electronic accelerators, as can be seen in Fig. 13.

In summary, this work showcases the effectiveness of cross-layer design of BNN accelerators with the emerging silicon photonics technology for energy-/area-efficient implementations and for performance-oriented designs. Overall, we can see

Table 3 Inference time on ROBIN-PO and Intel i7 desktop for the four models considered in evaluations

Model no.	Parameters	Datasets	Inference time (for one image)	
			ROBIN-PO (μs)	i7-4790 (ms)
1	60,642	Sign MNIST	0.0218	0.16
2	1,546,570	CIFAR10	0.28	1.75
3	13,570,186	STL10	2.3	2.5
4	552,362	SVHN	0.11	1.25

that our energy-efficient design (*ROBIN-EO*) exhibits EPB values ~4× lower than electronic BNN accelerators and ~933× lower than the photonic BNN accelerator, while the performance-oriented design (*ROBIN-PO*) shows ~3× and ~25× better FPS than the electronic and photonic BNN accelerators, respectively. With the growing maturity of silicon photonic device fabrication in CMOS-compatible processes, it is expected that the energy costs of device tuning, losses, and laser power overheads will go further down, making an even stronger case for considering optical-domain accelerators for deep learning inference.

6.5 Comparison to CPU-Based Inference

To highlight the advantage of dedicated inference acceleration, we have compared the performance of our *ROBIN* architecture against a standard desktop CPU performing inference on these models. The CPU we have considered is an Intel i7-4790, and we have used Tensorflow to analyze the latency for inference. The CPU, i7-4790, is reported to have an average power consumption of approximately 103 W. This power consumption is comparable to the ~90 W we report for the *ROBIN*-PO variant. The summary of observations for inference time is shown in Table 3. The *ROBIN* accelerator is observed to provide several orders of magnitude reduction in inference time for all the four models and datasets, compared to the Intel i7 system.

7 Conclusion

In this chapter we proposed *ROBIN,* an optical-domain BNN accelerator which utilizes device-level, circuit-level, and architecture-level optimizations to save on energy and area while improving overall throughput. Through our optimization efforts, we identified two variants of *ROBIN*: *ROBIN-EO*, which is optimized for energy and area efficiency, and *ROBIN-PO*, which exhibits higher FPS performance, at the expense of greater power consumption. Our simulation analysis showed that *ROBIN* exhibits significantly better EPB performance than the various state-of-the-art optical neural network accelerators. Owing to significantly lower power consumption reported by the electronic BNN accelerators considered, *ROBIN* variants are not able to obtain better FPS/Watt than them, but upon closer examination, both *ROBIN* variants can be seen to have better throughput than the electronic BNN accelerators. These results highlight the promise of our proposed *ROBIN* accelerator for accelerating BNN model execution for resource-constrained platforms.

The work described in this chapter is focused on BNN acceleration using photonic systems. In this work, we considered how photonic systems can be used to accelerate the partially binarized networks, with weights remaining binary, while activations being multi-bit parameters. To improve on this work, one may consider employing mixed quantization in the models considered, where different layers have different levels of quantization for their activation parameters. This can enable better accuracy for the considered models. The photonic system- and device-level optimizations discussed in this chapter are not limited to BNN inference accelerators. These techniques may also be considered for other non-BNN accelerators for DNN/CNNs as well.

References

1. Jouppi, N.P., Young, C., Patil, N., Patterson, D., Agrawal, G., Bajwa, R., Bates, S., Bhatia, S., Boden, N., Borchers, A., Boyle, R., Cantin, P., Chao, C., Clark, C., Coriell, J., Daley, M., Dau, M., Dean, J., Gelb, B., Ghaemmaghami, T.V., Gottipati, R., Gulland, W., Hagmann, R., Ho, C.R., Hogberg, D., Hu, J., Hundt, R., Hurt, D., Ibarz, J., Jaffey, A., Jaworski, A., Kaplan, A., Khaitan, H., Koch, A., Kumar, N., Lacy, S., Laudon, J., Law, J., Le, D., Leary, C., Liu, Z., Lucke, K., Lundin, A., MacKean, G., Maggiore, A., Mahony, M., Miller, K., Nagarajan, R., Narayanaswami, R., Ni, R., Nix, K., Norrie, T., Omernick, M., Penukonda, N., Phelps, A., Ross, J., Ross, M., Salek, A., Samadiani, E., Severn, C., Sizikov, G., Snelham, M., Souter, J., Steinberg, D., Swing, A., Tan, M., Thorson, G., Tian, B., Toma, H., Tuttle, E., Vasudevan, V., Walter, R., Wang, W., Wilcox, E., Yoon, D.H.: In-datacenter performance analysis of a tensor processing unit. In: ISCA. IEEE (2017)
2. Intel Movidius VPU. [Online]: https://www.intel.com/content/www/us/en/products/processors/movidius-vpu/movidius-myriad-x.html (2020)
3. Coubariaux, M., Hubara, I., Soudry, D., El-Yaniv, R., Bengio, Y.: BinaryNet: training deep neural networks with weights and activations constrained to +1 or −1. arXiv 2016, arXiv:1602.02830
4. Hubara, I., Courbariaux, M., Soudry, D., El-Yaniv, R., Bengio, Y.: Binarized neural networks. In: NIPS. Curran Associates, Inc. (2016)
5. Waldrop, M.M.: The chips are down for Moore's law. Nat. News. **530**(7589), 144 (2016)
6. Pasricha, S., Dutt, N.: On-Chip Communication Architectures. Morgan Kauffman (2008). ISBN 978-0-12-373892-9
7. Ziabari, A.K., Abellán, J.L., Ubal, R., Chen, C., Joshi, A., Kaeli, D.: Leveraging silicon-photonic noc for designing scalable GPUs. In: ACM ICS. ACM (2015)
8. Bahirat, S., Pasricha, S.: METEOR: hybrid photonic ring-mesh network-on-chip for multicore architectures. ACM Trans. Embed. Comput. Syst. **13**(3), 1–33 (2014)
9. Bahirat, S., Pasricha, S.: HELIX: design and synthesis of hybrid nanophotonic application-specific network-on-chip architectures. In: IEEE International Symposium on Quality Electronic Design (ISQED). IEEE (2014)
10. Bahirat, S., Pasricha, S.: 3D HELIX: design and synthesis of hybrid nanophotonic application-specific 3D network-on-chip architectures. In: Workshop on Exploiting Silicon Photonics for Energy efficient Heterogeneous Parallel Architectures (SiPhotonics). IEEE (2014)

11. Bahirat, S., Pasricha, S.: A particle swarm optimization approach for synthesizing application-specific hybrid photonic networks-on-chip. In: IEEE International Symposium on Quality Electronic Design (ISQED). IEEE (2012)
12. Bahirat, S., Pasricha, S.: UC-PHOTON: a novel hybrid photonic network-on-chip for multiple use-case applications. In: IEEE International Symposium on Quality Electronic Design (ISQED). IEEE (2010)
13. Bahirat, S., Pasricha, S.: Exploring hybrid photonic networks-on-chip for emerging chip multiprocessors. In: IEEE/ACM International Conference on Hardware/Software Codesign and System Synthesis (CODES+ISSS). IEEE (2009)
14. Chittamuru, S.V.R., Thakkar, I., Pasricha, S., Vatsavai, S.S., Bhat, V.: Exploiting process variations to secure photonic NoC architectures from snooping attacks. In: IEEE Transactions on Computer-Aided Design of Integrated Circuits and Systems, (TCAD). IEEE (2021)
15. Chittamuru, S.V.R., Thakkar, I., Pasricha, S.: LIBRA: thermal and process variation aware reliability management in photonic networks-on-chip. IEEE Trans. Multi-Scale Comput. Syst. 4(4), 758–772 (2018)
16. Chittamuru, S.V.R., Dharnidhar, D., Pasricha, S., Mahapatra, R.: BiGNoC: accelerating big data computing with application-specific photonic network-on-chip architectures. IEEE Trans. Parallel Distrib. Syst. 29(11), 2402 (2018)
17. Chittamuru, S.V.R., Thakkar, I., Pasricha, S.: HYDRA: heterodyne crosstalk mitigation with double microring resonators and data encoding for photonic NoC. IEEE Trans. Very Large Scale Integr. VLSI Syst. 26(1), 168–181 (2018)
18. Chittamuru, S.V.R., Desai, S., Pasricha, S.: SWIFTNoC: a reconfigurable silicon-photonic network with multicast enabled channel sharing for multicore architectures. ACM J. Emerg. Technol. Comput. Syst. 13(4), 1–27 (2017)
19. Chittamuru, S.V.R., Pasricha, S.: Crosstalk mitigation for high-radix and low-diameter photonic NoC architectures. IEEE Des. Test. 32(3), 29–39 (2015)
20. Thakkar, I., Chittamuru, S.V.R., Pasricha, S.: Mitigating the energy impacts of VBTI aging in photonic networks-on-chip architectures with multilevel signaling. In: IEEE Workshop on Energy-Efficient Networks of Computers (E2NC). IEEE (2018)
21. Pasricha, S., Chittamuru, S.V.R., Thakkar, I., Bhat, V.: Securing photonic NoC architectures from hardware trojans. In: IEEE/ACM International Symposium on Networks-on-Chip (NOCS). IEEE (2018)
22. Chittamuru, S.V.R., Thakkar, I., Pasricha, S.: SOTERIA: exploiting process variations to enhance hardware security with photonic NoC architectures. In: IEEE/ACM De-sign Automation Conference (DAC). IEEE (2018)
23. Thakkar, I., Chittamuru, S.V.R., Pasricha, S.: Improving the reliability and energy-efficiency of high-bandwidth photonic NoC architectures with multilevel signaling. In: IEEE/ACM International Symposium on Networks-on-Chip (NOCS). IEEE (2017)
24. Chittamuru, S.V.R., Thakkar, I., Pasricha, S.: Analyzing voltage bias and temperature induced aging effects in photonic interconnects for manycore computing. In: ACM System Level Interconnect Prediction Workshop (SLIP). IEEE (2017)
25. Dang, D., Chittamuru, S.V.R., Mahapatra, R.N., Pasricha, S.: Islands of heaters: a novel thermal management framework for photonic NoCs. In: IEEE/ACM Asia & South Pacific Design Automation Conference (ASPDAC). IEEE (2017)
26. Thakkar, I., Chittamuru, S.V.R., Pasricha, S.: A comparative analysis of front-end and back-end compatible silicon photonic on-chip interconnects. In: ACM/IEEE System Level Interconnect Prediction Workshop (SLIP). IEEE (2016)
27. Thakkar, I., Chittamuru, S.V.R., Pasricha, S.: Run-time laser power management in photonic NoCs with on-chip semiconductor optical amplifiers. In: IEEE/ACM International Symposium on Networks-on-Chip (NOCS). IEEE (2016)
28. Chittamuru, S.V.R., Thakkar, I., Pasricha, S.: PICO: mitigating heterodyne cross-talk due to process variations and intermodulation effects in photonic NoCs. In: IEEE/ACM Design Automation Conference (DAC). IEEE (2016)

29. Chittamuru, S.V.R., Thakkar, I., Pasricha, S.: Process variation aware cross-talk mitigation for DWDM based photonic NoC architectures. In: IEEE International Symposium on Quality Electronic Design (ISQED). IEEE (2016)
30. Chittamuru, S.V.R., Pasricha, S.: SPECTRA: a framework for thermal reliability management in silicon-photonic networks-on-chip. In: IEEE International Conference on VLSI Design (VLSI). IEEE (2016). [8] Pasricha, S., Nikdast, M.: A survey of silicon photonics for energy efficient manycore computing. In IEEE Design and Test, vol. 37, no. 4 (2020)
31. Miller, D.A.: Silicon photonics: meshing optics with applications. Nat. Photonics. **11**(7), 403 (2017)
32. Sunny, F., Mirza, A., Nikdast, M., Pasricha, S.: CrossLight: A cross-layer optimized silicon photonic neural network accelerator. In: ACM/IEEE Design Automation Conference (DAC). IEEE (2021)
33. Sunny, F., Nikdast, M., Pasricha, S.: SONIC: a sparse neural network inference accelerator with silicon photonics for energy-efficient deep learning. In: Asia and South Pacific Design Automation Conference (ASP-DAC). IEEE (2022)
34. Pintus, P., Hofbaurer, M., Manganelli, C.L., Fournier, M., Gundavarapu, S., Lemonnier, O., Gambini, F.: PWM-Driven thermally tunable silicon microring resonators: design, fabrication, and characterization. Laser Photonics Rev. **13**(9), 1800275 (2019)
35. Sunny, F., Mirza, A., Nikdast, M., Pasricha, S.: ROBIN: a robust optical binary neural network accelerator. ACM Trans. Embed. Comput. Syst. **20**(5), 1–24 (2021)
36. Sunny, F., Taheri, E., Nikdast, M., Pasricha, S.: A survey on silicon photonics for deep learning. arXiv 2021, arXiv:2101.01751
37. Shen, Y., Harris, N.C., Skirlo, S., Prabhu, M., Jones, T.B., Hochberg, M., Sun, X., Zhao, S., Larochelle, H., Englund, D., Soljacic, M.: Deep learning with coherent nanophotonic circuits. Nat. Photonics. **11**(7), 441–446 (2017)
38. Zhao, Z., Liu, D., Li, M., Ying, Z., Zhang, L., Xu, B., Yu, B., Chen, R.T., Pan, D.Z.: Hardware-software co-design of slimmed optical neural networks. In: IEEE/ACM ASPDAC. IEEE (2019)
39. Bangari, V., Marquez, B.A., Miller, H., Tait, A.N., Nahmias, M.A., De Lima, T.F., Peng, H.T., Prucnal, P.R., Shastri, B.J.: Digital electronics and analog photonics for convolutional neural networks (DEAP-CNNs). IEEE J. Sel. Top. Quantum Electron. **26**(1), 1–13 (2020)
40. Liu, W., Liu, W., Ye, Y., Lou, Q., Xie, Y., Jiang, L.: HolyLight: a nanophotonic accelerator for deep learning in data centers. In: IEEE/ACM DATE. IEEE (2019)
41. Shiflett, K., Wright, D., Karanth, A., Louri, A.: PIXEL: photonic neural network accelerator. In: HPCA. IEEE (2020)
42. Tait, A.N., De Lima, T.F., Zhou, E., Wu, A.X., Nahmias, M.A., Shastri, B.J., Prucnal, P.R.: Neuromorphic photonic networks using silicon photonic weight banks. Sci. Rep. **7**(1), 1–10 (2017)
43. Mourgias-Alexandris, G., Totovic, A., Tsakyridis, A., Passalis, N., Vyrsokinos, K., Tefas, A., Pleros, N.: Neuromorphic photonics with coherent linear neurons using dual-IQ modulation cells. IEEE J. Lightwave Technol. **38**(4), 811–819 (2020)
44. Pask, C.: Generalized parameters for tunneling ray attenuation in optical fibers. J. Opt. Soc. Am. **68**(1), 110 (1978)
45. Anderson, J., Sun, S., Alkabani, Y., Sorger, V., El-Ghazawi, T.: Photonic processor for fully discretized neural networks. In: IEEE ASAP. IEEE (2019)
46. Zokae, F., Lou, Q., Youngblood, N., Liu, W., Xie, Y., Jiang, L.: LightBulb: a photonic-nonvolatile-memory-based accelerator for binarized convolutional neural networks. In: IEEE/ACM DATE. IEEE (2020)
47. Totovic, A.R., Dabos, G., Passalis, N., Tefas, A., Pleros, N.: Femtojoule per MAC neuromorphic photonics: an energy and technology roadmap. IEEE J. Sel. Top. Quantum Electron. **26**(5), 1–15 (2020)
48. Nikdast, M., Nicolescu, G., Trajkovic, J., Liboiron-Ladouceur, O.: Chip-scale silicon photonic interconnects: a formal study on fabrication non-uniformity. IEEE J. Lightwave Technol. **34**(16), 3682–3695 (2016)

49. Stefan, A., Stöferle, T., Marchiori, C., Caimi, D., Czornomaz, L., Stuckelberger, M., Sousa, M., Offrein, B.J., Fompeyrine, J.: A hybrid barium titanate–silicon photonics platform for ultraefficient electro-optic tuning. IEEE J. Lightwave Technol. **34**(8), 1688–1693 (2016)
50. Bengio, Y., Léonard, N., Courville, A.: Estimating or propagating gradients through stochastic neurons for conditional computation. arXiv preprint, arXiv:13126199 (2013)
51. Lu, L., Li, X., Gao, W., Li, X., Zhou, L., Chen, J.: Silicon non-blocking 4× 4 optical switch chip integrated with both thermal and electro-optic tuners. IEEE Photonics J. **11**(6), 1–9 (2019)
52. Milanizadeh, M., Aguiar, D., Melloni, A., Morichetti, F.: Canceling thermal cross-talk effects in photonic integrated circuits. IEEE J. Lightwave Technol. **37**(4), 1325–1332 (2019)
53. Lumerical Solutions Inc.: Lumerical MODE. [Online]. Available: http://www.lumerical.com/tcad-products/mode/
54. Liu, Y., Sun, W., Xie, H., Zhang, N., Xu, K., Yao, Y., Xiao, S., Song, Q.: Adiabatic and ultracompact waveguide tapers based on digital metamaterials. IEEE J. Sel. Top. Quantum Electron. **25**(3), 1–6 (2019)
55. Duong, L., Nikdast, M., Le Beux, S., Xu, J., Wu, X., Wang, Z., Yang, P.: A case study of signal-to-noise ratio in ring based optical networks-on-chip. IEEE Des. Test. **31**(5), 55–65 (2014)
56. Bogaerts, W., Heyn, P.D., Vaerenburgh, T.V., De Vos, K., Selvaraj, S.K., Claes, T., Dumon, P., Bienstman, P., Thourhout, D.V., Baets, R.: Silicon microring resonators. Laser Photonics Rev. **6**(1), 47–73 (2012)
57. Su, Z., Hosseini, E.S., Timurdogan, E., Sun, J., Leake, G., Coolbaugh, D.D., Watts, M.R.: Reduced wafer-scale frequency variation in adiabatic microring resonators. In: OFC. Optica Publishing Group (2014)
58. Xu, Q., Fattal, D., Beausoleil, R.G.: Silicon microring resonators with 1.5-μm radius. Opt. Express. **16**(6), 4309–4315 (2008)
59. Little, B.E., Chu, S.T., Haus, H.A., Foresi, J., Laine, J.-P.: Microring resonator channel dropping filters. IEEE J. Lightwave Technol. **15**(6), 998–1005 (1997)
60. Xia, J., Bianco, A., Bonetto, E., Gaudino, R.: On the design of microring resonator devices for switching applications in flexible-grid networks. In: IEEE International Conference on Communications (ICC), pp. 3371–3376. IEEE (2014)
61. Chen, T., Du, Z., Sun, N., Wang, J., Wu, C., Chen, Y., Temam, O.: DianNao: a small-footprint high-throughput accelerator for ubiquitous machine-learning. In: ACM ASPLOS. ACM (2014)
62. QKeras: https://github.com/google/qkeras
63. Frandsen, L.H., Borel, P.I., Zhuang, Y.X., Harpøth, A., Thorhauge, M., Kristensen, M., Bogaerts, W., Dumon, P., Baets, R., Wiaux, V., Wouters, J.: Ultralow-loss 3-dB photonic crystal waveguide splitter. Opt. Lett. **29**(14), 1623–1625 (2004)
64. Tu, Y., Fu, P.H., Huang, D.W.: High-efficiency ultra-broadband multi-tip edge couplers for integration of distributed feedback laser with silicon-on-insulator waveguide. IEEE Photonics J. **11**(4), 1–13 (2019)
65. Bahirat, S., Pasricha, S.: OPAL: a multi-layer hybrid photonic NoC for 3D ICs. In: IEEE/ACM ASPDAC. IEEE (2011)
66. Jayatileka, H., Caverley, M., Jaeger, N.A.F., Shekhar, S., Chrostowski, L.: Crosstalk limitations of microring-resonator based WDM demultiplexers on SOI. In: OIC. IEEE (2015)
67. Timurdogan, E., Sorace-Agaskar, C.M., Hosseini, E.S., Leake, G., Coolbaugh, D.D., Watts, M.R.: Vertical junction silicon microdisk modulator with integrated thermal tuner. In: CLEO: Science and Innovations, OSA. IEEE (2013)
68. Shen, J., Shikata, A., Fernando, L.D., Guthrie, N., Chen, B., Maddox, M., Mascarenhas, N., Kapusta, R., Coln, M.C.W.: A 16-bit 16-MS/s SAR ADC with on-chip calibration in 55-nm CMOS. IEEE J. Solid State Circuits. **53**(4), 1149–1160 (2018)
69. Wu, B., Zhu, S., We, B., Chiu, Y.: A 24.7 mW 65 nm CMOS SARassisted CT modulator with second-order noise coupling achieving 45 MHz bandwidth and 75.3 dB SNDR. IEEE J. Solid State Circuits. **51**(12), 2893–2905 (2016)
70. Ruan, Z., Zhu, Y., Chen, P., Shi, Y., He, S., Cai, X., Liu, L.: Efficient hybrid integration of long-wavelength VCSELs on silicon photonic circuits. IEEE J. Lightwave Technol. **38**(18), 5100–5106 (2020)

71. Güngördü, A.D., Dündar, G., Yelten, M.B.: A high performance TIA design in 40 nm CMOS. In: IEEE ISCAS. IEEE (2020)
72. Wang, B., Huang, Z., Sorin, W.V., Zeng, X., Liang, D., Fiorentino, M., Beausoleil, R.G.: A low-voltage Si-Ge avalanche photodiode for high-speed and energy efficient silicon photonic links. IEEE J. Lightwave Technol. 38(12), 3156–3163 (2020)
73. Mirza, A., Sunny, F., Pasricha, S., Nikdast, M.: Silicon photonic microring resonators: design optimization under fabrication non-uniformity. In: IEEE/ACM Design, Automation and Test in Europe (DATE) Conference and Exhibition, pp. 484–489. IEEE, Grenoble (2020)
74. Bahadori, M., Nikdast, M., Rumley, S., Yuan Dai, L., Janosik, N., Van Vaerenbergh, T., Gazman, A., Cheng, Q., Polster, R., Bergman, K.: Design space exploration of microring resonators in silicon photonic interconnects: impact of the ring curvature. IEEE J. Lightwave Technol. 36(13), 2767–2782 (2018)
75. Qin, E., Samajdar, A., Kwon, H., Nadella, V., Srinivasan, S., Das, D., Kaul, B., Krishna, T.: SIGMA: a sparse and irregular GEMM accelerator with flexible interconnects for DNN training. In: IEEE HPCA. IEEE (2020)
76. Cass, S.: Taking AI to the edge: Google's TPU now comes in a maker-friendly package. IEEE Spectr. 56(5), 16–17 (2019)
77. Luo, T., Liu, S., Li, L., Wang, Y., Zhang, S., Chen, T., Xu, Z., Temam, O., Chen, Y.: DaDianNao: a neural network supercomputer. IEEE Trans. Comput. 66(1), 73–88 (2017)
78. Aimar, A., Mostafa, H., Calabrese, E., Rios-Navarro, A., Tapiador-Morales, R., Lungu, I.A., Milde, M.B., Corradi, F., Linares-Barranco, A., Liu, S.C., Delbruck, T.: NullHop: a flexible convolutional neural network accelerator based on sparse representations of feature maps. IEEE Trans. Neural Netw. Learn. Syst. 30(3), 644–656 (2016)
79. Guo, P., Ma, H., Chen, R., Li, P., Xie, S., Wang, D.: FBNA: a Fully binarized neural network accelerator. In: International Conference on Field Programmable Logic and Applications. IEEE (2018)
80. Umuroglu, Y., Fraser, N.J., Gambardella, G., Blott, M., Leong, P., Jahre, M., Vissers, K.: FINN: a framework for fast, scalable binarized neural network inference. In: ACM/SIGDA FPGA. ACM (2017)
81. Capra, M., Bussolino, B., Marchisio, A., Shafique, M., Masera, G., Martina, M.: An updated survey of efficient hardware architectures for accelerating deep convolutional neural networks. Future Internet. 12(7), 113 (2020)

Low-Latency, Energy-Efficient In-DRAM CNN Acceleration with Bit-Parallel Unary Computing

Ishan G. Thakkar, Supreeth M. Shivanandamurthy, and Sayed Ahmad Salehi

1 Introduction

Convolutional neural networks (CNNs) have achieved remarkable progress in recent years, and they are being aggressively utilized in real-world applications related to artificial intelligence (AI) and machine learning [1, 2]. In general, CNNs mimic biological neural networks and utilize compute-heavy arithmetic functions such as multiply-accumulate (MAC), nonlinear activation, and pooling. Although these CNN functions are amenable to acceleration because of a high degree of compute parallelism, their acceleration using traditional ASIC platforms (e.g., Dadiannao [1], EIE [3]) is challenging because of the need to avoid the memory wall while accessing their large number of operands [4]. To address this problem, several prior works have explored processing-in-memory (PIM) designs based on the emerging non-volatile memory (NVM) crossbar technologies (e.g., ISAAC [2], PRIME [5], XNOR-RRAM [6]) as well as the traditional DRAM technology (e.g., DRISA [7], SCOPE [8], DRACC [9], LACC [10]). Such PIM designs strive to avoid data movement to consequently achieve a balance between computational efficiency and memory performance while processing CNNs in situ.

However, it is challenging to support MAC operations in PIM designs. The NVM crossbar-based PIM designs, such as ISAAC [2] and PRIME [5], leverage the Kirchhoff's law to perform MAC operations in the analog domain. However, such analog-computing-based accelerators require power-hungry and sluggish digital-to-analog converters and analog-to-digital converters (DACs and ADCs), which

I. G. Thakkar (✉) · S. M. Shivanandamurthy · S. A. Salehi
University of Kentucky, Lexington, KY, USA
e-mail: igthakkar@uky.edu; supreethms@uky.edu; SayedSalehi@uky.edu

© The Author(s), under exclusive license to Springer Nature Switzerland AG 2024
S. Pasricha, M. Shafique (eds.), *Embedded Machine Learning for Cyber-Physical, IoT, and Edge Computing*, https://doi.org/10.1007/978-3-031-19568-6_14

diminishes the performance and energy-efficiency benefits of such accelerators. Alternatively, the DRAM-based PIM designs implement in situ MAC operations digitally, for which they break a single MAC operation into multiple functionally complete memory operation cycles (MOCs) that are serially run on a single subarray (the smallest logical cell array in a DRAM module). Multiple such subarrays typically work in parallel to achieve high processing throughput. Such designs require a very larger number of MOCs per MAC operation. For instance, DRISA [7] requires up to 222 MOCs per MAC. To reduce the required number of MOCs, SCOPE [8], DRACC [9], and LACC [10] employ light-weight optimizations that simplify the implementation of MAC operations. SCOPE adopts rate-coded unary (stochastic) computing to implement approximate multiplication, requiring a reduced number of up to 25 MOCs per MAC [8]. On the other hand, DRACC [9] eliminates most multiply operations by employing quantized CNNs that use ternary weights, whereas LACC [10] employs lookup table-based multiply operations. Because of these optimizations, DRACC and LACC require a reduced number of MOCs per MAC of up to 13 and 11, respectively. This can still incur very high latency and energy consumption as one MOC can incur up to 49 ns latency and up to 4nJ energy consumption [7, 9, 11], depending on the utilized DRAM technology node and subarray size (bitline length). The high latency and energy values per MAC operation have prevented the DRAM-based PIM designs from being immediately adopted for CNN inference.

In this chapter, we present a novel CNN accelerator called ATRIA. ATRIA employs bit-parallel rate-coded unary (stochastic) computing, which enables it to perform 16 MAC operations in only 2 consecutive MOCs. ATRIA is most related to SCOPE [8]. It significantly improves upon SCOPE in two ways. First, SCOPE uses rate-coded unary (stochastic) computing to perform only multiply operations, whereas it uses the conventional binary arithmetic to perform accumulate operations. In contrast, ATRIA performs both multiply and accumulate operations using bit-parallel rate-coded unary (stochastic) computing. Second, both SCOPE and ATRIA require expensive binary-to-stochastic (B-to-S) and stochastic-to-binary (S-to-B) conversions of operands, but ATRIA is better able to hide the latency of these conversions by successfully removing them from the critical processing path. Moreover, ATRIA restricts the precision errors induced due to the rate-coded unary (stochastic)-computing-based accumulate operations by employing stochastic operands that are $2\times$ larger in size. As a result, ATRIA exhibits only 3.5% drop in CNN inference accuracy on average compared to SCOPE. Despite this slight drawback, ATRIA substantially outperforms SCOPE as well as other in-DRAM accelerators such as DRISA and LACC in terms of the latency, throughput [frames per second (FPS)], and efficiency (FPS/W/mm^2) of processing state-of-the-art CNNs.

2   Concept of Bit-Parallel Rate-Coded Unary (Stochastic) Computing

The use of rate-coded unary (stochastic) computing simplifies the implementation of complex arithmetic functions, such as multiplication and accumulation, by reducing them to simple bit-wise logical operations [12]. To perform a multiplication of 2 N-bit stochastic operands (A and B in Fig. 1a) in the bit-serial manner, the bit streams of the operands are applied to an AND gate serially, and the bit-wise output of the AND gate is collected for total N clock cycles to generate the multiplication output bit stream (C in Fig. 1a). Similarly, to perform a scaled accumulation of 4 (or more) N-bit stochastic operands in the bit-serial manner (A, B, C, D in Fig. 1b), the bit streams of the operands are applied to a MUX, whose bit-wise output is selected by a 2-bit (or larger) random number (RND in Fig. 1b) every clock cycle for total N clock cycles, to generate the output bit stream that represents a scaled accumulation (E in Fig. 1b). To reduce the area and static power consumption of computing, such bit-serial implementation of rate-coded unary (stochastic) computing compromises the latency of computing.

In contrast, we observe that the latency of computing can be improved by N× if the rate-coded unary (stochastic) computing can be implemented in the bit-parallel manner. For example, if N copies of AND gates and MUX circuits are available (Figs. 2a and b), the N-bit outputs for the stochastic multiplication and scaled accumulation can be obtained in one clock cycle in the bit-parallel manner. In a nutshell, the idea for such bit-parallel implementation of rate-coded unary (stochastic) computing is to transform the input bit streams into bit vectors by striping them across the N copies of the AND gates and MUX circuits and then perform bit-wise AND and MUX operations to generate output bit vectors. For instance, the individual N bits a_1 to a_N, b_1 to b_N, c_1 to c_N, and d_1 and d_N of operands A, B, C, and D from Fig. 1b are striped across N copies of MUXs in Fig. 2b. As a result, the individual N bits of the scaled accumulation output E can be collected in a bit-parallel manner from N MUXs. For such bit-parallel scaled accumulation (i.e., MUX operation), total N RND signals (RND_1 to RND_N) are needed, which can be generated a priori and made available in a parallel manner

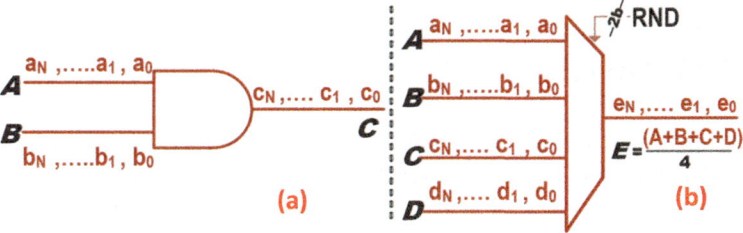

Fig. 1 Bit-serial rate-coded unary (stochastic) computing circuits for (**a**) multiplication (AND gate), (**b**) scaled accumulation (MUX)

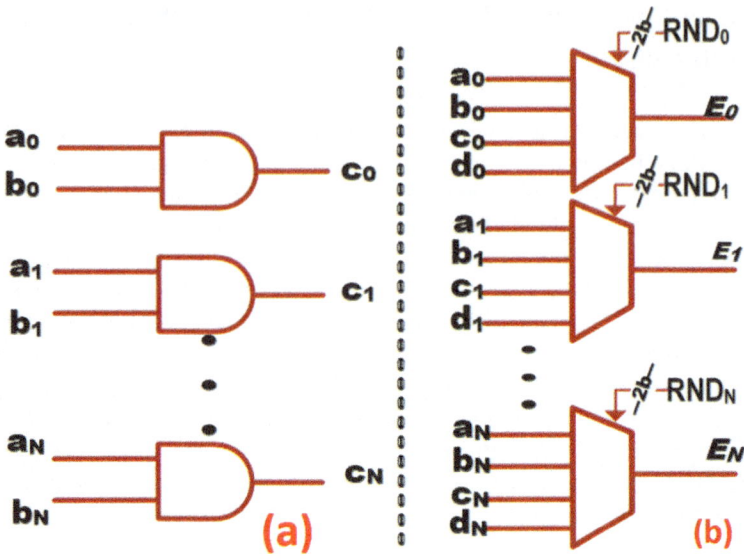

Fig. 2 Bit-parallel rate-coded unary (stochastic) computing circuits for (**a**) multiplication (an array of AND gates), (**b**) scaled accumulation (an array of MUXs). Here, the individual N bits of operands A, B, C, and D from Fig. 1 are striped across N copies of AND gates and MUXs

(Fig. 2b). Although Fig. 2b illustrates bit-parallel scaled accumulation for only four input stochastic operands (A, B, C, and D), this concept can be extended for more or less than 4 input stochastic operands as well.

Such bit-parallel rate-coded unary (stochastic) computing naturally fits well for in-DRAM processing of applications because the inherent parallelism of DRAM makes it fundamentally easy to provision data in the bit-parallel manner. Our proposed in-DRAM accelerator ATRIA employs such bit-parallel rate-coded unary (stochastic) computing to implement in-DRAM MAC operations for the first time and exploits the benefits of such implementation to substantially improve the latency and throughput of in-DRAM CNN processing, compared to the in-DRAM CNN processing accelerators from prior work.

3   ATRIA: Overview

Our ATRIA accelerator architecture employs an 8Gb DRAM module with 8 chips. Figure 3 illustrates the schematic of one such chip. Each chip has 8 banks, with 64 subarrays per bank, and 32 mats per subarray of 256×256 bits size each. Each row in a subarray is of 8Kb size; therefore, each subarray contains total 8Kb sense amplifiers (S/As) and write drivers (W/Ds). Each subarray acts as a processing element (PE), which is defined as the smallest independent cell-array structure that

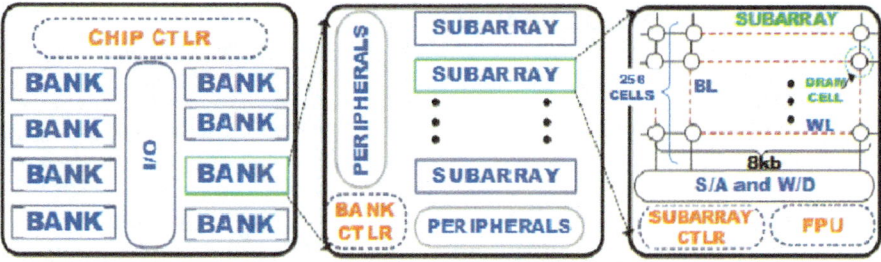

Fig. 3 The hierarchical structure of our ATRIA accelerator chip

can perform computing. Therefore, there are a total of 4096 PEs in ATRIA. Like the other in-DRAM accelerators from prior work (e.g., DRISA [7], SCOPE [8], LACC [10]), the PEs in ATRIA can also operate in parallel to process CNN inference in situ. To process CNN inference, each PE (i.e., subarray) in ATRIA employs a feature processing unit (FPU), as shown in Fig. 3. In addition, to orchestrate these in-parallel processing operations inside the PEs, ATRIA employs hierarchical controllers (chip, bank, and subarray controllers (CTLRs) in Fig. 3). The operation of these hierarchical CTLRs is described in Sect. 3.3. The structure and operation of each FPU in ATRIA support our concept of bit-parallel rate-coded unary (stochastic) computing for in situ processing of CNNs, as discussed next.

3.1 Structure of a PE in ATRIA

A PE of our ATRIA accelerator is basically a DRAM subarray that is integrated with an FPU and a subarray CTLR, as illustrated in Fig. 4. The subarray part of the PE is structured in the manner the conventional DRAM subarrays are organized [13, 14]. Therefore, in this section, we only provide details of the structure of the FPU. The role of the subarray CTLR is discussed in Sect. 3.3. The FPU consists of various hardware components that support the implementation of the following six functions: (i) bit-parallel stochastic multiply operation (MUL), (ii) bit-parallel stochastic accumulate operation (ACC), (iii) binary-to-stochastic (B-to-S) conversion, (iv) stochastic-to-binary (S-to-B) conversion through pop counter (PC), (v) nonlinear activation function ReLU, and (vi) max-pooling function. To support bit-parallel MUL, three 8Kb rows of the subarray (Row 1, Row 2, and Row 3 in Fig. 4a) are reserved and operated following the triple row activation and charge-sharing protocol of AAP memory operation cycle (MOC) from Ambit [11] (see Sect. 3.2).

The hardware components that support bit-parallel ACC consist of an array of 512 copies of 16:1 MUXs and their associated 512 copies of 4-bit registers (Fig. 4a). These 4-bit registers store the pre-determined random values that enable the output selection (16:1) for their respective MUXs. Each MUX has 16 inputs;

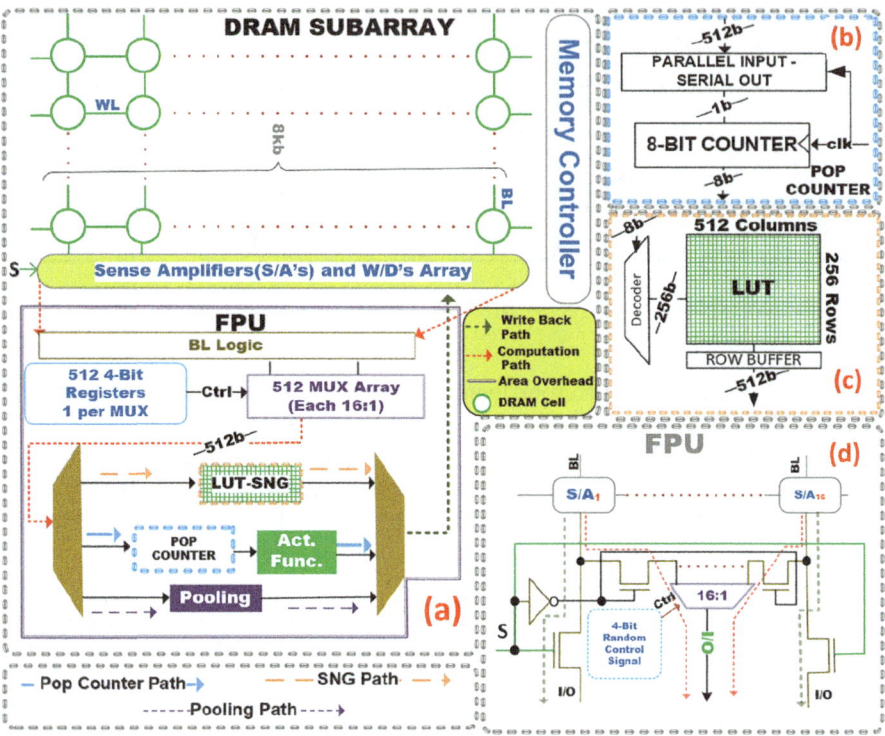

Fig. 4 Schematic of a processing element (PE) of ATRIA. (**a**) Schematic of a subarray and feature processing unit (FPU); (**b**) pop counter for S-to-B conversion [2]; (**c**) LUT for B-to-S conversion; (**d**) a 16:1 MUX and its connections with S/As as part of the FPU

therefore, the total number of inputs for the entire array of 512 MUXs is 8Kb. These 8Kb MUX inputs are connected to 8Kb S/As, with 16 adjacent S/As feeding one MUX and vice versa (Fig. 4d). Note that the S/As in the commodity DRAMs typically connect to I/O logic through signal S and related control transistors (M_1 to M_16) (Fig. 4d). To facilitate connections of S/As to MUXs, ATRIA employs one additional inverter (INV) and 16 transistor switches (T1 to T16) per MUX, which can be controlled by the same signal S (Fig. 4d). An 8Kb row from the subarray can be read into 8Kb S/As (Fig. 4a), which can hold in total 16 stochastic bit vectors of 512-bit size each ($16 \times 512 = 8Kb$). These 16 stochastic bit vectors can be striped across 512 MUXs, so that each individual bit of a bit vector is fed into a different MUX with each MUX having all its 16 inputs from 16 different bit vectors. This arrangement sets up the array of MUXs to perform a 16-operand scaled ACC in the bit-parallel manner, following our concept of bit-parallel rate-coded unary (stochastic) computing discussed in Sect. 2. The detailed functioning of this array of MUXs for performing scaled ACC is presented in Sect. 3.2.

In addition, to implement in-memory B-to-S conversion, each FPU in ATRIA employs a lookup table (Fig. 4c). Our idea of using lookup table-based B-to-S conversion is inspired from the design of SCOPE accelerator [8]. This enables ATRIA to employ the deterministic method for B-to-S conversion to eliminate correlation errors [8]. Moreover, each FPU in ATRIA employs an additional lookup table to perform ReLU (Fig. 4a). Further, it also incorporates a pop counter to perform in-memory S-to-B conversion (Fig. 4b), as well as logic to implement max-pooling function (Fig. 4a). ATRIA implements the max-pooling and ReLU functions in the binary domain. This mandates that the results of processing of every CNN layer's parameters always go through S-to-B, ReLU, and then B-to-S conversions before they can activate processing of the next CNN layer. This in turn eliminates the undesirable propagation of precision errors (which are very common in rate-coded unary (stochastic) computing [12]) between the stochastic operations of two consecutive CNN layers (see more on errors in Sect. 4.2. The overheads of incorporating FPUs in ATRIA PEs are discussed in Sect. 3.4. The next section describes the functioning of an FPU-enabled PE of our ATRIA accelerator.

3.2   Functioning of a PE in ATRIA

Each PE of our ATRIA accelerator can perform all essential functions required for processing CNNs, such as MAC, max pooling, and ReLU. In addition, since ATRIA employs rate-coded unary (stochastic) computing, each PE can also perform important functions for implementing rate-coded unary (stochastic) computing, such as B-to-S and S-to-B (pop count) conversions. On one hand, each PE performs B-to-S, S-to-B (pop count), ReLU, and max-pooling functions by relaying the related operands along the data processing path in the FPU through the corresponding hardware components (Fig. 4a). To orchestrate the relaying of the operands to perform these functions, the PE makes use of the subarray CTLR whose functioning along with the functioning of other hierarchical CTLRs in ATRIA is discussed in Sect. 3.3. On the other hand, each PE of ATRIA can perform a MAC function (F_{MAC}) of 16 stochastic operands of 512-bit size each, by employing a series of total five memory operation cycles (MOCs) (similar to the AAP/AP MOC from [7, 11]). These MOCs engage the reserved rows Row 1, Row 2, and Row 3 (Fig. 4a) and the MUXs in the FPU, as discussed next.

Figure 5 illustrates how ATRIA performs F_{MAC}. ATRIA performs F_{MAC} in two main steps.

Step 1 engages the reserved subarray rows Row 1, Row 2, and Row 3 to perform MUL. Step 2 engages the array of MUXs to perform ACC. Before performing F_{MAC}, ATRIA first makes the involved stochastic operands available in the reserved subarray rows Row 1 and Row 2. For that, it performs two MOCs similar to RowClone [15] to copy the contents of two source rows into Row 1 and Row 2, respectively. Consequently, Row 1 contains 16 512-bit operands N_1 to N_{16} (Fig. 5a). Similarly, Row 2 contains 16 512-bit operands M_1 to M_{16} (Fig. 5a). In addition,

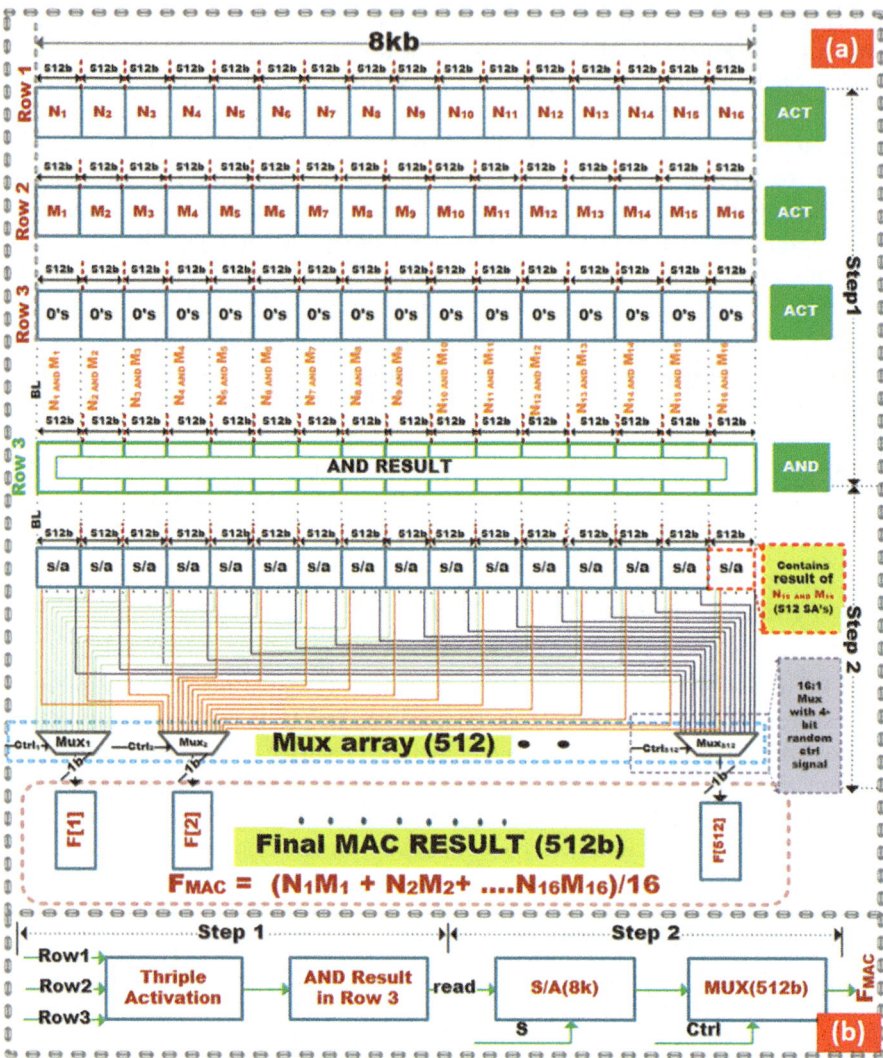

Fig. 5 A schematic showing the operation of a PE of ATRIA to perform a 16-operand multiply-accumulate (MAC) function (F_{MAC})

ATRIA initializes Row 3 with "0's" at system boot. After these initial steps, ATRIA schedules Step 1 of F_{MAC}, which employs the triple row activation and charge-sharing-based MOC from Ambit [11] to perform bit-parallel logical AND (i.e., stochastic MUL) of the involved operands N_1 to N_{16} and M_1 to M_{16}. At the end of the MOC for Step 1, Row 3 contains the results of bit-parallel logical AND, i.e., N_1 AND M_1 to N_{16} AND M_{16} (Fig. 5a and b). These results essentially represent the outcome of bit-parallel stochastic MUL, i.e., $N_1 M_1$ to $N_{16} M_{16}$. After this, ATRIA schedules Step 2 of F_{MAC}, where it performs a MOC to read the stochastic MUL results from Row 3 into S/As. These results from S/As are then pushed through the array of 16:1 MUXs, MUX1 to MUX512. The 512-bit output of this array of MUXs is selected using the pre-latched random control signals RND1 to RND512. This 512-bit output is the stochastic scaled ACC of the input operands $N_1 M_1$ to $N_{16} M_{16}$. In other words, this 512-bit output presents $F_{MAC} = (N_1 M_1 + N_2 M_2 + \ldots + N_{16} M_{16})/16$ (Fig. 5a and b). ATRIA then uses one more MOC to store the result of this F_{MAC} into a row in the subarray through W/Ds. Thus, ATRIA uses only 5 MOCs (2 MOCs for initializing Row 1 and Row 2, 1 MOC for MUL, 1 MOC for ACC, and 1 MOC for write back) to perform a scaled MAC function F_{MAC} (also called dot product) of 16 stochastic operands. In other words, if a MAC operation is conventionally defined as a MUL of two operands followed by an accumulate operation (i.e., A = A + $N_i M_i$), then ATRIA uses only 5 MOCs to perform 16 MAC operations in parallel. However, we find from our evaluation results in Sect. 4.4 that the use of bit-parallel rate-coded unary (stochastic) computing in ATRIA can increase precision errors. Nevertheless, we also find that the increased precision errors are worth tolerating for due to the substantial performance benefits of ATRIA.

3.3   System Integration and Controller Design

In this section, we describe how our ATRIA accelerator integrates with the host system and how the hierarchical controllers of ATRIA orchestrate the processing of CNNs. ATRIA integrates with the host system in the same way the conventional GPU- or FPGA-based accelerators do through PCIe bus. For a CNN processing using ATRIA, the host system stores the weighting parameters and inputs of the CNN in the individual PEs (subarrays) of ATRIA via direct memory access (DMA). We adopt the strategy from SCOPE [8], wherein the weighting parameters are stored in ATRIA in the stochastic format. This strategy ensures that in situ B-to-S conversions are required only for activation parameters, which dramatically reduces the number of in situ B-to-S conversions. As a result, the latency and energy of processing CNNs with ATRIA are dramatically reduced as well.

After storing the inputs and weighting parameters of a CNN in PEs of ATRIA, the host-side ATRIA CTLR (not shown in Fig. 3) orchestrates the processing of the CNN in conjunction with the hierarchical ATRIA CTLRs shown in Fig. 3. The host-side ATRIA controller generates a series of μ operations, which are received by the hierarchical ATRIA CTLRs. We adopt the designs from [7] for these CTLRs. These CTLRs support simultaneous multi-subarray/bank activation for better parallelism. The first chip-level CTLR is essentially a decoder, and it also helps with inter-bank data movement. The bank-level CTLRs decode the μ operations, convert them into addresses, vector lengths, and control codes, and then send them to subarray CTLRs in the active subarrays. The subarray CTLR consists of address latches, local decoders, and counters. The address latches are essential for multi-subarray activation [7]. The counters are used for continuously updating addresses to local subarray decoders. In addition, the subarray CTLR also contains buffers to support communication of operands.

Inter-bank and inter-subarray data communications in ATRIA are supported through the interconnects design adopted from LISA [16]. Data communications are carried out in binary format instead of stochastic format, which results in better energy efficiency [8]. Also, the inclusion of buffers in the subarray CTLRs enables pipelined data communications, which enables better use of resources and efficient hiding of long latencies, reducing the memory bottleneck to improve the throughput of CNN processing with ATRIA.

3.4 Overhead Analysis

Table 1 lists the latency, energy, and area overheads of various hardware components that are part of the FPUs inside the PEs of our ATRIA accelerator. These results are based on our logic synthesis analysis for 22nm node. We considered standard SRAM for LUT implementation. After accounting for the extra area overhead of these components from Table 1, the total area for 8Gb ATRIA accelerator becomes $77mm^2$. For comparison, DRISA-1T1C-NOR [7], DRISA-3T1C [8], SCOPE-Vanilla [8], SCOPE-H2D [8], and LACC [10] consume $55mm^2$, $64.6mm^2$, $259.4mm^2$, $273.4mm^2$, and $61mm^2$ area, respectively. Thus, ATRIA consumes larger area than DRISA-1T1C-NOR, DRISA-3T1C, and LACC. Nevertheless, ATRIA still achieves substantially better area and energy efficiency compared to these accelerators (Sect. 4.4). Similarly, despite the S-to-B pop counter in ATRIA incurring a long latency of 256ns (Table 1), the performance of ATRIA does not get much affected, as ATRIA manages to keep this latency out of the critical processing path (Sect. 4.3).

Table 1 Latency, energy, and area overhead values of various hardware components of the FPUs in the PEs of ATRIA

Component	Total area (mm^2)	Latency (ns)	Energy per PE (pJ)
16:1 MUXs for ACC	1.3×10^{-3}	2	10
4-bit registers for RND storage	1.1×10^{-5}	2	15.6
B-to-S LUT (512×256)	3.4	1	0.3
S-to-B pop counter (PC) (2GHz)	2.1×10^{-5}	256	153.6
ReLU LUT	1.2	1	0.3
Max-pooling logic	4.1	5	940

4   Evaluation

4.1   Modeling and Setup for Evaluation

We evaluate ATRIA and compare it with other in-DRAM accelerators from prior work such as SCOPE-Vanilla [8], SCOPE-H2D [8], DRISA-1T1C-NOR [7], DRISA-3T1C [7], and LACC [10]. We first evaluate the per-MAC latency, per-MAC energy, and total area values for our considered accelerators. We divide the evaluation of per-MAC latency/energy into two parts: latency/energy of a multiply operation (MUL) and latency/energy of an accumulate operation (ACC). All our considered accelerators follow the AAP/AP memory operation cycle (MOC) from Ambit [11]. Therefore, the latency and energy values per MOC and the total number of MOCs per MAC are evaluated first for all considered accelerators. Different accelerators have different latency and energy per MOC because they employ different lengths of local bitlines in their subarrays. For example, DRISA [7] and SCOPE [8] employ shorter local bitlines with only 64 cells per bitline. In contrast, LACC employs 512 cells per bitline, whereas ATRIA employs 256 cells per bitline. Shorter bitlines typically yield lower latency per MOC [13]. We evaluate latency using SPICE-based [17] modeling of local bitlines. To evaluate per-MOC energy as well as total accelerator area, we used CACTI [18]. We developed a custom simulator in Python to model the MOC-accurate transaction-level performance behavior of our considered accelerators, as well as to evaluate system-level performance metrics such as frames per second (FPS), latency, efficiency (FPS/W/mm^2), and memory bottleneck ratio. Memory bottleneck ratio is defined as the ratio of total stall time (time for which an accelerator needs to wait for the operands) over total inference processing time. We considered four state-of-the-art CNNs to evaluate these metrics. The quantized versions of these CNN models were trained using PyTorch for ImageNet dataset and 8-bit fixed precision of activation and weight parameters. These activation and weight parameters were extracted and provided as the input to our Python-based performance simulator, which also took our evaluated energy, latency, and area values for our considered accelerators as the input. Next, we present and discuss the results of our simulation-based study.

4.2   Precision Error and Accuracy Results

ATRIA has one caveat compared to SCOPE. The use of MUX-based bit-parallel stochastic accumulation in ATRIA can increase the absolute precision error (APE) of computing, as explained in [12]. An APE for an operation (i.e., MUL or ACC) is defined as the absolute difference between the expected result and the observed result of the operation. From [12] and [19], APE depends on the operand values, input size (i.e., the number of operands), and operand size (i.e., bit-stream length). For a MUX-based stochastic ACC with an input size of 16 (as is the case for

Table 2 Average APE (μAPE), standard deviation in APE (σAPE) and CNN testing accuracy (A) for SCOPE-Vanilla, SCOPE-H2D, and ATRIA for various CNNs

CNN Benchmarks	SCOPE-Vanilla			SCOPE-H2D			ATRIA		
	μAPE	σAPE	A(%)	μAPE	σAPE	A (%)	μAPE	σAPE	A(%)
AlexNet	0.23	0.04	93.6	0.09	0.01	96.7	0.33	0.05	92.2
GoogleNet	0.30	0.05	87.7	0.17	0.03	88.5	0.41	0.07	87.7
VGG16	0.35	0.05	91.9	0.21	0.03	95.1	0.53	0.09	90.2
ResNET-50	0.26	0.04	90.1	0.12	0.02	93.6	0.47	0.08	89.8

ATRIA), the average APE (μAPE) can be reduced to an acceptable value in the range between 0.2 and 0.54, if the operand size is kept 512 bits or longer [12, 19]. Therefore, we increase the operand size, i.e., bit-vector length, of the bit-parallel stochastic operands in ATRIA to 512 bits from their full-precision length of 256 bits (corresponds to 8-bit binary operands). The resultant μAPE values and the corresponding standard deviation in APE (σAPE) for four benchmark CNNs are listed in Table 2. The μAPE and σAPE values in Table 2 were obtained for the complete set of individual APEs for all MAC results required in the respective CNNs when the inferences of these CNNs are implemented on ATRIA, SCOPE-Vanilla, and SCOPE-H2D for the ImageNet dataset. Table 2 also lists the inference accuracy results. As evident, ATRIA exhibits 2.9$\times$ and 1.5$\times$ more μAPE, and 3.2$\times$ and 1.6$\times$ more σAPE than SCOPE-H2D and SCOPE-Vanilla, respectively, on average across the CNNs. Nevertheless, compared to SCOPE-H2D and SCOPE-Vanilla, ATRIA exhibits only 3.5% and 0.85% drop in inference accuracy on average across the CNNs, which we reason is acceptable due to the significant performance benefits of ATRIA, as evident from Sects. 4.3 and 4.4.

4.3 Per-MAC Latency Results

Table 3 lists our evaluated latency values and the number of Pes (#PEs) for ATRIA and other in-DRAM CNN accelerators. The latency values include values for MUL and ACC in the number of MOCs (#MOCs), latency per MOC in ns, as well as the latency values for LUT-based B-to-S conversion and pop-count (PC) operations (required for S-to-B conversion). From Table 3, ATRIA holds three crucial advantages. First, it exhibits smaller per-MAC latency over SCOPE, DRISA, and LACC (Table 3). This is because ATRIA performs 16 MAC operations in parallel. For that, ATRIA uses in total 5 MOCs (total 85ns latency with each MOC incurring 17ns latency) (Sect. 3.2), 2 MOCs to copy the operand rows, 1 MOC to perform 16 in-parallel MULs, 1 MOC to perform 16 in-parallel ACCs, and 1 MOC to store the MAC result. In Table 3, for ATRIA, 2 MOCs for operand row copy are counted in total MUL MOCs, and 1 MOC for MAC result store is counted in total ACC MOCs. Thus, by performing 16 MAC operations in parallel, ATRIA achieves shorter per-MAC latency.

Table 3 Comparison of various accelerators with ATRIA, in terms of the number of PEs (#PEs) and latency of MUL, ACC, MAC, binary to stochastic conversion (B-to-S), and pop-count (PC) operations

Various accelerators	Latency values						
	MUL #MOCs	ACC #MOCs	MOC (ns)	MAC (ns)	B-to-S (ns)	PC (ns)	#PEs
DRISA-3T1C [1]	200	11	8	1768	–	–	32,768
DRISA-1T1C-NOR [1]	200	22	10	2110	–	–	16,384
LACC [3]	1	10	21	231	–	–	16,384
SCOPE-Vanilla [2]	3	4	8	56	1	176	65,536
SCOPE-H2D [2]	21	4	8	200	1	176	65,536
ATRIA	3/16	2/16	17	5.25	1	256	4098

Second, ATRIA can better hide the latency for PC operations, compared to SCOPE. This is because SCOPE utilizes full adder-based PC operations that need to be performed inside PEs. Therefore, despite using the as-late-as-possible (ALAP) scheduling algorithm, PC operations in SCOPE inevitably stall the PEs. In contrast, ATRIA offloads PC operations to dedicated serial counters (operating at 2GHz) per PE (Sects. 3.2 and 3.3). As a result, ATRIA does not need to stall PEs for PC operations, enabling itself to better hide PC latency. Therefore, although ATRIA yields higher latency per PC operation than SCOPE (Table 3), ATRIA efficiently hides this higher latency, not letting it affect the performance.

Third, ATRIA exhibits smaller bottleneck ratio compared to SCOPE and DRISA (see Fig. 6d in Sect. 4.4). Bottleneck ratio is defined in Section IV.A. ATRIA achieves lower bottleneck ratio because the use of a massively large number of PEs in SCOPE and DRISA results in unavoidable inter-PE communication latency, a substantial portion of which remains on the critical processing path because of the inherently limited parallelism available for such inter-PE communications. In contrast, ATRIA is better at hiding the inter-PE communication latency, due to its smaller number of PEs and its LISA [16] substrate-based implementation of intra-bank, inter-bank, and inter-PE data communications (Sect. 3.2).

4.4 CNN Inference Performance Results

We evaluate the performance of ATRIA and compare it with the following inDRAM CNN accelerators from prior work: DRISA-3T1C [7], DRISA1T1CNOR [7], SCOPEVanilla [8], SCOPEH2D [8], and LACC [10]. We consider four CNNs: VGG16 [20], AlexNet [21], ResNET_50 [22], GoogleNET [23], with the ImageNet dataset. Using the setup described in Sect. 4.1, we evaluated latency, FPS, FPS/W/mm^2, and bottleneck ratio, for batch sizes of 1 and 64. Figure 6a shows efficiency (FPS/W/mm^2) results. For batch size 1, ATRIA is $18\times$, $64\times$, $98\times$, and $50\times$ more efficient than DRISA-1T1C-NOR, DRISA-3T1C, SCOPE-Vanilla, and SCOPE-H2D, respectively, on average across CNNs. However, ATRIA is 15% less efficient than LACC, due to the LACC's lower area (Sect. 3.4). Nevertheless, for batch size 64, ATRIA is more efficient than LACC as well. ATRIA is $136\times$, $522\times$, $3.4\times$, $71\times$, and $95\times$ more efficient than DRISA-1T1C-NOR, DRISA-3T1C, LACC, SCOPE-Vanilla, and SCOPE-H2D, respectively, on average across CNNs. In general, ATRIA is more efficient due to the following two reasons: (i) better FPS due to lower per-MAC latency (Table 3) and (ii) a reasonable average power consumption of 23.4W.

Figure 6b shows CNN processing latency results normalized w.r.t. ATRIA. For batch size 1, ATRIA achieves $7.4\times$, $18\times$, $3.3\times$, $6.5\times$, and $4.4\times$ lower latency than DRISA-1T1C-NOR, DRISA-3T1C, LACC, SCOPE-Vanilla, and SCOPE-H2D, respectively, on average across CNNs. Similarly, for batch size 64, ATRIA achieves $44\times$, $107\times$, $10\times$, $1.2\times$, and $2.6\times$ lower latency than DRISA-1T1C-NOR, DRISA-3T1C, LACC, SCOPE-Vanilla, and SCOPE-H2D, respectively, on average across CNNs. ATRIA achieves lower CNN processing latency because of its lower per-MAC latency and its ability of efficiently hiding its higher S-to-B conversion latency. Moreover, DRISA-1T1C-NOR, DRISA-3T1C, LACC, SCOPE-Vanilla, SCOPE-H2D, and ATRIA achieve $60\times$, $59\times$, $30\times$, $2\times$, $6\times$, and $10\times$ higher latency for batch size 64 than batch size 1. This is because the higher parallelism of SCOPE variants (more #PEs in Table 3) allows them to process larger batch size without saturating the latency benefits, by distributing the batch processing across multiple PEs.

Figure 6c shows FPS results. For batch size 1, ATRIA has on average $7.4\times$, $18\times$, $3.3\times$, $6.5\times$, and $4.4\times$ higher FPS than DRISA-1T1C-NOR, DRISA-3T1C, LACC, SCOPE-Vanilla, and SCOPE-H2D, respectively. For batch size 64, ATRIA has on average $44\times$, $107\times$, $10\times$, $1.2\times$, and $2.6\times$ higher FPS than DRISA-1T1C-NOR, DRISA-3T1C, LACC, SCOPE-Vanilla, and SCOPE-H2D, respectively. ATRIA has higher FPS due to the combined effects of lower per-MAC latency and lower memory bottleneck ratio (Sect. 4.3), as discussed next.

Finally, Fig. 6d gives memory bottleneck ratio (MBR) results. MBR for all accelerators reduces for batch size 64 than batch size 1 because increasing batch size to 64 does not substantially increase the stall time for weighting parameter accesses, but doing so increases CNN processing time due to the required time sharing of resources across multiple batch inputs, resulting in lower MBR. For batch size 64, ATRIA has lower MBR than all other accelerators, except for LACC. LACC has only 1% MBR for batch size 64, which corroborates the results from [10]. This is because the kernel mapping algorithm used in LACC enables better resource utilization. SCOPE variants have the highest MBR for both batch sizes because in SCOPE the latency for S-to-B conversions comes in the critical path (Sect. 4.3). In contrast, ATRIA is able to better hide this latency to achieve lower MBR.

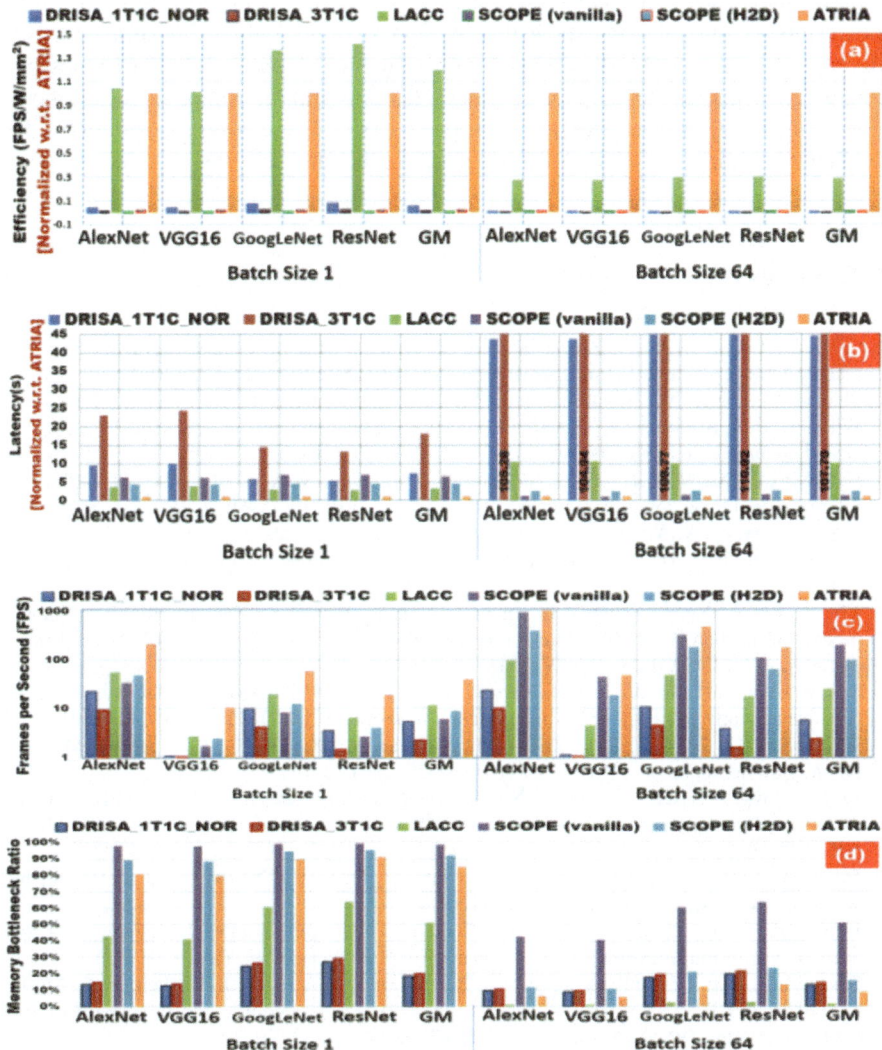

Fig. 6 (**a**) Efficiency (FPS/W/mm^2), (**b**) latency, (**c**) throughput (FPS), and (**d**) memory bottleneck ratio (MBR) results for various in-DRAM accelerators across CNNs. GM means geometric mean

5 Conclusions

In this chapter, we presented an energy-efficient and high-throughput CNN accelerator called ATRIA, which utilizes the novel concept of bit-parallel rate-coded unary (stochastic) computing to achieve ultra-low latency for multiply-accumulate (MAC) operations. We mapped four benchmark CNNs on ATRIA to compare its performance with five state-of-the-art in-DRAM accelerators from prior work. The results of our analysis show that ATRIA exhibits only 3.5% drop in CNN inference accuracy and still achieves improvements of up to 3.2× in frames per second (FPS) and up to 10× in efficiency (FPS/W/mm²), compared to the best-performing in-DRAM accelerator from prior work. These results corroborate the excellent capabilities of ATRIA for accelerating the inference tasks of deep CNNs.

References

1. Chen, Y., Luo, T., Liu, S., Zhang, S., He, L., Wang, J., Li, L., Chen, T., Xu, Z., Sun, N.: DaDianNao: A machine-learning supercomputer. IEEE MICRO, 609–622 (2014)
2. Shafiee, A., Nag, A., Muralimanohar, N., Balasubramanian, R., Strachan, J.P., Hu, M., Williams, R.S., Srikumar, V.: ISAAC: A convolutional neural network accelerator with in-situ analog arithmetic in crossbars. ACM ISCA **44**(3), 14–26 (2016)
3. Han, S., Liu, X., Mao, H., Pu, J., Pedram, A., Horowitz, M.A., Dally, W.J.: EIE: Efficient inference engine on compressed deep neural network. IEEE ISCA **44**(3), 243–254 (2016)
4. Eckert, C., Wang, X., Wang, J., Subramaniyan, A., Iyer, R., Sylvester, D., Blaaauw, D., Das, R.: Neural cache: Bit-serial in-cache acceleration of deep neural networks. ACM ISCA, 383–396 (2017)
5. Chi, P., Li, S., Xu, C., Zhang, T., Zhao, J., Liu, Y., Wang, Y., Xie, Y.: PRIME: A novel processing-in-memory architecture for neural network computation in ReRAM-based main memory. ACM ISCA **44**(3), 27–39 (2016)
6. Sun, X., Yin, S., Peng, X., Liu, R., Seo, J.-s., Yu, S.: XNOR-RRAM: A scalable and parallel resistive synaptic architecture for binary neural networks. IEEE DATE, 1423–1428 (2018)
7. Li, S., Niu, D., Malladi, K.T., Zheng, H., Brennan, B., Xie, Y.: DRISA: A DRAM-based reconfigurable in-situ accelerator. IEEE Micro, 288–301 (2017)
8. Li, S., Glova, A.O., Hu, X., Gu, P., Niu, D., Malladi, K.T., Zheng, H., Brennan, B., Xie, Y.: SCOPE: A stochastic computing engine for DRAM based in-situ accelerator. IEEE Micro, 696–709 (2018)
9. Deng, Q., Jiang, L., Zhang, Y., Zhang, M., Yang, J.: DrAcc: a DRAM based accelerator for accurate CNN inference. IEEE DAC, 1–6 (2018)
10. Deng, Q., Zhang, Y., Zhang, M., Yang, J.: LAcc: Exploiting lookup table-based fast and accurate vector multiplication in DRAM-based CNN accelerator. In: DAC (2019)
11. Seshadri, V., Lee, D., Mullins, T., Hassan, H., Boroumand, A., Kim, J., Kozuch, M.A., Mutlu, O., Gibbons, P.B., Mowry, T.C.: Ambit: In-memory accelerator for bulk bitwise operations using commodity DRAM technology. IEEE MICRO, 273–287 (2017)
12. Ren, A., Li, Z., Ding, C., Qiu, Q., Wang, Y., Li, J., Qian, X., Yuan, B.: SC-DCNN: Highly-scalable deep convolutional neural network using stochastic computing. ACM ASPLOS (2017)
13. Jacob, B., Wang, D., Ng, S.: Memory Systems: Cache, DRAM, Disk. Morgan Kaufmann (2010)
14. Thakkar, I.G., Pasricha, S.: 3D-ProWiz: An energy-efficient and optically-interfaced 3D DRAM architecture with reduced data access overhead. IEEE TMSCS **1**(3), 168–184 (2015)
15. Seshadri, V., Kim, Y., Fallin, C., Lee, D., Ausavarungnirun, R., Pekhimenko, G., Luo, Y., Mutlu, O., Gibbons, P.B., Kozuch, M.A.: RowClone: Fast and energy-efficient in-DRAM bulk data copy and initialization. IEEE MICRO, 185–197 (2013)
16. Chang, K.K., Nair, P.J., Lee, D., Ghose, S., Qureshi, M.K., Mutlu, O.: Low-cost inter-linked subarrays (LISA): Enabling fast inter-subarray data movement in DRAM. IEEE HPCA, 568–580 (2016)
17. Lou, Q., Pan, C., McGuinness, J., Horvath, A., Naeemi, A., Niemier, M., Hu, X.S.: A mixed signal architecture for convolutional neural networks. ACM JETC **15**(2), 1–26 (2019)

18. Balasubramonian, R., Kahng, A.B., Muralimanohar, N., Shafiee, A., Srinivas, V.: CACTI 7: New tools for interconnect exploration in innovative off-chip memories. TACO **14**(2), 1–25 (2017)
19. Li, Z., Ren, A., Li, J., Qiu, Q., Wang, Y., Yuan, B.: DSCNN: Hardware-oriented optimization for stochastic computing based deep convolutional neural networks. IEEE ICCD, 678–681 (2016)
20. Simonyan, K., Zisserman, A.: Very deep convolutional networks for large-scale image recognition. Preprint (2014). arXiv:1409.1556
21. Krizhevsky, A., Sutskever, I., Hinton, G.E.: ImageNet classification with deep convolutional neural networks. Adv. Neural Inf. Process. Syst. **25**, 1097–1105 (2012)
22. He, K., Zhang, X., Ren, S., Sun, J.: Deep residual learning for image recognition. IEEE CVPR, 770–778 (2016)
23. Szegedy, C., Liu, W., Jia, Y., Sermanet, P., Reed, S., Anguelov, D., Erhan, D., Vanhoucke, V., Rabinovich, A.: Going deeper with convolutions. IEEE CVPR, 1–9 (2015)

Index

SPRINGER NATURE

GPSR Compliance

The European Union's (EU) General Product Safety Regulation (GPSR) is a set of rules that requires consumer products to be safe and our obligations to ensure this.

If you have any concerns about our products, you can contact us on ProductSafety@springernature.com

In case Publisher is established outside the EU, the EU authorized representative is:

Springer Nature Customer Service Center GmbH
Europaplatz 3
69115 Heidelberg, Germany

The manufacturer's authorised representative in the EU is Springer
Nature Customer Service Centre GmbH, Europaplatz 3, 69115 Heidelberg,
Germany. If you have any concerns regarding our products, please
contact ProductSafety@springernature.com

Printed and bound by CPI Group (UK) Ltd, Croydon, CR0 4YY
27/04/2026
02097573-0007